Lecture Notes in
Earth Sciences

Edited by Somdev Bhattacharji, Gerald M. Friedman,
Horst J. Neugebauer and Adolf Seilacher

37

A. Armanini G. Di Silvio (Eds.)

Fluvial Hydraulics of Mountain Regions

Springer-Verlag
Berlin Heidelberg New York London Paris
Tokyo Hong Kong Barcelona Budapest

Editors

Prof. Aronne Armanini
Dept. of Civil and Environmental Engineering
University of Trent
I-38050 Mesiano di Povo, Italy

Prof. Giampaolo Di Silvio
Institute of Hydraulics, Faculty of Engineering
University of Padua
Via Loredan 20, I-35131 Padua

ISBN 3-540-54491-7 Springer-Verlag Berlin Heidelberg New York
ISBN 0-387-54491-7 Springer-Verlag New York Berlin Heidelberg

Printed in Germany

Typesetting: Camera ready by author
Printing and binding: Druckhaus Beltz, Hemsbach/Bergstr.
32/3140-543210 – Printed on acid-free paper

Preface

Following the economical and social development of the local communities, mountain regions of temperate climates are increasingly becoming the site of valuable infrastructures and important urban and industrial settlements. As the catastrophic events of last years in the European Alps have clearly shown, the vulnerability of these territories has correspondingly increased, in terms of both property damage and losses of human life.

Until recently, the hydraulic scientific community has paid little attention to mountain watersheds, except perhaps during the period if the hydropower development. Nevertheless attention was then focused on problems and methodologies somewhat different from the issues of actual environmental concern.

More recently, however, hydraulic engineers have joined their colleagues from forest and rural engineering, who have traditionally dealt with erosion control in mountain areas, to bring in their own methodology, already experienced in lowland rivers. At the same time, academic people focused an interest in some phenomena, like massive transport, which is typical of mountain environment.

To bring together all these contributions and to make the state of the art of the mountain river science (oropotamology) and technology, an International Workshop was called at the University of Trent (Italy), on October 1989, under the sponsorship of Fluvial Hydraulic Section of the IAHR.

Three main topics have been recognized as particularly relevant from the point of view of both research and professional people:

a) Hydrodynamics of steep channels and local scale process;

b) Sediment movement and sediment training, with special emphasis on massive transport;

c) Particular features of sediment transport related to non-uniform grain-size.

However, as it is the case in these circumstances, the contest of several contributions often spread over more than one topic. In the following Introduction to papers, the three topics were split into 11 Sections, each one devoted to a more particular aspect recurrently addressed during the discussion. The same paper, thus, may be mentioned in different Sections of the Introduction.

Trento, march 1991

the editors

Acknowledgements

The Organizing Committee of the Workshop and the Editors of this book are indebted to the International Advisory Board, who have selected the papers:

Prof. Selim Yalin (Chairman of the Fluvial Hydraulic Sec. I.A.H.R.)
Prof. James C.Bathurst (University of Newcastle upon Tyne, U.K)
Dr. Lianzhen Ding (I.R.T.C.E.S., Beijing, China)
Dr. Martin Jaeggi (E. T. H., Zurich,Switzerland)
Prof. Masanori Michiue (University of Tottori, Japan)

and to the Moderators of the Sessions of the Workshop:

Prof. Matheus deVries, (Delft University of Technology, The Netherlands)
Dr.Gerrit J. Klaassen, (Delft Hydraulics, De Voorst Laboratory, The Netherlands)
Prof. James C.Bathurst (University of Newcastle upon Tyne, U.K)

who have been fundamental for stimulating the discussion and for distilling the most relevant aspects in which the following Introduction was articulated.

The Workshop was organized in the framework of the activities of the Italian Groups for the Disaster Prevention (CNR-GNDCI)) and for Sediment Transport (MURST 40%).

Table of Contents

C Non-uniformity of Sediment: Grain Sorting; Bed Armouring; Transport of Fine-Sediment Suspensions.

List of Contributors

Aronne Armanini , *Department of Civil and Environmental Engineering, University of Trent, Trent, Italy*
Kazuo Ashida , *Disaster Prevention Research Institute, Kyoto University, Gokasho, Uji, Kyoto 611, Japan*
Wojciech Bartnik , *Department of Hydraulic Engineering, Academy of Agriculture, Cracow, Poland*
Wilhelm Bechteler , *Laboratory for Hydraulics and Hydraulic Structure, University of the Armed Forces, Munich, Germany*
Maurice Bouvard , *Institut de Mècanique de Grenoble, Grenoble, France*
Stefano Brunelli , *Hydraulics Institute "G.Poleni", University of Padua, Padua, Italy*
Tim R.H. Davies , *Natural Resources Engineering Department, Lincoln College, University of Canterbury, New Zealand*
Giampaolo Di Silvio , *Hydraulics Institute "G.Poleni", University of Padua, Padua, Italy*
Francesco Dellagiacoma , *Provincia Autonoma di Trento, Trent, Italy*
Shinji Egashira , *Disaster Prevention Research Institute, Kyoto University, Gokasho, Uji, Kyoto 611, Japan*
Luigi Ferrari , *Provincia Autonoma di Trento, Trent, Italy*
Masaharu Fujita , *Department of Civil Engineering, Tottori University, Tottori, Japan*
Zoran Gavrilovic , *Institute for the Development of Water Resources "Jaroslav Cerni", Jaraslava Cernog, Beograd, Jugoslavia*
Paolo Ghilardi , *Istituto di Idraulica, Politecnico di Milano, Milano, Italy*
Walter H. Graf , *Laboratoire de Recherches Hydrauliques, Ecole polytechnique Fédérale, Lausanne, Suisse*
Tadaharu Ishikawa , *Department of Civil Engineering, Tokyo Institute of Technology, Tokyo, Japan*
Koichi Kato , *Department of Civil Engineering, Ehime University, Matsuyama, Japan*
Alberto Lamberti , *Hydraulics Institute, University of Bologna, Italy*
Zivorad Matovic , *Institute for the Development of Water Resources "Jaroslav Cerni", Jaraslava Cernog, Beograd, Jugoslavia*
Giovanni Menduni , *Istituto di Idraulica, Politecnico di Milano, Milano, Italy*
Masanori Michiue , *Department of Civil Engineering, Tottori University, Tottori, Japan*
Takahisa Mizuyama , *Public Works Research Institute, Ministry of construction, Japanese Government, Tsukuba, Japan*
Luigi Montefusco , *Department of Civil Engineering, University of Florence, Italy*
Hiroji Nakagawa , *Department of Civil Engineering, Kyoto University, Kyoto, Japan*
Ennio Paris , *Department of Civil Engineering, University of Florence, Italy*
Gary Parker , *St. Anthony Falls Hydraulic Laboratory, University of Minnesota, Minneapolis, Minnesota, USA*
Massimo Peviani , *I.S.M.E.S. s.p.a. , Bergamo, Italy*
Domenico Pianese , *Department of Hydraulics, Water Resources Management and Environmental Engineering, University of Naples, Napoli, Italy*
Dieter Rickenmann , *Laboratory of Hydraulics, Hydrology and Glaciology, E.T.H.- Zentrum, Zürich, Switzerland*
Fabio Rossi , *Institute of Civil Engineering, University of Salerno, Penta di Fasciano, Italy*

Helmut Scheuerlein , *Obernach Hydraulics Laboratory, Technical University Munich, Munich, Germany*
Friedrich Schöberl , *Institut für Konstruktiven Wassebau und Tunnelbau, Universität Innsbruck, Austria*
Giovanni Seminara , *Hydraulics Institute, University of Genoa, Genoa, Italy*
Yoshihiko Shimizu , *Department of Civil Engineering, Ehime University, Matsuyama, Japan*
Lechostaw Suszka , *Institute of Hydroengineering, Polish academy of Sciences, Gdansk, Poland*
Tetsuro Tsujimoto , *Department of Civil Engineering, Kanazawa University, Kanazawa, Japan*
Marco Tubino , *Hydraulics Institute, University of Genoa, Genoa, Italy*
Koichi Suzuki , *Department of Civil Engineering, Ehime University, Matsuyama, Japan*
Tamotsu Takahashi , *Disaster Prevention Research Institute, Kyoto University, Kyoto, Japan*
G. Vogel , *Laboratory for Hydraulics and Hydraulic Structure, University of the Armed Forces, Munich, Germany*
H.J. Vollmers , *Laboratory for Hydraulics and Hydraulic Structure, University of the Armed Forces, Munich, Germany*
Hans Peter Willi , *Swiss Federal Office for Water Economy, Bern, Switzerland*

Introduction to the Papers

by

A. Armanini and G. Di Silvio

1. WATER FLOW IN MOUNTAIN STREAMS

One of the features that characterizes most water flow in mountain streams is the large relative roughness. In such a case the roughness elements of typical size k_s, say the intrusion within the main flow, have the same order of magnitude as the water depth h ; that is:

$$Z = \frac{h}{k_s} = O(1) \qquad 1)$$

and the velocity distribution deviates from the logarithmic law

When the relative submergence Z is small enough a *roughness sublayer* has been recognized near to the bottom, where the velocity distribution tends to become uniform and Reynolds stresses tend to be suppressed [W.H.GRAF in paper A1 and H.NAKAGAWA *et alii* in paper A2]. The reduction of measured Reynolds stresses near the bed apparently impairs the values of the bed shear stress $\tau_0 = \rho g h j$, j being the slope, calculated from the total force balance. H.NAKAGAWA *et alii* in paper A2 try to explain the characteristics of the roughness sublayer and the apparent discrepancies in terms of eddy shedding from individual roughness elements. As a consequence, hydraulic resistance also deviates from the normal law accepted in case large submergence:

Besides the velocity distribution (and the related estimation of resistance), this deviation also affects the incipient motion condition and sediment transport formulas (Sections 2 and 3).

In order to determine the parameters affecting the hydraulic resistance in a mountain stream in clear water conditions it is convenient to apply concepts from dimensional analysis. The shear stress exerted on the bed by the water depends on the following parameters: mean velocity U , water depth h , acceleration due to gravity g , fluid density ρ , representative diameter of bed material D_s, and its size and space distribution n :

$$f(\tau_0 \ , \ U, h, g, \rho, D_s, n, \mu) = 0 \qquad 2)$$

By means of π−theorem, equation 2) can be expressed as a function of the following dimensionless groups:

$$f(\frac{U}{u_*} \ , \ \frac{U}{\sqrt{gh}} \ , \ \frac{h}{D_s} \ , \ \frac{\rho U h}{\mu}, \ n \ ) = 0 \qquad 3)$$

where, in the first group, the shear velocity $u_* = \sqrt{\frac{\tau_0}{\rho}}$ has been introduced. Even if other combinations of different dimensionless groups are possible, the five assumed seem to be suitable to describe the phenomenon under consideration.

The first dimensionless group represents the relative resistance and can be expressed also by means of the Darcy-Weisbach friction factor f, or the Chèzy coefficient χ : $\frac{U}{u_*} = \sqrt{\frac{8}{f}} = \frac{\chi}{\sqrt{g}}$.

The second group represents the Froude number: $Fr = \frac{U}{\sqrt{gh}}$.

The third ratio is the relative submergence: $Z = \frac{h}{D_s}$.

The fourth group is the Reynolds number: $Re = \frac{\mu}{\rho U h}$. In the natural water courses the turbulent shear stresses prevail over the viscous forces, such that viscosity does not influence the resistance and consequently Reynolds number Re can be overlooked.

The fifth parameter n accounts for both the roughness size distribution and its spacing (packing and imbrication). In most cases, even for non-uniform material, the grain-size distribution allows for the definition of a unique *representative diameter* (e.g. D_{50} or D_{90}) which should be also chosen considering the spacing produced by the flow.

By including the n parameter in the definition of representative diameter, one has:

$$\frac{U}{u_*} = f(Z, Fr) \qquad 4)$$

Experiments [GRAF, paper A1] evidence that resistance follows the logarithmic law:

$$\frac{U}{u_*} = \frac{1}{\kappa} ln \frac{R_h}{D_s} + \bar{B}r \qquad 5)$$

where $\bar{B}r$ is a function of relative submergence and of Froude number :

$$\bar{B}r = f(Z, Fr) \qquad 6)$$

if the Kàrmàn constant κ is taken equal to 0.4 , $\bar{B}r$ ranges between 6.25 (Keulegan's value for large submergence) and 1.3. In the paper A1 this dependency is clearly shown, even if no one explicit relationship is proposed.

The problem is tackled in a similar way also by T.TSUJIMOTO in paper A7, where the resistance law is expressed in terms of relative submergence and of two more parameters α e β, introduced in order to best fit the experimental data. The author gives no precise indications on the nature of these parameters. A dependency of these parameters on Froude number, which does not appear in the proposed expression for resistance, seems to be possible.

One more parameter that influences the hydrodynamic resistance of mountain streams is focused by S.EGASHIRA and K.ASHIDA : an additional resistance introduced by *steep-pool o chutes and pools* in supercritical currents is presented in paper A3 as a function of Froude number and geometrical characteristics of bed forms.

2. INCIPIENT MOTION CONDITION

As a further consequence of small submergence on mountain streams, also the incipient motion criteria should be reviewed. With the same procedure adopted for hydraulic resistance, the dimensionless groups affecting incipient motion condition may be evidenced by means of dimensional analysis.

The relationship among nine parameters :

$$f(\ U, h, \tau_0, \rho_s, g, \rho, D_s, \mu, n) = 0 \qquad 7)$$

reduces to a relationship among 6 dimensionless groups:

$$f\left(\frac{(u_*^2)_{cr}}{D_s g\Delta}, \frac{\rho u_* D_s}{\mu}, \frac{h}{D_s}, \frac{U_{cr}}{u_*}, \frac{U_{cr}}{\sqrt{gh}}, n\right) = 0 \qquad 8)$$

where $\Delta = \dfrac{\rho_s - \rho}{\rho}$

The first group is the Shields mobility parameter: $\theta_{cr} = \dfrac{(u_*^2)_{cr}}{D_s g\Delta}$. If the bed is steep it is convenient [W.GRAF, paper A1] to modify this parameter, which represents the ratio between hydrodynamic forces and grain stresses, in order to account for the gravity force in the individual grain equilibrium:

$$\theta'_{cr} = \frac{(u_*^2)_{cr}}{D_s g\Delta(tan\Phi cos\alpha - sin\alpha)} \qquad 9)$$

where Φ is the angle of repose of the bed material and α is angle of channel bed. A similar relation has been proposed also by T.TSUJIMOTO in paper A7.

The second group is the grain Reynolds number: $R_* = \frac{\rho u_* D_s}{\mu}$.

The third ratio is the relative submergence: $Z = \frac{h}{D_s}$.

The fourth group is the relative resistance: $\frac{U}{u_*} = \sqrt{\frac{8}{f}}$.

The fifth group is the Froude number: $Fr = \frac{U}{\sqrt{gh}}$.

The last parameter n is related to roughness size distribution and spacing. Also in this case, as for flow resistance, the dependence on this parameter can be included in the definition of the representative diameter. The grain-size distribution, however, is very important in all cases where a selective removal of material occurs (static or dynamic armouring), as specified in Section 6.

The direct dependence on relative resistance $\frac{U}{u_*}$ can be omitted because it depends on submergence Z and Froude number Fr (Section 1).

Compared to Shields analysis, then, the critical shear stress (incipient motion) in mountain streams shows a further dependence on relative submergence Z and Froude number Fr; that is:

$$\theta'_{cr} = \frac{(u_*^2)_{cr}}{D_s g\Delta(tan\Phi cos\alpha - sin\alpha)} = f(\frac{\rho u_* D_s}{\mu}, \frac{h}{D_s}, \frac{U}{\sqrt{gh}}) \qquad 10)$$

The dependence of θ'_{cr} on submergence is clearly shown by W.GRAF. In paper A1, in fact, three zones are distinguished where the mobility parameter θ'_{cr} is affected in different ways by the relative submergence Z . L.SUSZKA in paper A4 gives a relation between θ'_{cr} and Z .

Different combinations of dimensionless groups are used in different papers. In paper A1 the modified Shields parameter θ'_{cr} is related to channel slope j instead of Froude number, but it should be remarked that channel slope is a combination of Froude number and relative resistance:

$$j = (\frac{u_*}{U} Fr)^2 \qquad 11)$$

W.BARTNIK in paper A6 introduces a *critical Froude number* $Fr_{cr} = \frac{U_{cr}}{\sqrt{gh}}$, to describe the incipient motion. This formulation, which does not include the Shields parameter, is not very convenient because the ratio between Shields parameter and Froude number depends on submergence:

$$\frac{U^2}{gh} = \frac{u_*^2}{g\Delta D_s} \cdot \frac{\chi^2\Delta}{g} \cdot \frac{D_s}{h} \qquad 12)$$

that is, critical Froude number is not constant even under large degrees of submergence.

Indeed, as the first factor (Shields parameter) is virtually constant and the Chèzy coefficient is proportional to $(h/D_s)^{1/6}$ (Strickler), one obtains that the critical Froude number, Fr_{cr}, should be proportional to $(h/D_s)^{-1/3}$, close to the exponent proposed in Fig. 4 of paper A6.

T.TSUJIMOTO, in paper A7, prefers to use a dimensionless *critical water discharge* ,(Schoklitsch approach) because this parameter can be measured much more precisely than shear stress and water depth:

$$q_c^* = \frac{q_{cr}}{D_s\sqrt{gD_s}} \qquad 13)$$

This dimensionless parameter is assumed to depend on bed slope trough two experimental parameters α e β , already introduced in the formulation of hydraulic resistance: however the expression of these parameters is not defined.

One may observe that the combination of ratio 13) with a uniform flow formula and Shields parameter gives:

$$q_c^* = \theta_{cr} \;\; \Delta \;\; Fr \;\; \sqrt{\frac{h}{D_s}} \;\; \frac{1}{j} \qquad 12)$$

In TSUJIMOTO approach, the dependence on relative density Δ is missing: however Δ is nearly constant in the natural streams. Besides, the dimensionless critical discharge q_c^* results to be nearly proportional to j^{-1}, which implies that the product $\left(\theta_{cr} \;\; Fr \;\; \sqrt{\frac{h}{D_s}} \right)$ is nearly constant: similar experimental results have been obtained by W.BARTNIK [paper A6] and by L.SUSZKA [paper A4].

3. SEDIMENT TRANSPORT FORMULAS

A small relative submergence may also modify the sediment transport formulas. With respect to the quantities considered in the previous sections, one more parameter relevant to sediment transport rate should be introduced in the dimensional analysis. The corresponding dimensionless group is the Einstein's transport parameter:

$$\Phi = \frac{q_s}{D_s \sqrt{g \Delta D_s}} \qquad 13)$$

In lowland streams Φ depends on θ and θ_{cr}. In the case of mountain streams with small submergence and strong slopes, the transport parameter Φ will depend also on channel slope j (or on Froude number). The explicit dependence on submergence Z usually does not appear because it is implicitly accounted for by θ_{cr}. A relation of this type for the bed-load is proposed by T.TSUJIMOTO in paper A7.

A specific transport formula to be applied where pool- and chute effects are present is given by S.EGASHIRA and K.ASHIDA [Paper A3].

In the case of massive transport both hydrodynamics and sediment transport problems must be approached from a different point of view: more details are given in Sections 4 and 5. In the present section the effect on the sediment transport of cohesive material dispersed in water will be considered. This problem is introduced by D.RICKENMANN [paper C7], as a modification of a ordinary sediment transport formula for steep channels (e.g. Smart-Jäggi 1983) in order to account for clay suspension. As a result, an increase of suspension density produces an increase of bed-load transport rate.

Finally a relationship between bed-load and water discharge is used by F.SCHÖBERL in paper A5, in the case of a glacierized watershed, where the bed-load discharge is determined by the characteristics of the liquid hydrograph.

A similar approach is also used by T.ISHIKAWA [paper B4] in order to establish a correlation between the flow rate and the sediment accumulation in an alluvial fun.

4. INTENSE SEDIMENT YIELD AND TRANSPORT

During severe meteorological events mountain rivers may be subject to a very intense sediment transport, many orders of magnitude larger than sediment transport in ordinary conditions. Intense transport in the rivers is invariably related to a very high lateral sediment input, basically under the form of **mass movement**: i.e. landslides from the basin slopes and/or debris flow from the steepest branches of the hydrographic network.

The huge amount of material suddenly fed into the river first settles down in correspondence to the input, then is rapidly removed and transported downstream with a complex process of differential deposition along its way. Deposition, basically controlled by the variations of river morphology, preferably occurs in the relatively less steep reaches. In some of these reaches bottom aggradation may reach several meters in few hours, with dramatic consequences as far as inundation is concerned.

Although the general trend is to have the largest quantities of material and the coarsest fractions of sediment deposited in the upper part of the river, not far from the location where sediment input takes place, fine particles may travel for a long distance during the same flood event, so that bottom aggradation may take place tens of kilometres downstream, where the river flows in relatively ample valleys. A rapid propagation of the amounts of fine sediments, following the occurrence of a landslide in the upper part of the basin, is reported by M.FUJITA et alii in paper B5.

While bottom aggradation in the lower parts of the river is definitely less dramatic than in the upper reaches, the increase of only one meter in the bottom elevation of a river with a relatively small slope may still produce substantial reduction of its conveyance capacity. As the river flows in a more vulnerable area, moreover, the effects of an inundation here are also generally more serious than in the upper part of the watershed.

A general review of the risks posed by mountain streams and of the difficulties of assessing them is given in the paper B8 by T.R.DAVIES. Practical suggestions are also given in the paper to mitigate the risk of death and destruction, as a consequence of catastrophic erosion and related inundation. In fact, although landslides, debris flow and other similar phenomena may present *per se* a hazard to the inhabitants inasmuch as they may destroy buildings and infrastructures, the main disaster in mountain regions is still represented by inundations.

Inundations related to intense sediment transport and the subsequent overaggradation of river bottom are relatively frequent in some geographical areas. According to historical records, in many valleys of the Alps they typically occur in correspondence to particularly severe meteorological events with an average frequency of 50-100 years. For example, the last occurrence in central Alps, July 1987, that produced catastrophic inundations in many valleys of Switzerland and Italy [paper C7 by G.DI SILVIO and A.PEVIANI] had been preceded by similar disasters in the same valleys, recorded by chronicles back to 1300.

Overaggradation due to excessive sediment input has been frequently recorded in other areas. In his key-note lecture [paper B1], T.MIZUYAMA reports the riverbed changes (up to 20 meters) which occurred in the Inamata valley (Japan) in 1983, while Z.GAVRILOVIC and Z.MATOVICH [Paper B7] describe the overaggradation in the Vlasina river (Yugoslavia) of 1988.

Intense sediment transport in mountain rivers presents two fundamental aspects that must be properly investigated: (i) the excessive sediment input in the streams produced by mass movement; (ii) the movement of sediment along the streams and corresponding erosion-deposition processes.

The first aspect is essentially related to the triggering of landslides and debris-flow, as well as to the movement of these masses (either as dry material or high-concentrated mixtures of sediment and water) down into the river: it will be developed in the next Section 5.

The second aspect is basically connected with the amount of sediments that can be conveyed in the river by a given water-flow (transport formulae). Since the transport in mountain rivers is strongly controlled by its grain-size composition and by the composition of the river bed , this second aspect will be treated in Section 6.

5. LANDSLIDES AND DEBRIS FLOW

The problem of mass movement (landslides and debris-flow) may be treated from different points of view. The first, definitely empirical, approach intends to single-out the lower and upper conditions (thresholds) above which mass movement may or, respectively, should occur in a given area. Most authors provide thresholds criteria simply in terms of minimum height of rain that is to fall, with different durations, before generalized collapses take place. Unfortunately these criteria (like the ones suggested by Campbell, 1975, for the area of Santa Monica, California, or by Mizuyama, 1983, for a Japanese watershed) have no general validity but are limited to specific geological and morphological conditions.

A second, more deterministic approach attempts to simulate the instability conditions corresponding to the initiation of mass movement. A model for the simulation of shallow landslides from the slopes of a given basin has been proposed by T. MIZUYIAMA in his already mentioned keynote lecture to the Workshop [paper B1]; the model includes the reproduction of vertical unsaturated seepage (rainfall infiltration) and horizontal saturated flow (groundwater table), as well as an analysis of slope stability. Although quite realistic and very useful for sensitivity analysis, the model is not generally applicable to a real watershed, where other types of mass movement besides shallow landslides (e.g., debris flow) take place.

Triggering conditions of debris flow, in fact, are different from these of landslides: while landslides are not necessarily saturated, debris flow is always related to the presence of both surface flow and underground flow. A classification of different types of massive flow and a discussion of the criteria for their existence is made by T. TAKAHASHI in paper B9. In this paper the well-known contributions of the author to the analysis of flow of sediment-water mixtures are resumed and developed. The approach is typical of continuum mechanics, as it especially investigates the relations between stresses and deformation (constitutive equations) of these mixtures, while the dynamic equilibrium conditions considered in the analysis are necessarily very simple (uniform flow).

Mathematical models describing intense sediment transport in uniform flow are obtained under different hypothesis, regarding both the constitutive equations and the boundary conditions to be prescribed at the bottom. A rather general model has been presented to the Workshop by A.LAMBERTI and L.MONTEFUSCO [paper B2]. While the required computations are not trivial, the numerical overall results are apparently in line with those provided by empirical formulae.

A simple uniform-flow formula for water-sediment mixtures, developed at the Institute Jaroslav Cerni of Belgrade (Yugoslavia), has been proposed by Z. GAVRILOVIC and Z. MATOVIC [paper B7]. This empirical, Chèzy-type formula gives the mixture velocity as a function of bottom slope, flow depth and mixture density; an empirical criterion is also given to estimate the density (i.e., the sediment concentration) of a certain debris flow. In addition to the grain-size of the transported material and the slope of the stream, the criterion is based on the effects produced by the passage of debris flow so it cannot be used for predictive purposes.

Sediment concentration, in fact, is probably the most crucial quantity to predict the rate of sediments fed into the river by possible debris flows occurring in the steepest tributaries during a given storm. Indeed, as soon as the hydrograph $q_w(t)$ of the tributary is computed via a rainfall-runoff model, the corresponding sediment discharge $q_s(t)$ is:

$$q_s = \frac{C}{1-C} q_w \qquad 14)$$

where C is the volume concentration of the mixture.

According to the experiments reported by TAKAHASHI [paper B9, Fig. 4], a quasi-equilibrium flow (no erosion or deposition) takes place along a flume, with a volume concentration:

$$C \simeq C_{s\infty} = \frac{\rho tan\theta}{(\rho - \sigma)(tan\phi - tan\theta)} \qquad 15)$$

where ρ and σ are the densities of water and sediments, $tan\theta$ is the flume slope and $tan\phi$ the repose slope of the material. The value $C_{s\infty}$ is also equal to the initial concentration of the mixture, when massive slippage of the saturated bed occurs under the action of surface water ("incipient instability").

In a real steep tributary, then, debris flow occurs when an initial concentration is controlled by the bed slope (eq. 15), and as soon as: *(i) bed saturation occurs* and *(ii) the bed contains enough material to load the water discharge according to eq.14* . The debris flow concentration, in principle, is expected to change along the tributary , if it changes its slope or discharge; as a first approximation, however this concentration may be assumed to remain equal to $C_{s\infty}$ down to the main river, if the tributary is reasonably short and regular.

6. NON-UNIFORM GRAIN-SIZE DISTRIBUTION

The sediments fed into the river by landslides and debris-flows are rapidly removed and conveyed downstream by the river discharge. In order to simulate correctly the selective process of entrainment, transport and deposition of this material, it is of paramount importance to take into account its grain-size distribution. Mountain rivers, in fact, are characterized by a widely extended grain-size curve (ranging from boulders to fine sand) both for the transported material and for the material that constitutes the bed. As a consequence, the behaviour of such different grain-size classes can hardly be described by only one *representative diameter.*

When particles of different diameter are considered, many problems arise, related to the mobility of each grain-size class. Let us consider, first, a number of different materials with uniform grain-size; common sense suggests that coarse materials are less mobile than fine materials, and indeed all the experimental formulae show that transport rate $T(d)$ decreases when grain-size, d, increases.

By increasing the flow discharge, however, the mobility of all the material tends to be more and more the same especially as far as bed-load transport is concerned; both Meyer-Peter and van Rijn formulas, for example, indicate that this mode of transport becomes independent from the particle diameter when the shear stress is much larger than the critical value corresponding to incipient motion.

If now we consider a mixture of different materials, the mobility of each grain-sizes is even more equalized by the fact that coarse particles protrude more in the velocity profile than with a uniform material and that fine particles are protected by the larger ones. To take this equalizing effect into account , different correction procedures *(hiding- and-exposure- coefficients)* have been proposed (Egiazaroff, Day, Ranga-Raju, etc.).

In the limit case of equal mobility for all the grain-size classes, the composition of the river bottom in equilibrium conditions should be equal to the composition of the transported material. The bottom of real rivers, however, is always coarser than the transported material. In fact, a perfect equal mobility of all the particles conveyed by the stream is never achieved, basically for two reasons: during dry periods of the year the water-flow (the shear stress) is below the critical value for many grain-sizes, so that for most of the time bed-load is limited to relatively fine gravel and sand; during floods, on the other hand, equal mobility is probably reached for all the particles moving as bed-load, but not for the finer particles abundantly transported in suspension in these occasions.

The *non-equal mobility* hypothesis is not only supported by many laboratory experiments, but also by the simple observation than the material forming the bed of mountain rivers (parent material) is invariably coarser than the material of the landslides feeding the transport, as well as coarser than the material trapped in a reservoir.

It should be noted that many phenomena of mountain river morphology may only be explained by a different mobility of different grain-size classes. The typical concave profile and downstream decreasing of particle diameters, as displayed by most rivers in absence of lithological controls, is invariably due longitudinal sorting based on differential transport, although it may be partly produced by abrasion in the case of soft rocks. In paper C1 G. PARKER proposes a model for evaluating the relative importance of either mechanism and applies it to Red Deer River, Alberta, Canada, where the transported material is composed of both limestone and quartz. Assuming that selective sorting is the only active mechanism, it is shown that the longitudinal distribution of both slope and grain-size, progressive decreasing downstream, can be considered a *quasi-equilibrium* configuration: in fact, the evolution towards the final equilibrium configuration would take an extremely long time.

As it will be seen again in one of the next Sections, the "aggradation" and "coarsening" of the bottom that represent this almost imperceptible adaptation process may be interpreted in terms of a kinematic double wave (a bottom profile and a grain-size profile) propagating very slowly in the flow direction.

Another aspect related to the different mobility of different grain classes is *vertical sorting.* Vertical sorting (characterized by a coarser surface layer above finer particles underneath) should be distinguished in:

- **static armour**, in which all the particles in the bed are immobile (for a given discharge); static armour is the final equilibrium stage (after a partial removal of the smaller particles) in which the residual fine material is protected by the coarse one. Static armouring is eventually reached when clear water is fed in the stream;

- **semi-static armour**, in which only the particles smaller than a certain diameter can still move (for a given discharge). Semi-static armouring is eventually reached when only fine particle are fed in the stream;

- **dynamic armour** (called also *pavement*), in which all the particles of the bed are susceptible to motion. Dynamic armour occurs when the grain composition of the bottom is properly related to the gradation curve of the material fed in the stream, as in equilibrium or quasi-equilibrium conditions.

It is obvious that the immobile coarser particles in the static and semi-static armour impede or reduce the entrainment of the smaller ones. Although they are not permanently at rest, the same role is also played by the coarser particles of the dynamic armour, or pavement. In this respect, the effect of the pavement is a sort of *vertical hiding* of the smaller particles of the sub-pavement. Should the vertical hiding be sufficiently strong, the grain-size composition of the sub-pavement would be equal to the composition the transported material (*equal mobility* conditions), although the pavement composition is coarser.

Problems arising from the development of static and semi-static armouring are discussed in paper C3 by M. BOUVARD, with special reference to the results of an undistorted scale (1:12.5) hydraulic model of a torrent of Southern French Alps.

The formation of a static armour has been investigated on a laboratory flume by A. LAMBERTI and E. PARIS [paper C5]. The paper provides the experimental data of five runs carried on in the following way: in the first part of each test, feeding the flume with a constant discharge of clear water, the time history of bottom composition and output solid discharge has been recorded until the output solid discharge vanishes (static armour completely developed); in the second part of the same test the static armour is first removed by a sudden increase of water discharge and subsequent evolution of the bottom is then recorded.

While the *static armour* is the final configuration of the bottom when the solid discharge becomes zero, a *dynamic armour* may also be present in equilibrium conditions with a solid transport different from zero. This phenomenon has been investigated by K. SUZUKI and K. KATO [paper C4] in a laboratory flume, where the sediment input is adjusted to be equal to the sediment output (*equilibrium conditions*). Experimental data regard the transport of a non-uniform sand with an average diameter of 0.41 cm and a standard deviation of 3. The pavement results to be always coarser than the sand fed into the flume, provided that bed shear stress is not excessively large.

The paper of SUSUKI and KATO does not give any quantitative information on the sub-pavement (called substrate in the paper). This does not, in any case, require the same composition of the supplied sand (transport), as such a circumstance occurs only if *equal mobility* applies. In fact, similar experiments reported in Paper C7 by G. DI SILVIO and S. BRUNELLI, show that sub-pavement (substratum) is always coarser than the transported material, although finer than the pavement.

The vertical structure of the bed (pavement and sub-pavement), in the presence of non-uniform material, is also important in the computation of the transport rate. If the movement of each grain-size class would be completely unaffected by the movement of the others, the transport rate of the $i-th$ class should be simply given by the product:

$$\beta_i \; T(d_i) \tag{16)}$$

where $T(d_i)$ is the transport of that class computed by any transport formula as if the bottom were made of uniform material and β_i is the percentage of the same class present in the bottom. We have seen before, however, that a certain interaction among the grain classes (hiding and exposure effect) does occur as it is revealed, among others, by the formation of a coarser pavement over a finer sub-pavement. As a result one should decide whether β_i in eq. 16) is the percentage of the $i-th$ class in the pavement or in the sub-pavement and, in addition , a certain correction should be introduced in the expression of $T(d_i)$ in order to take this interaction into account . In general, β_i is supposed to refer to the bottom surface (pavement) as the one which is more subject to the water-flow; in some instances however, it has been referred to the sub-pavement. As for the hiding-and-exposure correction, two procedures have been proposed to increase the inherent lower mobility of larger diameters; the first one (Day, Misri, etc.) consists of a coefficient that augments the *effective* shear stress that water-flow transmit to larger particles; the second one (Egiazaroff, Ashida-Michiue, etc.) reduces the *critical* shear stress that defines the incipient motion of larger particles. In some cases (Ranga-Raju), both procedures have been applied together.

It should be stressed that the hiding-and-exposure correction depend on whether the percentage β_i in eq. 16) refers to the pavement or to sub-pavement. In the first case, as the pavement composition is definitely coarser than the transport composition, the required correction is relatively small; in the second case, by contrast, the sub-pavement composition tends to be finer and more similar to the composition of the transport (quasi equal-mobility) so that the correction should be much larger.

Some of the available procedures to evaluate sediment transport rate for non-uniform grain-size material have been tested by both K. SUZUKI and K. KATO [paper C4] and A. LAMBERTI and E. PARIS [paper C5], who have utilized the experimental data for this purpose obtained from their respective experiments. Apparently not all the procedures apply equally well to every set of data.

Other procedures have been applied in the mathematical morphological models discussed in the next section.

7. MATHEMATICAL MORPHOLOGICAL MODELS

The morphology of rivers is controlled by the interaction of the water with the sedimentary system on which the water flows. The forms arising from such an interaction regard both the planimetric and altimetrical characteristics of the river at different spatial scales.

At the spatial scale of the river width, it is particularly interesting to note the tendency of a channel flowing in a non- cohesive material to deviate from a rectangular cross section and straight axis by developing a system of *bars* and *meanders*, i.e., respectively, altimetrical and planimetric deviations. A thorough review of the most recent work in this field is given by G. SEMINARA in paper A9, where the problem is approached from the point of view of linear and non-linear stability. The basic model for the analysis is represented by the two-dimensional depth-averaged water-flow equations, plus a secondary flow component accounting for the three-dimensional effects of the curvature. The sediment is supposed to have a uniform grain-size and the (bed-load) transport is expressed in terms of local shear stresses and bottom slope.

The linear theories are able to predict the conditions under which free bars occur, while non linear theories provide an interesting insight on their development and final configuration. It is to be noted that in these analyses the water discharge is considered to be steady. An attempt to evaluate the effects of flow unsteadiness in the development of alternate bars has been reported by M. TUBINO in paper A8; according to his experiments, the final bar height after a sequence of typical flood events may be evaluated in terms of a properly selected reference-value of the water discharge.

All the hypothesis assumed in the above mentioned stability analysis are better satisfied by lowland rivers or by large upland rivers than by relatively small rivers in mountain regions; grain sorting, for example, can hardly be neglected in the last case, as particles often range from sand to boulders. A typical morphology of mountain rivers, essentially related to the non- uniform grain-size of transported material, is the so-called *steep-pool bed*. Paper A3 by S. EGASHIRA and K. ASHIDA is devoted to the genesis and geometry of this peculiar formation.

Only when the stream is relatively wide and regular, are conditions such that multiple bars occur. Mountain river morphology, by contrast, is very often controlled by geological constraints (unerodable banks and sills) that determine local channel width and elevation. When the stream flows on a very ample alluvial deposit, on the other hand, multiple channel braiding takes place in dry conditions, while during a flood the number of channels generally becomes smaller, depending upon the water discharge; during these changes, the erosion-deposition rate of the channel banks is comparable with that of the bottom.

These situations present insurmountable difficulties for an analytical approach and even a numerical model is very hard to manage if three-dimensional phenomena are to be reproduced in detail. For mountain rivers, then, a number of one-dimensional models have been developed, especially intended to reproduce *cross- section averaged* aggradation and degradation. From the practical point of view, in fact, the average bed elevation is probably the most important information as the bottom may vary by several meters during a single storm. Aggradation or deposition occurs whenever the sediment fed into the stream from the watershed is, respectively, larger or smaller than the stream transport capacity. Depending upon the time-scale one is interested in, one may resort to either long-term evolution models (for simulating slow aggradation or degradation over decades or centuries) or to short-term evolution models (for simulating sudden aggradation or deposition during a single flood).

8. LONG-TERM EVOLUTION MODELS

The time-step of long-term evolution models is typically one year or more. A reasonable hypothesis is that the morphological characteristics of the stream (width, slope, bed composition, etc.) remain almost constant during the hydrological cycle and that only the fluctuating water discharge (floods) controls the *intra-annual* variations of the solid discharge. In this way, by integrating the sediment transport formula over the hydrological year (via the water-flow duration curve) the annual sediment transport is obtained, so that only the slow evolution of the stream (*inter-annual variations*) is simulated by the model.

This hypothesis is definitely acceptable, even for mountain streams, in ordinary years; only if some catastrophic event takes place; as mentioned in Section 4 (Intense Sediment Yield and Transport), the intra-annual variations of width, slope and bed composition cannot be ignored and a short-term model should be used (see next Section 9). In the long-term evolution model proposed by D. PIANESE and F. ROSSI [paper C2], the annual solid discharge of different grain-size classes is obtained from the Parker's (1982) transport equation (including a hiding coefficient) integrated over the flow duration curve. The evolution of bed composition is taken into account by considering a bottom *"mixing zone"* (according to Hirano scheme), corresponding to that part of the cross-section involved in the intra-annual pulsations (i.e., the erosion-deposition phenomena occurring, on the average, over the year); the definition of the mixing zone includes, of course, the transversal migration of the braiding channels.

A similar long-term morphological model has been also applied by G. DI SILVIO and M. PEVIANI [paper C7] in order to compute the quasi- equilibrium initial conditions (bottom grain-size composition along the stream) at the beginning of an exceptional flood.

In principle, every stream evolves towards an *"equilibrium configuration"* , namely, it tends to assume a bottom slope and a bottom composition such that the annual transport of each grain-size class is equal to the average amount of sediment yielded every year from the watershed slopes. Indeed, what will eventually occur in a laboratory flume subject to a constant input of water and sediment, should also take place in a natural stream under constant morphoclimatic conditions.

In a natural stream, however, the time required to reach the *"full-equilibrium"* conditions is extremely long. At local scale, instead, *"quasi-equilibrium"* conditions will be holding; in these conditions the (annually averaged) bottom gradation displays almost imperceptible time-variations, while the changes of bottom elevation (although more appreciable) are also very small.

In a mountain natural stream exempt from lithological constraints (non erodable sills) *"quasi-equilibrium"* conditions correspond to a concave profile and to a stream-wise decrease of bottom granulometry. As shown in paper C1 by G. PARKER, the slow profile aggradation may be approximately described by a kinematic wave propagating with an extremely low celerity.

In fact, the aggradation rate of a mountain stream is equal to the horizontal celerity of the bottom profile times the local slope. According to the long-term evolution models mentioned above, the propagation celerity increases with the ratio between the sediment transport rate and the area of the "mixing zone", but it also depends on the differential mobility of each grain-size class, and therefore on the hiding-and-exposure effect. Simplified expressions for the celerity of the bottom profile have been obtained, under different approximations, in paper B11 and paper C1; numerical applications show that the aggradation rate in *"quasi-equilibrium"* conditions results to be extremely small (centimetres per year) even for very active mountain rivers.

Most of the long-term morphological models of mountain streams take into account the non-uniform grain-size distribution of the material, by disaggregating it into grain-size classes and by writing the transport and conservation equations of each class. A different and very promising way to deal with this problem consists in summing-up the respective equations for all the grain-size classes; in this way the mean diameter as well as the second (and possibly the third) moment of the grain-size distribution can be used as dependent variables instead of the percentage of each grain-size class. If the second (or third) moment of the curve may be assumed to be constant in space and time, the number of equations and variables are substantially reduced. A description of this model is provided by A. ARMANINI in paper C8, together with a discussion on the limits and validity of the procedure.

9. SHORT TERM EVOLUTION MODELS

While the annual sediment transport in ordinary years can be computed by assuming that the morphological characteristics of the stream (width, slope, bed composition, etc.) are virtually constant during the hydrological cycle, this is no longer possible if large and sudden inputs of sediment occur. Where a landslide reaches the stream, in fact, the local slope increases and - above all - the bottom granulometry decreases in such a way that the transport capacity becomes much larger than in ordinary years. This circumstance has frequently been observed in mountain rivers, as previously mentioned in Section 4 (intense Sediment Yield and Transport).

M. FUJITA *et alii* [paper B5] report the large increase of sedimentation rate observed in Ontake reservoir (Japan), after a large landslide produced by an earthquake fell in the upper part of the watershed. The reservoir sedimentation process has been simulated by a short-term one-dimensional model, prescribing the upstream boundary condition at short distance from the reservoir itself; the day by day simulation covers different periods, each one of several months duration, after the landslide occurrence. Comparison between measurements and computations shows that the bed grain-size distribution to be prescribed at the upstream boundary of the model is fundamental for a good simulation; in fact, the bottom granulometry in this Section became very fine just after the landslide occurrence, but it has progressively coarsened in the following years. According to the authors, in order to obtain a good estimate of grain-size distribution in this Section, the movement of the material should be simulated starting from the landslide location, several kilometres upstream.

A short-term evolution model that simulates the movement of material along the hydrographic network, after the occurrence of landslides and debris flows in the upper part of the watershed, is reported by G. DI SILVIO and M. PEVIANI [paper B11]. The simulation covers a period of two or three days, during a catastrophic storm in the southern slopes of the Alps. The overfeeding and consequent overaggradation of the streams have been reproduced in the model and compared with measurements. As in the previous paper the model shows that the configuration of the river bottom becomes much finer after the falling of the landslides, with a subsequent increase of the transport rate.

Morphological evolution of a small river in a very short time-scale has been numerically simulated by G.GHILARDI and G:MENDUNI in paper B10. Some interesting features peculiar to upland rivers (rapid change of boundary conditions, interaction of waterflow and sediment transport in supercritical conditions, effect of unerodible bottom under a granular layer) have been considered in the model.

10. RESERVOIRS

One of the most urgent problems of mountain region hydraulics is the filling process of reservoirs by debris and sediments. This natural process, in fact, has the double negative consequence of reducing the storage capacity and efficiency of the reservoirs and damaging the water intake ability, the bottom outlet and their control devices.

Flushing of reservoirs, on the other hand, must be very carefully conducted both because the large amount of stored water required and for the environmental consequences that the mud discharge produces.

In paper B5 M.FUJITA *et alii*, propose a mathematical model for simulating a reservoirs sedimentation processes. The model is the usual multi-fractions Hirano model, accounting for both bed-load and suspended-load. The gradation curve is divided in 11 classes. The model has been applied in the simulation of the sedimentation process after a catastrophic landslide, the volume of which was estimated to be approximately $36 \cdot 10^6\ m^3$. The paper points out that the upstream sediment transport rate is decreasing in time due to the progressive armouring of the material deposited on the reach of the river upstream the reservoirs.

The opposite problem of the sluicing of a mountain reservoir is tackled in the next paper B6 by H.SCHEUERLEIN. The creation of an artificial channel at the bottom of the reservoir is proposed with the purpose of facilitating the flushing. A simplified model is proposed in order to valuate the efficiency of flushing remedy.

11. PREVENTION AND CONTROL

Although the 'art' of torrent control and soil protection in mountain areas is very old, it has been not object of research as valley river training. The hydraulic engineering community, and especially the fluid mechanics community, has not paid equal attention to these problems, focusing much more on the latter which truly developed into a science.

Frequently the research activity in this field has been limited to the investigation of sediment transport formulas in steep channels or to some armouring phenomena or, just very recently, to the massive sediment transport, debris flow and mud flow.

Little attention has been devoted to the problems related to the hydraulic structures for torrent control. Very often the hydraulic criteria for the design of such works are rather empirical in nature, and in some case subjective or based on the personal experience of the designer.

In paper B13, by A.ARMANINI *et alii*, an effort is presented to draw the historical progress of designing torrent control works and, in particular, of check dams according to the present configuration.

As far as *open check dams* are concerned, the functional difference between filtering dam and dosing dam is described: this different function is less evident in the open dams having an intermediate configuration: hydraulic design criteria for such a kind of open dams are less precise.

In paper B12 H.P.WILLI offers a review of the most important typological intervention by the Swiss federal authority for water control. The review includes general guidelines for torrent training but, as in the previous case, the hydraulic and structural criteria for designing the works do not seem be supported by well developed theoretical foundations.

The traditional procedure of physical modelling is sometimes the most reliable feature for the analysis of complex hydraulic situations. In paper B3 W.BECHTELER *et alii* describe a mobile bed model investigation of a re-naturalization of an alpine river.

12. REFERENCES

Bathurst,J.C. (1978) Flow resistance of large-scale roughness. , Jour. of Hydr.Div., ASCE, Vol.104, No.HY12, December.

Campbell,R.H. (1975). Soil slips, debris flows and rainstormes in the Santa Monica mountains and vicinity, southern California, U.S. Geol.Survey, Prof. Paper 851.

Day,T.J.(1980). A Study of the Transport of Graded Sediments. HRS Wallingford Rep. No.IT 190,April.

Egiazaroff,I.V.(1965). Calculation of Non-Uniform Sediment Concentrations, Jour. of Hydr. Div., ASCE,Vol 91,HY4.

Einstein,H.A.(1950). The Bed-Load Function for Sediment Transport in Open Channel Flows. Tech. Bull. No.1026, U.S.Dept. of Agr., Soil Cons.Serv., Sept.

Mayer-Peter,E. and Müller,R. (1984), Formulas for bed-load transport, 2nd Meeting IAHSR, Stockholm, Sweden, pp 1-26.

Ranga Raju,K.G.(1985). Transport of Sediment Mixture, Ippen Lecture, XXI IAHR Congr., Melbourne, Australia.

Rijn,L.C. van (1984). the prediction of sediment transport and alluvial roughness, Jour. of Hydr. Div., ASCE, Vol.110.

Smart,G.M., Jäggi M.N.R.(1983). Sediment transport in steep slopes, Mitteilung der Versuchsansalt für Wasserbau, Hydrlogie und Glaziologie, N. 64, ETH Zürich.

Mizuyama,T. and Uehara,S. (1983). Experimental study of the depositional process of debris flow. Trans. Jap. Geomorph. Union, 41.

Schoklitsch,A. 1962, Handbuch des Wasserbaues, 3dt ed., Springer-Verlag, Vienna.

Hydrodynamics of Steep Channels and Local-scale Processes

FLOW RESISTANCE OVER A GRAVEL BED: ITS CONSEQUENCE ON INITIAL SEDIMENT MOVEMENT

Walter H. Graf

Laboratoire de Recherches Hydrauliques
Ecole polytechnique Fédérale, Lausanne, Suisse

ABSTRACT

The first part concerns itself with the friction factor of a gravel bed. Velocity distributions are measured in a gravel-bed flume having large slopes of $0.2 < S_o$ (%) < 2 and total water depths of $7 < D(\mathrm{cm}) < 23$. Two uniform gravel sizes of $d_s = 2.35$ cm and $d_s = 1.35$ cm are investigated. It is shown that : (i) the velocity distribution (see Fig.1) can be described by the logarithmic law, eqn (2) (see Fig.2a) in the inner region and by a parabolic law, eqn (4) (see Fig.2b) in the outer region; (ii) the friction velocities are reasonably equal to the ones computed from the energy slope; (iii) the position of the reference level, y_o, can be established. The flow-resistance relations, eqns (6) and (6a), were researched and rendered : (i) the numerical constants, Br (see Fig.3) and $\bar{B}r$, depend upon the relative roughness (see Fig.4); (ii) where 3 zones can be identified; (iii) zone 1 being for small relative roughness with $\bar{B}r \approx 6.25$, proposed by Keulegan (1938); (iv) zone 3 being for large relative roughness with $\bar{B}r \approx 3.25$, proposed by Graf (1984). Two independent laboratory experiments and one set of field data (see Fig.6) are used to demonstrate the validity of the proposed flow-resistance relation.
The second part deals with the consequence of the above-developed flow-resistance relation on the initiation of grain movement on the bed. The results for steep-sloped and gravel-bed channels do not seem to agree with the well-accepted Shields diagram (see Fig.8). The understanding of the hydrodynamics of the turbulent flow over rough surfaces, expressed with the flow resistance, eqn (6a), and an appropriate constant, $\bar{B}r$ (see Fig.4), help to explain the deviation from the Shields diagram if relative roughness are of importance, i.e. : $(d_s/D) > 0.04$. Data, now available in the literature, are used (see Fig.10) to present in a simple way for the determination of initial sediment movement.

FRICTION FACTOR OF GRAVEL BED

Turbulent Flow over Rough Surface

In turbulent flow over a rough, rigid boundary the turbulence is directly affected by forces resulting from flow around the roughness elements.

It is common to distinguish two regions : (i) a region close to the boundary, the wall or *inner* region, where the wall shear stress and the wall roughness are the important parameters and (ii) a region farther away from the boundary, the core or *outer* region, where the wall stress is the important parameter.

Inner Region (Fig. 1) : According to the mixing length concept of Prandtl a logarithmic distribution is given as :

$$\frac{u}{u_*} = \frac{1}{\kappa} ln \frac{y + y_o}{d_s} + Br \tag{1}$$

or written as a velocity-deficit distribution as :

$$\frac{u_{max} - u}{u_*} = \frac{1}{\kappa} ln \frac{D + y_o}{y + y_o} + B \tag{2}$$

where $u_* = \sqrt{\tau_o/\rho}$ is the friction velocity, Br is a numerical constant of integration, B is an additional correction term and κ is Karman's universal constant. If $\kappa = 0.4$, a value of $B = 3.7$ should be taken (Hinze, 1975 p. 631). Br depends on the size, shape and distribution of the roughness; the most probable value for walls with sand-grain roughness is given as $Br \approx 8.5 \pm 15\%$ (Reynolds, 1974, p. 187). There is some uncertainty in the definition of the position of y_o.

Mentioned should be that eqn (1) applies to this part of the boundary layer, where the shear stress stays reasonably constant, being limited to $y/D \lesssim 0.15$.

Outer Region (Fig. 1) : The conditions which led to the logarithmic velocity no longer are valid, and no theory is available. However, with experimental evidence available, self similarity is given according to Karman, with a velocity-defect distribution such as :

$$\frac{u_{max} - u}{u_*} = fct\left(\frac{y + y_o}{D + y_o}\right) \tag{3}$$

This equation applies to the entire boundary layer, including the inner region, given with eqn (2). In this equation, the roughness parameter, d_s , does not appear; thus it is valid for flow over rough and smooth wall roughness.

An empirical relationship for eqn (3) is given by Hama (Hinze, 1975, p. 631) as :

$$\frac{u_{max} - u}{u_*} = C\left(1 - \frac{y + y_o}{D + y_o}\right)^2 \tag{4}$$

being valid in the region of y/D > 0.15 and independent on the flow's Reynolds number; C = 9.6 is taken.

Another relationship for eqn (3) is given by Coles (Nezu and Rodi, 1986, p. 337) as :

$$\frac{u_{max} - u}{u_*} = \frac{1}{\kappa}\, ln\left(\frac{D + y_o}{y + y_o}\right) + \frac{2\Pi}{\kappa}\cos^2\frac{\pi}{2}\left(\frac{D + y_o}{y + y_o}\right)^{-1} \tag{5}$$

where Π is Coles' parameter expressing the strength of the wake function. Eqn (5) is valid throughout the entire boundary layer. For the inner region, limited to $y/D \lesssim 0.15$, Cole's parameter vanishes and eqn (5) merges into the usual log law.

Flow resistance : Integrating eqn (1) across the boundary layer gives according to Tu et al. (1988) :

$$\frac{\bar{u}}{u_*} = \frac{1}{\kappa}\, ln\, \frac{D + y_o}{d_s} + \left(Br - \frac{1}{\kappa}\right) \tag{6}$$

or written in its usual form as :

$$\frac{U}{u_*} = \sqrt{\frac{8}{f}} = \frac{1}{\kappa}\, ln\, \frac{R_h}{d_s} + \overline{Br} \tag{6a}$$

where R_h is the hydraulic radius, $\bar{u}$ and U are the depth averaged and the cross-sectional averaged velocity, and f is the friction factor.

The numerical constant is taken usually as $\overline{Br}$ = 6.25 (Keulegan, 1938, p. 727) for a range of $7.4 < R_h/d_s < 1030$, including gravel-size data of $7.4 < R_h/d_s < 20$. For gravel-size data it was proposed that $\overline{Br}$ = 3.25 (Graf, 1984) for a range of $1.0 < R_h/d_s < 10$.

Experimental Data

Velocity distributions have been measured with either micropropellers or with pitot-static tubes at the EPFL installation (Graf et Suszka, 1987). The flow rate covers a range of $21 < Q$ (ℓ/s) < 148, having bottom slopes of $0.2 < S_o$ (%) < 2 and total water depths of $7 < D$ (cm) < 23. Uniform gravel of d_s = 2.35 cm (set No. 1) and d_s = 1.35 cm (set No. 2) was used to make a moveable bed. 50 EPFL runs have been performed; the hydraulic conditions were kept to avoid sediment transport. A typical velocity distribution for run STEM4 is given in Fig. 1.

The EPFL data (set No. 1 only) are used to obtain from eqn (2) and eqn (4) simultaneously, according to a technique proposed by Vedula and Achanta (1985) and further developed by Tu, Tsujimoto, and Graf (1988), - firstly for $y_o = 0$, and subsequently for $y_o \neq 0$, using a technique proposed by Perry, Schofield et Joubert (1969) - the following results :

(i) The friction velocities, u_* , obtained from the velocity distribution, are reasonably equal to the ones calculated directly by $u_* = \sqrt{gR_hS_o}$.

(ii) The coefficient C in eqn (4) fluctuates between $7.9 < C < 11.4$, being on the average comparable to C = 9.6, as given by Hinze (1975).

(iii) The correction term B in eqn (2) fluctuates between $1.0 < B < 3.3$ being on the average smaller than B = 3.7 as given by Hinze (1975).

(iv) The limit of the separation between the inner and outer region fluctuates between $0.13 < y/D < 0.2$, being on the average comparable to y/D = 0.15, as given by Hinze (1975).

(v) The position of y_o in the velocity-distribution equations varies between $-0.3 < y_o/d_S < 0.4$ being on the average comparable to the values cited by Zippe et Graf (1983) as $y_o/d_S \cong 0.5$ and by Hinze (1975) as $y_o/d_S \approx 0.25$

The data for run STEM4 presented in Fig. 2 are shown before (· : y_o = 0) and after (/ : $y_o \neq 0$) the adjustment of the position; eqn (2) and eqn (4) are plotted.

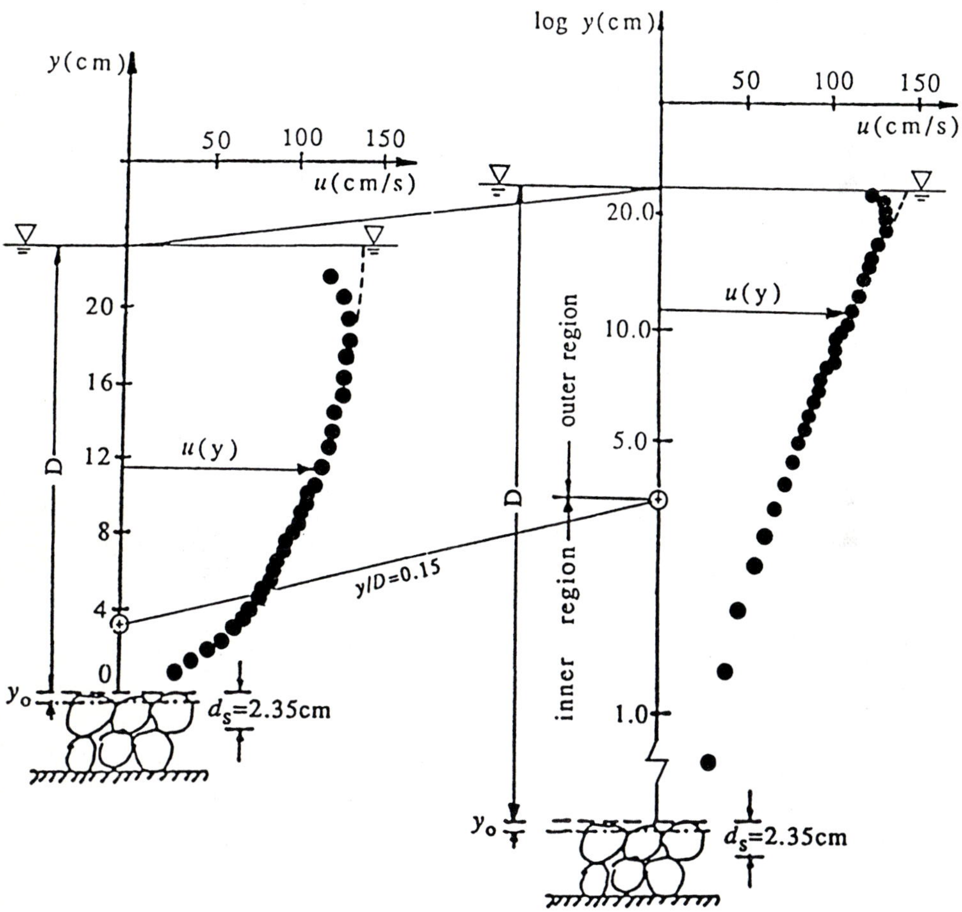

Fig. 1 Velocity distribution over gravel bed; run STEM4

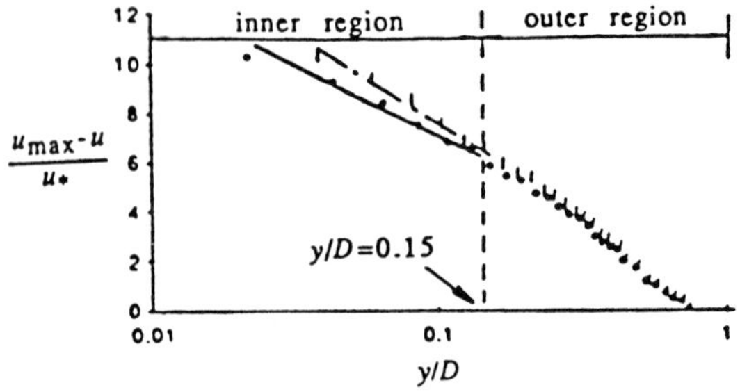

Fig. 2a Measured Velocity Distribution and eqn (2); run STEM4

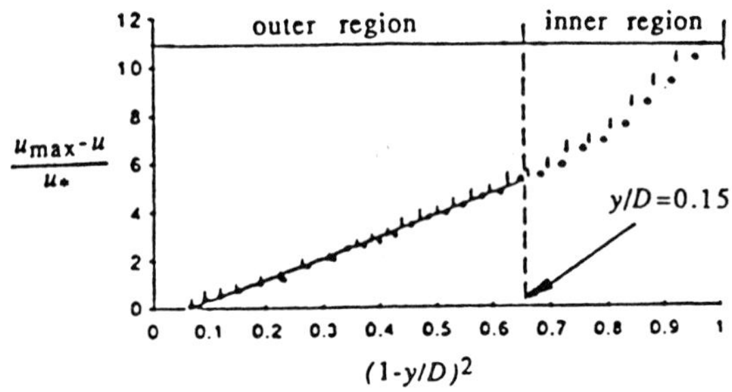

Fig. 2b Measured Velocity Distribution and eqn (4); run STEM4

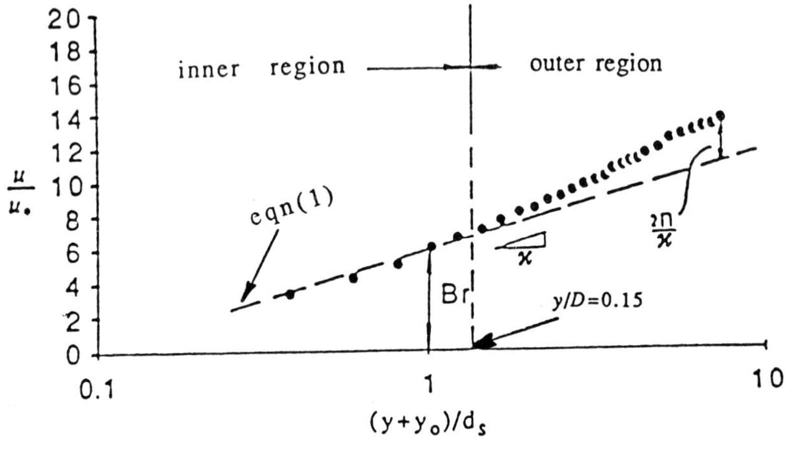

Fig. 3 Velocity Distribution; run STEM4;
y_o/d_s=0.17, Br=5.8, κ=0.4, Π=0.58

Above results will be used to investigate eqn (1). The velocity distribution for run STEM4 is given in Fig. 3. The EPFL data (sets No. 1 and No. 2) are used to obtain the following results :

(i) The numerical constant, Br , in eqn (1) and eqn (6) fluctuates between $3.8 < Br < 8.7$, being on the average smaller than $Br = 8.5$, as given by Keulegan (1938). A dependency of Br, but also of $\overline{Br}$ on the relative roughness, $d_s/(D + y_o)$ is evident in Fig. 4, where 3 zones might be distinguished :

Zone 1 : $Br = 8.5$ as quoted in the literature, being valid for small relative roughness, say $d_s/(D + y_o) < 0.05$; this value is in agreement with Keulegan's (1938, p. 727) research.

Zone 2 : $5.5 < Br < 8.5$ being a transition zone with a noticeable dependency on the flow's Froude number.

Zone 3 : $Br \cong 5.5$ being reasonably constant but showing a certain dependency on the flow's Froude number.

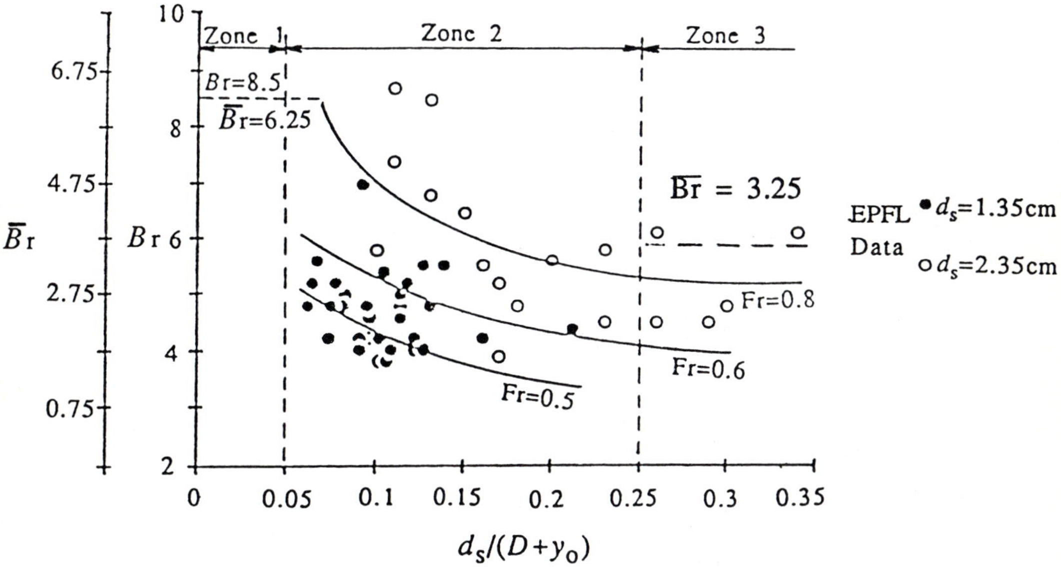

Fig. 4 Numerical Constants in eqn (6) versus Relative Roughness; EPFL Data

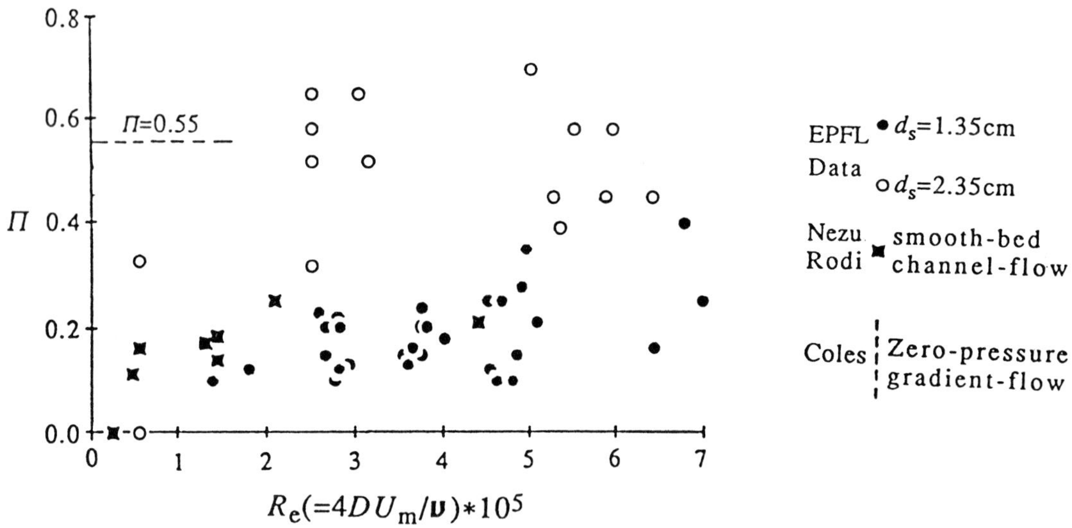

Fig. 5 Wake Strength, Π , in eqn (5) versus Reynolds Number; EPFL Data

(ii) The wake strength, Π , defined in eqn (5) and given in Fig. 5 fluctuates between $0 < \Pi < 0.7$, being on the average larger than $\pi \cong 0.2$, as given by Nezu and Rodi (1986) and Cardoso, Graf, and Gust (1989) for smooth channels, but about the same, $\Pi = 0.55$, as for zero-pressure gradient-flow (Hinze 1975, p. 697). A limited Reynolds number dependency may be detected.

Thus we reach a preliminary conclusion from our experimental data : that the friction factor, f , can be predicted with eqn (6a) together with Br, as given in Fig. 4. It remains to be seen if this prediction has general validity.

Two laboratory data sets obtained with the same EPFL installation are given in Cao (1985) and Suszka (1987). Only those data for flat beds but with and without sediment transport were used and are plotted in Fig. 6. Agreement between the measured (∘) and the with eqn (6a) calculated (+) values is reasonable within the to-be-expected spread.

Field data have been earlier investigated (Graf 1984, Table 2 and Fig. 5) and show reasonably agreement. Recently another set of field data by Colosimo et al. (1988) became available. As is to be seen from Fig. 6, the agreement is reasonable; d_{84} is used since the d_{50} value was not reported.

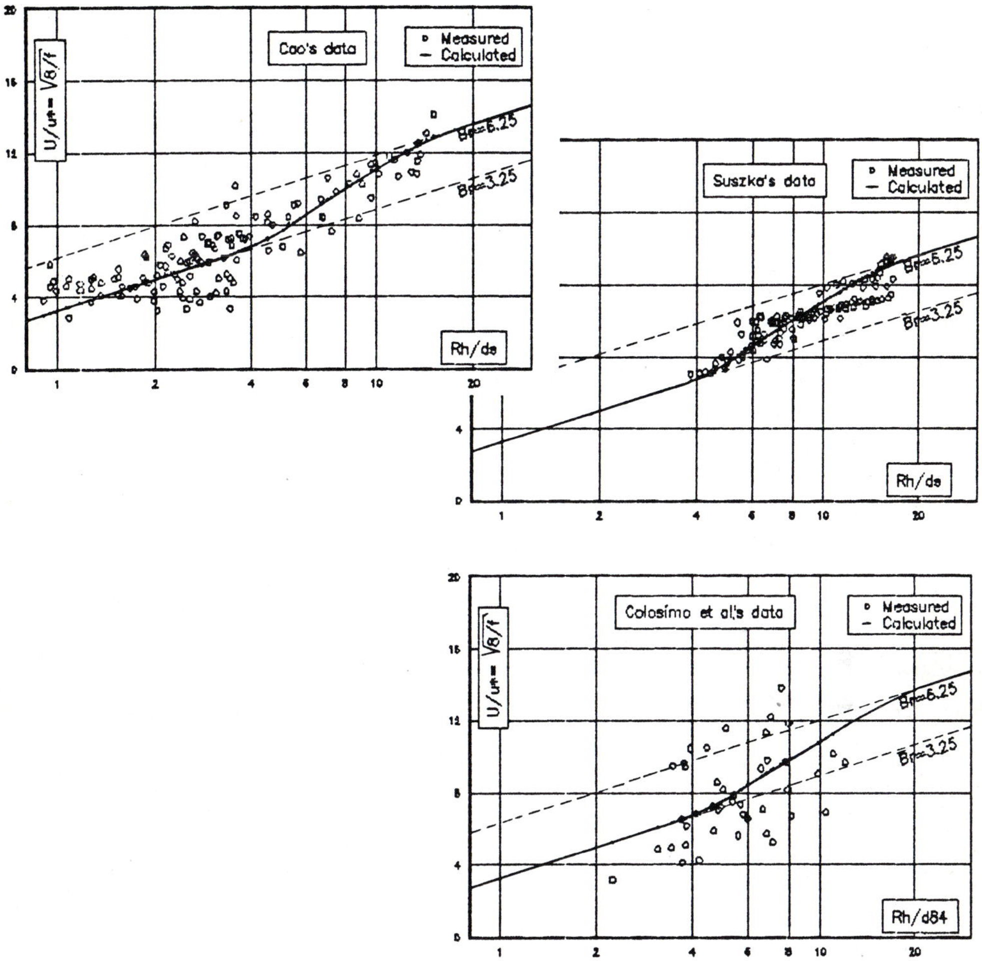

Fig.6 Friction factor versus Relative Submergence; data from Cao (1985), Suszka(1987) and Colosimo et al. (1988).

CRITICAL SHEAR STRESS

Influence of Slope

The condition of incipient motion under the action of the hydrodynamic drag, F_D, on the particle and its submerged weight, W, but neglecting the lift force, F_L, can be written (Graf, 1984 p.84) as:

$$\tan \Phi = \frac{W \sin \alpha + F_D}{W \cos \alpha} \tag{7}$$

where Φ is the angle of repose and α is the angle of the channel bed. Appropriate substitution for F_D and W renders:

$$\frac{(\tau_o)_{cr}}{(\gamma_s - \gamma)\, d_s} = A''' (\tan \Phi \cos \alpha \ - \sin \alpha) \tag{8}$$

where $(\tau_o)_{cr}$ is the critical shear stress, and A''' is a sediment coefficient. For channels with weak slopes, the relation, eqn (8), reduces to (Graf, 1984, p.91):

$$\frac{(\tau_o)_{cr}}{(\gamma_s - \gamma)\, d_s} = A'' \tag{9}$$

According to the Shields (1936) relation, it can be rewritten as:

$$\frac{(\tau_o)_{cr}}{(\gamma_s - \gamma)\, d_s} = \text{fct}\left(\frac{d_s\, u_*}{\nu}\right) \tag{10}$$

This well-known relation must be modified for strong slopes, according to eqn (8), and becomes (Bathurst et al., 1982):

$$\frac{(\tau_o)_{cr}}{(\gamma_s - \gamma)\, d_s\, (\tan \Phi \cos \alpha \ - \sin \alpha)} = \text{fct}\left(\frac{d_s\, u_*}{\nu}\right) \tag{11}$$

It can be shown (see Fig.7), that eqn (11) becomes identical with eqn (10), only for slopes being smaller than $S_o < 1\%$. Thus for larger slopes

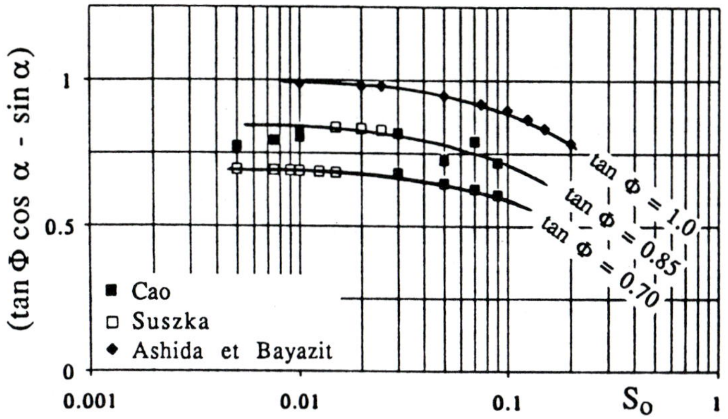

Fig.7 Plot of (tan Φ cos α - sin α) versus S_o

of $S_o > 1\%$, eqn (10) should be modified to read:

$$\frac{(\tau_o)_{cr}}{(\gamma_s - \gamma)\, d} = \text{fct}\left(\frac{d\, u_*}{\nu},\, S_o\right) \tag{11a}$$

Eqn (11) was compared with laboratory and field data in Fig.8, which does, however, not fully explain the phenomenon. Nevertheless, from Fig.8 we retain that the particle Reynolds number is of minor importance, if the data fall in a region of $Re_* = \frac{d\, u_*}{\nu} > 200$.

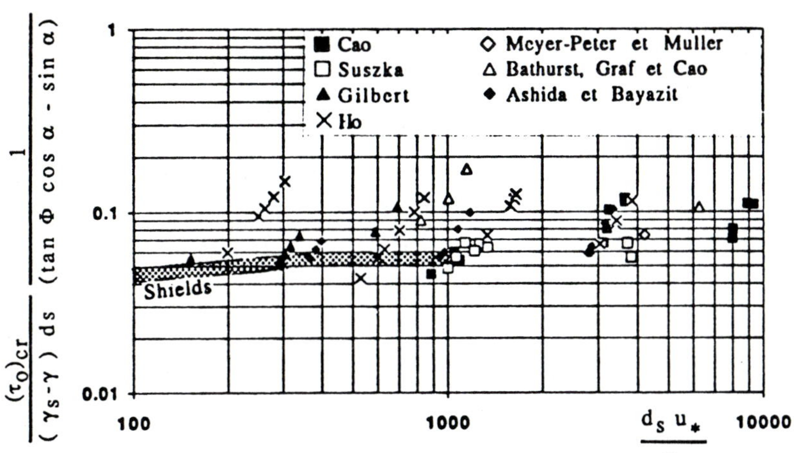

Fig.8 Plot of $\frac{(\tau_o)_{cr}}{(\gamma_s-\gamma)\, d_s} \frac{1}{(\tan \Phi \cos \alpha - \sin \alpha)}$ versus $\left(\frac{d_s\, u_*}{\nu}\right)$

For deep channels on weak slopes with small grain-sized material, when the relative submergence, D/d_s, is large, the Shields diagram has shown to be valid (Bettess,1984). However, for shallow channels on steep slopes with large grain-sized material, deviation from the Shields curve are evident (Bathurst et al.,1982 and Graf and Suszka, 1987) as seen here with Fig.8.

This fact was already remarkably well realized by Shields (1936) who observed that the slope appears in "only the product, $R_h S_o$, . . . , though an important exception must be mentioned, namely, in the case the slope becomes very appreciable. In this case, $R_h S_o$ is no longer constant . . ., but increases with increasing slope and decreasing depth". By plotting the dimensionless shear against the slope, for steep channel data, this dependency was shown in Fig.9 by Graf and Suszka(1987). " The explanation of this occurrence lies in the fact that with decreasing depth the relative roughness, d_s/R_h, increases . . . , the deviations are first noticeable with d_s/R_h values greater than ~ 1/25."

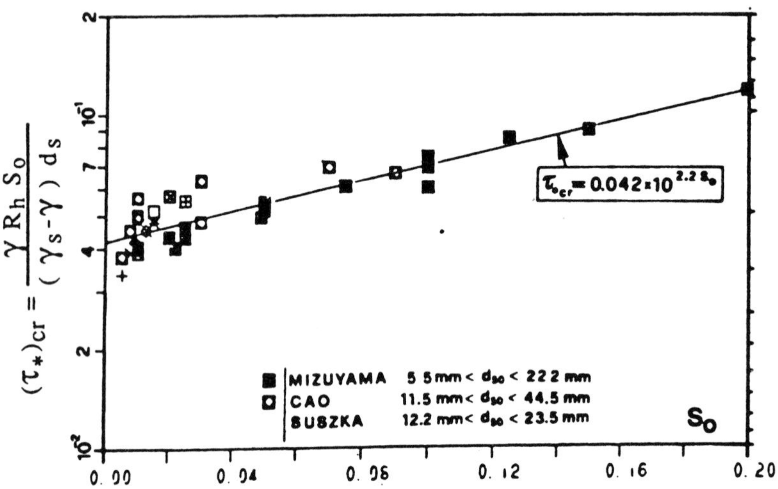

Fig.9 Dimensionless critical shear stress, $(\tau_*)_{cr}$, versus bed slope, S_o.

Influence of Relative Depth

In other words, Shields (1936) drew attention to the fact that his diagram renders only correct values if the relative roughness is $d_s/R_h < 0.04$ or the relative submergence is $R_h/d_s > 25$.

The importance of the relative roughness was already evidenced in the description of the flow resistance, given with eqn (6a). With the definition of the shear velocity, $u_*^2 = \tau_o/\rho$, eqn (6a) can be rewritten, assuming $R_h \approx D$, as :

$$\tau_o = \rho \left(\frac{U}{\frac{1}{\kappa} \ln \left(\frac{D}{d_s}\right) + \bar{B}r} \right)^2 \tag{12}$$

Thus the shear stress, τ_o, as well as the critical shear stress, $(\tau_o)_{cr}$, depend on the velocity distribution.
The data for steep channels, which exist in the literature, are now plotted in Fig. 10 according to :

$$\frac{(\tau_o)_{cr}}{(\gamma_s - \gamma)\, d_s\, (\tan \Phi \cos \alpha - \sin \alpha)} = \text{fct}\left(\frac{d_s}{D}\right) \tag{13}$$

Three zones might be distinguished :

Zone 1 : for $d_s/D < 0.04$, the dimensionless critical shear stress, $(\tau_*)_{cr} \approx 0.06$, is essentially constant. According to Shields (1936), the Shields diagram is valid. The velocity distribution of eqn (6a) is described by taking $\bar{B}r = 6.25$ after Keulegan (1938), as given here in Fig.4 (zone1).

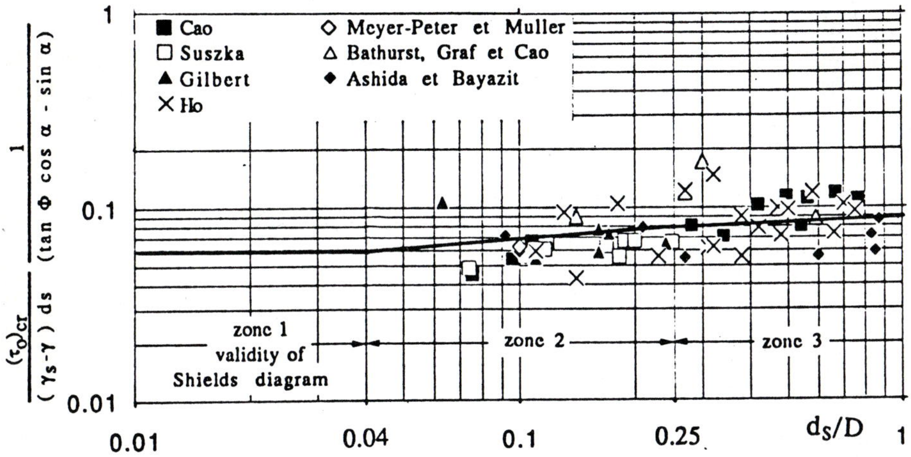

Fig.10 Dimensionless critical shear stress, $(\tau_*)_{cr}$, versus relative roughness, (d_s/D). Eye-fitted lines are drawn through experimental data for the steep-slope channels.

Zone 3 : for $d_s/D > 0.25$, the dimensionless shear stress, $(\tau_*)_{cr}$ = fct (d_s/D) , depends weakly on the relative roughness. The velocity distribution of eqn (6a) is described by taking Br = 3.25 after Graf (1984), as given here in Fig.4 (zone 3). A dependency on the flow's Froude number cannot be ruled out.

Zone 2 : for $0.04 < d_s/D < 0.25$, the dimensionless shear stress, $(\tau_*)_{cr}$ = fct (d_s/D), depends on the relative roughness. The velocity distribution of eqn (6a) is described by taking 3.25 < Br < 6.25 , as given in Fig.4 (zone 2)

Above preliminary conclusions are indicated with eye-fitted lines in Fig.10. The verdict is still out, and other data are still necessary to check the generality, but Bettess (1984) has similarly identified an existence of zone 3 in Fig.10.

CONCLUSIONS

Careful measurements together with a knowledge of theoretical hydrodynamics allowed to describe rather accurately the turbulent flow over rough surfaces.

Subsequently, this permitted to explain existing data on the initial sediment motion for different relative roughnesses. The Shields diagram was found to be only valid for small relative roughnesses, while for large relative roughnesses a useful diagram is proposed with Fig.10.

I wish to express my appreciation to my collaborators, B. Kironoto, H. Tu, and T. Tsujimoto, who participated at different stages in this research .

BIBLIOGRAPHY

Bathurst, J.C, Graf, W.H, and Cao, H.H. (1982); "Initiation of Sediment Transport in Steep Channels with Coarse Bed Material"; Mechanics of Sediment Transport, Euromech 156, Istanbul, A. Balkema, Rotterdam, NL.

Bettess, R. (1984); "Initiation of sediment transport in gravel streams"; Proc., Instn. Civ. Engg., 77, Part 2, Tech. Note 407, pp.79-88.

Cao, H.H. (1985); "Résistance hydraulique d'un lit de gravier mobile à pente raide; étude expérimentale"; doctoral dissertation, No. 589, Ecole Polytechnique Fédérale , Lausanne, Suisse.

Cardoso, A.H., Graf, W.H., and Gust, G. (1989); "Uniform Flow in Smooth Open Channel"; J. Hydr. Res., Vol 27, No.4

Colosimo, C., Copertino, V.A., and Veltri, M. (1987); "Friction Factor Evaluation in Gravel-Bed Rivers"; Proc., Am. Soc. Civ. Engg., J. Hydraulic Engg., Vol.114, No.8.

Graf, W.H. (1984); "*Hydraulics of Sediment Transport*"; Water Resources Publ., Littleton, Colorado, USA.

Graf, W.H. (1984),"Flow Resistance for Steep, Mobile Channels"; Seminar "Idraulica de Territorio Montano", Bressanone, Univ. Padua, Italy, Oct.

Graf, W.H., Cao, H.H. and Suszka,L. (1983); "Hydraulic of Steep, Mobile-Bed Channels"; Proc., XXI Congress, Int. Ass. Hydr. Res., Melbourne, Australia.

Graf, W.H., and Suszka, L. (1987); "Sediment Transport in Steep Channels"; Proc., Jap. Soc. Civ. Engg., J. Hydroscie. and Hydr. Engg., Vol.5, No.1.

Hinze, J.O. (1975); "*Turbulence*", McGraw Hill Book Company, New York.

Keulegan, G.H. (1938); "Laws of Turbulent flow in Open Channels"; J. Res. Nat. Bureau Stand., U.S. Dept. of Commerce, Vol.21, pp. 707 - 741.

Nezu, L and W. Rodi (1986), "Open Channel Flow Measurements with A Laser Doppler Anemometer"; Proc., Am. Soc. Civ. Engg., J. Hydraulic Engg., Vol.112, No.HY5, pp.335-355.

Reynolds, J.A. (1974); "*Turbulent Flows In Engineering*"; John Wiley & Sons Ltd, London.

Shields, A. (1936); "Anwendung der Aehnlichkeitsmechanik und Turbulenzforschung auf die Geschiebebewegung"; Mitteil., PVWES, Berlin, no. 26.

Suszka,L. (1987); "Sediment transport at Steady and Unsteady Flow; a Laboratory Study", doctoral dissertation, No. 704, Ecole Polytechnique Fédérale , Lausanne, Suisse.

Tu, H., Tsujimoto, T., and Graf, W.H. (1988); "Velocity Distribution in A Gravel-Bed Flume"; Proc., 6th Congress, Asian and Pacific Regional Div., Int. Ass. Hydr. Res., Kyoto, Japan.

Vedula, S and R.R. Achanta (1985); "Shear from Velocity Profiles: A New Approach"; Proc., Am. Soc. Civ. Engg., J. Hydraulic Engg., Vol.111, No.HY1, pp.131-143.

Yen, B.C. et al. (1989); "Incipient Sediment Transport in Steep Channels"; paper in preparation.

Zippe, H.J., and Graf, W.H. (1983); "Turbulent Boundary-Layer Flow over Permeable and Non-Permeable Rough Surfaces"; J. Hydr. Res., Vol 21, No.1

TURBULENT FLOW WITH SMALL RELATIVE SUBMERGENCE

Hiroji Nakagawa[(1)], Tetsuro Tsujimoto [(2)] and Yoshihiko Shimizu [(3)]

[(1)]Department of Civil Engineering, Kyoto University, Kyoto, Japan
[(2)]Department of Civil Engineering, Kanazawa University, Kanazawa, Japan
[(3)]Department of Civil Engineering, Ehime University, Matsuyama, Japan

ABSTRACT

The turbulent structure of flow with small relative submergence, which should be clarified for advanced study of sediment transport in mountain rivers, is different from that with sufficient relative submergence which is familiar to us according to the recent development of research of turbulent open channel flow. In this study, in order to clarify and understand the fundamental mechanism of flow with small relative submergence, turbulence measurements are conducted for flow over a bed composed of glass beads under small relative submergence. The "roughness sublayer" is recognized near the bed where the velocity profile becomes more uniform than the log-law profile and the turbulence is suppressed. These characteristics are attempted to be phenomenologically explained as an influence of eddy sheddings from individual roughness elements on the ordinary wall-turbulence.

1. Introduction

The turbulent structure of flow with small relative submergence is quite different from that of flow with sufficient submergence, and it must affect the mechanics of sediment transport and fluvial processes in mountain regions. In spite of many previous research works on mean flow over large roughness, the detailed structure of flow including turbulence characteristics has not been yet investigated.

As for the mean flow velocity, some researchers (Iwagaki 1955, O'Loughlin & Annambhotla 1969, Mizuyama 1977, Raupach et al. 1980) already pointed out that the velocity distribution near the bed becomes more uniform than the log-law profile, and some attempts were proposed to explain such degenerated velocity profile simply by modifying the mixing length or eddy kinematic viscosity but none related to the detailed turbulence structure. This near-bed region is often called "roughness sublayer".

The turbulence characteristics of flow with large submergence have been recently investigated to have several universal expressions on the statistical properties of them and to accumulate knowledges about mechanics of turbulence by focussing to their organized behavior such as bursting (Nakagawa 1984). Nevertheless, few have been done for flow with small relative submergence or flow over large roughness.

In this study, the flow measurements on glass-beads beds have been conducted mainly to clarify the turbulence characteristics which must degenerate the velocity profile.

2. Experiments

In the flume experiments, rough beds were made of closely packed spherical glass beads instead of natural gravel beds in order to avoid the irregularities involved in individual exposure and orientation of bed materials which actually affect much the hydraulic resistance but less the turbulence characteristics. Three sizes of glass beads were used: d=1.25, 1.7 and 2.97cm (d=diameter of bed materials).

Some runs were conducted with a hot-film anemometer in a 9m long 0.33m wide flume, and the other with a lazer Doppler anemometer (LDA) in a 20m long 0.4m wide flume. On the flow measurements, the sampling frequency was 200Hz, and the sampling period for each position was about 45s.

The experiments were conducted under uniform flow condition and covered the following range: i_b=0.0009~0.005; Fr=0.23~0.64; Re=3800~30000; and h/d=1.3~4.2, in which i_b=bed slope; h=flow depth; $Fr \equiv U_m/\sqrt{gh}$; U_m=depth-averaged velocity; h=flow depth; g=gravity acceleration; $Re \equiv U_m h/\nu$; and ν=kinematic viscosity.

3. Velocity Distribution, Reynolds Stress and Turbulence Intensity

Some examples of mean velocity distribution, U(y), is shown in Fig.1, and it is obviously recognized that the velocity distribution near the bed becomes more uniform than the log-law profile with the universal Kármán constant (κ=0.4). The "reference bed" (y=0) is here assumed to be the tangential surface of the roughness glass-beads. One would claim that the reference bed should be shifted down so as the measured distribution follows the log-law with κ=0.4. Certainly such procedure is less difficult in these cases, but the reason why the authors insist to avoid such modification will be referred to with the measured Reynolds-stress distribution soon.

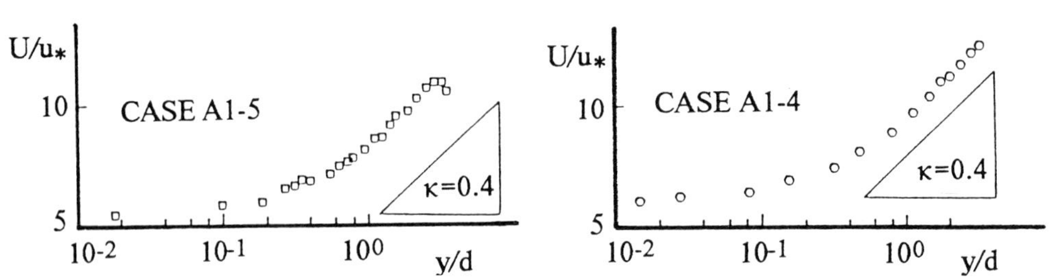

Fig.1 Mean velocity distribution.

Figure 2 shows some examples of the measured distribution of the Reynolds stress ($-\rho\overline{u'v'}$; ρ=mass density of fluid; and u', v'=longitudinal and vertical components of velocity fluctuation). What should be emphasized is that the Reynolds stress is obviously suppressed in the region near the bed. Though one could modify the velocity profile by changing the reference bed, one could not do the Reynolds-stress distribution. One cannot help recognizing the existence of a special region near the bed. The degenerated region of the Reynolds-stress distribution and that of velocity profile are quite consistent to each other. This region is here called "roughness sublayer. While, the region where the log-law is valid is called "inertial sublayer". If the Reynolds-stress distribution is certainly degenerated, the velocity distribution must be degenerated. When the vertical change of the mixing length is assumed to be invariant, the measured Reynolds-stress distribution well reproduces the measured velocity profile (see Fig.3). Any correction of the reference bed to insist the log-law is no longer significant. That is the reason why the authors reject the correction of the reference bed.

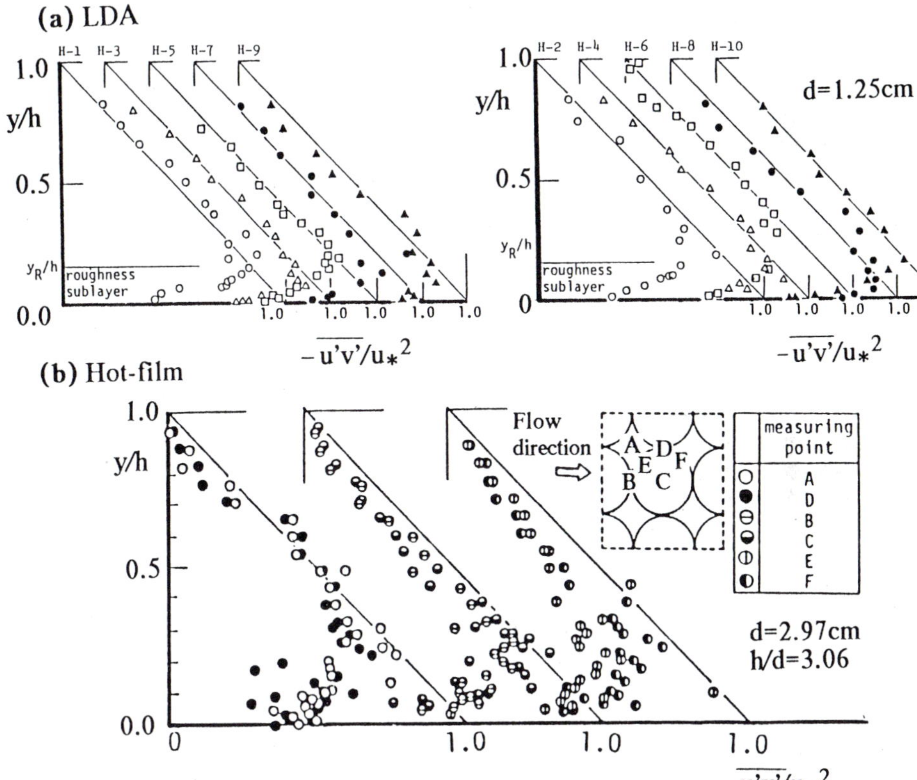

Fig. 2 Distribution of Reynolds stress.

Figure 2(a) show the data obtained with LDA, while Fig.2(b) the data with a hot-film. The suppression of the Reynolds stress is independent of the instruments. Moreover, the data obtained from the different position relative to the individual roughness (see Fig.2(b)) have no essential differences, and thus the suppression of the Reynolds stress near the bed is almost independent of the horizontal location. What fills the gap between this degenerated distribution of the Reynolds stress and the ordinary triangular distribution? It will be discussed in the next chapter.

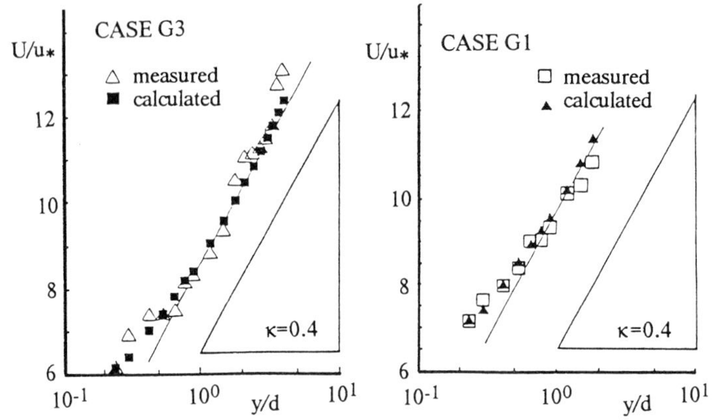

Fig.3 Calculated velocity distribution from degenerated Reynolds-stress distribution.

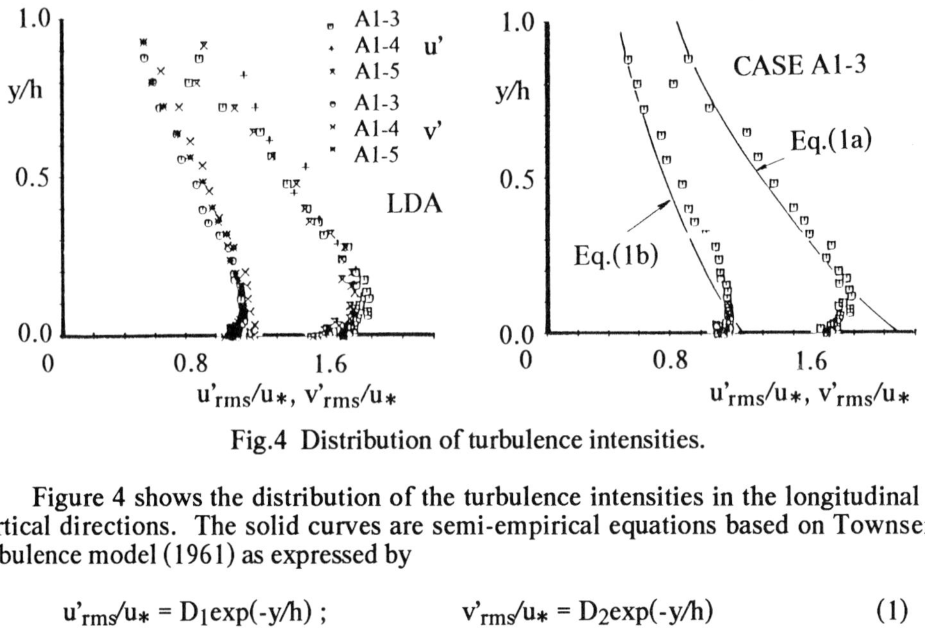

Fig.4 Distribution of turbulence intensities.

Figure 4 shows the distribution of the turbulence intensities in the longitudinal and vertical directions. The solid curves are semi-empirical equations based on Townsend's turbulence model (1961) as expressed by

$$u'_{rms}/u_* = D_1 \exp(-y/h) \; ; \qquad v'_{rms}/u_* = D_2 \exp(-y/h) \qquad (1)$$

in which u'_{rms}, v'_{rms}=turbulence intensities in the longitudinal and vertical directions; u_*=shear velocity; and the numerical values are determined based on the many laboratory data for flow with sufficient submergence: D_1=2.3 and D_2=1.27 (Nezu 1977). Obviously in the roughness sublayer, the turbulence intensities are smaller than those for flow with sufficient submergence. Fig.5(a) shows that the relation between the dimensionless correlation coefficient of the longitudinal and the vertical turbulence intensities, $r_T \equiv \overline{u'v'}/(u'_{rms}v'_{rms})$, and the relative height from the bed (y/h) is still preserved even in the roughness sublayer. The curve drawn in Fig.5(a) expresses the following equation deduced from the triangular Reynolds-stress distribution and Eq.(1).

$$-r_T \equiv \frac{-\overline{u'v'}}{u'_{rms}v'_{rms}} = \frac{1}{D_1 D_2}\left(1-\frac{y}{h}\right)\exp\left(\frac{2y}{h}\right) \tag{2}$$

Eq.(2) well represents the experimental data. This fact implies that the turbulence is suppressed in the roughness sublayer preserving the universal correlation structure between the longitudinal and vertical components. Although the distribution of turbulence intensities is degenerated from Eq.(1), the ratio of v'_{rms} to u'_{rms} is kept constant ($k_{uv}=D_2/D_1$) as shown in Fig.5(b).

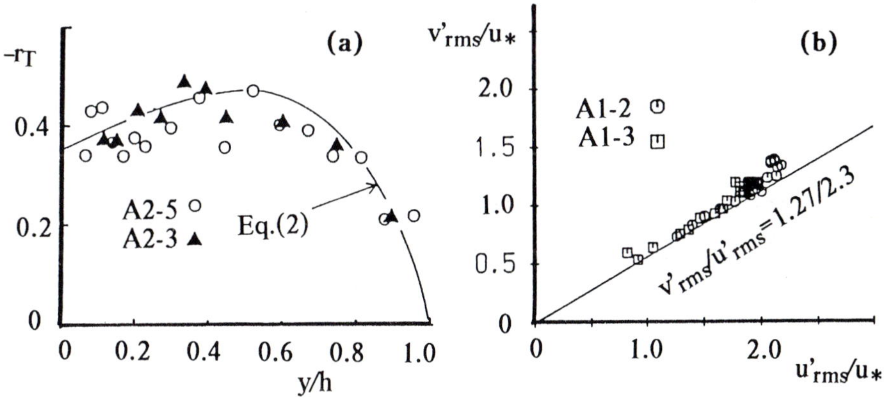

Fig.5 Correlation of velocity fluctuation between longitudinal and vertical directions.

In the roughness sublayer, the turbulence is suppressed and subsequently the Reynolds-stress distribution is degenerated to bring about a uniformalization of the velocity profile. The thickness of the roughness sublayer might be measured from any of Figs.1, 2 and 4, and those obtained from them are consistent to each other. The thickness of the roughness sublayer is almost (0.2~0.4)d. Thus, the roughness sublayer with the above-mentioned characteristics is emphasized particularly in the case of flow with small relative submergence.

The above-mentioned characteristics of the roughness sublayer have been also recognized for gravel-bed flow with small relative submergence (Tsujimoto 1989), though the the thickness of the roughness sublayer of gravel-bed flow is slightly larger (≅d) because of the higher irregularity of the gravel-bed surface than that of flow over the most packed glass beads.

4. Organized Structure in Roughness Sublayer

An inspection of the temporal auto-correlation of velocity fluctuation clarifies a distinguished difference of the fluctuating properties between the inertial sublayer and the roughness sublayer. Fig.6 shows the change of the feature of the temporal auto-correlation $r_{uu}(\tau)$ (τ=lag time) with the relative height. In the region far from the bed (outside of the roughness sublayer), $r_{uu}(\tau)$ decreases with the lag time as sometimes conveniently approximated by an exponential function. In the roughness sublayer, several sub-peaks are recognized in $r_{uu}(\tau)$, and it suggests an intensive periodic behavior. In other words, the velocity fluctuation in the roughness sublayer is somewhat "organized".

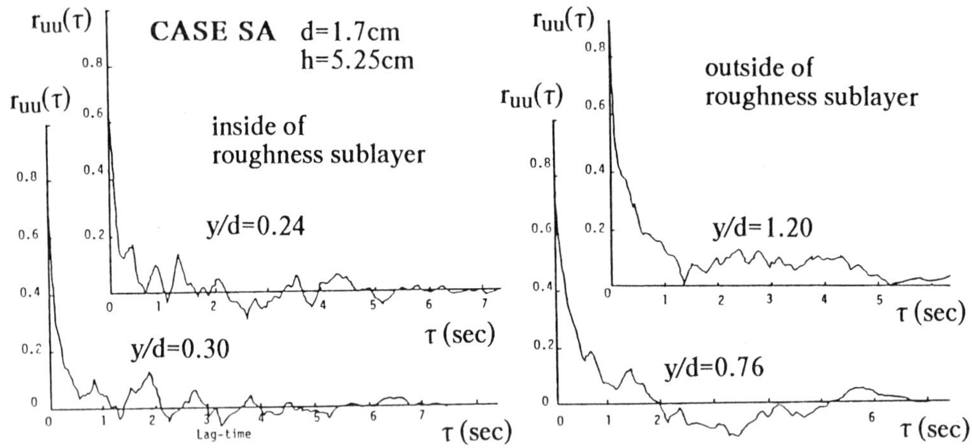

Fig.6 Temporal auto-correlation of velocity fluctuation.

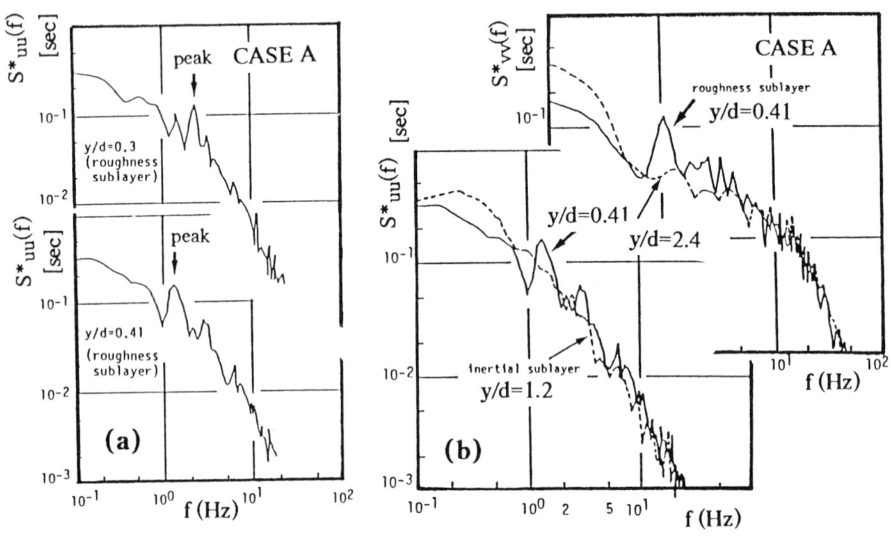

Fig.7 Spectra of velocity fluctuation.

Figure 7 shows the frequency spectra of the velocity fluctuation. In Fig.7(a), the spectra of the longitudinal velocity fluctuation in the roughness sublayer, and a peak is recognized at the frequency of the order of 1~3Hz. Such a peak is also recognized in the vertical component of fluctuation as shown in Fig.7(b). In the region far from the bed, no significant spectral peaks appear and the spectra are rather smooth as expressed by well-known universal forms (see the dashed curves in Fig.7(b)). The predominant frequency appearing in the roughness sublayer might correspond to the eddy-sheddings intervals from the individual roughness elements. Actually, the Strouhal number of the predominant frequency of the spectra is almost consistent to that of eddy-sheddings from the individual roughness elements.

According to the investigations of auto-correlation and spectral characteristics, we can guess that the momentum exchange in the roughness sublayer is indebted to the eddy sheddings from the individual roughness elements, the velocity fluctuation due to which is less random than the "ordinary turbulence" but somewhat periodic or "organized".

When an obvious "organized structure", which is horizontally fixed, is involved in the total velocity fluctuation, the velocity fluctuation might be favorably or formally divided into the fluctuating component due to such an organized motion and that due to turbulence, which are to be represented as (u_S, v_S) and (u_T, v_T), respectively. Then, the instantaneous velocity components in the longitudinal and vertical directions, u and v, are written as follows if a two-dimensional approximation is here adopted for simplicity.

$$u(x,y,t) = U(y)+\overline{u}_S(x,y)+u_S'(t)+u_T'(t) \tag{3}$$

$$v(x,y,t) = \overline{v}_S(x,y)+v_S'(t)+v_T'(t) \tag{4}$$

in which - means a time average; ' means a temporal fluctuating component from the time average; and U=horizontally and temporally averaged velocity. Therefore, the Reynolds stress obtained by a point measurement, τ_R, is written as

$$\tau_R \equiv -\rho\overline{(u-\overline{u})(v-\overline{v})} = -\rho(\overline{u_S'v_S'}+\overline{u_T'v_T'}) \tag{5}$$

in which a correlation between the organized component and the turbulence component has been reasonably neglected. By the way, the total momentum flux in the vertical direction, τ, is expressed by

$$\tau \equiv -\rho\overline{\overline{uv}} = -\rho\overline{\overline{(U+\overline{u}_S+u_S'+u_T')(\overline{v}_S+v_S'+v_S')}} = -\rho(U\overline{v}_S+\overline{u}_S\overline{v}_S)+\tau_R \tag{6}$$

Under uniform flow condition, $\tau(y)$ shows a triangular distribution, and the above equation implies that $\tau_R>\tau$ with upward secondary motion while $\tau_R<\tau$ with downward motion.

Take an example of the flow with stable cellular currents. Fig.8 shows the experimental results by Nakagawa et al. (1981). They conducted the turbulence measurements for the flow with cellular secondary currents brought about the lateral variations of bed roughness. Corresponding to the distribution of the vertical velocity component by cellular motion (Fig.8(a)), the distribution of the Reynolds stress shifts from the triangular one (Fig.8(b)), and then the velocity profile also shifts from the logarithmic law (Fig.8(c)). A convex profile of the Reynolds stress brings a velocity profile as shown by the curve (A) of Fig.8(c); while a concave profile of the Reynolds stress brings a velocity profile as shown by the curve (B). Thus, the flow with stable organized structures is characterized by an intensive "horizontal inhomogeneity".

For less stable cellular currents, which is brought about by unfavorable aspect ratio, the above-mentioned feature and "horizontal inhomogeneity" fade away more or less as shown in Fig.9 (the experimental results by Nakagawa et al. 1981), because instability of the location of cells makes the data from a point measurement somewhat spatially averaged.

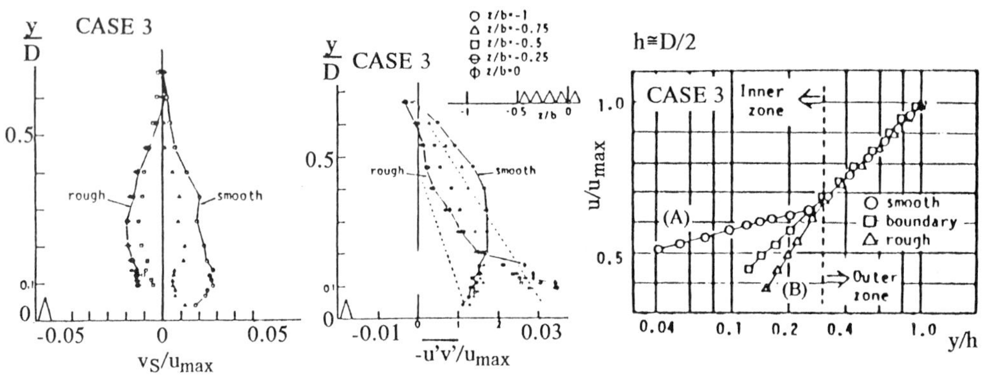

Fig.8 Flow with stable cellular secondary currents.

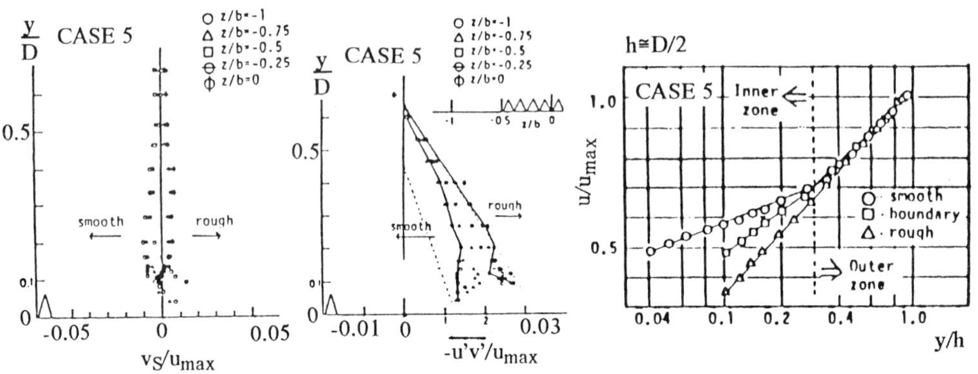

Fig.9 Flow with unstable cellular secondary currents.

The organized structure of flow in the roughness sublayer is further unstable or less fixed spatially, and thus "horizontal inhomogeneity" becomes no longer significant through a spatial averaging process in spite of the point measurement. A horizontal average of Eq.(6) yields

$$\langle\tau\rangle = -\rho\langle(U+\bar{u}_S)\bar{v}_S+\tau_R/\rho\rangle = -\rho\langle\bar{u}_S\bar{v}_S\rangle-\langle\tau_R\rangle \qquad (7)$$

$\bar{u}_S$ is positive when $\bar{v}_S$ is negative; while, $\bar{u}_S$ is negative when $\bar{v}_S$ is positive. Hence, $\bar{u}_S\bar{v}_S$ is always negative, and $\langle\bar{u}_S\bar{v}_S\rangle$ becomes negative. It means the suppression of the measured Reynolds stress ($\langle\tau_R\rangle$), because $\langle\tau\rangle$ must be balanced with the gravity.

The vertical component of flow, v_S, exists only the vicinity of the roughness elements, and it might have a maximum near the bed because it might be brought about by eddy-sheddings from individual roughness elements. Hence, the decrease of the Reynolds stress takes place only in the roughness sublayer, and the distribution of the Reynolds stress through the flow depth falls rather a convex profile independently of the measuring points. This is consistent with the measured distribution of the Reynolds stress of flow with small relative submergence shown in Fig.2, and this tendency is consistent to the fact that the measured velocity profile is uniformalized in the roughness sublayer.

Figure 10 shows a comparison among the flow with stable cellular motion, that with unstable cellular motion and the flow affected by the eddy-sheddings from the roughness. The view of this chapter well explains these different flow characteristics.

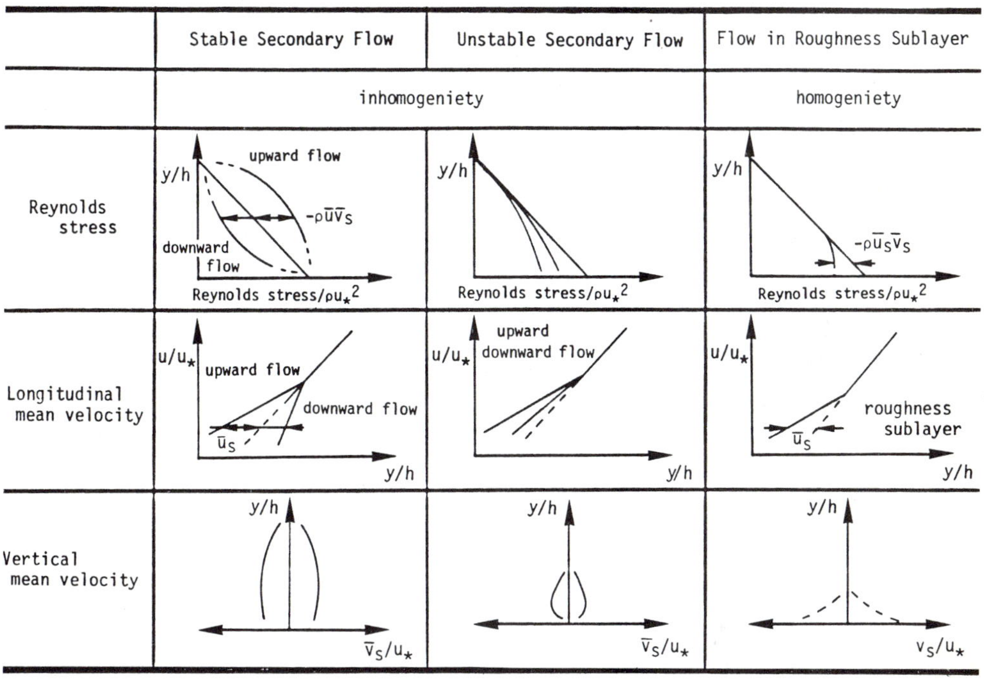

Fig.10 Comparison among flow with stable cellular secondary currents, flow with unstable cellular currents and flow with small relative submergence.

Further Discussion on Turbulence Characteristics

The characteristics of turbulence in the roughness sublayer are distinguished from those in the inertial sublayer. The difference are characterized by a suppression of the Reynolds stress in the roughness sublayer, which will be further discussed.

Recently, the contribution of the "bursting" phenomena on the Reynolds stress is focussed. The bursting is constituted by the gradual "sweep" ($u'>0$, $v'<0$) and the rapid "ejection" ($u'<0$, $v'>0$), and these cause a negative correlation of the velocity fluctuation. That is the "Reynolds stress". In order to realize the bursting, the following contribution of each event to the Reynolds stress is often inspected by turbulence measurements.

$$RS_i = \frac{1}{\overline{u'v'}} \lim_{T\to\infty} \frac{1}{T}\int_0^T u'(t)v'(t)I_i(t,H)\, dt \tag{8}$$

$$I_i(t,H) = 1 \;\; \left(|u'v'| \geq H|\overline{u'v'}| \right); \qquad I_i(t,H) = 0 \;\; \left(|u'v'| \leq H|\overline{u'v'}| \right) \tag{9}$$

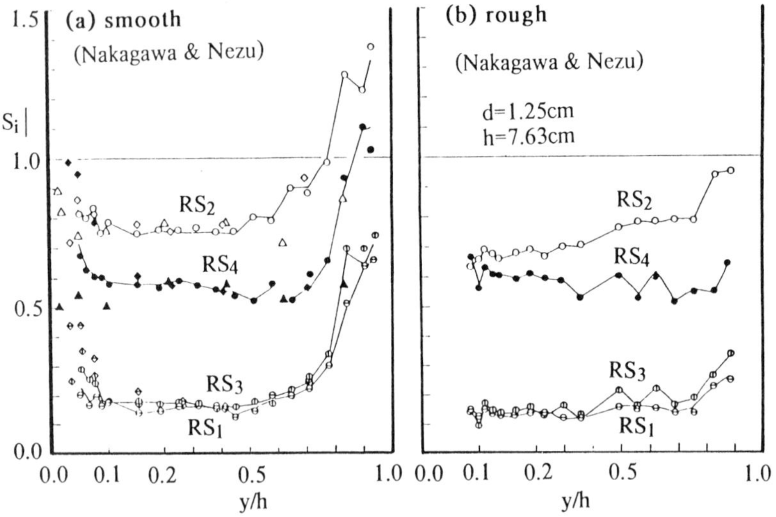

Fig. 11 Contribution of each event of velocity fluctuaion to the Reynolds stress for flow over smooth androuhg beds (Data obtained by Nakagawa & Nezu).

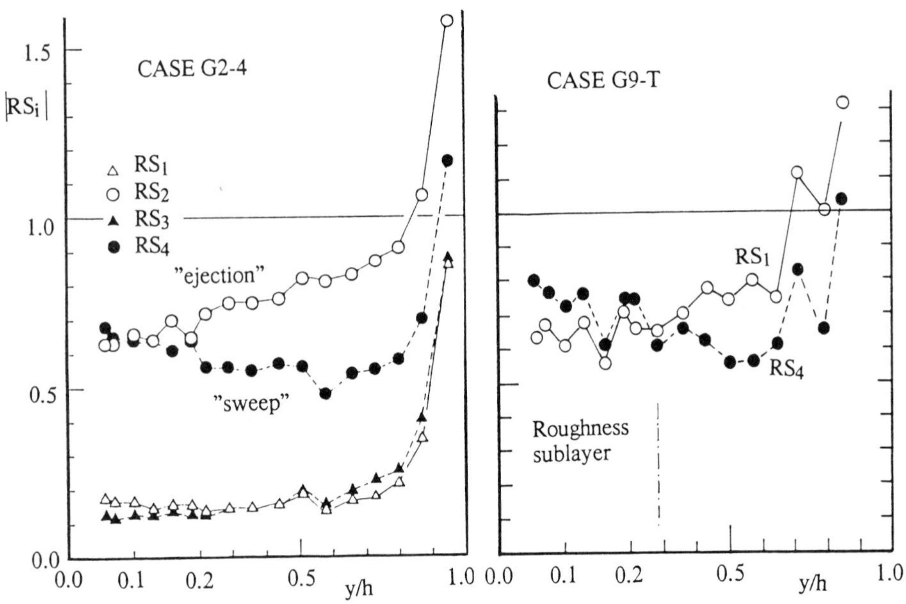

Fig. 12 Contribution of each event of velocity fluctuation to the Reynolds stress for flow with small relative submergence.

in which H is the threshold to distinguish the bursting; and the subscript i means as follows: i=1: $u'>0$ and $v'>0$; i=2: $u'<0$ and $v'>0$ (ejection); i=3: $u'<0$ and $v'<0$; i=4: $u'>0$, $v'<0$

(sweep). It is obvious that the events i=2 and 4 dominate and bring about the Reynolds stress. Fig.11 shows the experimental results (H=0) by Nakagawa & Nezu (1977), and it demonstrates that the "ejection" dominates particularly for the flow over a smooth bed. This fact makes the ejection often simply called bursting rather than the repetition of sweep and ejection. For the flow over a rough bed, the ejection is slightly suppressed near the bed, and it might be the reason why the bursting on the rough bed is not so distinguishable.

The data of the present study on the flow with small relative submergence are shown in Fig.12 (H=0). The suppression of the ejection event is more obvious. It suggests that the "bursting" might be degenerated in the case of flow over large roughness. The bursting might be no longer characterized as the repetition of gradual sweep and rapid ejection. One might guess that active momentum exchange due to turbulence is no longer required because the eddy-shedding pattern might be contributive to it. As a reference, for flow with cellular secondary currents, the upward motion suppresses the ejection and the downward motion promotes it apparently.

Conclusion

The experiments clarified the uniformalization of the velocity profile and the suppression of the turbulence intensities and the Reynolds stress in the roughness sublayer. The characteristics of the roughness sublayer are emphasized in the case of flow with small relative submergence.

In the roughness sublayer, the momentum exchange is indebted to somewhat organized flow pattern subjected to the eddy-sheddings from the individual roughness elements. It is suggested by an appearance of the relatively periodical component of low frequency in the velocity fluctuation as clarified by spectral analysis, and such an organized behavior is restricted in the roughness sublayer. Moreover, it is so unstable that even a point measurement detects a somewhat horizontally averaged behavior. The horizontally averaging process explains a deficit of the Reynolds stress in the roughness sublayer. An inspection of the structure of the correlation of the fluctuating velocity suggests that the suppression of the Reynolds stress might depend particularly on that of the "ejection", and this fact might make the bursting on the rough bed indistinguishable.

References

1. Christensen, B.A. (1971) : Incipient motion on cohesionless channel banks. *Proc. Sedimentation Symposium*, Berkeley, California, USA, Chapter 4.
2. Iwagaki, Y. (1955) : Study on surface erosion by sheet flow due to rainfall. *Doctoral Thesis*, Kyoto University (in Japanese).
3. Mizuyama, T. (1977) : Bedload transport in steep channels. *Doctoral Thesis*, Kyoto University, 96p. (in Japanese).
4. Nakagawa, H. and I. Nezu (1977) : Prediction of the contributions to the Reynolds stress from bursting events in open-channel flows. *Journal of Fluid Mechanics*, Vol.80, Part 1, pp.99-128.
5. Nakagawa, H., I. Nezu and A. Tominaga (1981) : Turbulent structure with and without cellular secondary currents over various bed configurations. *Annuals*, Disaster Prevention Research Institute, Kyoto University, No.24B-2, pp.315-338 (in Japanese).
6. Nakagawa, H., T. Tsujimoto and Y. Shimuzu (1988a) : Velocity profile of flow over rough permeable bed. *Proc. 6th Congress of APD-IAHR*, Kyoto, Japan, Vol.II, pp.449-456.

7. Nakagawa, H., T. Tsujimoto, Y. Shimuzu and S. Murakami (1988b) : Interaction between flow over rough permeable bed and flow in porous medium composed of bed materials. *Proc. of the 3rd International Symposium on Refined Flow Modelling and Turbulent Measurements*, Tokyo, Japan, pp.627-634.
8. Nezu, I. (1977) : Turbulent structure in open-channel flow. *Doctoral Thesis*, Kyoto University, 118p. (in Japanese).
9. O'Loughlin, E.M. and V.S.S. Annambhotla (1969) : Flow phenomena near rough boundaries. *Journal of Hydraulic Research*, IAHR, Vol.7, No.2, pp.231-250.
10. Raupach, M.R., A.S. Thom and I. Edwards (1980) : A wind-tunnel study of turbulent flow close to regularly arrayed rough surface. *Boundary-Layer Meteorology*, Vol.18, pp.373-397.
11. Townsend, A.A. (1961) : Equilibrium layers and wall turbulence. *Journal of Fluid Mechanics*, Vol.11, pp.97-120.
12. Tsujimoto, T. (1989) : Bed-load transport in steep channels. *International Workshop on Fluvial Hydraulics of Mountain Regions*, Trent, Italy.

FLOW RESISTANCE AND SEDIMENT TRANSPORTATION IN STREAMS WITH STEP-POOL BED MORPHOLOGY

Shinji Egashira and Kazuo Ashida

Disaster Prevention Research Institute
Kyoto University, Gokasho, Uji, Kyoto 611, Japan

ABSTRACT

Sediment transportation is changeable during a flood both in quantity and quality due to the interactions among the bed morphology, flow structure and bed material. Three subjects, which are associated with sediment phenomena in mountain streams, are chosen for discussion in the present study. (1) Studies on step-pool bed morphology, typical of mountain rivers, is reviewed with attention focused on the mechanism of its birth and geometric feature. (2) A resistance equation of the flow over step-pool bed forms is presented. In the derivation of this formula, the energy loss due to a strong eddy formed in the separation zone downstream of the wave crest is assumed and evaluated in terms of entrainment velocity. (3) Discussions are devoted to a mathematical model for predicting the sediment discharge rate of through puts or over-passing loads on stable step-pool bed systems.

INTRODUCTION

Mountain streams are characterized by such features as steep slope, a wide range of grain sizes, and small drainage area. Sediment motion, therefore, takes place in two regimes. The first of these, which occurs in a very steep region, is a massive movement such as debris flows. The second which is more familiar is manifested in terms of the mode of bed or suspended loads. Mechanisms of sediment transportation, therefore, are very complex.

With attention focused on reaches where the birth of massive movement cannot occur but the massive flow of sediment may continue, a step-pool bed system with an armored bed surface is generally formed there. Step-pool morphology is characterized by step-like geometric features spaced longitudinally down the channel, and interlocking structures composed of large imbricated stones or cobbles on the river bed. On the step surface, an armor coat is formed. Such armored bed forms may be produced, destroyed and reformed during floods, or on the other hand may occasionally survive a large flood without change. Flow regime and flow resistance may change coincidently with these. Sediment transport, therefore, is affected very much both quantitatively as well as qualitatively by the step-pool topography.

In the present paper three subjects, i.e. the birth and the geometric features of step-pool morphology, friction law of the flow over step-pool bed forms and sediment transportation on the beds with stable step-pool bed forms, will be chosen for discussion. The first of these is concerned with the genesis and geometric features of step-pool topography. The studies of these will be reviewed briefly. The second subject, which is concerned with the flow over the step-pool morphology, will be discussed with attention focused on the energy dissipation of mean flow. In the third, discussion will be made on the mechanisms of sediment transport on stable step-pool systems without changes during a flood.

GENESIS AND GEOMETRY OF STEP-POOL MORPHOLOGY

This chapter will be devoted to reviewal studies. Whittaker and Jaeggi [1], who may be a pioneer in this subject, studied the step-pool bed topography by means of flume experiments, and presented results on the hydraulic conditions for its genesis as well as for the wave length. According to them, the step-pool topography is produced for the condition;

$$0.075 < i_b \tag{1}$$

in which i_b is the bed slope. Furthermore, the wave length is closely related to those of antidunes.

The flume study performed by Whittaker and Jaeggi affected subsequent researches [2,3]. Ashida, Egashira and Ando [2] conducted a large number of flume tests as follows: The flume employed in the experiments is 6m long and 20cm wide. Three kinds of sediment were used; Sand-A (d_{max} = 3.8cm, d_m = 0.71 cm, $\sqrt{d_{84}/d_{16}}$ = 4.28), Sand-B (d_{max} = 2.54 cm, d_m = 0.53cm, $\sqrt{d_{84}/d_{16}}$ = 3.26), and Sand-C (d_{50} = 0.6cm, $\sqrt{d_{84}/d_{16}}$ = 1.24). Sand-C which is approximately uniform is employed to investigate the possibility of the genesis of step-pool morphology. Flow discharge were supplied at a constant rate in each test, with no sediment supply. In these experiments, the condition of the genesis, the developing process and the geometric features were investigated for step-pool morphology.

Photo 1 shows the final stage of step-pool topography obtained from the experiment; Q(flow discharge)=7.5 ℓ/s, Sand-A, i_b (initial bed slope)=0.05 and initial Froude Number=1.36. The flow discharge in this case is a little above the condition for the critical shear stress of the largest particle. At the initial stage of this test, untidunes appeared with active sediment transport. Then the transport capacity or tractive force decreases a little with bed degradation. Some largest particles stop at the upstream side of the crest of untidunes with active armoring and sorting of bed sediment. In the sorting process, part of sediment transported to the largest particles stopped already can be trapped there and the downstream side of these, on the other hand, is eroded. Through such processes, untidunes come to halt and change into step-pool morphology as shown in photo 1.

Photo 1 Step-pool topography produced by flume test.

In figure 1, the temporal changes of mean wave length and height, which are the results obtained from the test with $i_{bo} = 0.075$, $Q = 2.23\ell/s$, $Fro = 1.67$ and Sand-B, are shown. According to the results, the wave length is determined at an initial stage during active sediment movement, which emphasizes that the step-pool topography is not only caused by antidunes but also, as shown by Whittaker and Jaeggi, its wave length can be determined by these. With respect to the wave height, it may increase gradually to a statical equilibrium if sediment sorting or armoring phenomenon continues. It should be noted that step-pool topography cannot be produced, so long as using the uniform sand grains.

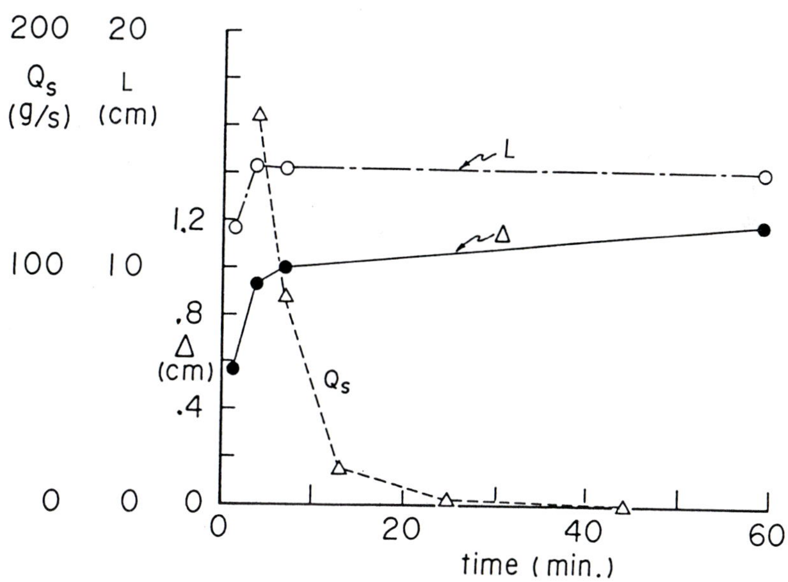

Figure 1 Temporal changes of wave length, Λ, and height, Δ, and sediment discharge, Q_s.

From experimental observations and results, Ashida, Egashira and Ando concluded that step-pool bed systems were produced only if the three conditions were satisfied; sediment mixture, upper flow regime and active armoring or sorting. The second and third conditions can be formulated;

$$\frac{h}{d_{mo}} \leq \left(\frac{\sigma}{\rho} - 1\right)\tau_{*mo}\left(6.0 + 2.5\ln\frac{h}{k_s}\right)^2 \tag{2}$$

$$\left.\begin{array}{l} u_*^2/u_{*cmo}^2 > 1 \\ u_*^2/u_{*cmax}^2 < 1 \end{array}\right\} \tag{3}$$

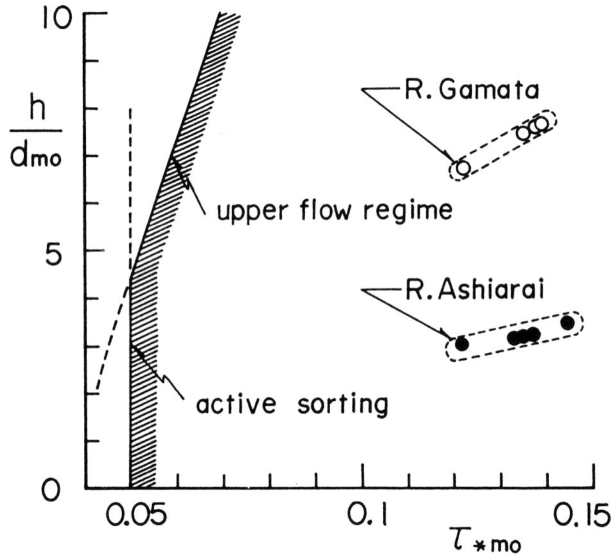

Figure 2 Domain for the birth of step-pool morphology.

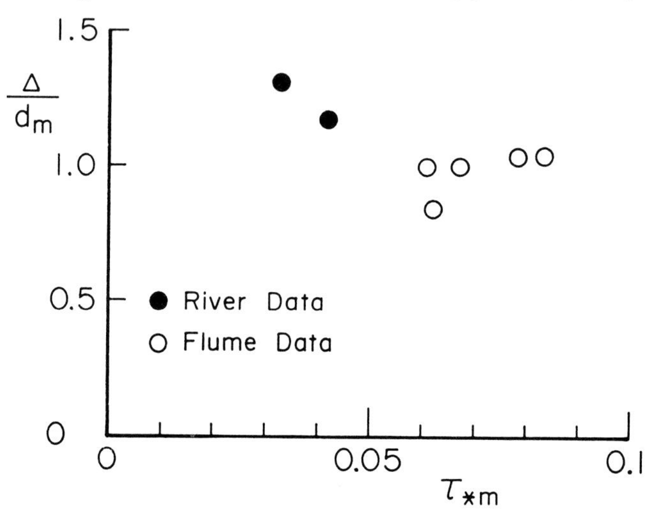

Figure 3 Mean wave heights of step-pool morphology.

in which h is the flow depth, d_{mo} is the mean size of initial sediment, σ is the mass density of the sediment, ρ is the mass density of water, u_* is the shear velocity, u_{*cmo} is the critical shear velocity for d_{mo}, u_{*cmax} is the critical shear velocity for the maximum grain size, τ_{*mo} is the non-dimensional shear stress in terms of d_{mo}, and k_s is the equivalent roughness.

In figure 2, the above relations are shown with using $k_s = d_{mo}$. Data collected by Egashira, Ashida, et al. [4] in mountain streams with step-pool topography are plotted in the formative region of the bed forms. Those of the flume tests [2], which are not shown herein, were successfully explained in the formative region.

The wave heights of step-pool morphology were investigated by Egashira et al. [2,4] in terms of dimensional analysis. The results are shown in figure 3 with data from flume tests as well as from mountain streams. Therein, Δ is the mean wave height and d_m is the mean particle size of armor coats.

According to the results, wave heights could be explained in terms of d_m;

$$\Delta / d_m = \text{const} \approx 1 \tag{4}$$

The above relation is very suggestive, taking the varieties of mean particle size of armor coats from 1.0cm to 50cm into consideration.

FLOW RESISTANCE

DERIVATION OF FRICTION FACTOR

The flow depths in mountain streams are generally so shallow that the flow structure might be affected directly by the irregularities of flow boundary or grains composing the bed. The flow in mountain rivers, therefore, is very complex. There have been many researches on the friction low for such flows [5, 6, 7]. Most of these emphasize that the friction factor may be expressed in terms of h/d (flow depth/representative grain size) and one associated parameter. These may be useful for such flows as in the reaches without a stable step-pool bed form.

But, the flow over a step-pool bed form takes a variety of flow regime such as subcritical flow, supercritical flow and chutes and pools. Both in sub-and super-critical flows, the rate of energy dissipation in the separation zone downstream of the crest plays an important role in the flow resistance, and in the flow of chutes and pools the energy dissipation due to hydraulic jump may control the friction law in a sense of longitudinally spatial mean. We will discuss friction factor from a view point mentioned above.

In figure 4, three flow patterns over step-pool topography are shown with notations used herein. The stream line along the bed surface separates at the crest, and reattaches to a rear of the crest. A strong eddy motion which may be called an "entrained eddy" is formed in the separation zone. This eddy may be produced by turbulence into which kinetic energy is transferred from the mean flow. Such kinds of eddy motion are very similar to those formed in the regions downstream of a step-down [8] as well as of crests of general sand waves [9]. According to the reports by Raudkivi [9] and Imamoto, et al. [8], it is found that turbulent intensity or turbulent energy is very large outside the demarcation line between the main flow and the entrained eddy and, on the other hand, very weak inside it. This brings about that part of the mean flow energy causes a turbulence and dissipates due to the formation of an entrained eddy.

Such an energy cascade occurs in the flows of turbulent jet [10] in homogeneous media as well as in density stratified field. The flow rate of mass and energy at the boundary can be generally specified in terms of the "entrainment velocity".

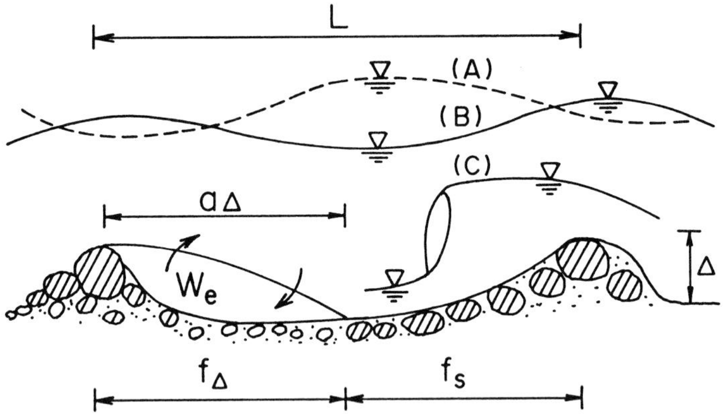

Figure 4 Schematics of flow over a step-pool bed form.

The process of energy dissipation mentioned above can be roughly formulated as follows. Referring to the notations shown in figure 4, the energy dissipation of mean flow in the region between the crest and the reattachment point is expressed in terms of mean velocity and friction factor;

$$\frac{1}{8} f_\Delta v^3 a\Delta = \int_A -\overline{u'w'}\,\frac{d\overline{u}}{dz} dA \tag{5}$$

in which f_Δ is the friction factor of the region defined in figure 4, v is the mean flow velocity, $a\Delta$ is the length of separation zone, Δ is the wave height, a is the empirical factor associated with the length of separation zone, $-\overline{u'w'}$ is the Reynolds stress and A is the area of integration which is specified as

$$A = k_1 a\Delta^2/2 \tag{6}$$

Herein, k_1 is the coefficient of order one. Reynolds stress may be expressed, using the entrainment velocity, W_e;

$$-\overline{u'w'} = k_2 v W_e \tag{7}$$

in which k_2 is the coefficient of order one. Entrainment velocity can be evaluated in terms of entrainment coefficient, E_n.

$$W_e = E_n v \tag{8}$$

The velocity gradient can be roughly estimated as

$$\frac{d\overline{u}}{dz} = k_3 \frac{v}{h} \tag{9}$$

in which k_3 is the coefficient of order one and h is the flow depth longitudinally averaged.

Substituting equations from (6) to (9) into (5) yields

$$f_\Delta = 4KE_n\frac{\Delta}{h} \;, \quad (K = k_1k_2k_3) \tag{10}$$

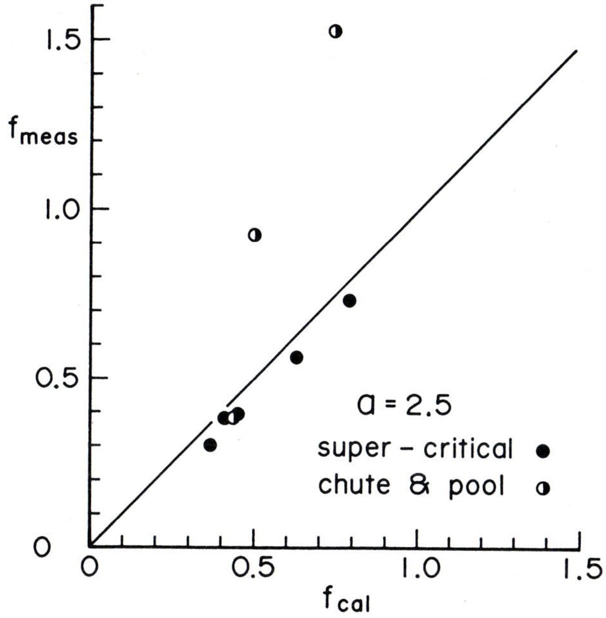

(a) Flow over an artificial step-pool bed form

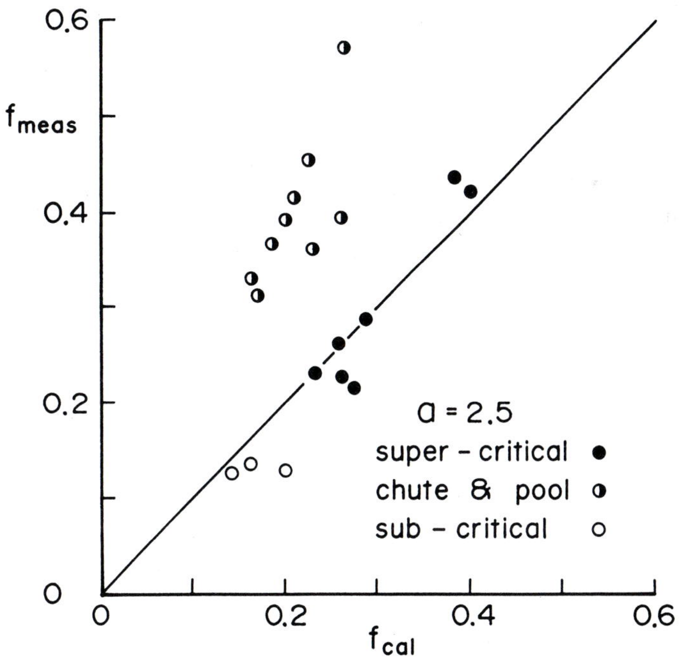

(b) Flow over step-pool bed forms produced experimentally

Figure 5 Comparison between flume data and calculated values with equation (12) for friction factors.

Two empirical coefficients, K and E_n, in equation (10), should be determined from world-widely known data. But herein, the factor of

$$E_n = 0.08$$

will be assigned to the entrainment coefficient according to the results for the flow of a half jet.

With respect to K, it was determined by the flow over artificial, two-dimensional dunes;

$$K = 6.0$$

These two factors will be hold so long as some contradictions do not occur.

The friction factor, f_s, in the region between the reattachment point and the wave crest or in the wall region can be estimated by the logarithmic law for velocity distribution. Consequently,

$$f_s = \frac{8}{\{6.0 + 2.5\ln(h/k_s)\}^2} \tag{11}$$

From equations (10) and (11), one can obtain the friction factor averaged in the region from one crest to the next;

$$f = \frac{a\Delta}{L} f_\Delta + \left(1 - \frac{a\Delta}{L}\right) f_s \tag{12}$$

except the flow in chutes and pools.

In case of chutes and pools, we should take the energy dissipation due to hydraulic jump into consideration. Subsequently, the friction factor is expressed as

$$f = \frac{a\Delta}{L} f_\Delta + \left(1 - \frac{a\Delta}{L}\right) f_s + \frac{8}{F_r^2}\frac{\delta H}{L} \tag{13}$$

in which F_r is the Froude Number and δH is the energy loss due to hydraulic jump.

APPLICATION OF THE RESISTANCE LAW TO FLUME DATA

Flume tests were performed to collect data of friction factors for flows [11,12] over an artificial step-pool bed form with the wave length 20cm and the wave height 2.27cm which is made of glass beads with the diameter 2.4cm as well as over step-pool bed forms experimentally produced, using sands and gravels with the maximum size 3.80cm, mean size 1.31cm and $\sqrt{d_{84}/d_{16}}$=2.27.

In figures 5(a) and (b), experimental data are compared with those calculated by equation (12) for friction factors. As shown in the figures, experimental data are well predicted by equation (12) except those of chutes and pools. Data of chutes and pools are under estimated because the equation (12) does not account for the energy loss due to hydraulic jump. It should be noted that the value, a = 2.5, somewhat smaller than that in the case of two-dimensional dunes, is used in these calculations, because the length of separation zone shortens due to the effects of flow around particles composing the crest.

SEDIMENT TRANSPORT ON THE BED WITH STEP-POOL MORPHOLOGY

MATHEMATICAL MODEL

Sediment transport in mountain streams sometimes is very scarce or occurs much less than that expected in an equilibrium state even if the flow discharge exceeds over some critical values.

Such occasions prevail in the channels with stable step-pool systems or fully armored beds where the sediment transportation is controlled by the rates of sediment supply from upstream region. In such conditions sediment particles may be transported as over-passing loads or through puts, filling pools and pores in armored bed surface with themselves or eroding themselves from the bed. But original step-pool systems with armored bed surface remain unchanged although the concentration of each grain fraction in bed surface changes temporally and spatially due to sediment depositions and erosions. A mathematical model to evaluate such a sediment transportation will be discussed.

(a) Sediment Transport on Flat, Armored Bed

Let us consider, at first, the transport process of sediment particles on the armored bed without step-pool bed forms. In figure 6, it is shown that sediment grains which are finer than those of armored bed surface are transported, filling the pores among larger grains composing the bed surface with themselves. The sediment storage, V, among the pores is written in unit area as

$$V = kd_a A_s / A_{se} \tag{14}$$

in which d_a is the mean diameter of the armor coat, A_s is the area covered by the sediment to be transported, A_{se} is the area covered all over by the transported sediment and k is the empirical constant estimated to be 1/6.

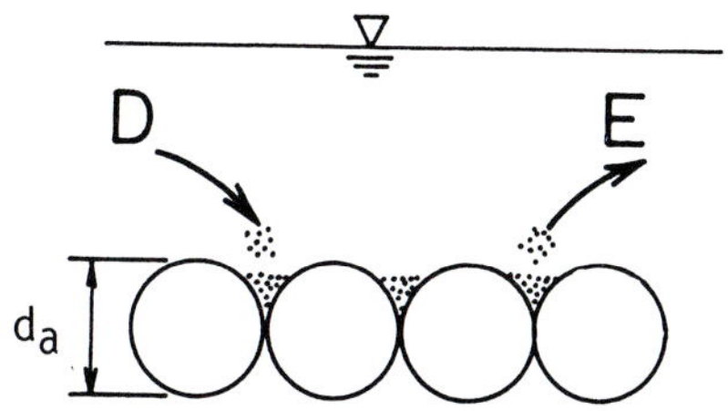

Figure 6 Sediment transport process on flat bed with armored surface.

The rate of change of sediment storage can be expressed in terms of erosion and deposition velocities, E and D;

$$kd_a \frac{\partial}{\partial t}\left(\frac{A_s}{A_{se}}\right) = \frac{1}{1-\lambda}(D - E) \tag{15}$$

in which λ is the porosity of sediment. Spatial change of sediment transport rate is also written as

$$\partial q_b / \partial x = E - D \tag{16}$$

in which q_b is the sediment transport rate on the armored bed. Erosion velocity is considered to be proportional to the ratio of coverage, A_s/A_{se}, and, therefore, takes a form;

$$E = (k_3/k_2)(A_s/A_{se}) d P_s \tag{17}$$

in which k_3/k_2 takes the value of 2/3 in case of spherical grains, d is the grain size of sediment and P_s is the pick-up rate, and it is given by Nakagawa and Tsujimoto [13].

$$P_s = F_O \tau_* \left(1 - \frac{\tau_{*c}}{\tau_*}\right)^3 \left\{\frac{(\sigma/\rho - 1)g}{d}\right\}^{1/2} \tag{18}$$

Herein, F_o is the empirical coefficient of 0.03 to 0.035, τ_* is the nondimensional tractive force, σ is the mass density of sediment, ρ is the mass density of water and g is the gravitational acceleration. Deposition velocity may be proportional to the sediment transport rate and in inverse proportion to the step length, Λ, of sediment particle;

$$D = q_b/\Lambda \tag{19}$$

We can calculate the sediment transport rate on flat bed with armored surface, using equations (15) and (16) with supplementary equations from (17) to (19). But, a further discussion is desirable, in order to investigate the transport process of through puts. So, let us introduce nondimensional parameters with "prime" defined as follows.

$$t = \sqrt{d/\{(\sigma/\rho - 1)g\}}\, t' \qquad x = \Lambda x'$$

$$A_s = A_{se}A'_s \qquad q_b = q_{be}q'_b$$

Herein, A'_s should be called a "coverage ratio" and, in additions, Λ and q_{be} can be written respectively;

$$\Lambda = \Lambda' d \tag{20}$$

$$q_{be} = (k_3/k_2)P_s\Lambda' d^2 \tag{21}$$

Substitution of the relations defined above into equations (15) and (16) yields the formulas;

$$\frac{\partial A'_s}{\partial t'} = C(q'_b - A'_s) \tag{22}$$

$$\frac{\partial q'_b}{\partial x'} = A'_s - q'_b \tag{23}$$

in which C is the parameter which controls the propagation of sediment particles and it is defined as;

$$C = \frac{k_3}{k_2}\frac{F_o}{1-\lambda}\frac{1}{k}\frac{d}{d_a}\tau_*\left(1 - \frac{\tau_{*c}}{\tau_*}\right)^3 \tag{24}$$

Equations (22) and (23) bring about the differential equations with hyperbolic type for A'_s and q'_b,

$$\frac{\partial^2 A'_s}{\partial t'\partial x'} + \left(1 - \frac{1}{C}\frac{\partial C}{\partial x'}\right)\frac{\partial A'_s}{\partial t'} + C\frac{\partial A'_s}{\partial x'} = 0 \tag{25}$$

$$\frac{\partial^2 q'_b}{\partial t'\partial x'} + \frac{\partial q'_b}{\partial t'} + C\frac{\partial q'_b}{\partial x'} = 0 \tag{26}$$

In equation (25), the condition,

$$1 - \frac{1}{C}\frac{\partial C}{\partial x'} \doteq 1$$

can be generally accepted.

(b) Sediment Transport on Step-pool Topography with Armored Bed

In channels with step-pool topography, sediment transportation depends on the sediment volume stored in the pool as well as on the coverage ratio, A'_s, in the region of flat, armored bed. Such an occasion is shown in figure 7.

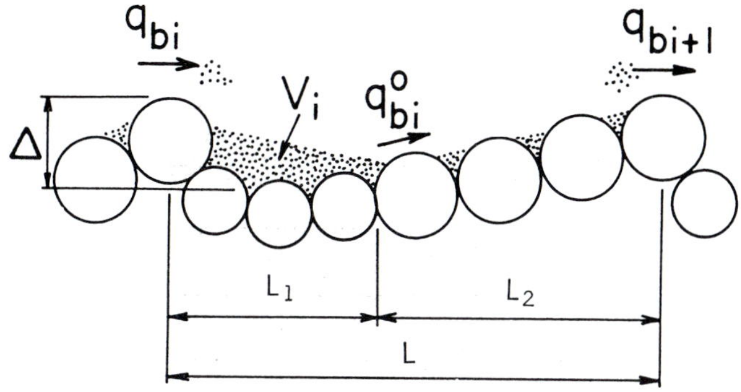

Figure 7 Sediment transport process on step-pool topography.

Therein L is the wave length, L_1 is the length of pool region which is approximately $L_1 = a\Delta$, and L_2 is the length of the wall region.

With attention focused on the sediment stored in i-th pool, the rate of temporal change can be described as

$$\frac{\partial V_i}{\partial t} = \frac{1}{1-\lambda}(q_{bi} - q_i^o) \tag{27}$$

in which V_i is the sediment volume stored in i-th pool, q_{bi} is the inflow rate of sediment into i-th pool and q_{bi}^o is the outflow rate of sediment from it.

We introduce an equilibrium volume of sediment storage, V_{ei}, which means that sediment transport rate attains to an equilibrium rate, q_{be}, if the sediment storage comes up V_{ei}. In additions, q_{bi} is supposed to take a form;

$$q_{bi}^o = (V_i/V_{ei})\, q_{be} \tag{28}$$

Herein, some non-dimensional parameters are introduced;

$$V_i = V_{ei}\, V_i' \ , \quad q_{bi} = q_{be} q_{bi}' \ \text{ and } \ q_{bi}^o = q_{be}\, q_{bi}^{o'}$$

Substitution of these into equation (27) yields

$$\frac{\partial V_i'}{\partial t'} = K_p(q_{bi}{}' - V_i') \tag{29}$$

in which

$$K_p = \frac{q_{be}}{V_{ei}(1-\lambda)} \sqrt{\frac{d}{(\sigma/\rho - 1)g}} \tag{30}$$

In the above equations, the equilibrium storage volume can be estimated with the formula;

$$V_{ei} = \frac{1}{2} k_e L_1 \Delta = \frac{1}{2} k_e a \Delta^2 \tag{31}$$

in which k_e is the empirical coefficient of order one, and $a\Delta$ is the length of separation zone defined before. q'_{bi} appeared in equation (29) is calculated with the formulas shown in subsection (a). Sediment transportation associated with over passing loads or through puts, therefore, can be predicted in reaches with step-pool systems by using either equations (22), (23) and (29) or equations (25), (26) and (29), with supplementary equations shown above.

APPLICATION OF MATHEMATICAL MODEL TO FLUME TESTS

Experiments were performed to obtain data for over passing loads, using the flume 12m long and 27.5cm wide [12]. At first, step-pool morphology was produced under the condition with flow discharge of 11.82 ℓ/s, initial bed slope of 0.08, no sediment supply and bed material with maximum size of 3.8cm, mean size of 1.31cm and $\sqrt{d_{84}/d_{16}} = 2.77$. Flow discharge continued at constant rate of 11.82 ℓ/s until the rate of sediment discharge came up negligibly small. The bed degraded due to active sediment motions from its initial slope of 0.08 to the statical equilibrium slope of 0.033. Then, a step-pool bed form of which the mean wave length and height were respectively 30.0cm and 3.0cm was formed. The mean particle size of the armored bed was 2.63cm, which was two times larger than that of original material. A stable step-pool system with armored bed surface was prepared by such a manner.

Two kinds of flume experiment were made; The one (Run-A) was for testing the transport model shown in subsection (a) and the other (Run-B) for testing the model in subsection (b). Run-A was carried out after the step topography employed in Run-B was smoothed by pressing it. Run-B, therefore, was performed before Run-A. Both in these tests, the flow discharge was 5.90 ℓ/s, and sediment with mean size of 0.063 cm and $\sqrt{d_{84}/d_{16}} = 1.32$, for over passing loads which was much finer than that of the bed surface, was supplied with the constant rate 44.6 gf/sec during 21 minutes at the section 4m upstream from the flume end.

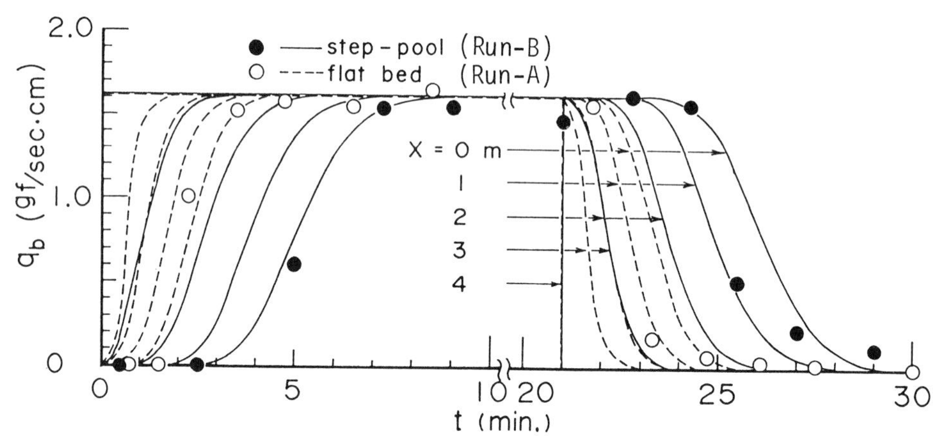

Figure 8 Results computed with the present method for sediment discharge curves at the sections 0, 1, 2, and 3m from downstream end vs. flume data for sediment discharge rates at the down stream end (x=0m).

Results computed at four sections with the present method and experimental data are shown for over-passing loads in figure 8. Flume data show that sediment particles on the flat bed with an armor coat propagate faster than those on the bed with step-pool topography. Such an experimental fact is predicted well by the present method.

Empirical factors necessary for calculating the over passing loads were determined according to either the theoretical tools or the experimental facts as follows; $k_3/k_2 = 2/3, F_o = 0.035, \lambda = 0.4, k = 1/6, L = 30\text{cm}$, $\Delta = 3\text{cm}$, $V_{ei} = 22.5\text{cm}^2$ (k_e=2, a=2.5), and the equivalent roughness was specified, taking the result performed by Muramoto, Kawata and Nunomura [14] into consideration;

$$k_s = (1 - A_s')d_a + A_s'd \qquad (32)$$

CONCLUSIONS

Three subjects associated with sediment phenomena in mountain streams were chosen for consideration in the present paper, and discussed from the view point of sediment hydraulics. The results may be summarized as follows:

(1) According to the reviewal studies on the genesis and geometric features of step-pool bed morphology, such bed forms may be produced whenever the three conditions, *i.e.* upper flow regime, bed material with sediment mixture and active, selective entrainment, are satisfied, and those wave heights are of the same order of magnitudes as the mean size of sediment particles composing the armor coat.

(2) A friction law of flow over step-pool morphology is presented, taking both the energy dissipation of mean flow due to entrained eddies in separation zones and the energy dissipation in wall regions into consideration. The friction law shows that the flow resistance depends on the relative wave height, Δ/h, the relative wave length, L_1/L, and the roughness, k_s, in the wall region.

(3) Discussions are made on the sediment transportation such as over-passing loads or through puts on the bed with stable step-pool topography. These emphasize that the transport processes can be devolved into two parts; the one in the pool region and the other in the wall region. In the pool region, the sediment transport rate depends on the temporal storage volume of sediment to be transported and in the wall region it is affected by the coverage ratio of transported sediment to that of parent, armored bed surface.

The method for predicting the over-passing loads, which is applicable for the case of step-pool morphology as well as for the case of flat bed with an armor coat, is presented by formulating the two transport processes. It is found that the results obtained from numerical analyses with the present method accord well with flume data.

ACKNOWLEDGMENT

Flume data employed in this paper were obtained by the authors and their coworkers-Miss N. Ando, Mr. N. Nishimoto and Mr. T. Nishino-when they were graduated students. We are grateful for their help with the flume tests and relating discussions. In addition, Mr. Y. Yoshida, technician, helped with flume tests and computer processing, and in drawing the figures. We should be grateful for his help.

REFERENCES

1. Whittaker, J. F. and Jaeggi, M. N. R.: Origin of Step-pool Systems in Mountain Streams, Proc. ASCE, Jour. Hydraulic Div., Vol. 108, No. HY6, 758-773, 1982.

2. Ashida, K., Egashira, S. and Ando, N.: Generation and Geometric Features of step-pool Bed Forms, Annuals, Disaster Prevention Research Institute, Kyoto Univ., No.27B-2, 341-353, 1984. (in Japanese)

3. Hasegawa, K. and Fujita, M.: Flow Resistance Caused by Step-pool Systems in Mountain Streams, Proc. 30th Japanese Conference on Hydraulics, 79-84, 1986. (in Japanese)

4. Egashira, S., Ashida, K., Sawada, T. and Nishimoto, N.: Geometric Features of Step-pool Bed Forms in Mountain Streams, Proc. 29th Japanese Conference on Hydraulics, 537-542, 1985. (in Japanese)

5. Bathurst, J.C.: Flow Resistance of Large-Scale Roughness, Proc. ASCE, Jour. Hydraulic Div., Vol, 104, No. Hy. 12, 1578-1603, 1978.

6. Hey, R. D.: Flow Resistance in Gravel-Bed Rivers, Proc. ASCE, Jour. Hydraulic Div., Vol. 91, No. HY4, 365-379, 1979.

7. Graf., W. H., Cao Hun Hai, and Bathurst, J. C.: Blench Equilibrium Parameters; Supplementary Data, Proc. ASCE, Jour. Hydraulic Eng. Vol. 109, No. 10, 1369-1374, 1983.

8. Imamoto, H., Fujii, G., Nishio, N. and Tanaka, K.: On the characteristics of an Open Channel Flow Through Passage of Cross-Sectional Variation (3), Annuals, Disaster Prevention Research Institute, Kyoto Univ., 331-353, 1978. (in Japanese)

9. Raudkivi, A. J.: Loose Boundary Hydraulics, Pergamon Press, 206-208, 1967.

10. Tsubaki, T. and Komatsu, T.: Flow Properties and Turbulent Entrainment of Two Dimensional Stratified Surface Jet, Proc. Japan Society of Civil Eng., No. 273, 69-81, 1978. (in Japanese)

11. Ashida, K., Egashira, S. and Nishino, T.: Structure and Friction Law of Flow over a Step-pool Bed Form, Annuals, Disaster Prevention Research Institute, Kyoto University, No. 29B-2, 391-403, 1986. (in Japanese)

12. Egashira, S., Ashida, k. and Nishimoto, N.: The Role of Step-pool Bed Form in Sediment Transport, Proc. 30th Japanese Conference on Hydraulics, 223-228, 1986. (in Japanese)

13. Nakagawa, H. and Tsujimoto, T.: Sand Bed Instability due to Bed Load Motion, Proc. ASCE, Jour. Hydraulic Div. Vol. 106, No. HY12, 2029-2051, 1980.

14. Muramoto, Y., Kawata, Y. and Nunomura, A.: Basic Study on the Bed Load in Gravel River, Annuals, Disaster Prevention Research Institute, Kyoto Univ., No. 20B-2, 451-474, 1977. (in Japanese)

MODIFICATION OF TRANSPORT RATE FORMULA FOR STEEP CHANNELS

Lechostaw Suszka

Institute of Hydroengineering
Polish academy of Sciences, Gdansk, Poland

ABSTRACT

The sediment transport formula for steep channels was modified basing on the experimental data obtained by five authors. The constant value of the dimensionless critical shear stress was replaced by the function of the relative depth. Initiation of motion was determined by means of the probability concept of the sediment transport, where the probability is constant and equals $p = 0.0001$.

1. INTRODUCTION

Recently, when the knowledge on sediment transport in plain rivers is fairly extensive, the hydraulic engineers are trying to get to know the phenomenon of the sediment transport in mountain rivers better. Since such rivers have steep slopes $S_o > 0.005$, and small relative depths, $h/d_{50} < 20$, the scale of the bed particle diameter is of the same order of magnitude as the scale of depth. Hence, the influence of bed roughness on flow becomes more significant; velocity distribution and bed resistance also differ distinctly from those for greater relative depths, when $h/d_{50} > 100$. The question arises, whether the dimensionless critical shear stress, τ_{*cr}, for particle Reynolds number $Re_* > 600$ is constant or it depends on other parameters. Another problem is subjectivity in determining the initiation of motion. The opinions of authors differ as to what should be treated as the beginning of the sediment motion. It seems therefore useless to compare the data of the critical shear stress taken from observations and measurements by various authors. It is proposed to use Pazis's (1977) probability concept to determine the critical shear stress. Several sets of data were examined to verify whether, assuming the same degree of the probability of one particle motion, the critical shear stress is constant or variable.

The main purpose of this research was to modify the bed-load formula for steady, uniform flow and small relative depth as proposed by Graf and Suszka (1987) and Suszka (1987) (see Fig.1). The scatter of their data in the lower sediment-transport range is not random but systematic rather, as regards the bed slope, S_o, or relative depth, h/d_{50}. The relative depth was proposed therefore as an additional parameter.

2. INITIATION OF SEDIMENT MOTION

One of the difficulties in sedimentation processes is the determination of the point of initiation of sediment motion, in other words, the value of the shear stress at which sediment starts to move.

Shields (1936) was the first to determine the critical shear stress by extrapolating a plot of measured sediment discharge against shear stress to the zero transport rate. The original Shields diagram for initiation of sediment motion (Fig. 2) shows that for a particle Reynolds number $Re_* = u_* d_{50}/\nu$ greater than 500, the dimensionless critical shear stress, τ_{*cr} equals 0.06 and does not change for a higher Re_*. The Shields method of determining the critical shear stress by extrapolating was not precise and rather subjective. Several studies have been carried out to find a more objective criterion for the critical shear stress. One of the methods based on the probability concept was presented by Pazis and Graf (1977).

3. PROBABILITY CONCEPT

The authors suggest that the probability p of one particle motion is equal

$$p = N'/N_t \tag{1}$$

where: p is the probability of one particle motion,

N' is the number of particles passing the unit width cross section per second,

N_t is the number of particles that occur on bed area F, which corresponds to the surface of the unit width and u_L length,

u_L is the velocity of flow near the bed at level $z = 0.35d_{50}$ above the theoretical bed level (Einstein, El-Samni (1949)) according to the logarithmic low of the velocity distribution, $u_L = 5.88u_*$.

The N' value can be calculated from the relation

$$q_s = N'\gamma_s V \tag{2}$$

where: q_s is the sediment transport expressed by weight,
V is the volume of one particle, $V = \Pi d^3/6$,
γ_s is the specific weight of the bed particles.

Hence

$$N' = 6q_s/\Pi d^3\gamma_s \tag{3}$$

then

$$N_t = F/\xi(\Pi d^2)/4 \tag{4}$$

where: ξ is the factor that replaces the voids in the upper layer particles, $\xi = 1.103$.

From (1), (3) and (4)

$$p = 0.281q_s/\gamma_s d_{50}u_* \tag{5}$$

If the sediment transport is given in volume units (q_s' (m^3/sm)) then

$$q_s = q_s'\gamma_s \tag{6}$$

This gives

$$p = \frac{0.281q_s'}{d\ u_*} \tag{7}$$

where: u_* is the shear stress velocity, $u_* = \sqrt{gR_b S_o}$,
g is the gravity,
R_b is the hydraulic radius after correction of the wall effects
S_o is the bed slope.

4. SHIELDS PARAMETER AS A FUNCTION OF THE RELATIVE DEPTH

Dimensionless shear stress for steep channels should be slightly modified as compared with the classical Shields parameter, $\tau_* = \tau_o/gd_{50}(\rho_s - \rho)$ and according to Mizuyama (1977) it should be as follows:

$$\tau_{*s} = \tau_* /(\tan\phi\cos\theta - s\ \sin\theta/(s - 1)) \tag{8}$$

where: θ is the bed slope of the channel,
ϕ is the friction angle of gravel,
s is the ratio of the sediment and water density.

Five sets of data from laboratory flume experiments were used to find the critical shear stress τ_{*Scr}, corresponding to the probability of one particle motion, $p = 10^{-4}$, as a function of the relative depth, h/d_{50} (see Figs. 3a-e). All data are obtained from experiments with a gravel bed flume. The ranges of the hydraulic parameters used in the experiments are given in Tabl.1.

Tabl.1 Ranges of hydraulic data used in the analysis.

Author Year	Particle diameter d_{50} [mm]	Bed slope S_o	Relative depth h/d_{50}	Particle Reynolds Numb. $Re_* = u_* d_{50}/\nu$
Suszka	12.2	0.005 - 0.015	5.74-20.1	868 - 1788
(1987)	23.5	0.015 - 0.025	4.00-8.94	3712 - 4383
Cao	22.2	0.01 - 0.09	0.98-11.4	2400 - 5500
(1985)	43.5	0.01 - 0.09	0.90-6.00	2980 -14000
Meyer-Peter	3.3	0.0027	42.0-73.3	147 - 200
and Muller	28.6	0.003 - 0.018	11.9-38.8	3758 - 6900
(1948)				
USWES	4.1	0.003 - 0.0045	18.0-32.0	180 - 288
(1935)				
Ho Pang	4.1	0.0017- 0.0033	16.9-50.7	170 - 258
Yung (1939)	6.1	0.0033- 0.005	16.7-25.7	288 - 414

For each kind of bed material and bed slope (in each set of data) critical Shields parameters, τ_{*Scr}, corresponding to the probability of one particle motion, $p = 10^{-4}$, were found. Altogether, 26 points which describe the initiation of the motion as a function of the relative depth, h/d_{50}, were obtained (Tabl.2).

Table 2. Critical shear stress corresponding to the probability of one particle motion, p = 10^{-4}, from experiment measurements given Tabl.1.

Source	τ_{*s}	Re_*	h/d_{50}	d_{50}	S_o
Suszka	0.0335	1020.0	13.5	12.2	0.0050
	0.0440	1150.0	11.0	12.2	0.0075
	0.0430	1250.0	8.65	12.2	0.0090
	0.0450	1300.0	8.20	12.2	0.0100
	0.0525	3800.0	6.50	23.5	0.0150
	0.0570	3700.0	5.00	23.5	0.0200
	0.0620	3900.0	4.10	23.5	0.0250
Cao	0.0510	3000.0	8.60	22.2	0.0100
	0.0690	3300.0	3.50	22.2	0.0300
	0.0760	3400.0	2.40	22.2	0.0500
	0.0960	3600.0	1.80	22.2	0.0700
	0.0980	3600.0	1.40	22.2	0.0900
	0.0490	7800.0	3.00	44.5	0.0300
	0.0640	8900.0	1.90	44.5	0.0500
	0.0700	8200.0	1.30	44.5	0.0700
	0.0900	9000.0	1.20	44.5	0.0900
Meyer-Peter and Muller	0.0300	140.8	30.0	3.3	varied
	0.0460	3800.0	17.0	28.6	varied
	0.0410	3800.0	30.0	28.6	varied
	0.0500	4000.0	40.0	28.6	varied
USWES	0.0315	185.0	19.0	4.1	0.0030
	0.0360	190.0	15.0	4.1	0.0040
	0.0410	210.0	15.0	4.1	0.0045
Ho Pang Yung	0.0340	190.0	50.0	4.1	0.0017
	0.0320	310.0	20.0	4.1	0.0033
	0.0400	340.0	15.0	6.1	0.0050
	0.0325	193.0	19.0	6.1	0.0033

All the points obtained were approximated in the log-log coordinate system by the method of least squares (Fig.4). It may be noted that the critical shear stress, τ_{*Scr}, can be shown as a power function of the relative depth, and the best approximation of this function can be given by the following equation:

$$\tau_{*Scr} = 0.0851\ (h/d_{50})^{-0.266} \qquad (9)$$

Furthermore, fifteen experimental points of dimensionless shear stress τ_{*Scr}, obtained by Mizuyama (1977) and presented by Ashida and Bayazit (1973) were shown additionally in Fig.4. All the points were obtained in circumstances in which sediment transport did not occur during three-minute experiments. The same set of points can be presented in the plot τ_{*Scr} versus Re_*. If the relative depth is used as the third paramęter a family of parallel lines can be obtained (Fig.5). This tendency is similar to that obtained by Neill (see Bogardi (1977) p.221) in Fig.6.

Why is it that the critical dimensionless shear stress increases with decrease in relative depth? The effect of the reduction of the U/u_*, where U is the mean velocity in the cross-section, for the decreasing relative depth does, of course, becomes important. Furthemore, Ashida and Bayazit's (1973) results showed that the nondimensional velocity u/u_* at a certain level above the bed with constant shear stress, decreases significantly as the slope increases and therefore greäter shear stress is required to move particles. This effect can be explained by greater energy dissipation in the separation zone downstream of gräins for shallower flows. Ashida and Bayazit suggest that the turbulence characteristic of the flow may be expected to change considerably in case of small relative depth, but this is only a hypothesis.

5. SEDIMENT TRANSPORT

Now, let us return to the bed-load formula proposed by Graf and Suszka (1987)

$$\Phi = 10.4\ \Psi^{-1.5}(1 - 0.045\ \Psi)^{2.5} \tag{10}$$

where: Φ is the sediment transport parameter

$$\Phi = CUR_h/\sqrt{(\rho_s/\rho - 1)gd_{50}^3} \tag{11}$$

with C as a volumetric concentration of the transported particles and Ψ as the flow intensity being the reciprocal of the dimensionless shear stress

$$\Psi = 1/\tau_* = gd_{50}(\rho_s - \rho)/\gamma R_h S_o \tag{12}$$

The value 0.045 in the formula (10) seems to represent the Shields parameter at the beginning of the sediment transport. Since the relation between dimensionless critical shear stress and the relative depth has been established, it is possible to replace the constant value by the variable (eq.(9)). All the experimental points (Tabl.1) were divided into six categories according relative depth. The limiting values of h/d_{50} were 2, 4, 7, 10 and 20. Points which belong to the same range were indicated identically (see Fig.7). When all points were plotted between Φ versus Ψ coordinates, it was obvious that the points with smaller relative depth were situated below those with greater relative depths.

When the critical shear stress constant value, 0.045, in eq.(1) is replaced by τ_{*Scr} from eq.(9) the lines representing constant relative depths can be drawn in plot Φ versus Ψ. All experimental points and the lines corresponding to $h/d_{50} = 2$ and 20 and also the line taken for constant critical shear stress $\tau_{*cr} = 0.045$ are given in Fig.7. The lines for $h/d_{50} = 2$ and 20 separate the data which belong to appropriate ranges of corresponding h/d_{50} quite well. To show the advantage of the modified relation (13)

$$\Phi = 10.4\,\Psi^{-1.5}(1 - \tau_{*Scr}\,\Psi)^{2.5} \tag{13}$$

where τ_{*Scr} is calculated by equation (9), the author compared Φ values measured with Φ calculated with the former formula (10) and those calculated by eq.(13) proposed in this paper.

6. CONCLUSIONS

The analysis of five sets of sediment data was carried out in order to modify the transport rate formula for steep channels. The Shields parameter was used in formula (13) as a function of the relative depth basing on Pazis's probability concept.

The folowing conclusions can be drawn on the basis of the analysis of the experimental data of the author and others.

1) Conventional determination of the critical Shields parameter is subjective.
2) The probability of one particle motion can be calculated on the basis of Pazis's probability concept.
3) The constant value of the probability of one particle motion $p = 10^{-4}$, can be used as impartial criterion to calculate the Shields parameter.
4) In the case of gravel steep streams the Shields parameter for particle Reynolds number, $Re_* > 600$, is not constant but it increases with diminishing relative depth.
5) The author agrees with the explanation of Ashida and Bayazit concerning increase of the Shields parameter for decreasing relative depth.
6) The sediment transport formula (1) proposed by Graf and Suszka (1977) for steady uniform flow in gravel streams is modified taking the relative depth as a parameter.

REFERENCES

ASHIDA K., BAYAZIT M. (1973),"Initiation of Motion and Roughness of Flows in Steep Channel", Proc. 15th Congress Intl. Ass. Hydraul. Res., Istanbul, Turkey, Vol.1, pp.475-484.

BOGARDI J. (1977),"Sediment Transport in Alluvial Streams", Akademiai Kiado.

CAO H.H. (1984),"Etude Experimentale des Ecoulement dans un Canal a Pente Raide et avec Lit a Gravier", These No 589, Ecole Polytechnique Federale de Lausanne, Lausanne, Switzerland.

EINSTEIN H.A, EL-SAMNI E.S. (1949),"Hydrodynamics Forces on a Rough Wall", Review of Modern Phys., Vol.21, No.3.

GRAF W.H., SUSZKA L. (1987),"Sediment Transport in Steep Channels", Journal of Hydrosciences and Hydraulic Engrg., Japan Soc. Civ. Engs., Vol.5/1, pp. 11-26.

HO PANG-YUNG (1939),"Abhangigkeit der Geschiebebewegung von der Kornform und der Temperature", Mitteil. Preuss. Versuschsanst. fur Wasser. und Schiffbau, Berlin, Vol.37.

MEYER-PETER E., MULLER R. (1948),"Formulas for Bed Load Transport", Proc. 2nd Meeting of Int. Assoc for Hydraulic Structures Res., Stockholm, pp.39-64.

MIZUYAMA T. (1977),"Bedload Transport in Steep Channels", Doctoral Thesis, Kyoto University, Kyoto, Japan.

PAZIS, GRAF W.H. (1977),"Erosion et Deposition; un Concept Probabiliste", Proc. 17th Congress Intl. Ass. Hydraul. Res., Baden-Baden., Vol.1, pp.39-46.

SHIELDS A. (1936),"Anwendung der Ahnlichkeits-Mechanik und Turbulenzforschung auf die Geschiebebewegung, Mitteil. Preuss. Versuchsanst. Wasser, Erd, Schiffsbau, Berlin, No.26.

SUSZKA L. (1987),"Sediment Transport at Steady and Unsteady Flow; A Laboratory Study", These No 704, Ecole Polytechnique Federale de Lausanne, Lausanne, Switzerland.

U.S. Army Corps Of Engineers (1935),"Studies of River Bed Materials and their Movement with Special References to the Lower Mississippi River, U.S. Waterways Expt. Sta., Vicksburg, Miss., paper 17, pp.161.

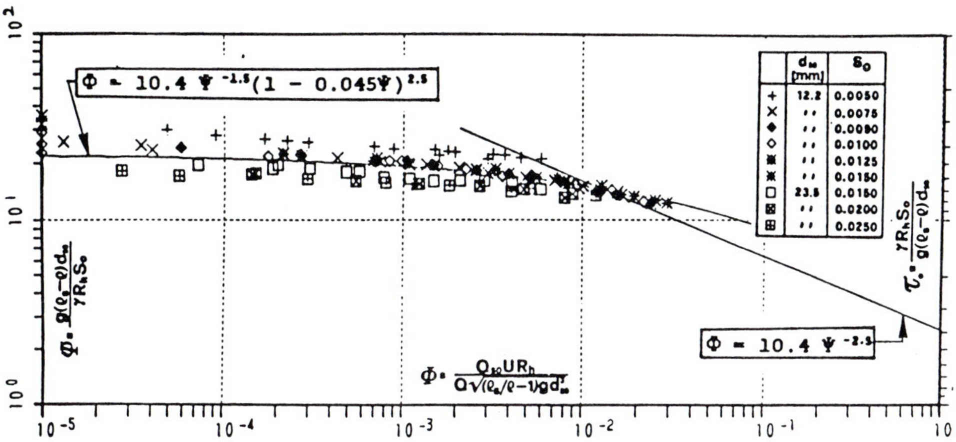

Fig.1 Shear-intensity parameter, Ψ, vs. sediment parameter, Φ, according to Graf and Suszka (1987).

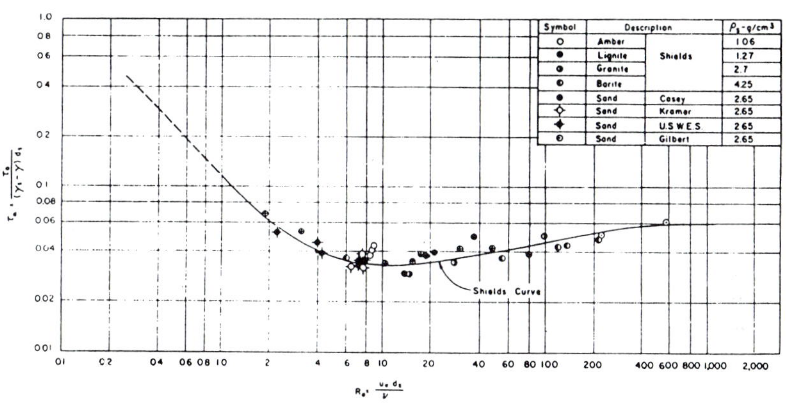

Fig.2 Shields diagram for initiation of sediment motion.

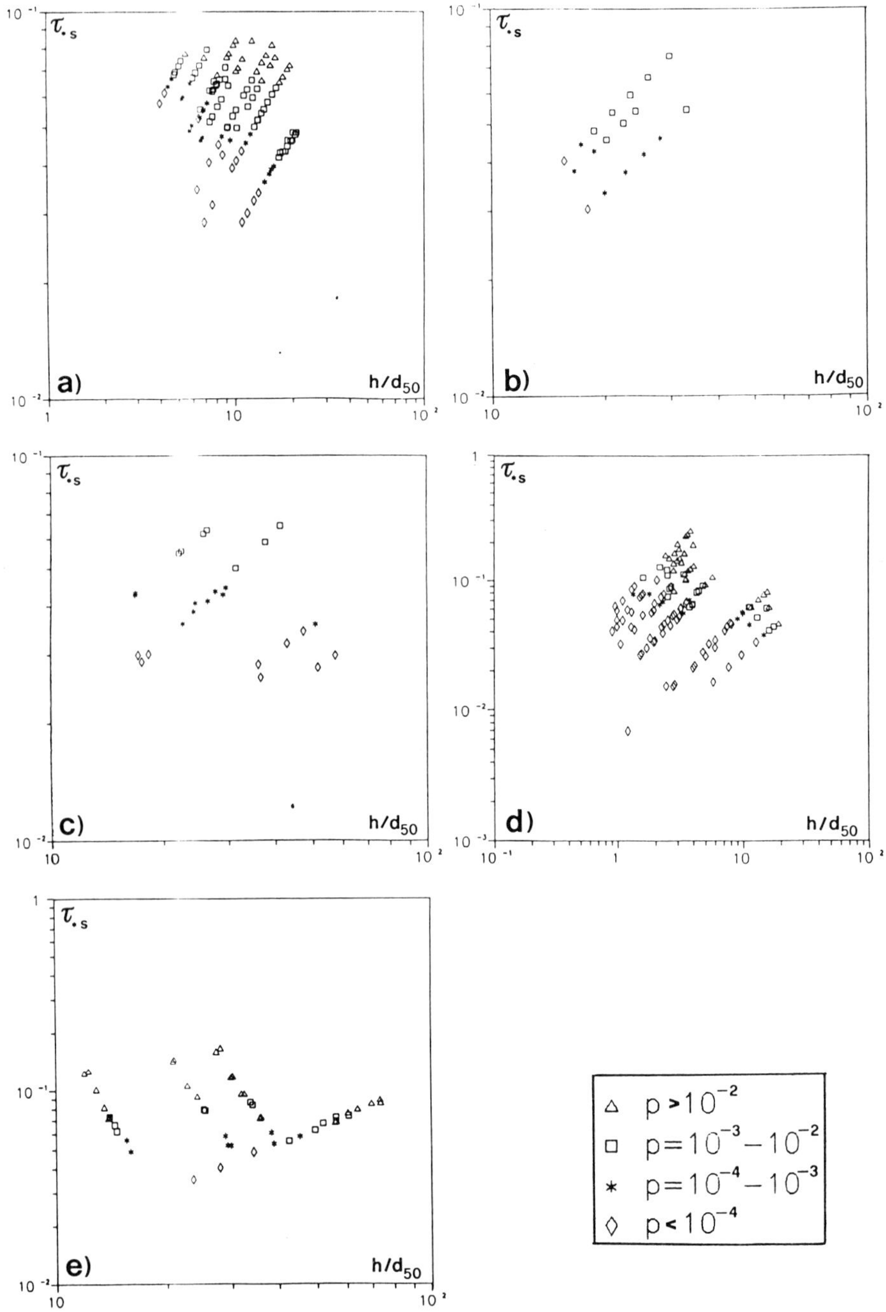

Fig.3 Dimensionless shear stress, τ_{*s}, vs. relative depth, h/d_{50}, with probability of sediment motion, p, as a parameter. Sources of data: a) Graf and Suszka, b) USWES, c) Ho Pang Yung, d) Cao, e) Meyer-Peter and Muller (see Tabl. 1).

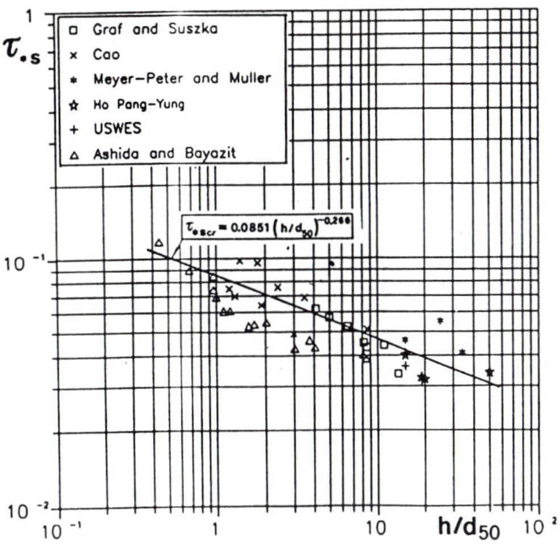

Fig.4 Shields parameter, τ_{*Scr}, given for probability of motion, $p = 10^{-4}$, vs. relative depth, h/d_{50}.

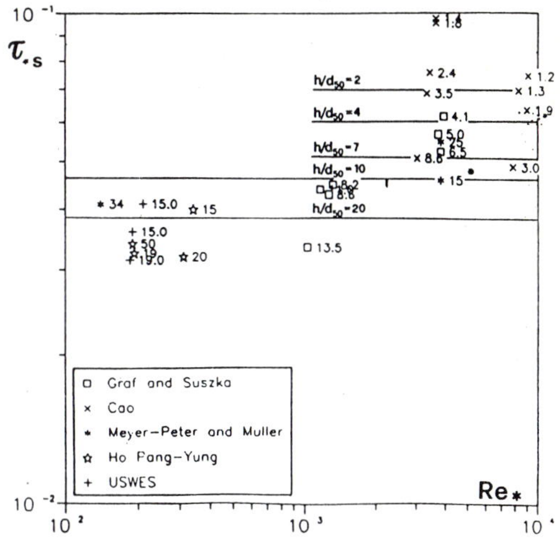

Fig.5 Shields parameter, τ_{*Scr}, vs. particle Reynolds number Re_*. Values at right of each point indicates relative depth, h/d_{50}.

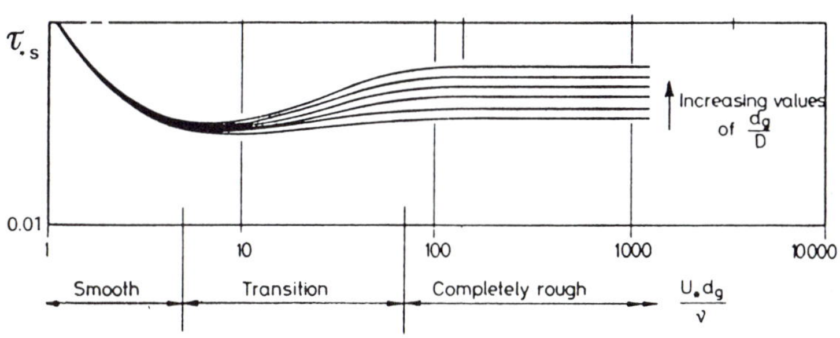

Fig.6 Shields parameter, τ_{*Scr}, vs. particle Reynolds number Re_*. according to Neill. D is the depth, d_g is the particle diameter.

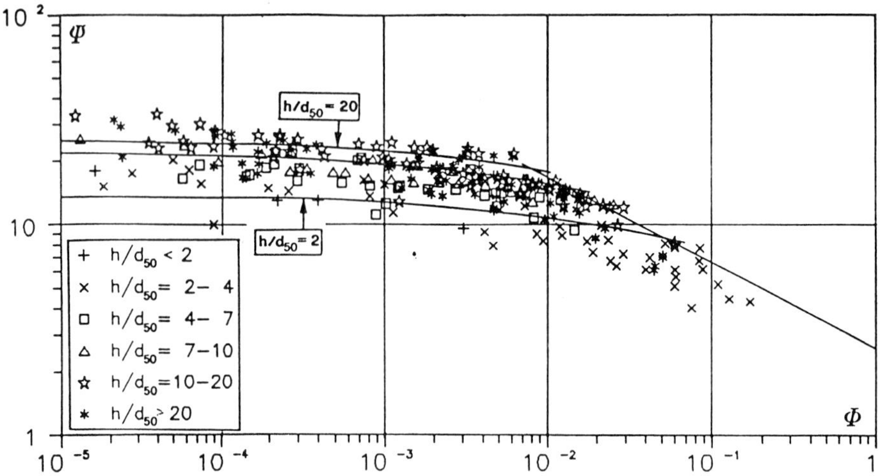

Fig.7 Shear-intensity parameter, Ψ, vs. sediment parameter, Φ. Lines indicate equation modified (13) with h/d_{50} = 2 and 20 and former equation (10).

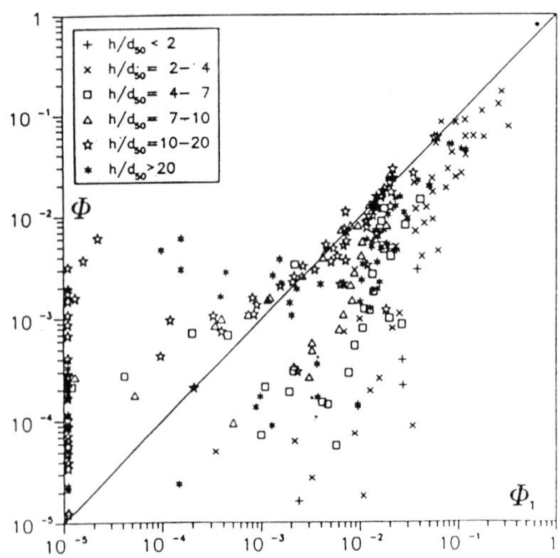

Fig.8a Sediment parameter measured, Φ, vs. sediment parameter calculated, Φ_1.

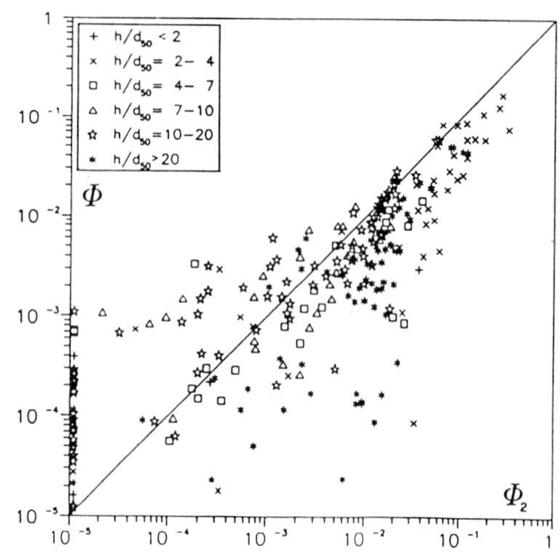

Fig.8b Sediment parameter measured, Φ, vs. sediment parameter calculated, Φ_2.

CONTINUOUS SIMULATION OF SEDIMENT TRANSPORT IN THE CASE OF A GLACIERIZED WATERSHED

Friedrich Schöberl

Institut für Konstruktiven Wassebau
und Tunnelbau, Universität Innsbruck, Austria

ABSTRACT

Sediment supply in mountain streams is permanently fluctuating. Under natural conditions, the transport is mainly caused by storm flood waves and snow or glacial melting processes. The information about the variation of the transported material at these events and about the change in size distribution and rate is very scarce, since only few direct observation stages exist, e.g. Lauffer-Sommer (1982), Zeller (1985), Tacconi-Billi (1987).

At the Pitzbach, a small tributary of the river Inn, a particular device has been prepared in form of a tyrolian weir to continuously monitor bed load and suspended matter.

Resulting load graphs permit the evaluation of relations that govern the sediment supply. This paper continues and extends the work presented earlier by Hofer, (1985) and Schöberl, (1988). It can be shown that characteristic features of the discharge hydrograph dominate the bed load discharge, while for the suspended material additional factors are involved. Due to the evaluation of characteristic paramters a continous simulation of the bed load is possible. The principle of this method is outlined for the case of the Pitzbach and the results gained are presented.

CHARACTERISTIC OF THE SITE

The sediment yield of mountainuous areas can be estimated, amoung other methods, by the help of intake weirs as done in various cases by the TIWAG, Lauffer - Sommer (1982). These intakes are built in form of tyrolian weirs furnished with bottom trash racks. Sediments up to a diameter of 150 mm are withdrawn with water and trapped in an annexed settling chamber. Nearly all sediments are caught at lower flows, at high stages only about 5 % of the coarsest material is bypassed over the weir.

At one of these sites, special efforts were made to improve the equipment and to intensify the observations. This catchment area belongs to the upper reach of the Pitzbach, a small tributary of the river Inn. The area of 26,8 km² is 59,5 % glaciated. This fact impacts strongly the flow and transport regime of this creek. In cooperation with the hydraulic laboratory of the University of Innsbruck continuous monitoring of the deposition pattern at different positions along the settling chamber were arranged to allow a more accurate reading of the deposited volume. The settling chamber is automatically flashed if a certain deposition level is exceeded. Moreover a continuous measuring of the suspended load was possible by installing a turbidity meter supported by short-term sampling to control changes of the transported grain material.

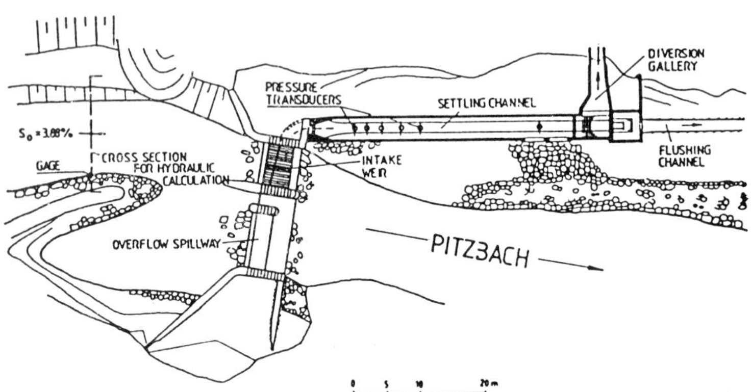

Fig.1 situation at the oberservation site

The bed layer of this reach consists of coarse material up to 300 mm and causes a state of latent erosion. Beneath a stage of 0.5 m the flow becomes extremely rough. Fig.2 illustrates the grain size distribution of the top layer and the basic mixture lying underneath. Except at extreme floods, the continuously trapped material is of lesser size than that of the bed composition.

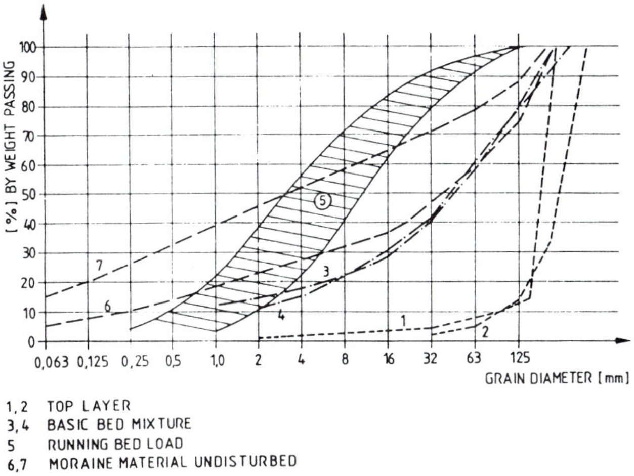

Fig.2 characteristic grain size distributions of the Pitzbach

RUN OFF CONDITIONS

During the summer period, glaciarized watersheds produce periodically fluctuating discharges, caused by melting processes activated daily. Therefore, in general, the observed hydrographs belong to two different flow phenomena. While the basis wave is due to the diurnal glacier run off, each rain- fall superposes the glacier run off, causing sharp rising translation waves.

BED LOAD ESTIMATION

For the same discharge, the sediment load can differ significantly, depending on peak or base flow conditions. The high sensitivity of sediment movement refers to the fact that in nature, particularly in paved beds, the observed load cannot match the transport capacity normally totalling much less. The transport is regulated by erosion processes in the supplying tributaries and not related to the movement of the entire bed. Only extreme floods can provide a motion of all components of the bed.

According to many investigations, a simple relation of discharge and sediment rate under unsteady natural conditions does not exist. Lacking more precise estimations, a poor medium power law is accepted in many cases. Some investigators as i.e. Remy Berzencovich (1960) tried to describe the stochastic behavior by means of statistical distributions for different discharge intervals.

In the case of the Pitzbach a first evaluation of transport rates succeeded in relating the sediment load to characteristic differences of discharge and rain intensity, Hofer, (1985).

For the basic glacial wave, the daily minimum rate Qg_{min} was regressed to the minimum discharge Q_{min}

$$Qg_{min} = f(Q_{min})$$

and the increase of load dQg due to the swelling of discharge dQ could be approximated in form

$$dQg = f(dQ, Q_{min}).$$

Rain events with rain intensities exceeding 0,04 mm/min affect the load increase additionally. Considering the exeeding rain intensity as a further influencing parameter we arrive at

$$dQg = f(dQ, Q_{min}, i_m)$$

with Qg as sediment load, Q_{min} as base flow, dQ discharge difference between peak and base flow and i_m rain intensity in (mm/min). In each case the relationships f() were approximated in terms of e-functions.

Beside permanent observations of hydrographs, the analysed relations demand continuous rain gauging too. Moreover the results yield only term values for the peak and base conditions and don't allow time dependence in detail. In comparision, continuous calculations require particular knowledge of the wave recession features, since rainfalls occur independently to the melting wave.

Therefore, the key role for the superposition of rain and melting waves relates to the descending part of the wave, because the time dependent recession controls the starting position of every new event.

Further investigations about the unsteady character of flow conditions, Schöberl-Hofer, (1987) and Schöberl, (1988) proved the opportunity to describe the decreasing transport rate and also the increasing rate exclusively by features of the observed hydrographs.

WAVE RECESSION

Independently of the kind of event and of the drop of shear level for the difference between peak and base flow, all waves together follow an uniform recession law in the form :

$$- dQg = - Qgmax.[1 - e^{- 1,75.(dt/dto)^{0,2}}]$$

with Qgmax the peak sediment rate, dt the actual time interval, dto the base interval of a compared glacial wave recession totalling 16 ^{h} .

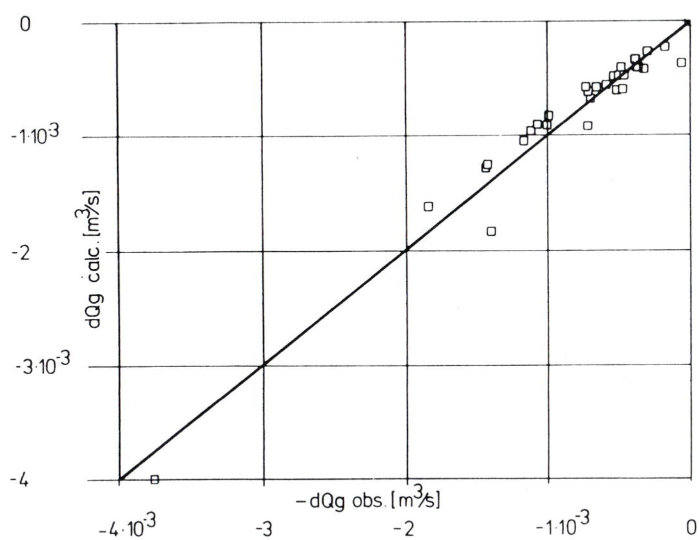

Fig.3 observed load differences dQg (peak-base flow) versus calculated dQg_{calc} for the descending limb of the hydrographs

In comparison with the rate recession, a different behavior emerges for the increasing limb of the waves.

INCREASING WAVE LIMB

At first, an attempt was made to normalize the observed rates with numbers based on shear velocities. But strong differences in the flow conditions for stages below 0,5 m brought worse results compared to discharge based relations. So the latter were finally preferred.

For the glacial wave the relation of the increase of rate has been slightly simplified, for convenience, compared to the former approach of 1985 :

$$dQg = f(dQ) = 0{,}08 \cdot e^{0{,}8 \cdot dQ}$$

In the event of rains the effect of the rain intensity is replaced by a ratio comparing the time of the rising part with that of the glacial wave. Furthermore additional influence has to be considered referring to the starting discharge Q_{min}.

In that case the gradient between the base and peak flow is expressed by:

$$dQg = d_1 \cdot (e^{0{,}8 \cdot dQ}) \cdot e^{d_2} \quad \text{with}$$

$$d_1 = (0{,}05 + 0{,}08 \cdot T_*^{1{,}5})$$

$$d_2 = 0{,}45 \cdot [e^{(-0{,}6 \cdot T_* \cdot Q_{min}/Qo)}]$$

$$T_* = [(dto/dt) - 1] \cdot (Qo/Q_{min})$$

Qo denoting the threshold discharge for the bed load movement (2 m^3/s), Q_{min} the starting level due to the rain event, dto the average time interval of a compared glacial wave and dt the interval of the rising part of the hydrograph.

The regression for the investigated waves is demonstrated by Fig.4

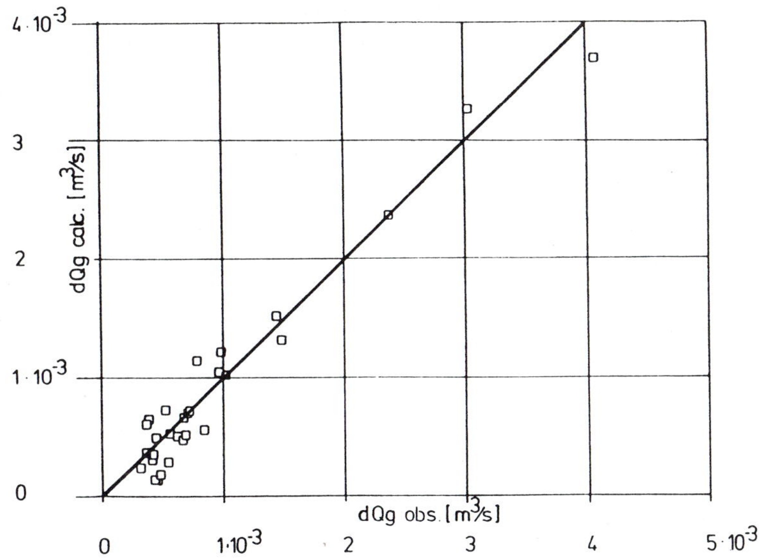

Fig.4 observed load differences dQg (base-peak flow) versus calculated dQg_{calc} for the rising limb of the hydrographs

FULL WAVE SIMULATION - BED LOAD GRAPH ESTIMATION

Laws for both wave limbs combined, formulate the pattern of a simple load graph. To meet the sligthly different shapes of hydrographs, the primary relation units are extended with a discharge based form factor : $\{dQi(dti)/dQ(dto)\}^{\alpha}$, dQi denoting the actual change of discharge and dQ = the total difference between peak and base flow condition, and the exponent α figuring 1,0 for the recession part and 2,5 for the rising section of the hydrograph.

Due to these characteristic parameters a continuous simulation of material transport is possible. As an example, Fig. 5 compares the measured and calculated rates for a certain period.

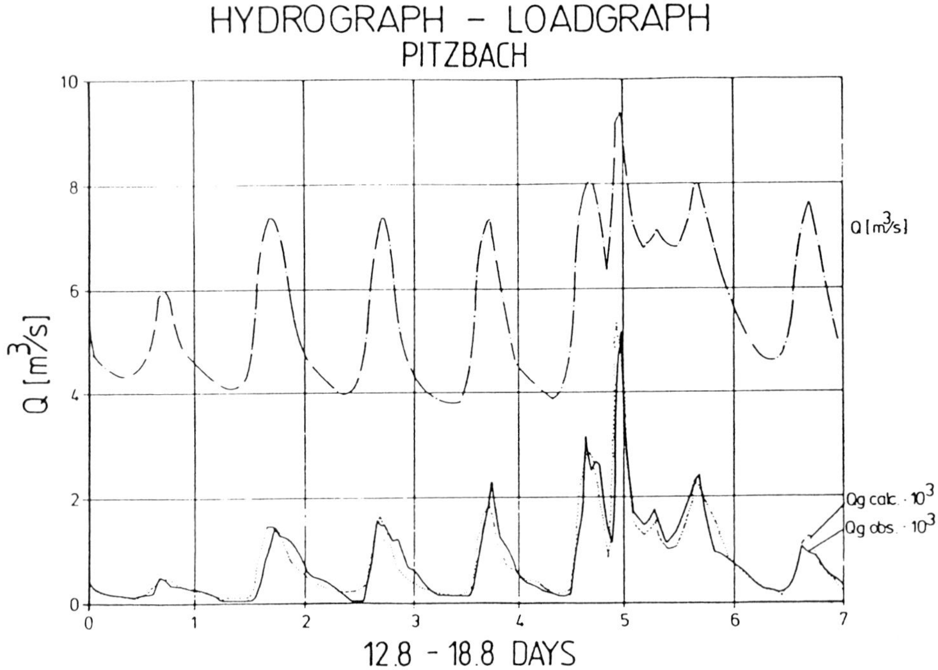

Fig.5 Example comparing the measured load graph with the calculated graph

SUSPENDED MATERIAL TRANSPORT

The equations for the bed load behavior are only partly transformable to the transport of suspended material. The fluctuating transport of this component turns out to be much more complex and is determinded by various supplemental factors, also see Hofer (1985).

Although the change of the concentrations of suspended material and the change of bed load discharge follow a mean similar trend, the principal relations are distorted by following factors :

* seasonal influences determining the minimum daily concentration. The flow evolution of melting snow-covered areas, typical for spring and early summer, differs from that of glacial melting in summer and early autumn.

* the suspended wave fails to coincide with the runoff wave, being ahead about 2 hours. The differences in the speed of both waves vary.

* "Föhndays", a particular meteorologic phenomenon of the Eastern Alps with strong warm winds on the northern side provide progressive peaks of transport rates.

* pre-history effects, - well known from analogous phenomenons like varying flood peaks due to different pre-moistening of the river basins - i.e. long-term colder periods with low melting activities act as trapping agents, while first days with higher temperatures and higher solar radiation cause progressive transport peaks compared with runoff waves of similar height in other periods. So the concentration gradient is also affected by the duration of preceding colder periods.

* Not all erosion areas react uniformly and higher flow peaks can dissolve in several subwaves of suspension transport

  and eventually

* opposite to bed load, the recession curve of the load graph of rain events differ from the depletion characteristic of the glacial wave.

To formulate relations of equal quality to bed load, more detailed investigations are necessary to determine the impact of all the influences summarized above. The governing relations must be put forward in a following step.

CONCLUSIONS

For bed load, characteristics of the hydrograph allow a continuous simulation of the material transport. The lack of information data, needed to describe the complex field situations with the randomly interferred sediment supply (with rates mostly less than the transport capacity as usually calculated), is the reason why strictly hydraulically based methods suffer severe difficulties.

The hydrological appraoch presented herein, establishing the concept of a discharge related load graph, seems to be convenient in describing the bed load evolution of glacial melting and rain-runoff waves in a simple way. For the suspended material, problems arose in formulating similar relationships, since additional factors complicate the transport process. To accurately pursue the wave translation a more complex model is necessary to enable equally satisfactory results.

Of course, these results are only a first step in analyzing sediment supply of a glacierized basin. But it seems likely that due to the observed bed load similar relationships hold in analogous basins.

REFERENCES

BATHURST J.C., **GRAF** W.H., **CAO** H.H. (1987)
Bed Load Discharge Equations for Steep Mountain Rivers
(SEDIMENT TRANSPORT IN GRAVEL-BED RIVERS, ED. BY THORNE C.R.,BATHURST J.C. AND HEY R.D.)

HOFER B. (1985)
Der Feststofftransport von Hochgebirgsbächen am Beispiel des Pitzbaches
(DISS. INSTITUT F. KONSTR. WASSERBAU UND TUNNELBAU, UNIVERISTÄT INNSBRUCK)

HRISSANTHOU V. (1989)
Oberflächenerosion, Bodenabtrag
(4.DVWK FBL FÜR TECHN. HYDRAULIK, BERECHNUNG DES FESTSTOFFTRANSPORTES FÜR DIE INGENIUERÜRAXIS, MÜNCHEN NEUBIBERG)

JÄGGI M., **RICKENMANN** D. (1987)
Applications of Sediment Transport Formulae in Mountain Streams
(22 ND IAHR CONGRESS, TOPICS IN FLUVIAL HYDRAULICS, LAUSANNE)

KARIM M. **KENNEWOO** H. **JULIEN** P. (1986)
Washload and Fine Sediment Load
(JOURNAL OF HYDRAULIC ENGINEERING -ASCE.
VOL 112, NO.6, JUNE 1986,P 541-545)

LAUFFER H., **SOMMER** N. (1982)
Studies on Sediment Transport in Mountain Streams of the Eastern Alps
(COMMISSION INTERNATIONALE DES GRANDS BARRAGES RIO DE JANEIRO 1982 Q 54, R 28)

REMY-BERZENKOVICH E. (1960)
Analyse des Feststofftriebes fließender Gewässer, SCHRIFTENREIHE DES ÖSTERR. WASSERWIRTSCHAFTSVERBANDES, HEFT 41, WIEN)

SCHÖBERL F. (1983)
Statistische Untersuchungen über die Geschiebeverfrachtung in Hochgebirgsbächen
(FORSCHUNGSGESELLSCHAFT FÜR VORBEUGENDE HOCHWASSERBEKÄMPFUNG, SCHRIFTENREIHE 1 - SEMINARE 3/V - 1983, HOCHWASSERSCHUTZ UND EROSIONSSCHUTZ)

SCHÖBERL F., **HOFER** B. (1987)
Influence of Unsteady Flow Conditions on Sediment Transport in a Mountain Water Course
(DISCUSSION, 23 IAHR CONGRESS LAUSANNE)

SCHÖBERL F. (1988)
Bed Sediment Transport in Relation to Unsteady Flow Conditions in Mountain Rivers
(INTERNATIONAL CONFERENCE ON FLUVIAL HYDRAULICS, BUDAPEST 1988)

TACCONI P., **BILLI** P. (1987)
Bed Load Transport Measurements by the Vortex-tube Trap on Virginio Creek, Italy
(SEDIMENT TRANSPORT IN GRAVEL-BED RIVERS, ED. BY THORNE C.R.,BATHURST J.C. AND HEY R.D.)

WILLIAMS J.R. 1978)
A Sediment Graph Model Based on an Instantanuous Unit Sediment Graph
(WATER RESOURCES RESEARCH, VOL. 14, NO. 4)

ZELLER J. (1985)
Feststoffmessung in kleinen Gebirgseinzugsgebieten (WASSER ENERGIE LUFT, JG. 77, HEFT 7/8

DETERMINATION OF THE CRITICAL CONDITIONS OF INCIPIENT NOTION OF BED LOAD IN MOUNTAIN RIVERS

Wojciech Bartnik

Department of Hydraulic Engineering
Academy of Agriculture, Cracow, Poland

ABSTRACT

The paper analyzes experimental data on incipient motion in mountain rivers. Studies of the incipient motion of bed load in rivers are of great importance for a large number of existing formula to calculate the bed load transport rate based on the knowledge of parameters of beginning of motion.

The data of four Polish rivers are discussed. The analysis shows that incipient motion of a grain is controlled by its relative submergence. The effects of relative submergence on the critical Froude number and on the resistance coefficient are also analysed.

INTRODUCTION

Radioactive tracer method of measuring incipient motion of bed particles was adopted in a few Polish mountain rivers [1] [Table 1]. In the experiments the tracer used was Jr-192 and Ta- 182. About 30 grains of each chosen fraction were introduced in the upstream region of each detectors. The detectors used were connected to the counting and recording system of a field isotope laboratory located on a bus.

The depth at which 90% of the total amount of marked grains move from the detector has been considered as *critical depth*. Furthermore, hydraulic measurements were carried out to define the mean velocities and the slope under incipient motion condition of the individual fractions of the bed material.

The studies and the experiments of Shields [5] experiments showed that the beginning of motion can be expressed as a simple relation between Fr_* and Re_*, but for coarse and non homogeneous materials determination of the critical conditions is rather indefinite. On the basis of radioisotopic measurements of the beginning of motion functional relationships to determine the critical conditions were found.

parameters	symbols		Wisłoka river	Dunajec river	Raba river
slope	J	‰	0.2-0.5	0.2-0.5	2.0-3.5
discharge	Qmax	m^3/s	750	730	330
depth of flow	h	m	1.0-3.1	1.0-4.7	0.55-2.0
width of flow	B	m	30.0	65.0	60.0
grain size	dm	m	0.0031	0.0274	0.070
standard deviation of mixture diam.	δ		4.0	2.77	1.95

Table 1: Parameters of characteristics rivers

During radioactive tracer experiments the critical mean velocity and critical depth of flow were observed for bed materials of given particles size d_i and specific weight of bed load. Because the critical mean velocity depends on particle size, depth of flow and other hydraulic factors, defining the parameter of Shields as:

$$f_c = \frac{\tau_c}{(\gamma_s - \gamma)d} \qquad 1)$$

and by introducing the expression $\tau = \rho v_*^2$, and according to the Chèzy formula:

$$\frac{v_c}{v_{*c}} = \frac{\chi}{\sqrt{g}} \qquad 2)$$

one obtains:

$$\frac{\rho v^2}{g\ (\rho_s - \rho)d} = \frac{\chi^2}{g} f_c \qquad 3)$$

The parameter on the left was termed *mobility number* M_n by Neille [2] and was found to be a function of d_i/h. The equation describing the straight line on the Fig.1 is:

$$M_n = 1.54(\frac{d_i}{h})^{-0.24} \qquad 3)$$

By introducing the relative density of the particles: $\frac{(\rho_s - \rho)}{\rho} = 1.65$, one finds:

$$v_c = 1.59\sqrt{g d_i}(\frac{h}{d_i})^{0.12} \qquad 4)$$

Fig.1 shows also the influence on the critical mean velocity of the ratio d_i/d_m. The mean critical velocity is higher for a fraction d_i when the ratio d_i/d_m decreases.

FLOW RESISTANCE

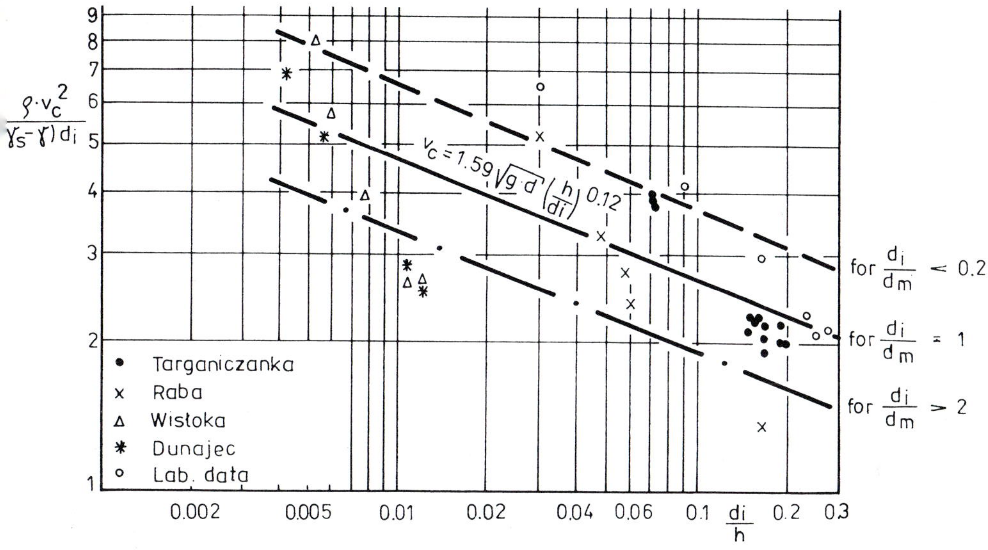

Fig.1 Mobility number M_n versus d_1/d_m

Flow resistance in fully developed turbulent flow is described by the Prandtl-von Karman equation::

$$\frac{v}{v_*} = \frac{1}{\sqrt{f}} = C + 2.03 lg \frac{h}{k_s} \tag{5}$$

Determination of the value k_s presents a major problem, especially when the roughness is irregular. In this paper the equation of resistance was found on the basis of the size of roughness element d_i, for which motion under critical conditions occurs. The basic data for this study are obtained from natural gravel bed river and laboratory experiments. The equation, representing resistance of flow under critical condition with beds coarse materials, has been found as (Fig. 2):

$$\frac{v}{v_*} = 2.53 + 5.75 lg \frac{h}{d_i} \tag{6}$$

MODIFIED SHIELDS EQUATION

Many objections have been recently directed to Shields approach. The last modification of Shields diagram is given by Gessler [4]:

$$\tau_c \left(\frac{1}{\rho \gamma \nu^2 (\gamma_s - \gamma)} \right)^{0.33} = \Phi \left(d \left(\frac{g(\rho_s - \rho)}{\rho \nu^2} \right)^{0.33} \right) \tag{7}$$

The data from the gravel bed Polish mountain rivers (*Dunajec, Raba, Wisłoka and Iarganiczanka*) don't agree with the curve given by Gessler, probably because the standard deviation of the mixture is large ; the difference, however, is not pronounced. The results are helpful for channel analysis and

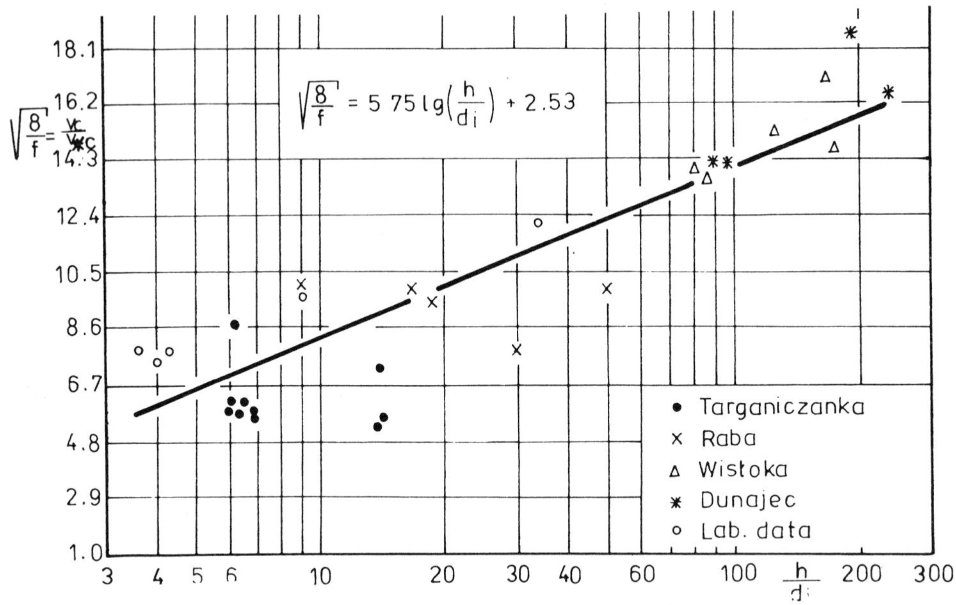

Fig.2 - Flow resistance relation vs. h/d_i.

stability design of cross- section of rivers, although coordinate parameters do not have a clear physical meaning.

CRITICAL FROUDE NUMBER

The problem of critical condition of coarse bed materials may also be described by critical Froude number. Expressing the experimental data for incipient motion as function of the critical Froude number by ,$v_c/\sqrt{gh_c}$, and the relative submergence, $Z = h_c/d_i$, the following relationship is obtained (Fig.4):

$$Fr_c = 1.35 \left(\frac{h_c}{d_i}\right)^{-0.35} \tag{8}$$

This relationship is very suitable for mountain rivers of wide ($B >> h$) and regular cross section.

This *critical Froude number* will be compared to Naill's *mobility number* M_n. According to Yalin [6] for the case of rough turbulent flow the incipient motion can be given by a relation between *mobility number* M_n and *critical dimensionless submergence* Z_{cr} . An empirical relation may be expressed as:

$$\sqrt{\frac{M_n}{Z_c}} = \frac{v_c}{1.65\sqrt{gh_c}} \tag{9}$$

As mobility number is only functions of dimensionless submergence (eq. 3), one finds:

$$\sqrt{\frac{M_n}{Z_c}} = 1.24\ Z_c^{-0.38} \tag{10}$$

and on base of it, the *critical Froude number* is:

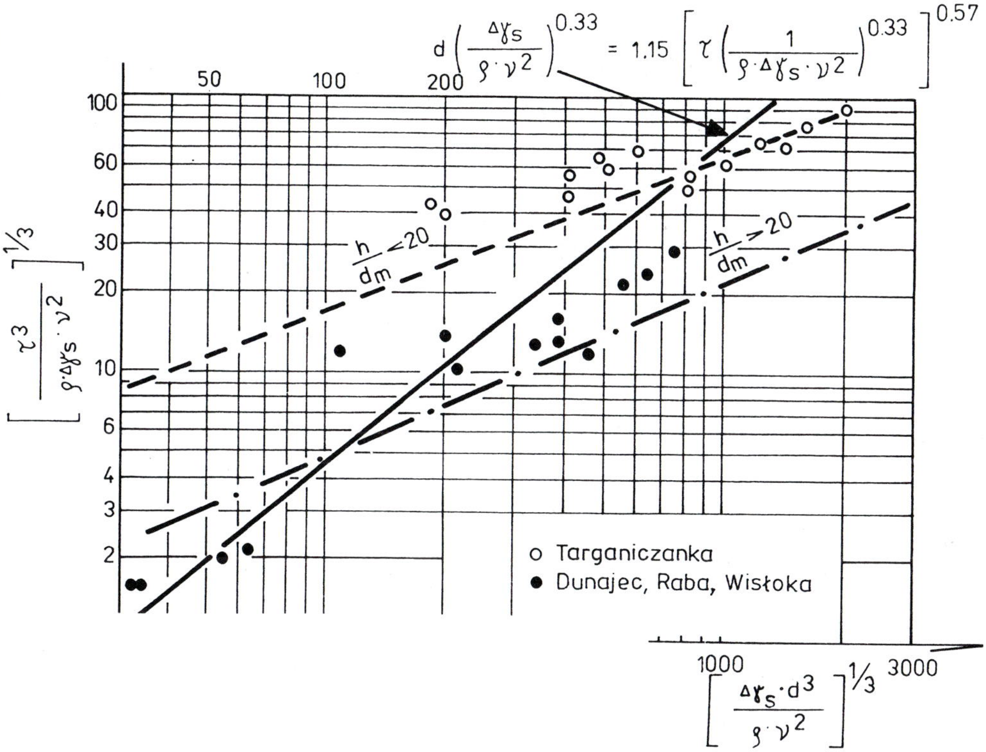

Fig.3 - Modified Shields diagram

$$Fr_c = 1.56\ Z_c^{-0.38} \qquad 11)$$

which is quite similar to eq. (8) found above.

LIST OF SYMBOLS

d_i	-fraction diameter
d_m	-typical grain size of mixture
f	-Shields parameter
f_m	-parameter of bed stability for dm
g	-acceleration due to gravity
k_i	-parameter of grain size stability
h	-depth of low
h_c	-critical depth of flow
v	-average velocity
v_c	-critical velocity
v	-shear velocity
n	-coefficient of Manning

j	-slope
k_s	-roughness height
Fr_c	-critical Froude number
F_d	-Froude number for particles
M_n	-mobility number
Re	-Reynolds number for flow
Z_c	-h/d critical dimensionless flow depth

REFERENCES

[1] Bartnik W.,Michalik A., Bednarczyk T., Radiotracer Investigations of Initial Parameters of Bed Load in Rivers, 6-th International Conference Transport and Sedimentation of Solid Particles June 88, Wroclaw, Poland

[2] Bogardi J., Sediment Transport in Alluvial Streams, Akademia Kiado , Budapest 1978

[3] Gladki H.,Michalik Al.,Bartnik W.,Measurement Bed Load Transport in Mountain Streams Using the Radioactive Tracer Method, IAHR - Workshop Rapperswill April 6-8, 1981

[4] Simons B.D., Stenturk F., Sediment Transport Technology, Fort Collins, Colorado 1977

[5] Wang Y.F., Bed-load transport in open channels, IAHR Baden-Baden 1977

[6] Yalin M.S., MECHANICS OF SEDIMENT TRANSPORT, Pergamon Press 1977

[7] Yalin M.S., Dimensional Approach to the Formulation of Regime Channels, International Conference on Fluvial Hydraulics Budapest Hungary 1988

BED-LOAD TRANSPORT IN STEEP CHANNELS

Tetsuro Tsujimoto

Department of Civil Engineering
Kanazawa University, Kanazawa, Japan

ABSTRACT

In this paper, hydraulic resistance, critical tractive force and bed-load discharge in steep gravel-bed channels are investigated. If the velocity distribution, which is much affected by small relative submergence, is formulated, the resistance law, dimensionless critical tractive force and bed-load formula can be reasonably predicted, and they are quite different from those established in alluvial sand bed channel which are familiar to us.

1. Introduction

Estimations of hydraulic resistance and sediment discharge are firstly important in river hydraulics, and many results have been already obtained for alluvial bed streams composed of sand. With recent development of mountain regions, hydraulics of mountain streams should be more investigated because the properties of flow and sediment transport in mountain streams characterized by steep slope and small relative submergence might be much different from those in alluvial rivers which are familiar to us. In the case of flow with very small relative submergence, however, the accuracy of measurements of flow depth is so low mainly due to the ambiguous definition of the hypothetical plane assumed to be a bed and thus the several data show so wide scattering that the peculiarity of flow and sediment transport in steep gravel-bed channels is scarcely taken account of. Then, the resistance law and sediment transport formula obtained for flow with sufficiently large submergence are often adopted as approximations. However, since the properties of flow and sediment transport in mountain streams are essentially different from those for alluvial streams, they should be reasonably clarified to achieve a better accuracy in estimating them in mountain regions. In fact, the critical tractive force, which is one of the important parameters of bed-load transport, increases with the bed slope as pointed out by Ashida & Bayazit (1973), Mizuyama (1977), Bettes (1984), Bathurst et al. (1987), Graf & Suszka (1987) and others; and hydraulics resistance increases with increase of relative roughness as pointed out by Griffiths (1981), Cao (1985) and others. Furthermore, the velocity profile of flow with small relative submergence might deviates from the logarithmic law as pointed out and attempted to be explained by Christensen (1971), O'Loughlin & Annamhotla (1972), Mizuyama (1977) and Nakagawa et al.(1988b)

In this study, the velocity profile, resistance law, critical tractive force for bed-load movement and bed-load discharge are investigated to obtain more reasonable and accurate estimations of them in mountain regions, particularly keeping the conformity of the accuracy of treatment among them. For simplicity, two-dimensional steady uniform flow over a flat bed composed of gravels of uniform diameter is considered as a first step of the research for fluvial hydraulics in mountain regions, though wide gradation of bed materials, various kind of bed forms and unsteadiness of flood flow are actually representative characteristics of flow in mountain streams.

2. Flume Experiments of Flow with Small Relative Submergence

Flume experiments were conducted in a 0.4m wide and 12m long flume with adjustable slope. Three kinds of gravels were used for preparing flat beds in the flume. The diameters of the used gravels are 7.29, 9.5 and 12.0mm. The flow measurements were carried out with a propeller currentmeter (the diameter of the propeller is 3mm). Flow was introduced to the flume from a constant-head tank through an adjustable valve, and the flow discharge was measured by a triangular wier at the end of the flume. The tail gate at the end of the flume could adjust the back water, and uniform flow condition was approximately accomplished for each run. The experimental conditions ranged as follows: S=bed slope=0.001~0.01; Re=1000~40000; Fr=0.12~0.82; and h/d=0.6~15.0, in which $Re \equiv Uh/\nu$; $Fr \equiv U/\sqrt{gh}$; U=depth-averaged flow velocity; h=flow depth; d=diameter of gravels; g=gravity acceleration; and ν=kinematic viscosity of water. The propeller currentmeter was connected to a FFT-analyzer and the time-averaged velocity and the turbulence intensity (defined as the root-mean square of the velocity fluctuation in the longitudinal direction) were calculated. The reference bed or the hypothetical bed (y=0) is conventionally, for simplicity, assumed to be 0.2d below the tangential top-surface of gravels.

The distributions of mean flow velocity and turbulence intensity of several runs of the present experiments, in which the relative submergence is systematically changed, are shown in Figs.1 and 2, respectively. In Fig.1, the present data are compared with the logarithmic law which is familiar for flow over a rough bed with sufficient submergence. In Fig.2, the data are compared with the following semi-empirical relation established for flow with sufficient submergence based on large amount of the previously conducted turbulence measurements.

$$\frac{u'_{rms}}{u_*} = D_1 \exp\left(-\frac{y}{h}\right) \qquad (1)$$

in which D_1=2.3 (Nezu 1977); $u_* \equiv \sqrt{ghS}$=shear velocity; and the subscript rms means a root-mean-square value. Both Figs.1 and 2 suggest that the distributions of velocity and turbulence intensity in the region near the bed are no longer consistent with those well-known for flow with sufficient submergence. The velocity profile becomes rather uniform, and the turbulent intensity is suppressed in this region, which is often called "roughness sublayer". The thickness of the roughness sublayer, y_R, is 0.3~1.2 times gravel diameter. The scattering of the values of y_R/d may depend on the respectively different surface arrangement of gravels for each run.

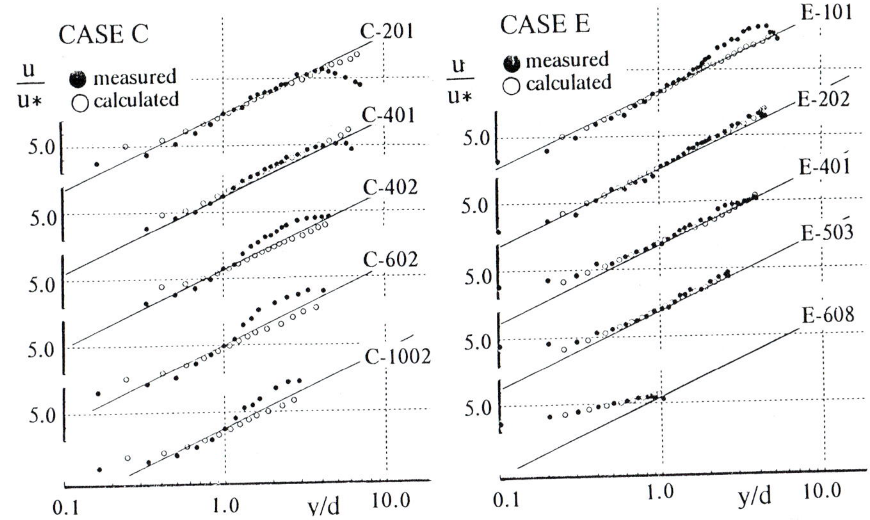

Fig. 1 Velocity distribution of flow over gravel bed.

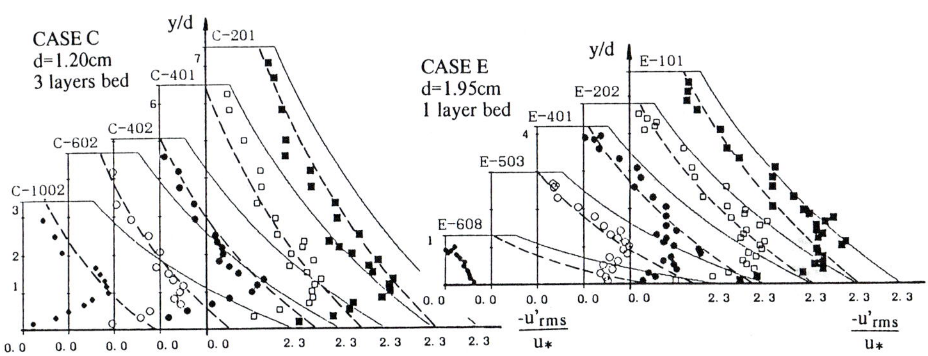

Fig. 2 Distribution of turbulence intensities of gravel-bed flow.

The Reynolds-stress distribution, $-\rho\overline{u'v'}$ (ρ=mass density of water), is more closely related to the velocity profile. In the present experiments without simultaneous measurements of the longitudinal and the vertical velocity fluctuations, the Reynolds stress was not directly obtained. However, the following indirect estimation might be possible: Based on the turbulence measurements of flow over large roughness (glass beads: d=12.5mm) with a lazer Doppler anemometer by Nakagawa et al. (1988b) where not only the turbulence intensities in both longitudinal and vertical directions but also the Reynolds stress were measured, the correlation of the longitudinal and the vertical fluctuations, r_T, is preserved as a universal function of the relative height from a bed ($\eta \equiv y/h$) even in the roughness sublayer, and the ratio of the turbulence intensity in the vertical direction to that in the longitudinal one is kept constant (k_{uv}=1.27/2.3) even in the roughness sublayer. The universal function $r_T(\eta)$ is written as

$$-r_T(\eta) \equiv \frac{-\overline{u'v'}}{u'_{rms} \cdot v'_{rms}} = \frac{1}{k_{uv} D_1^2}(1-\eta)\exp(2\eta) \tag{2}$$

When Eq.(2) is valid even in the roughness sublayer, the Reynolds stress distribution can be estimated only from the data of the turbulence intensity in the longitudinal direction, as follows:

$$\frac{-\overline{u'v'}}{u_*^2} = \frac{1}{D_1^2}(1-\eta)\cdot\exp(2\eta)\cdot\left(\frac{u'_{rms}}{u_*}\right)^2 \tag{3}$$

The estimated Reynolds-stress distribution is shown in Fig.3, and it demonstrates the suppression of the Reynolds stress in the roughness sublayer. This fact is consistent to that the velocity distribution is degenerated to be more uniform than the logarithmic law in the roughness sublayer. If the vertical distribution of the mixing length is invariant, the Reynolds-stress distribution estimated by Eq.(3) well explains the degeneration of the velocity profile as also shown in Fig.1. Further discussion is referred to in another paper (Nakagawa et al. 1989).

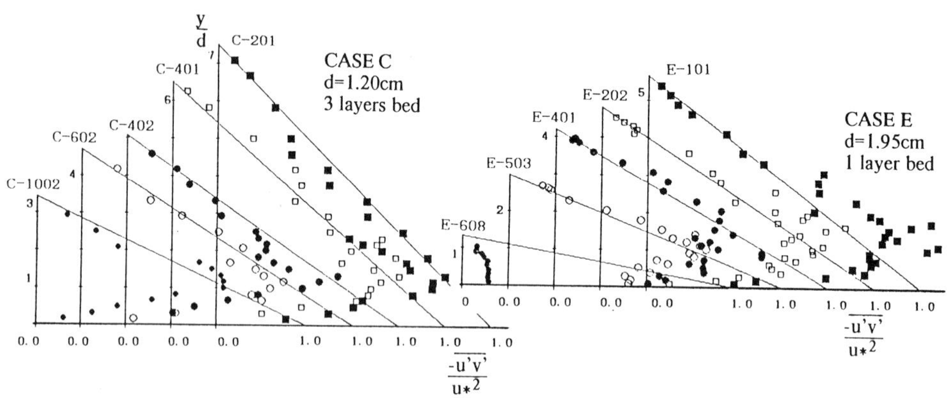

Fig.3 Reynolds-stress distribution of gravel-bed flow.

What is more important as for bed-load motion is the fact that the flow velocity near the bed-load particles becomes smaller than that for flow with sufficient submergence at the same shear velocity. In Fig.4(a), the values of $\varphi_d \equiv u_d/u_*$ (u_d=flow velocity at the height y=d; y=height from the reference bed) are plotted against the relative submergence. φ_d decreases with decrease of the relative submergence from φ_{d0}=8.5 (well-known value for hydrodynamically rough regime of Nikradse's experiments, if d is identified with the equivalent sand roughness) to around φ_{dmin}=3.0, in spite that the grain size Reynolds number ($Re_* \equiv u_* d/\nu$) is sufficiently large. In Fig.4(b) are plotted also the values of φ_d approximately evaluated from the data of the friction factor f (Cao 1985), by using the following relation which is obtained by an assumption of the logarithmic velocity distribution.

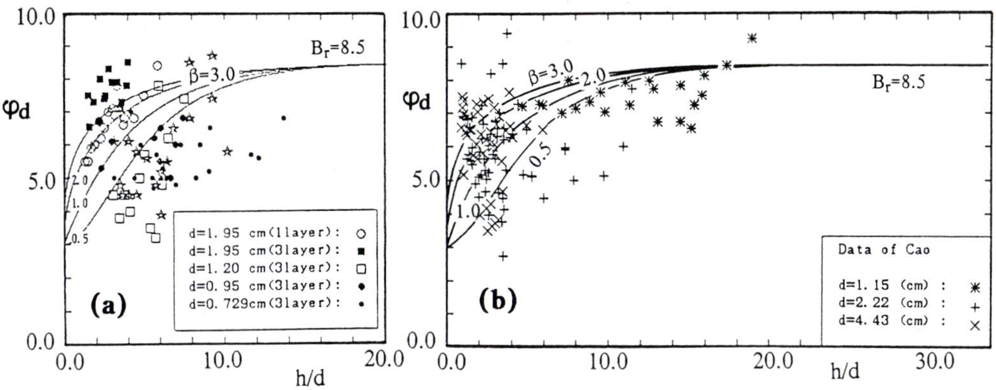

Fig.4 Relation between φ_d and h/d.

$$\varphi_d \cong \sqrt{\frac{8}{f}} - \frac{1}{\kappa}\left[\ln\frac{h}{d} - 1\right] \tag{4}$$

in which κ=Kármán constant. Although the data in Fig.4 are considerably scattering perhaps according to the differences of other hydraulic parameters, four curves indicated in Fig.4 are here chosen as representative of the relation between φ_d and h/d. These curves are expressed by the following equation.

$$\vartheta\left(\frac{h}{d}\right) \equiv \frac{\varphi_d}{\varphi_{d0}} + \frac{\varphi_{d0}-\varphi_{dmin}}{\varphi_{d0}}\left[1-\exp\left(-\frac{1}{\alpha}\frac{h}{d}\right)\right]^{1/\beta} + \frac{\varphi_d}{\varphi_{d0}} \tag{5}$$

in which α and β are empirical constants. φ_{dmin} is fixed to be 3.0 according to the data shown in Fig.4. α and β are here related to each other so that (φ_d-φ_{dmin}) becomes 99% of (φ_{d0}-φ_{dmin}) at h/d=20, and thus only β is an independent empirical constant. The four curves in the figure are obtained by changing the value of β as 0.5, 1.0, 2.0 and 3.0.

Simply, the logarithmic law is here adopted but φ_d changes following to Eq.(5). Then, the resistance law is obtained as follows:

$$\sqrt{\frac{8}{f}} \equiv \frac{U}{u_*} = \frac{1}{\kappa}\ln\frac{h}{d} + \varphi_d\left(\frac{h}{d}\right) - \frac{1}{\kappa} \tag{6}$$

Except the flow with too small relative submergence, the resistance law is more subjectively affected by the change of the value of φ_d with h/d than the change of the velocity profile due to the degeneration of the Reynolds-stress distribution. In Fig.5, the present data as for $\sqrt{8/f}$~(h/d) are shown with Eq.(6) and Keulegan's equation (φ_{d0} is used instead of φ_d as a function of h/d in Eq.(6)). The calculated curve of Eq.(6) with Eq.(5) in general describes a discrepancy of the $\sqrt{8/f}$~(h/d) relation from the Keulegan's equation in the range of h/d smaller than 20. In Fig.5, also the data of Cao (1985) are added.

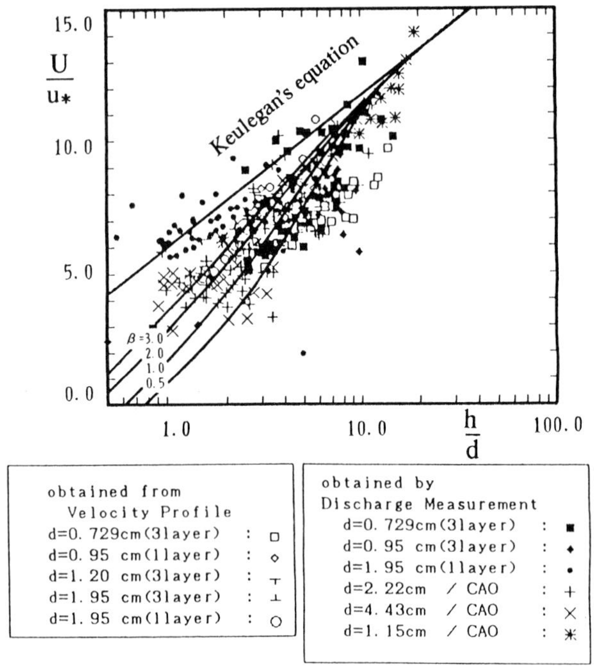

Fig.5 Relation between $\sqrt{8/f}$ and h/d.

3. Critical Tractive Force of Bed-Material Particles in Steep Channels

The forces acting upon a gravel particle resting on a rough bed with slope θ are the drag force D; lift force L ($=k_L D$); the weight of a gravel particle W; the buoyant force B_f; and the frictional force f_r (see Fig.6). These are expressed as follows:

$$D = \varepsilon_0 C_D A_2 d^2 \frac{\rho u_b^2}{2} \tag{7}$$

$$W = \sigma g A_3 d^3 ; \qquad B_f = \rho g A_3 d^3 \tag{8}$$

in which ε_0=sheltering coefficient ($\cong$0.4); C_D=drag coefficient; A_2, A_3=two- and three-dimensional geometrical coefficients of gravels; u_b=flow velocity near a bed; and $k_L \cong 0.85$ (Nakagawa & Tsujimoto 1986).

The magnitude of the resultant force acting on a gravel in the longitudinal direction, F, is expressed as follows:

$$\frac{F}{W} = (1+k_L\mu_f)\frac{D}{W} - [\mu_f(\cos\theta - \frac{B_f}{W}) - \sin\theta] \tag{9}$$

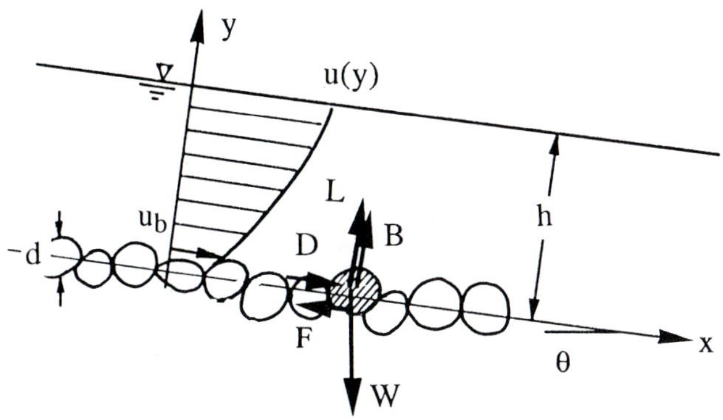

Fig.6 Schematic figure.

in which μ_f=static frictional coefficient of gravels ($\mu_f \cong 1.0$). If the sliding motion is considered as an initiation of motion of a gravel particle from a bed, F=0 corresponds to the so-called critical tractive force, and the dimensionless critical tractive force, τ_{*c}, can be expressed as follows:

$$\tau_{*c} = \frac{[(\sigma/\rho)\cos\theta - 1]\mu_f - \sin\theta}{(\sigma/\rho - 1)(1 + k_L\mu_f)\Xi_1\varphi_d^2} \tag{10}$$

in which $\tau_* \equiv u_*^2/[(\sigma/\rho - 1)gd]$; $\Xi_1 \equiv \varepsilon_0[C_D A_2/(2A_3)]$; σ=mass density of gravels; and the subscript c indicates the value for critical condition of sediment movement.

In the above, a sliding is assumed as a type of motion on dislodgement of a particle from a bed, but the equation of motion (a force-balance equation) is slightly but not essentially different according to a type of incipient motion. Furthermore, it is difficult to know true values of many coefficients introduced in the above treatment, and the numerical values introduced here are rather hypothetical. However, with the same equation of force-balance and numerical values for respective coefficients as the present study, not only the ordinary critical tractive force for mild slope bed as represented by Shields curve but also the effects of gradation of bed materials (Nakagawa et al. 1977), transpiration (Nakagawa et al. 1988a) and transverse bed-slope (Nakagawa et al. 1986) on the pick-up rate as well as the critical tractive force could be reasonably well evaluated. Hence, these formulation and numerical evaluations can be regarded significantly available. When the value of τ_{*c} in the case of $\theta \to 0$ is represented by τ_{*c0}, the following equation is obtained.

$$\gamma_c \equiv \frac{\tau_{*c}}{\tau_{*c0}} = \frac{\Psi_1(S)}{[\vartheta(h/d)]^2} \tag{11}$$

in which

$$\Psi_1(S) \equiv \frac{[(\sigma/\rho)\cos\theta - 1]\mu_f - (\sigma/\rho)\sin\theta}{(\sigma/\rho - 1)\mu_f} \tag{12}$$

$S \equiv \tan\theta$=bed slope; and the subscript 0 indicates the values for the case of S=0 with sufficient submergence. By the way, dimensionless bed shear stress, relative submergence and bed slope are interrelated as follows under uniform flow condition:

$$\tau_* = \frac{S}{(\sigma/\rho-1)} \cdot \frac{h}{d} \tag{13}$$

Hence, at the critical condition of bed-load transport, S and (h/d) are reversely proportional each other, and thus $1/[\vartheta(h/d)]^2$ is rewritten by a function of S as $\Psi_2(S)$. It means that $\gamma_c \equiv \tau_{*c}/\tau_{*c0}$ is written as a function of bed slope, and the effect of bed slope is composed of the two parts: $\Psi_1(S)$ as an effect of the gravity itself; and $\Psi_2(S)$ as an effect of degeneration of velocity distribution due to small relative submergence. The former is a decreasing function of S, while the latter an increasing one; and resultantly, τ_{*c}/τ_{*c0} falls an almost increasing function.

$$\gamma_c(S) \equiv \frac{\tau_{*c}}{\tau_{*c0}} = \Psi_1(S) \cdot \Psi_2(S) \tag{14}$$

In Fig.7, four curves of the relation between $\gamma_c \equiv \tau_{*c}/\tau_{*c0}$ and S are shown corresponding to the four curves representing the relation between $\varphi_d \equiv u_d/u_*$ and (h/d) in Fig.4, with the flume data of Mizuyama (1977), Cao (1985) and Suszka & Graf (1987). It is here assumed that τ_{*c0}=0.035. An empirical formula proposed by Graf & Suszka (1987), which is written as follows, is also drawn in the figure.

$$\frac{\tau_{*c}}{\tau_{*c0}} = 10^{2.2S} = \exp(5.06S) \tag{15}$$

They chose the value 0.042 as τ_{*c0}. The presently derived relation can well explain the data somewhat with the physical background.

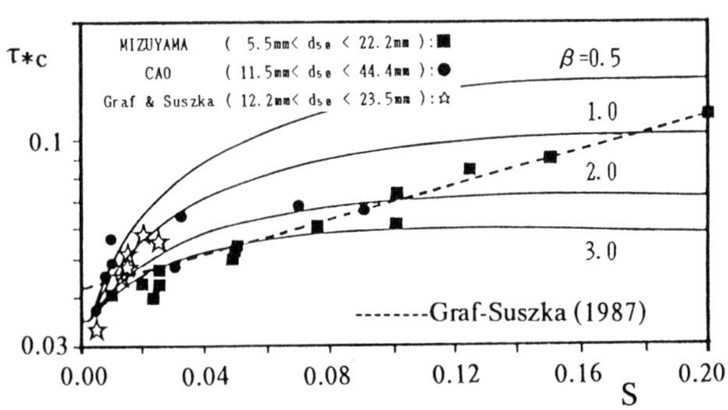

Fig.7 Variation of dimensionless critical tractive force against bed slope.

Dimensionless shear stress may be reasonable on consideration of the critical condition of bed-load movement from the view point of mechanics just as adopted in the above analysis. However, the use of depth and therefore shear stress as a criterion for initiation of movement is less appropriate because of difficulties in measuring depth due mainly to that in determination of "true" bed. From this reason, Bettes (1984) recommended that the water discharge per unit width q should be used instead of the bed shear stress and proposed the following empirical relation.

$$q_{*c} \equiv \frac{q_c}{\sqrt{gd^3}} = \frac{k_A}{S}\log\left(\frac{k_B}{S}\right) \tag{16}$$

in which k_A=0.134 and k_B=1.221. A more satisfactory agreement is achieved with an empirical equation by eye-fitting by Bathurst, Graf & Cao (1987) as follows:

$$q_{*c} = 0.15S^{-1.12} \tag{17}$$

Eqs.(16) and (17) are shown by dashed lines with the flume data in Fig.8. The present analysis also provides a prediction of q_{*c}. The curves based on the present model are drawn by a solid curve also in Fig.8 corresponding to the four curves in Fig.4.

$$q_{*c} = \sqrt{\tau_{*c0}}\sqrt{\sigma/\rho-1}\sqrt{\frac{8}{f}}\sqrt{\frac{h_c}{d}} \tag{18}$$

in which Eqs.(5), (6) and (12)~(14) have been used to calculate the relation between q_{*c} and S. The present model might explain the data slightly better than Eqs.(16) and (17), and it is more important that the present model is not empirical relation but has a reasonable background.

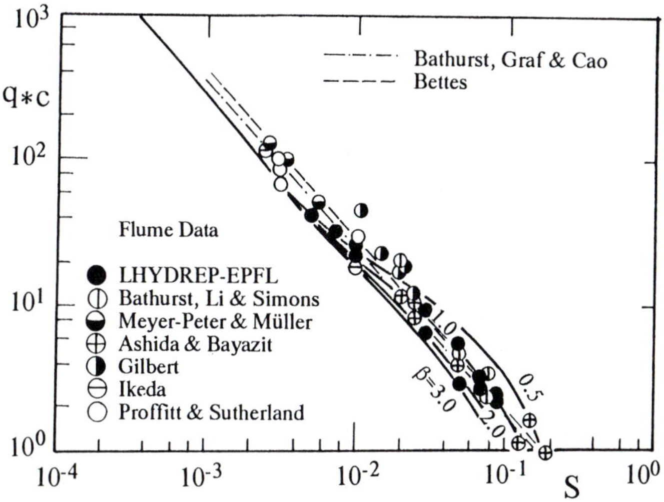

Fig.8 Critical flow discharge for bed mobility versus bed slope.

4. Bed-Load Transport Rate

The formulation of the critical tractive force contributes to establishing the transport formula, because many formulae previously proposed for alluvial streams have a similar form as

$$q_{B*} = A\tau_{*c}^{3/2}\left(1-\frac{\tau_{*c}}{\tau_*}\right)^m\left(1-\frac{\sqrt{\tau_{*c}}}{\sqrt{\tau_*}}\right)^n \tag{19}$$

in which $q_{B*} \equiv q_B/\sqrt{(\sigma/\rho-1)gd^3}$; q_B=bed-load transport rate; and A, m and n=empirical parameters. The exponent (3/2) of τ_* in Eq.(19) is reasonable from the energetical view point (Bagnold 1965), and the parameters chosen in the previously proposed formulae for alluvial streams are as follows: (1) A=8, m=3/2, n=0 for Meyer-Peter & Müller's (1949); (2) A=4.25, m=1, n=0 for Bagnold's (1965); and (3) A=17, m=n=1 for Ashida & Michiue's (1972).

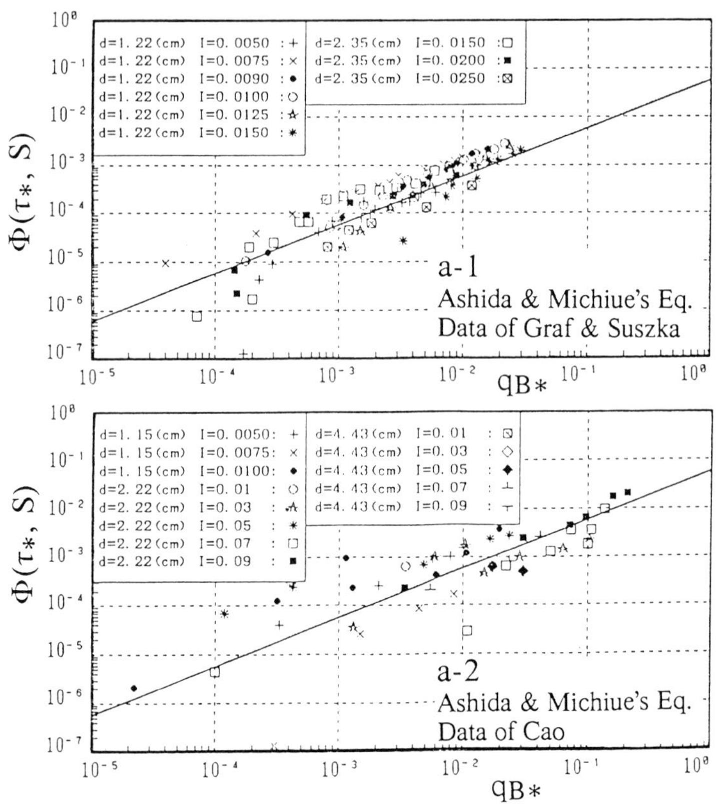

Fig.9 Determination of empirical constants in bed-load formula (a).

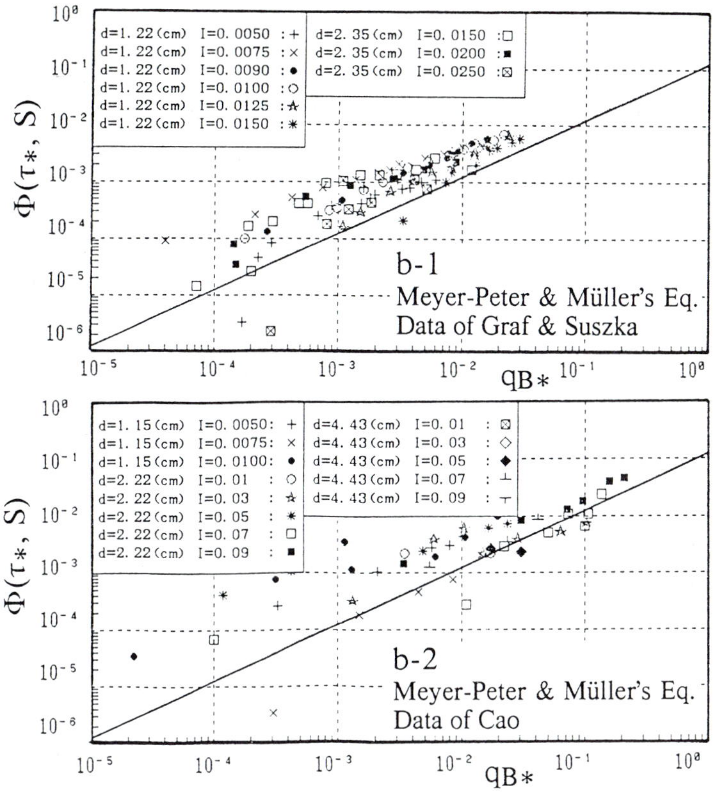

Fig.9 Determination of empirical constants in bed-load formula (b).

As aforementioned, the dimensionless critical tractive force is actually a function of bed slope, and thus, the relationship between q_{B*} and τ_* for steep streams with different bed slope differs from each other particularly in the range of lower transport intensity. If the dimensionless critical tractive force was reasonably evaluated by the aforementioned method for each bed slope, the bed-load transport rate would be more accurately predicted. Furthermore, A is also modified as $A=A_0/[\Psi_1(S)]^{3/2}$ because τ_* should be corrected by taking account of the gravity effect. Then, a bed-load transport formula for steep channels is proposed as follows:

$$q_{B*} = A_0 \cdot \Phi(\tau_*, S) \equiv A_0\left[\frac{\tau_*}{\Psi_1(S)}\right]^{3/2}\left(1-\frac{\gamma_c \tau_{*c0}}{\tau_*}\right)^m\left(1-\frac{\sqrt{\gamma_c \tau_{*c0}}}{\sqrt{\tau_*}}\right)^n \tag{20}$$

The flume data of bed-load discharge (Cao 1985; Graf & Suszka 1987), were used to determine the empirical constants. Based on the above mentioned three formulae, the linearity of the relation between q_{B*} and $\Phi(\tau_*, S)$ was inspected. Two examples are compared in Fig.p. As a result, the empirical constants are here determined as follows: $A_0=17$, $m=1$, $n=1$ and $\beta=1.0$. The transport formula with the presently determined parameters is shown with the flume data (Graf & Suszka and Cao) in Fig.10, and the agreements between the data and the proposed transport formula are fairly well. Another point to be demonstrated is that the value of β is certainly determined though it has not been numerically determined but it ranges β=0.5~3.0 before inspecting the bed-load discharge. The asymptote of the present model for $S \to 0$ falls Ashida & Michiue's formula for alluvial streams.

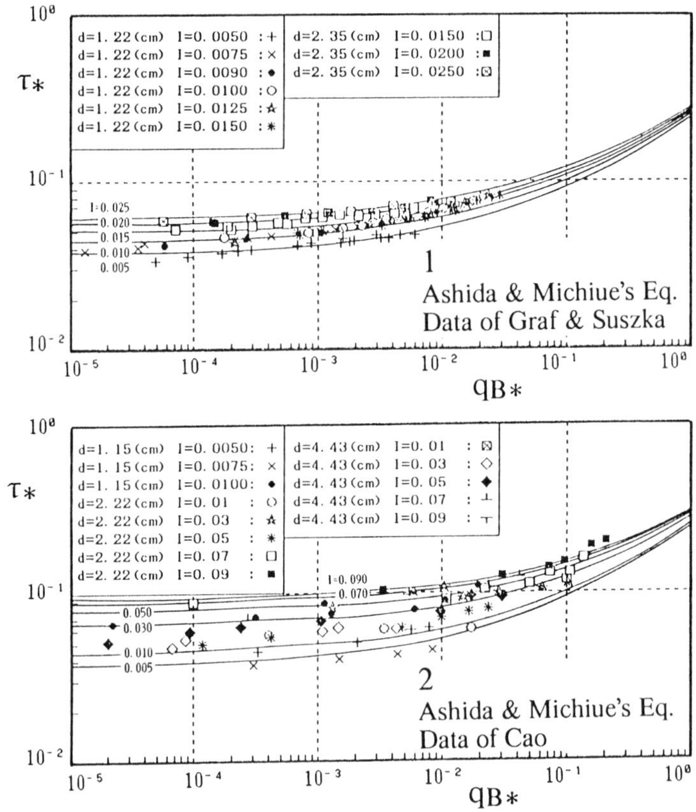

Fig.10 Bed-load function for each bed slope.

5. Conclusions

The following results are obtained in this study:

(1) In steep channels, flow has in general small relative submergence, and the velocity distribution is degenerated from that of flow with sufficient submergence. The velocity distribution in the near-bed region called "roughness sublayer" becomes more uniform than the log-law profile corresponding to the suppression of the turbulence there. Moreover, the value of φ_d, which corresponds to the ratio of the flow velocity at y=d (d=gravel diameter) to the shear velocity, systematically decreases with decrease of relative submergence. Though

the data of φ_d shows a considerable scattering, a representative curve is proposed for the relation between φ_d and h/d (relative submergence) with undetermined empirical constant (β). A discrepancy of resistance law from Keulegan's equation can be explained by using the established relation between φ_d and h/d.

(2) The critical tractive force in steep channels is affected by direct gravity effect and indirect effect through the change of velocity distribution. Based on the force-balance equation for bed-material particle, these two effects are formulated. The latter is evaluated based on the presently proposed relation between φ_d and h/d. A resultant expression gives an accurate and reasonable prediction of dimensionless critical tractive force in steep channels. In order to avoid the poor accuracy in determination of flow depth for flow with small relative submergence, the critical condition for mobile bed expressed by a dimensionless flow discharge is also proposed, and it can explain the flume data more favorably than the previous empirical formula.

(3) A reasonable evaluation of dimensionless critical tractive force gives a more accurate prediction of bed-load discharge in steep channels. Using the flume data of bed-load discharge in steep channels, empirical constants involved in the previous bed-load formula have been synthetically determined.

(4) On determination of bed-load discharge, the provisionally estimated value of β as an index of the velocity distribution degenerated by the small relative submergence of flow has been definitively determined. The velocity profile, the hydraulic resistance, the critical condition for mobile bed, and the bed-load discharge have been formulated in the forms including one empirical parameter β. The comparisons between the respective experimental data and them have determined a reliable numerical value of β uniquely.

Unsteady aspects of bed-load transport must be important particularly in mountain regions in general with relatively sharp hydrograph of floods, on which interesting studies have already started (Suszka & Graf, 1987; Tsujimoto, Graf & Suszka, 1988). Furthermore, the bed-load transport of heterogeneous materials and the subsequent sorting phenomena (Tsujimoto, 1989a, b) are also essentially important problem in gravel-bed streams. The present study must help the advanced step of these studies.

This study started at Ecole Polytechnique Fédérale de Lausanne (E.P.F.L.) when the author was staying there for research cooperation with Prof. Walter H. Graf. The author expresses his gratitude to Prof. Graf for his encouragement and scientific advise. At E.P.F.L., the author had a chance to discuss with Mr. L. Suszka, the data obtained by whom at E.P.F.L. were used in this study.

References

1. Ashida, K. and M. Bayazit (1973) : Initiation of motion and roughness of flows in steep channels. *Proc. 15th Congress of IAHR*, Istanbul, Turkey, Vol.1, pp.475-484.
2. Ashida, K. and M. Michiue (1972) : Study on hydraulic resistance and bedload transport rate in alluvial streams. *Proc. JSCE*, No.206, pp.59-69 (in Japanese).
3. Bagnold, R.A. (1965) : An approach to the sediment transport problems from general physics. *U.S.G.S. Professional Paper*, 422-J.
4. Bathurst, J.C., W.H. Graf and H.H. Cao (1987) : Bed load discharge equations for steep mountain rivers. *Sediment Transport in Gravel-Bed Rivers*, edited by C.R.

Thorne, J.C. Bathurst and R.D.Hey, John Wiley & Sons Ltd., pp.453-477.

5. Bettes, R. (1984) : Initiation of sediment transport in gravel streams. *Proc. Institution of Civil Engineers*, No.77, Part 2, pp.79-88.
6. Cao, H.H. (1985) : Resistance hydraulique d'un lit de gravier mobile a pent raide, etude experimentale. *Ph.D. Thesis*, Ecole Polytechnique Fédérale de Lausanne.
7. Christensen, B.A. (1971) : Incipient motion on cohesionless channel banks. *Proc. Sedimentation Symposium*, Berkeley, California, USA, Chapter 4.
8. Graf, W.H. and L. Suszka (1987) : Sediment transport in steep channels. *Journal of Hydroscience and Hydraulic Engineering*, JSCE, Vol.5, No.1, pp.11-26.
9. Griffiths, G. (1981) : Flow resistance in coarse gravel bed rivers. *Journal of the Hydraulics Division*, ASCE, Vol.107, HY7, pp.899-918.
10. Meyer-Peter, E. and R. Müller (1949) : Formulas for bed-load transport. *Proc. 2nd Congress of IAHR*, Stockholm, Sweden.
11. Mizuyama, T. (1977) : Bedload transport in steep channels. *Doctoral Thesis*, Kyoto University, 96p. (in Japanese).
12. Nakagawa, H and T. Tsujimoto (1986) : *Mechanics of Sediment Transport and Alluvial Hydraulics*, Gihodo Shuppan, Tokyo, Japan, 310p. (in Japanese).
13. Nakagawa, H., T. Tsujimoto and T. Hara (1977) : Armoring of alluvial bed composed of sediment mixtures. *Annuals*, Disaster Prevention Research Institute, Kyoto University, No.20B-2, pp.355-370 (in Japanese).
14. Nakagawa, H., T. Tsujimoto and S. Murakami (1986) : Non-equilibrium bed load transport along side slope of an alluvial streams. *Proc. 3rd International Symposium on River Sedimentation*, Jackson, Mississippi, USA, pp.885-893.
15. Nakagawa, H., T. Tsujimoto and S. Murakami (1988a) : Effect of suction or injection through bottom of a stream on bed load transport process. *Proc. IAHR International Conference on Fluvial Hydraulics*, Budapest, Hungary, pp.70-75.
16. Nakagawa, H., T. Tsujimoto, Y. Shimizu (1988b) : Velocity profile of flow over rough permeable bed. *Proc. 6th Congress of APD-IAHR*, Kyoto, Japan, Vol.II, pp.449-456.
17. Nakagawa, H., T. Tsujimoto, Y. Shimizu (1989) : Turbulent flow with small relative submergence. *International Workshop on Fluvial Hydraulics of Mountain Regions*, IAHR, Trent, Italy.
18. Nezu, I. (1977) : Turbulent structure in open-channel flow. *Doctoral Thesis*, Kyoto University, 118p. (in Japanese).
19. O'Loughlin, E.M. and V.S.S. Annambhotla (1969) : Flow phenomena near rough boundaries. *Journal of Hydraulic Research*, IAHR, Vol.7, No.2, pp.231-250.
20. Suszka, L. and W.H. Graf (1987) Sediment transport in steep channels at unsteady flow. *Proc. 22nd Congress of IAHR*, Fluvial Hydraulics, Lausanne, Switzerland, pp.166-170.
21. Tsujimoto, T. (1989) : Instability of bed-surface composition due to sorting process in a stream composed of sand and gravel. *Proc. International Symposium on Sediment Transport Modeling*, New Orleans, U.S.A.,1989.
22. Tsujimoto, T. (1989) : Longitudinal stripes of alternate lateral sorting due to cellular secondary currents. *Proc. 23rd Congress of IAHR*, Ottawa, Canada,1989.
23. Tsujimoto, T., W.H. Graf and L. Suszka (1988) : Bed-load Transport in unsteady flow. *Proc. 6th Congress of APD-IAHR*, Kyoto, Japan, Vol.II, pp.15-22.
24. Tu, H.Z., T. Tsujimoto and W.H. Graf (1988) : Velocity distribution in a gravel-bed flume. *Proc. 6th Congress of APD-IAHR*, Kyoto, Japan, Vol.II, pp.425-430.

FLUME EXPERIMENTS ON ALTERNATE BARS IN UNSTEADY FLOW

Marco Tubino

Hydraulics Institute
University of Genoa, Genoa, Italy

ABSTRACT

- Flume experiments are performed to investigate the process of development of alternate bars under unsteady flow conditions. The bed response to unsteady flow is analyzed for different values of the period of the flood and the observed temporal behaviour of the amplitude of bars, which appears to be strongly affected by the unsteady character of the flow, is compared with the theoretical predictions of Tubino and Seminara [1987].

1 . INTRODUCTION

The process of formation and development of migrating alternate bars has been the subject of several investigations starting from the cornerstone contributions by Kinoshita [1961] and Leopold and Wolman [1957]. This problem bears both a conceptual and practical interest. It is well known that, under appropriate conditions, the flat cohesionless bottom of a turbulent stream flowing in a straight channel loses stability and bottom perturbations arise and develop spontaneously as a result of an instability process leading to an alternating sequence of deep and shallow reaches. The formation of these bedforms is one of the basic processes controlling the channel development on a spatial scale of the order of the channel width and may play a relevant role in the process of meander formation (see Blondeaux and Seminara [1985] and Seminara and Tubino [1989]).

The practical motivation for investigating this process refers to the problem of formation of migrating bars in straight reaches of rivers as a result of river regulation works such as channelization and artificial straightening. This feature needs be controlled since the scouring and depositional effects related to the development of these bedforms can affect several aspects of fluvial engineering like navigation, bank protection and design of structures.

Many data have been collected on the occurrence of these bedforms both in the field and in the laboratory and on the geometrical and hydraulic characteristics of alternate bars under steady flow conditions, starting from the first observations of Kinoshita [1961] until the recent and detailed investigations of Fujita and Muramoto [1985]. Empirical criteria for alternate bar formation and predictors for the equilibrium length and height of bars are also available in the literature (see among others Ikeda [1984] and Kuroki and Kishi [1985]).

On the theoretical side a rational interpretation of the physical process has been developed through a large number of linear studies (Hansen [1967], Callander [1969], Engelund and Skovgaard [1973], Parker [1976], Fredsoe [1978]). Linear theories seek the conditions for incipient bar formation, the linear growth rate of perturbations and their wavelengths and wavespeeds under steady flow conditions. In particular they show that alternate bars develop in a supercritical regime where the width ratio β (i.e. the ratio between the half channel width and the flow depth) exceeds a critical value β_C, for given values of the Shields parameter θ and of the roughness parameter d_s (grain size scaled by flow depth) (Tubino [1986], Colombini et al. [1987]).

Linear studies result in an exponential growth of the bar amplitude and are obviously unable to predict a 'finite equilibrium amplitude' of the bed perturbations. The latter was determined by Colombini et al. [1987] by means of a weakly non linear analysis developed in a neighbourhood of the threshold conditions for bar formation. The development in time of the bar amplitude was found to be governed by a non linear ordinary differential equation (of Landau-Stuart type) the solution of which shows that non linear effects inhibit the exponential growth predicted by linear theory and lead to an equilibrium value asymptotically reached.

The non linear analysis of Colombini et al. [1987] has been found to predict successfully the equilibrium value and the temporal behaviour of the bar height during the process of development under steady flow conditions. The experimental evidence suggests that the wavelength of bottom perturbations rapidly attains a fairly stable value and then undergoes minor variations such that the formative process of migrating alternate bars is mainly characterized by the temporal increase of the wave height until the bar geometry reaches an equilibrium state: this condition allows one to define an appropriate time scale of the development process of bars.

The question then arises of how the above process is affected by the unsteady character of the flow, which is always characteristic of rivers. In gravel bed rivers the actual formative conditions for alternate bars are typically unsteady and alternate bars form and develop at flood stage and then undergo minor variations at lower stage. When comparing the characteristic time scale for bar development and the time scale of the basic unsteady flow it is found that in many cases the two scales are of the same order of magnitude, thus leading to the two processes strongly affecting each other. In this case the temporal behaviour of the amplitude of bars will not be predictable on the basis of the equilibrium values related to instantaneous flow conditions. In other words the bed response to unsteady flow is expected to exhibit a time lag with respect to the variable hydraulic conditions. What is then the significance of a steady equilibrium amplitude of bars since formative conditions are typically unsteady?

The latter problem was tackled theoretically (Tubino and Seminara [1987]) by means of a weakly non linear analysis suitable to describe the temporal behaviour of the amplitude of bars during flood propagation. The theory shows that the bar response to unsteady flow mainly depends on a parameter $\hat{U}$, which is a measure of the ratio between the time scale of flow unsteadiness and the time scale of bar instability, such that if $\hat{U}$ is O(1), as it occurs in nature, flow unsteadiness affects the instantaneous growth rate and phase of bar perturbations and controls the final amplitude reached by the bed configuration.

The above findings suggest that the concept of a 'finite equilibrium amplitude' of bars established in Colombini et al. [1987] can only keep meaningful provided it be referred to some 'reference flow stage' (i.e. to some reference amplitude) which will have to be defined.

In order to substantiate the main results of the theory (since no detailed data on alternate bars development under unsteady flow conditions were available) and to ultimately ascertain the validity limits of the weakly non linear approach some laboratory experiments have been performed both under steady and unsteady conditions, i.e. using a step like unsteady discharge simulating a continuous hydrograph.

The aim of these investigations was to ascertain the role played by the parameter $\hat{U}$ in controlling the effect of unsteadiness on bar development, to further verify the theoretical estimate of the time scale of the process of bar growth and to obtain detailed data on the temporal behaviour of the amplitude of bars during a flood event or during a sequence of similar flood events. The latter point is twofold since it allows one to study the effect of a given initial amplitude of bed configuration and to ascertain whether bars develop as a result of a single flood event or as a cumulative effect of typical flood events. In the latter case it is of interest to investigate the possible trend of the final amplitude reached by the bed configuration towards some upper bound depending on the flow conditions characterizing typical hydrographs.

The experimental procedures and results are reported in section 2. The main features of the theoretical treatment are briefly outlined in section 3. Finally section 4 is devoted to a comparison between theoretical and experimental results along with some concluding remarks.

2 . EXPERIMENTS

The experiments were performed in a flume of the Laboratory of the Institute of Hydraulics of Genoa University. The flume is 19 m long, 60 cm wide, and is equipped with a recirculating flow system for both water and sediment; the flume has smooth rigid walls and a movable bed. An adjustable weir controls the flow depth at the downstream end to minimize backwater effects and determine conditions of uniform flow. Water discharge is measured by an orifice plate flow meter in the recirculating pipe and is regulated by a pneumatic valve which is automatically driven in order to impose a prescribed discharge. A rather uniform sand with median diameter of 1.1 mm was used (see figure 1) and the surface of the bed was molded at the beginning of each experiment in order to get a prescribed initial slope. The flume was filled with sand in the reach between 1 m and 19 m from the upstream end and a by-pass system was arranged on the sediment recirculating system in order to get weighted samples during the experiments.

Measurements of flow depth and bed topography were performed with a point gauge and a video camera device was also used to determine the planimetric bed configuration. The flow was interrupted during the experiments in order to follow the temporal development of bar geometry: bed elevations were measured after the flow stoppage at 9 points in each cross section with a longitudinal spacing selected on the basis of the observed wavelength. Additional measurements were performed to obtain a more refined information about maximum and minimum bed elevations within each bar front. Some experiments were repeated without intermediate stoppages until a prescribed stage configuration in order to confirm the reproducibility of the experiments. No significant changes in the slope with respect to the initial configuration were observed during the experiments since the sediment load carried by the stream was trapped at the downstream end and recirculated in the upstream section.

In table 1 the hydraulic conditions and measured values of the bar wavelength W_L* and height H_B* (difference between maximum and minimum bed elevations within a bar front) are reported for some of the experiments performed under steady conditions in which the temporal development of bed configuration has been recorded (Q* is water discharge, D_o* is the average depth and t* is time). Figure 2 shows the time development of the wavelength and height of bars as observed in run ST72.

In tables 2 and 3 data are reported of two series of experiments performed under unsteady conditions. In the experiments of each series the total flow duration T_F* (i.e. the period of the flood event considered) was changed and the plane bed initial configuration was re-established except for run D721 where the bed configuration obtained at the end of experiment D72 was used as initial configuration.

Table 1

Run n.	Q^* (l/s)	Initial slope	D_o^* (cm)	Free surface slope	t^* (min)	W_L^* (cm)	H_B^* (cm)
ST70	6.44	.007	2.54	.00662	30 60 90	344 390 443	2.05 3.3 3.5
ST71	4.91	.007	2.20	.00702	30 60 90	400 465 506	1.5 3.2 3.87
ST72	5.89	.007	2.33	.00669	15 30 45 60 75	242 339 405 426 444	1.53 1.73 2.3 3.14 3.67

Table 2

		Initial slope = .005 Runs D51, D52, D53, D54		Initial slope = .007 Runs D71, D72, D721, D73	
Step	t^*/T_F^*	Q^* (l/s)	D_o^* (cm)	Q^* (l/s)	D_o^* (cm)
1	1/24	5.48	2.38	4.98	2.11
2	1/24	5.94	2.47	5.68	2.36
3	1/24	7.12	2.77	6.75	2.60
4	1/24	8.29	3.03	8.53	3.08
5	1/6	9.71	3.55	9.64	3.42
6	1/6	8.30	3.19	8.35	3.06
7	1/6	7.12	2.83	6.73	2.59
8	1/6	6.08	2.48	5.73	2.36
9	1/6	5.56	2.38	3.84	1.85

In each series of experiments the same initial slope was used and the same triangular shape hydrograph was simulated using a step like unsteady discharge with the same relative flow duration t^*/T_F^* in each step. In table 3 averaged values of W_L^*, H_B^* and C_B^* (wavespeed of bar fronts) observed in the last steps of the experiments are given. Figure 3 illustrates the development process of alternate bars observed in runs D72 and D721 which simulated the effect of a sequence of two identical flood events.

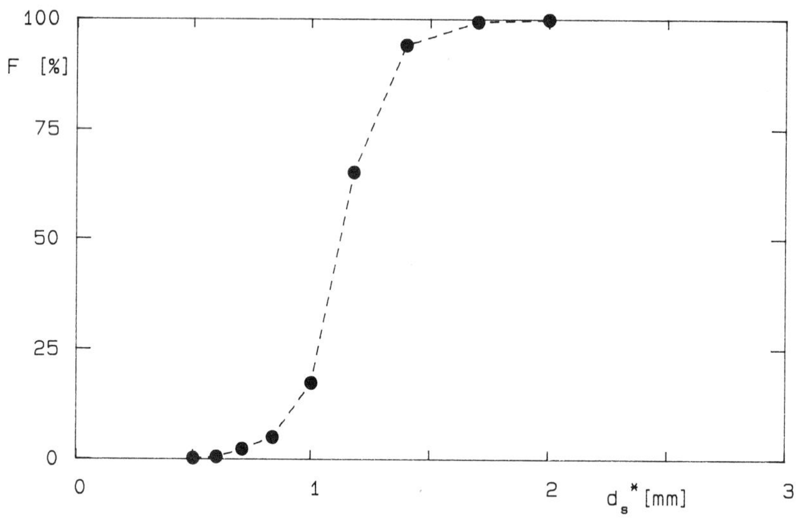

Figure 1 - Grain size accumulation curve of sand used

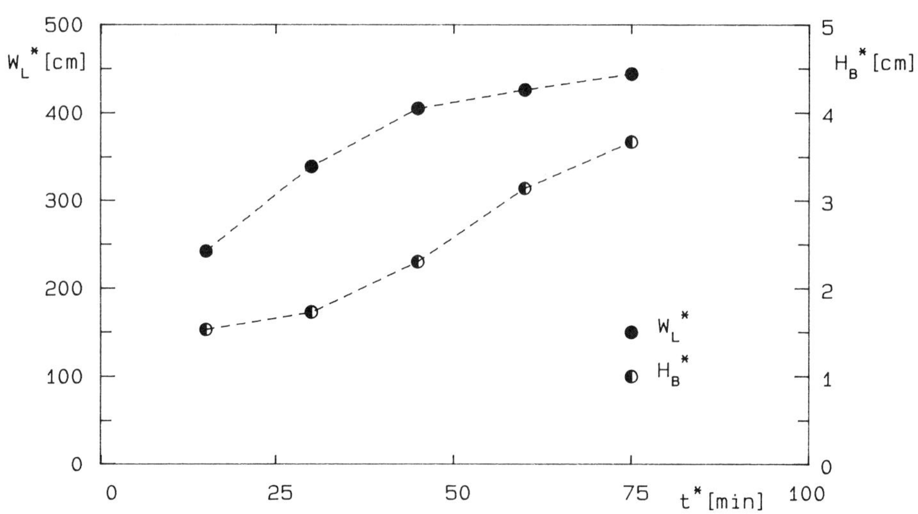

Figure 2 - Time development of the wavelength and height of bars as observed in run ST72

Table 3

Run	T_F^* (min)	Step	W_L^* (cm)	H_B^* (cm)	C_B^* (cm/min)
D51	24	9	225	0.9	---
D52	48	8	255	0.8	---
		9	321	1.2	---
D53	96	7	287	1.5	---
		8	293	1.86	5.7
		9	306	2.08	4.8
D54	144	6	254	1.5	---
		7	443	2.12	---
		8	481	2.8	---
		9	530	3.42	3.26
D71	24	9	346	1.33	---
D72	48	8	393	1.33	---
		9	393	2.7	6.9
D721(*)	48	5	417	3.28	12.0
		6	421	3.28	10.2
		7	423	3.9	8.1
		8	441	4.09	6.3
		9	451	4.24	3.1
D73	96	8	360	2.97	---
		9	430	4.0	---

(*) Run D721: initial configuration as at the end of Run D72

3 . OUTLINE OF PREVIOUS THEORY

Tubino and Seminara [1987] analyzed the effect of unsteadiness of the flow on the process of non linear growth of bars. They considered a straight reach of a channel with constant width 2B* and characteristic length much smaller than a typical length of a flood wave. Within this reach a flood wave is felt as a locally uniform unsteady flow. In other words, by referring to a reference basic (steady) state such that $\overline{U}_o^*$ and $\overline{D}_o^*$ are average speed and flow depth, the basic unsteady flow within the reach examined was written in the form:

$$U_o^* = \overline{U}_o U_o(\tau) \qquad D_o^* = \overline{D}_o D_o(\tau) \tag{1}$$

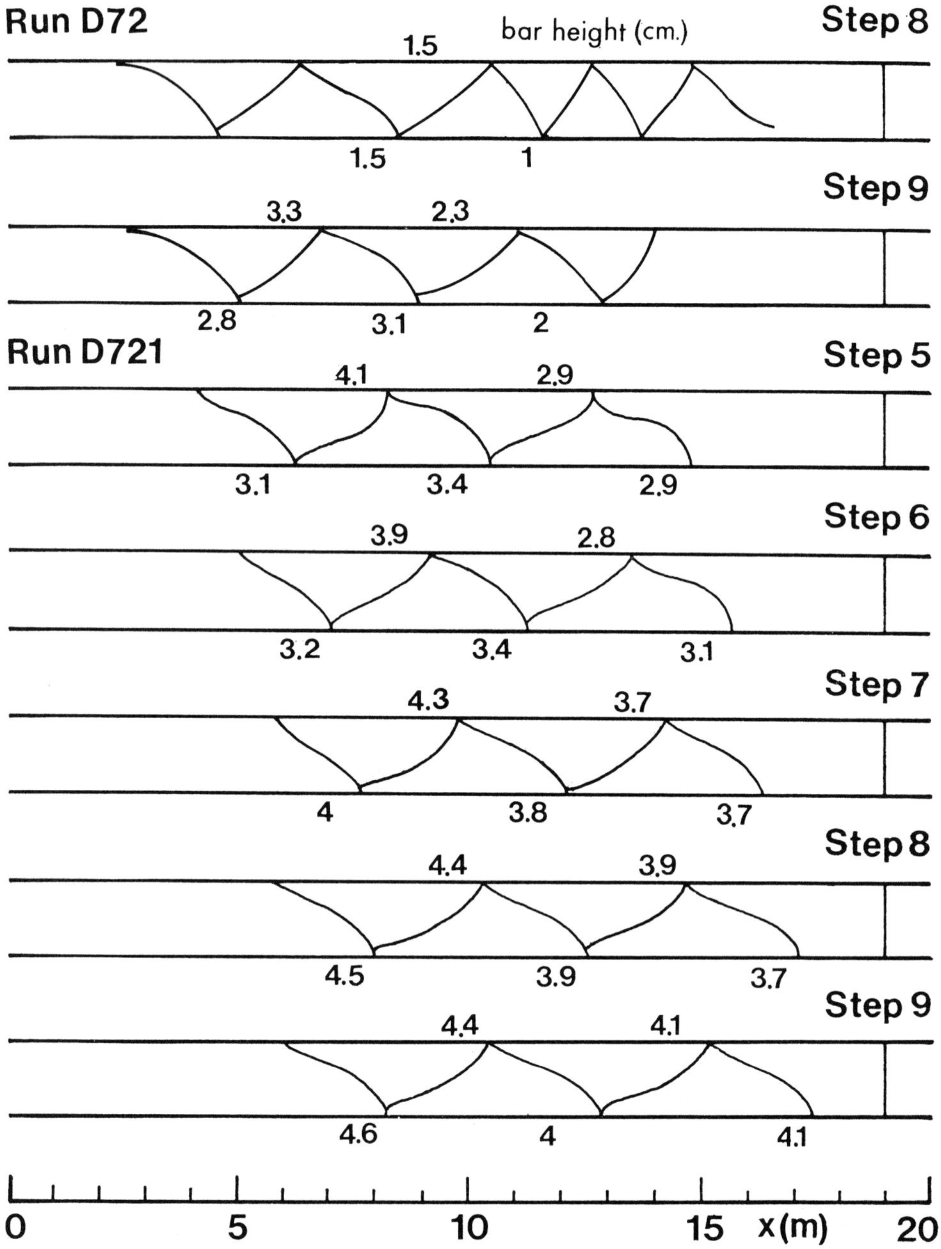

Figure 3 - Development process of alternate bars observed in runs D72 and D721

where U_o*, D_o* are the instantaneous average speed and flow depth respectively and τ is the dimensionless time variable appropriate to describe the basic unsteady flow which is given by

$$\tau = \sigma^* t^* \tag{2}$$

with σ^* characteristic frequency of the flood wave.

Assuming the half width of the channel B* as the spatial scale of the channel reach, they defined a 'small' dimensionless parameter σ in the form:

$$\sigma = \frac{\sigma^* B^*}{\overline{U}_o^*} \tag{3}$$

The latter can be thought of as the ratio between the average time required for the flow to travel along the channel reach and the characteristic period of the flood. Typical values of σ^* may range between $(10^{-3} \div 10^{-5}) s^{-1}$ in gravel bed rivers. The assumption made above, namely that within the spatial scale of bars a flood wave is felt as a locally uniform unsteady flow, is equivalent to the assumpion

$$\sigma \ll 1 \tag{4}$$

Tubino and Seminara [1987] developed a weakly non linear analysis assuming that the average value $\overline{\beta}$ of the width ratio $\beta(=B^*/D_o^*)$ (which is a function of τ when a uniform unsteady flow defined as in equation (1) is considered) was relatively close to the critical value β_c, i.e. to the threshold condition above which alternate bars can develop accordingly with the results of a linear stability theory. Thus

$$\overline{\beta} = \beta_c(\overline{\theta}, \overline{d}_s)[1 + \epsilon] \qquad \epsilon \ll 1 \tag{5}$$

where $\overline{\theta}$ and $\overline{d}_s(=d_s^*/\overline{D}_o^*)$ are reference values of Shields parameter and roughness parameter respectively, d_s^* being the grain size.

Furthermore also unsteadiness was assumed to be weak and the flow discharge was expressed in the form

$$Q^*(\tau) = 2B^* U_o^* D_o^* [1 + \delta q_1(\tau)] \qquad \delta \ll 1 \tag{6}$$

The assumption of small δ is representative of a typical situation of gravel bed rivers which become active at relatively high flows, thus having only a portion of the hydrograph such that Shields parameter exceeds the threshold value for sediment motion.

The non linear theory leads to derive an amplitude equation, describing the temporal behaviour of the amplitude of bars, which shows that the unsteadiness of the flow is likely to affect the process of development of bars and to control the final amplitude reached by the bed configuration at the end of a flood event provided

$$\hat{U}=\frac{\sigma}{\epsilon\alpha_{1R}}\sim O(1) \tag{7}$$

The parameter $\hat{U}$ represents the ratio between the time scale of the basic unsteady flow and the time scale of bar growth, ($\epsilon\alpha_{1R}$) being the linear growth rate of perturbations. When $\hat{U} >> 1$ the period of the flood is too short to allow for an effect of unsteadiness on bar development (the amplitude of bars cannot change during the flood event) whereas when $\hat{U} << 1$ the bed configuration can adapt to the instantaneous hydraulic conditions. An estimate of the parameter $\hat{U}$ can be readily obtained for a given reference state ($\overline{\beta}$, $\overline{\theta}$, $\overline{d}_s$) using figure 4, which gives the critical width ratio below which bars would not form as predicted by linear theory, and figure 5, where the function α_{1R} is plotted. Since typical values of σ for gravel bed rivers fall in the range ($10^{-2}\div10^{-4}$) it appears that there is a wide range of values of $\overline{\theta}$ and $\overline{d}_s$ such that $\hat{U}\sim O(1)$.

On the basis of theoretical results a procedure was introduced able to determine the final bar height for a given flood event; an asymptotic equilibrium value H_{TF}* for the latter height was introduced and predicted in terms of the reference dimensionless parameters: H_{TF}* coincides with the final value of bar height corresponding to an instantaneous bed response to hydraulic conditions and was obtained slightly modifying the equilibrium height corresponding to the reference state to account for the effect of unsteadiness.

4 . DISCUSSION

Some interesting conclusions can be drawn by comparing the main results of the theory outlined above with the experimental data. A glance at table 3 and figure 3 suggests that the unsteady part of the basic flow can play an important role in controlling bar development. The first effect is related to the time duration of the flood, (for gravel bed rivers the active portion of the flood): indeed for a given shape of the hydrograph (i.e. for a given series of experiments) the final amplitude reached by bar perturbations (step 9) shows a sharp dependence on T_F*; the latter dependence confirms the predicted relevance of the relative magnitude of the time scales of bar instability and of flow unsteadiness in controlling the process of bar development. This is accounted for in the theory by means of the parameter $\hat{U}$ (where $\sigma^* = (1/T_F^*)$): an estimate of the accuracy of the theoretical prediction can be obtained from figures (6)a-b in which the ratio between the final value H_{BF}* of bar height actually reached by bed configuration (step 9) and the value H_{TF}*, i.e. the final value of bar height corresponding to an instantaneous bed response to hydraulic conditions, is plotted versus $\hat{U}$ for each series of experiments. The theoretical prediction (solid line) shows a satisfactory agreement with experimental observations even though the theoretical trend appears sharper than the experimental one: indeed also in the experiments the interaction between bar development and flow unsteadiness appears to emerge in the range of values of $\hat{U}$ close to unity.

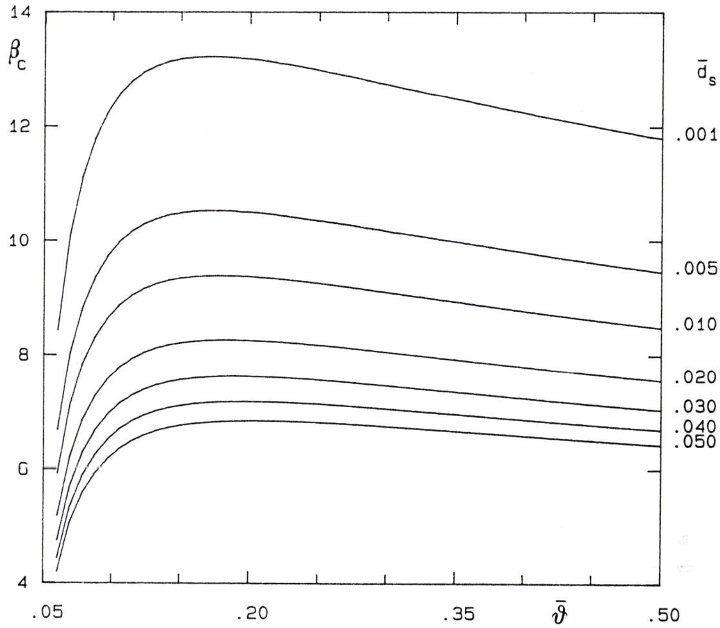

Figure 4 - The critical width ratio β_c is plotted versus $\bar{\vartheta}$ for some values of $\bar{d}_s$

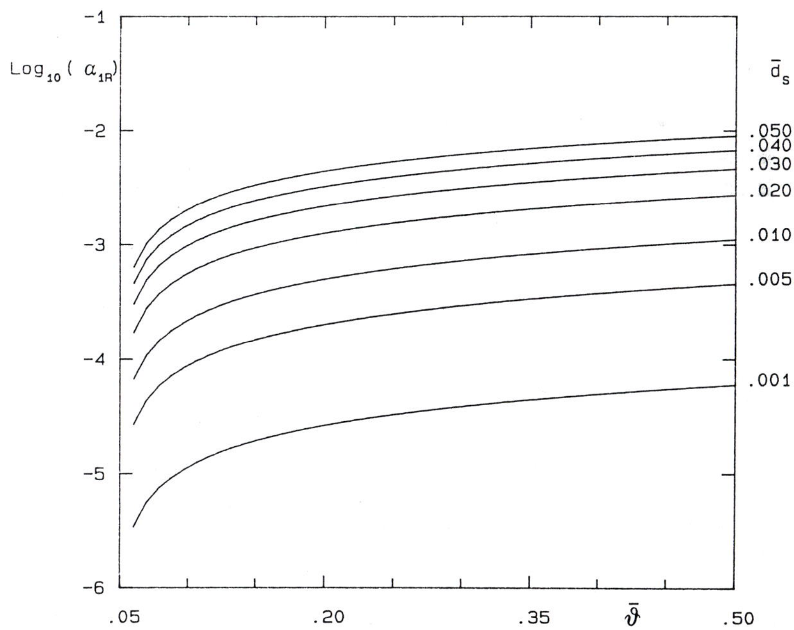

Figure 5 - The parameter α_{1R} is plotted versus $\bar{\vartheta}$ for some values of $\bar{d}_s$

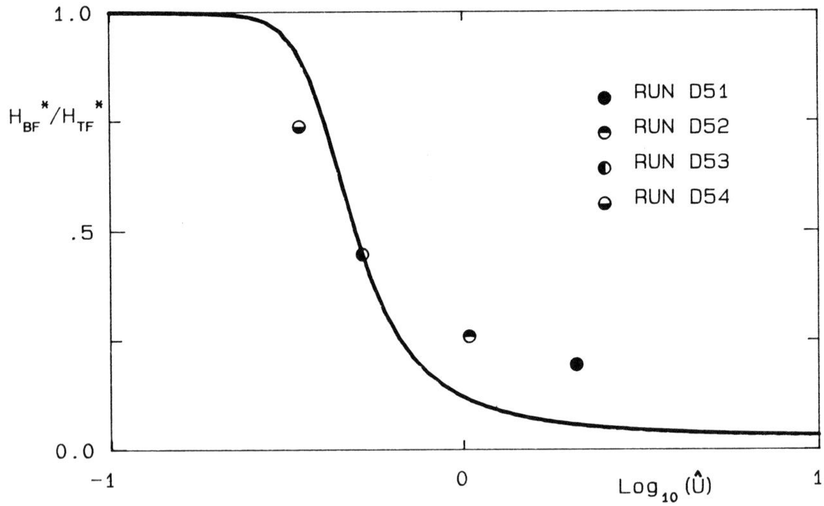

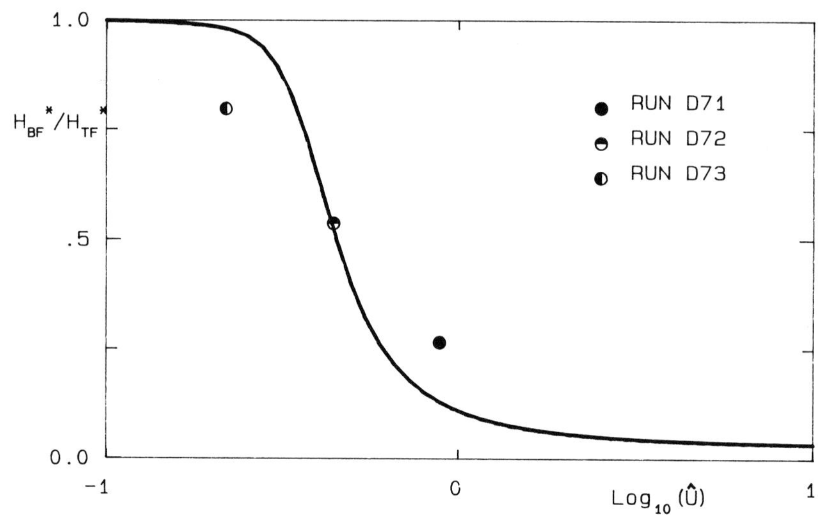

Figures 6(a,b) - The observed values of the ratio between the final amplitude reached by bar perturbations and H_{TF}^* are plotted versus $\hat{U}$: solid lines represent Tubino and Seminara (1987) predictions

The temporal behaviour of the bar height H_B* experimentally detected seems to be fairly well reproduced by the theory: a comparison is reported in figure 7a where the theoretical curve (solid line) and experimental data referring to the sequence of experiments D72 and D721 are reported. Also shown (dashed line) is the equilibrium bar height H_T* corresponding to the instantaneous hydraulic conditions. In the plot both H_B and H_T are dimensionless values of bar height (scaled by the reference flow depth $\overline{D}_o$*) and the dimensionless time variable τ ranges from 0 to 2 since two flood events are considered. The theory shows that, except for the initial stage, the process of bar development is mainly controlled by non linear effects. A damping in the process of bar growth is exhibited both by the experimental trend and by the theoretical predictions in the upper part of the flood (the peak of the flood may even fall within the stable regime $\beta < \beta_c$), the latter feature being consistent with the fact that during the rising stage the critical value β_c increases whence the actual width ratio decreases. The process of bar growth appears to be mainly concentrated in the longer falling stages which are to be considered as the formative portions of the floods. Figure 7a shows a clear trend of the bar height reached at the end of each flood event (τ=1, τ=2) towards an upperbound which appears to be well represented by the final value H_{TF}* of the bar height H_T* corresponding to instantaneous response. This feature confirms the theoretical conclusions reported above thus suggesting that, even though in actual situations the process of bar development appears to be strongly affected by unsteadiness and during a flood event the bed response is found to exhibit a phase lag with respect to the instantaneous hydraulic conditions, the final bar height reached after a sequence of typical flood events may be predicted in terms of a reference asymptotic value. Figure 7b shows a comparison between the measured values of the dimensionless wavespeed of bars C_B (scaled by the reference speed $\overline{U}_o$*) and theoretical predictions (solid line) for the same sequence of experiments as in figure 7a. The observed trend is satisfactorily predicted by the theory which appears to slightly overestimate the experimental values. From experimental and theoretical findings two main effects appear to affect C_B: the former is associated with the unsteadiness of the flow, producing an increase of C_B for increasing values of the discharge, the latter is related to finite amplitude effects which give rise to a non linear correction which tends to damp the value of C_B as the bar amplitude increases.

The results presented herein are quite encouraging and show that the theoretical analysis of Tubino and Seminara [1987] compares satisfactorily with experimental findings though the limits under strongly non linear situations will have to be ascertained. Some further features may originate in the latter case. Particularly coexistence of various bedforms may arise and complicate the experimental observations and theoretical interpretation. Indeed in the present experiments multiple rows bars sometimes appeared at low stages whereas small scale bedforms may arise and coexist with bars at high stages.

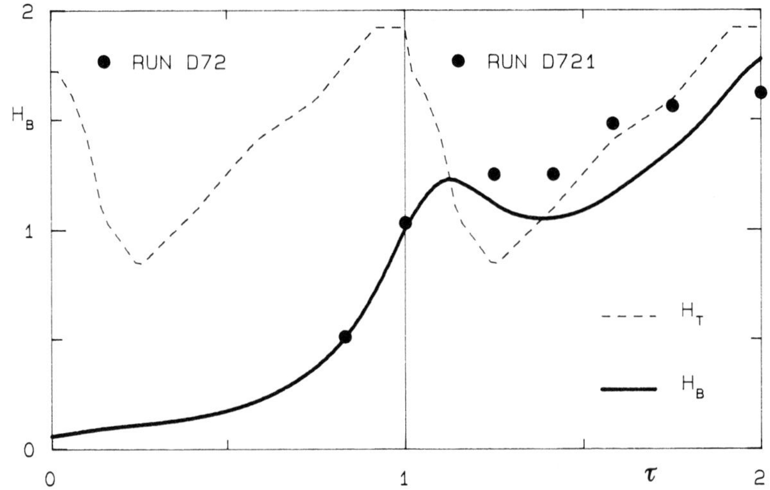

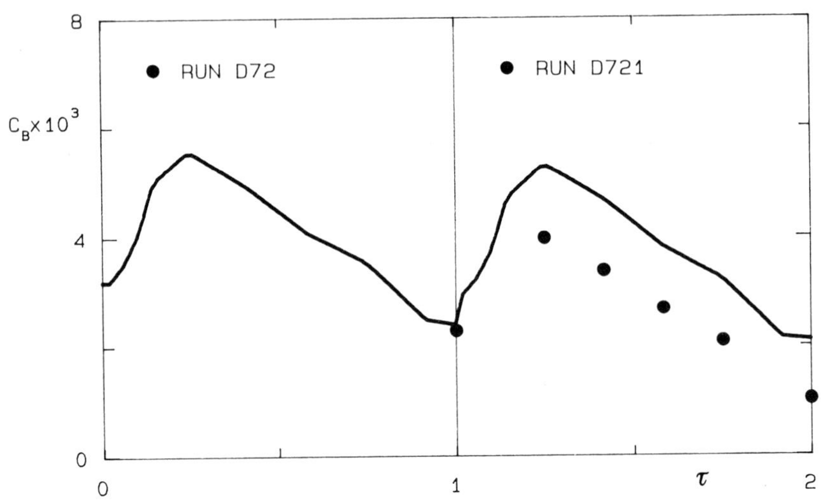

Figures 7(a,b) - The temporal behaviour of the dimensionless bar height H_B and wavespeed C_B of bars predicted by Tubino and Seminara (1987) is compared with experimental results of runs D72 and D721

Acknowledgements - This work was supported by the Italian Ministry of Public Education under grant MPI (40%) "Fenomeni di trasporto solido". The author expresses his gratitude to Luigi Stagi, University of Genoa, and Marco Taccini for their assistance in the experiments.

REFERENCES

Blondeaux, P., and G. Seminara, A unified bar-bend theory of river meanders, J. Fluid Mech., 157, 449-470, 1985.

Callander, R.A., Instability and river channels, J. Fluid Mech., 36, 465-480, 1969.

Colombini, M., G. Seminara, and M. Tubino, Finite - amplitude alternate bars, J. Fluid Mech., 181, 213-232, 1987.

Engelund, F., and O. Skovgaard, On the origin of meandering and braiding in alluvial streams, J. Fluid Mech., 57, 289-302, 1973.

Fredsoe, J., Meandering and braiding of rivers, J. Fluid Mech., 84, 609-624, 1978.

Fujita, Y., and Y. Muramoto, Studies on the process of development of alternate bars, D.P.R.I. Bull. 35, Kyoto Univ., Japan, 55-86, 1985.

Hansen, E., On the formation of meanders as a stability problem, Hydr. Lab. Tech. Univ. of Denmark Basic Res., Prog. Rep. 13, 9-13, 1967.

Ikeda, S., Prediction of alternate bar wavelength and height, J. Hydraul. Engng., ASCE, 110 (4), 371-385, 1984.

Kinoshita, R., Investigation of channel deformation in Ishikari River, Rept. Bureau of Resources, Dept. Science and Technology, Japan, 1-174, 1961.

Kuroki, M., and T. Kishi, Regime criteria on bars and braids, Hydraulics Papers 9, Hokkaido Univ., Japan, 1985.

Leopold, L.B., and M.G. Wolman, River channel patterns: braided, meandering and straight, U.S. Geol. Survey Prof. Paper 282-B, 1957.

Parker, G., On the cause and characteristic scales of meandering and braiding in rivers, J. Fluid Mech., 76, 457-480, 1976.

Seminara, G., and M. Tubino, Alternate bars and meandering: free forced and mixed interactions, in River Meandering, edited by S.Ikeda and G.Parker, Water Res. Monogr. 12, AGU, Washington, D.C., 1989.

Tubino, M., Linear and non linear aspects of large scale bedforms in erodible channels, Ph.D. dissertation, Univ. of Genoa, 1986 (in Italian).

Tubino, M., and G. Seminara, Unsteady alternate bars, Proceedings of the MPI meeting 'Leggi Morfologiche e loro Verifiche di Campo', University of Cosenza (Italy), June 25-26, 1987.

RIVER BARS AND NON LINEAR DYNAMICS

Giovanni Seminara

Hydraulics Institute
University of Genoa, Genoa, Italy

ABSTRACT

Some recent work on the mechanics of formation and non linear development of bars is reviewed. It is shown that many characteristic features of non linear systems, namely non linear dispersion, damped oscillations, resonance, instabilities and bifurcations are displayed by bars. Some novel results on non linear behaviour of resonant meanders are outlined, which appear to open new paths to model the development of meander bends of finite amplitudes.

CONTENT

1 . INTRODUCTION

The decade of the 80's has seen a considerable progress in research on river bars, a topic which is rapidly reaching the status of a subject fully belonging to the realm of non linear mechanics.

The approach followed in the present paper is based on the above premise. We wish to review some recent contributions which have improved our understanding of the mechanics of formation and non linear development of bars. It will appear that a variety of features characteristic of non linear systems have been brought up by the recent research.

River bars develop as a result of one of the most fascinating characteristics of rivers, namely their ability to determine their own shape through the interaction of the fluid with its, mostly non cohesive, container. Such interaction may take a variety of forms leading to different macroscopic effects (variation of bed topography, variation of channel alignement, cross section widening, sediment sorting, etc.). The spatial scales on which such interactions may occur are made available by the channel geometry and by the flow field: the relevant spatial scale of bars is channel width (2B). The rate at which the above interaction takes place is essentially associated with the erosion-deposition process which, specially in gravel bed rivers, is also affected by the highly unsteady character of the flow field.

The literature on river bars is very wide, encompassing contributions spreading among different areas (sedimentology, geomorphology, sediment transport, river engineering, etc.). Hence various classifications of bars have been proposed (see for instance Church & Jones (1982) or the ASCE Committee on Sedimentation (1966) for morphologically based classifications, the former specifically related to gravel bars). Following Seminara & Tubino (1989a), hereafter referred to as ST, we will simply distinguish between the following two classes of bars: free bars, spontaneously developing as a result of an instability of channel bottom, and forced bars, arising from some physical constraint imposed on the channel, such as curvature, river confluences, width variations. Free bars thus include single and multiple row bars, the former often called alternating bars (ASCE classification, 1966) being characteristic of straight or not too sinuous channels, the latter being the distinctive feature of braided rivers. Free bars are most often observed to migrate downstream (but see Ikeda & Ohta (1986) for observations of stationary bars). Forced bars, which include 'point bars' usually located at the convex sides of channel bends and 'tributary bars' which occur immediately downstream from points of lateral inflow into a channel, do not move relative to the channel. Free and forced bars are observed to coexist under conditions that will be discussed in sect. 3.2.

Research developed in the 70's focused mainly on linear theories of bars which will be reviewed in section 2. Non linear theories have been brought up by the recent research and will be reviewed in sect. 3. Possible lines of development of future research will be discussed in sect. 4.

2 . LINEAR THEORIES

2.1 Formulation

We wish to investigate the linear development of bed topography in cohesionless channels in response to any action causing the formation of bars, i.e. of relatively large scale perturbations with respect to the straight uniform configuration. The latter will be taken in the following as the basic state. This aim can be pursued fairly satisfactorily by means of a quasi 2-D model of the type recently developed by various authors including the writer and his coworkers. The main idea of the model, which was originally put forward by Engelund (1974), is to decompose secondary flow associated with curvature of the channel axis into a 'slowly varying' component with zero depth-average determined from the fully developed solution in constant curvature channels and a depth-averaged component determined from the 2-D de Saint-Venant equation modified as to account for the dispersive effect associated with transverse transport of longitudinal momentum due to the former component. The reader is referred to ST for a detailed discussion of the mathematical and physical basis of the model. Here it suffices to point out the main ingredients required for the model to properly account for the crucial mechanisms which control bar development:

i) the spatial scale of channel topography investigated herein is of the order of channel width;

ii) the time scale investigated is that associated with bottom erosion, i.e. orders of magnitude larger than the flow time scale in the steady case such to allow a quasi-steady description of the evolution of channel topography;

iii) a directional description of sediment transport is required, able to account for the deviation of bed load transport from the direction of average bottom stress due to the action of gravity on sediment particles (see eqn. (13) of ST);

iv) the flow field being slowly varying in space and time, the turbulent structure is assumed in equilibrium with the local conditions, with spatial and velocity scales determined by the local values of depth and friction velocity respectively: whence bottom stress is modelled as aligned with the mean bed velocity vector and bed load intensity is assumed to be in quasi equilibrium with the local value of the Shields stress (see eqns. (11), (12) of ST).

The linear character of the analysis obviously restricts its validity to small amplitude perturbations from the basic state. Further features play an important role in the development of gravel bars, namely bed sorting, armouring and unsteadiness of the basic flow. Some of them have been recently investigated and are given some attention below.

Let us first restrict our attention to the case of a steady basic state, and of sediment with uniform grain size. Following Kalkwijk & De Vriend (1980) we decompose the velocity field in the form

$$u = u_o(z)U(s,n) \tag{1}$$

$$v = \Gamma_o(z)\nu\mathscr{C}(s)U(s,n) + u_o(z)V(s,n) \tag{2}$$

where:

- (s,n,z) is a right-handed orthogonal coordinate system such that s is the arclength of channel axis, n is a horizontal transverse coordinate and z is directed upward: s,n and z are dimensionless, s and n being scaled by the channel half-width B and z by the unperturbed uniform depth D_o;
- (u,v,w) is the three-dimensional velocity vector averaged over turbulence and scaled by the unperturbed uniform average velocity U_o;
- (U,V) is the 2-D depth averaged velocity vector again scaled by U_o;
- $u_o(z)$ is the uniform logarithmic velocity distribution;
- $\Gamma_o(z)$ is the vertical distribution of the secondary flow component with zero depth-average;
- ν is a dimensionless measure of curvature of the channel axis defined in the form

$$\nu = \frac{B}{R_o} \tag{3}$$

with R_o typical curvature radius of the axis;
- $\mathscr{C}(s)$ is a dimensionless O(1) function describing the longitudinal distribution of channel curvature.

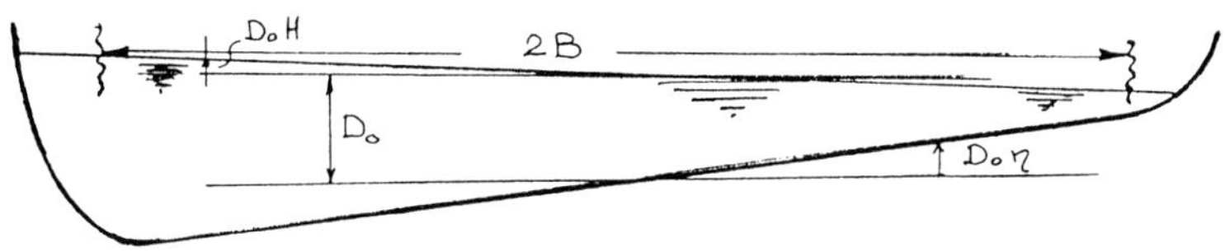

Fig. 1 - Sketch of channel cross-section.

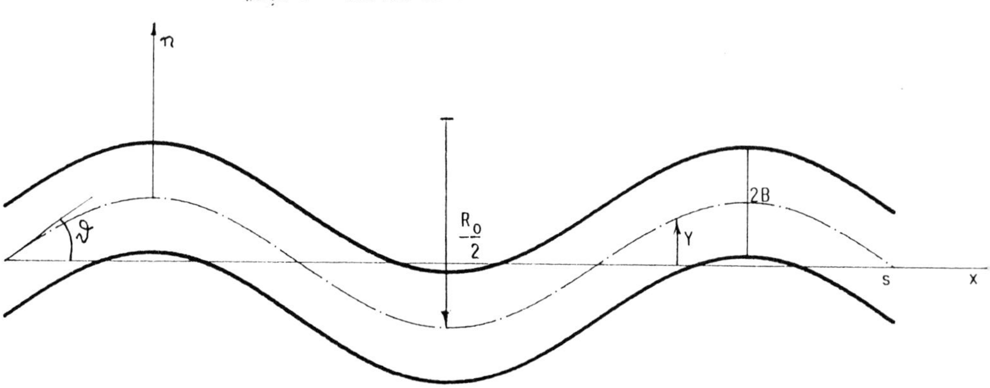

Fig. 2 Sketch of channel plane view.

In the following we assume:

$$\nu = \frac{B}{R_o} \ll 1 \qquad\qquad \beta = \frac{B}{D_o} \gg 1 \tag{4$_{a,b}$}$$

Assumption $(4)_a$ suggests the opportunity to first investigate the effects of curvature at a linear level; assumption $(4)_b$ allows us to ignore, as a first approximation, the effects of side wall boundary layers required for the flow field to adjust to the no-slip condition (but see Johannesson & Parker (1989) for an analysis of side wall effects).

Within the framework of the above theoretical scheme, the mathematical problem governing the linear spatial and temporal development of bars in arbitrarily curved channels reads:

$$U_{,s}+a_1U+(\eta+D)_{,s}+a_2D=a_3n\mathcal{C}(s)\nu \tag{5}$$

$$V_{,s}+a_4V+(\eta+D)_{,n}=[a_5\mathcal{C}(s)+a_6\mathcal{C}'(s)]\nu \tag{6}$$

$$U_{,s}+V_{,n}+D_{,s}=0 \tag{7}$$

$$a_7U_{,s}+V_{,n}+a_8\eta_{,nn}+a_9D_{,s}+\eta_{,t}=0 \tag{8}$$

$$\eta_{,n}=[a_{10}\mathcal{C}(s)]\nu \qquad (n=\pm 1) \tag{9}$$

$$V=0 \qquad (n=\pm 1) \tag{10}$$

where:

(U,V,η,D) are perturbations of depth average velocity, bed elevation and flow depth with respect to the basic state the latter two dimensionless quantities being scaled by the uniform depth D_o

t is dimensionless time scaled by $[(1-p)(BU_o^*)]/\sqrt{(\rho_s/\rho-1)gd_s^3}$ with ρ_s,p,d_s density, porosity and grain diameter of sediment respectively.

Finally coefficients a_{1-10} are functions of the unperturbed friction coefficient C_o, of the unperturbed bed load function ϕ_o, of their derivatives with respect to Shields stress θ and flow depth D, and of dispersive coefficients. Their structure can be found in detail in ST.

Eqns. (5-8) are linearized forms of longitudinal momentum equation (5), transverse momentum equation (6), flow continuity (7) and sediment continuity (8) respectively. Boundary conditions (9) and (10) impose the side walls to be impermeable to sediment and flow respectively.

It can be shown that the differential system (5-8) is parabolic both in the general unsteady case and in the steady case.

2.2 The natural response of the system : free bars

Let us first consider the unforced case defined by the condition of vanishing curvature

$$\mathcal{C}(s)=0 \tag{11}$$

The linear differential problem (5-8) becomes homogeneous and in the absence of any forcing non trivial solutions may only arise from an instability of the basic uniform state. The existence of such instability was first conceived by Callander (1968) and was then the subject of several investigations mainly in the 70's (see ST for a review). The possible growth of various modes is displayed by simply expanding the solution for (U,V,η,D) in Fourier series in the transverse direction in the form

$$(U,V,D,\eta)=\sum_{m_{odd}}(f_mS_m,g_mC_m,d_mS_m,e_mS_m)+\sum_{m_{even}}(f_mC_m,g_mS_m,d_mC_m,e_mC_m) \quad (12)$$

where

$$S_m=\sin\left(\frac{\pi}{2}mn\right) \quad , \quad C_m=\cos\left(\frac{\pi}{2}mn\right) \quad (13)$$

Linearity of the system (5-10) implies independency of the various modes which are readily shown to be governed by the following differential equation:

$$L_mg_m=A_og_m^{IV}+A_1g_m^{III}+A_2g_m^{II}+A_3g_m^{I}+A_4g_m+\frac{\partial}{\partial t}(b_og_m^{III}+b_1g_m^{II}+b_2g_m^{I}+b_3g_m)=0 \quad (14)$$

where the coefficients A_{o-4}, b_{o-3} are easily expressed in terms of the coefficients a_{1-10} of the original system (5-10) and an apex ' denotes differentiation with respect to the longitudinal coordinate. Eqn. (14) allows for simple dispersive waves solutions of the form

$$g_m\propto\exp i(\lambda s-\omega t)\exp\Omega t \quad (15)$$

provided a complex dispersion relationship be satisfied between the growth rate Ω, the angular speed ω and the wavenumber λ of perturbations for given values of the width to depth ratio β, the unperturbed Shields stress θ_o and the roughness ratio d_s/D_o (see eqn. (50) of Blondeaux & Seminara (1985)).

Two classes of free bars are described by the solution (15):

- if λ is assumed to be real and Ω does not vanish, temporally growing and migrating bars are described: these are the perturbations usually called 'alternating bars' (m=1) or 'multiple row bars' (m>1) in the literature;
- if λ is assumed to be complex and Ω is assumed to vanish, spatially growing and migrating bars are described by (15): an important role is played by a subset of these perturbations, namely those further characterized by vanishing migration speed ($\omega\equiv0$).

i) Temporally growing bars ($\Omega\neq0$, λ real).

The dispersion relationship is easily found by substituting from (15) into (14).

Analysis of this relationship allows one to determine, for each mode m stable ($\Omega<0$) and unstable ($\Omega>0$) regions in the space of parameters of the problem and to select the preferred mode as the most unstable one (maximum Ω) for any given set of parameters. Fig. 3 shows stability regions and marginal stability curves (such that $\Omega=0$) in the plane (β,λ) for the first few modes for typical values of θ_o and d_s/D_o. Fig. 4 shows the competition between various modes as the width to depth ratio increases, for the same values of θ_o and d_s/D_o as in fig. 3.

If we denote by subscript cm the critical conditions for mode m to become unstable we can readily show that $\lambda_{cm}=m\lambda_{c1}$ $\quad \beta_{cm}=m\beta_{c1}$

Linear theory predicts the wavelength and wavespeed of free bars as those corresponding to the critical conditions. Bars are invariably found to migrate downstream at criticality. Comparison between predicted and observed values of bar wavelengths and wavespeeds are reported by Colombini, Seminara & Tubino (1987) (see also ST). Research developed in the 70's assumed as a

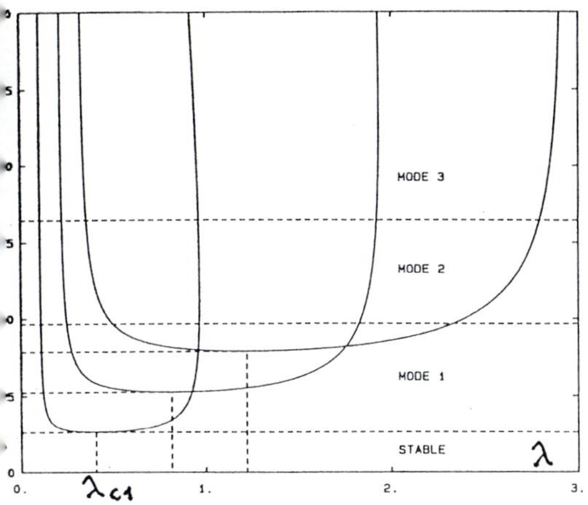

Fig. 3 – Marginal stability curves in the plane (β, λ) for the first three free modes (θ_o = .3 d_s/D_o = .01).

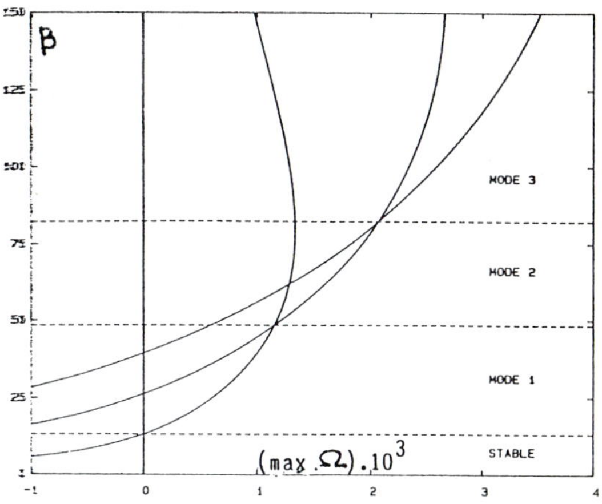

Fig. 4 – Competition between the first three free modes as the width-depth ratio increases (θ_o = .3 d_s/D_o = .01).

criterion to predict the morphological regime of rivers the order of preferred mode, whence rivers would keep straight provided they are narrow enough for β not to exceed β_{c1} for given values of θ_o and d_s/D_o, whereas meandering would occur provided mode 1 is selected, braiding taking over provided mode m with m>1 is preferred. The main assumption implicit in this approach is that the formation of alternate bars would imply incipient meandering, while the formation of multiple row bars would imply incipient braiding. Comparison with observed morhological behaviour is performed by various Authors (see Fredsøe (1978), Kishi (1980) and Blondeaux & Seminara (1983)). The above picture has been questioned in the 80's as will appear in the following.

ii) Spatially growing bars (λ complex, Ω=0).

The behaviour of spatially growing bars is dual with respect to that of temporally growing bars in that the marginal conditions for the former type of perturbations are identical with those of the latter (see Gaster (1962) for a complete discussion of this dual behaviour within the context of hydrodynamic stability).

The subset of spatially growing perturbations characterized by vanishing migration speed was first analyzed by Olesen (1983) and is of great relevance to the problem of forced bars, as it will appear that this type of perturbations may be excited by the forcing effect of curvature leading to the phenomena of resonant forced bars in sinuous channels (Blondeaux & Seminara (1985)) and overdeepening in the entrance region of curved channels (Struiksma & al. (1985)).

The complex values of λ corresponding to each mode can be obtained as the solutions of the algebraic equation:

$$-A_0\lambda^4 - iA_1\lambda^3 - A_2\lambda^2 + iA_3\lambda + A_4 = 0 \qquad (16)$$

Plots for the real (λ_r) and imaginary (λ_i) parts of λ versus β are given in fig. 5 for mode 1 and for given θ_o, d_s/D_o.

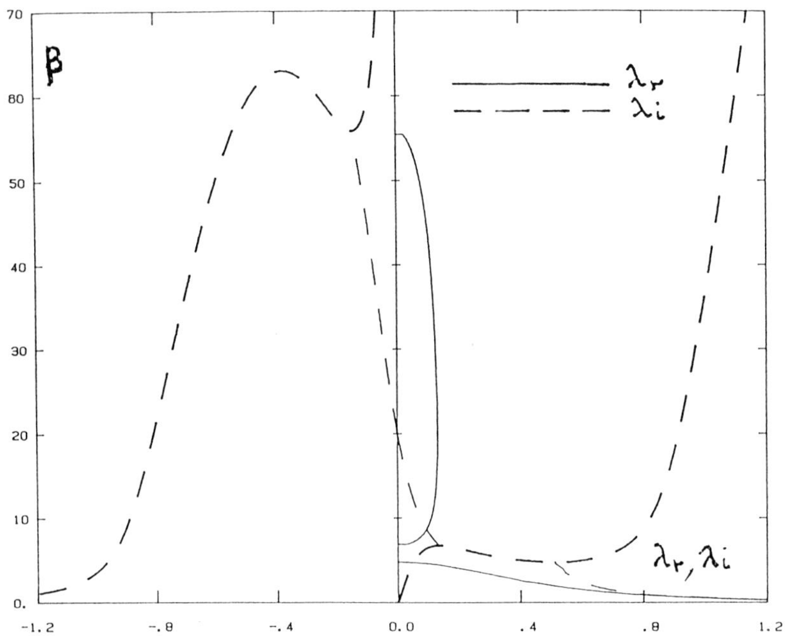

Fig. 5 – Wave number and growth rate of spatially growing first mode.

(θ_o = .1, d_s/D_o = .01)

It can be shown that for the general mode m the same plots as in fig. 5 apply, provided β and λ be replaced by β/m and λ/m respectively. In particular this implies that the growth rate of modes increases linearly with their order.

From figs. 5, 6 the following observations arise:

i) Among the possible natural responses of the system a particular class of steady bars which do not amplify either in space (λ_i = 0) or in time (Ω = 0) exists for particular values of β and λ_r, that will be denoted in the following by (β_R, λ_R).

ii) As pointed out by Struiksma & al. (1985), eqn. (16) has in general 4 complex solutions: two of them are always purely imaginary, corresponding to exponentially damped or growing perturbations, the remaining two are of the form ($\pm\ \lambda_r$ + i λ_i) and describe damped oscillations.

2.3 The forced response of the system : forced bars

Let us move to the case of bottom topography perturbed with respect to the uniform basic state due to the forcing effect of a non vanishing curvature of the channel axis. Though the complete system (5-10) can be solved exactly for a few special distributions of curvature (constant or sinusoidal) it proves instructive to express the solution in the following general form:

$$(U,V,D,\eta)=\Big\{(U_f(s)n,0,D_f(s)n,\eta_f(s)n)+\sum_{m_{odd}}(U_mS_m,V_mC_m,D_mS_m,\eta_mS_m)+$$

$$+\sum_{m_{even}}(U_mC_m,V_mS_m,D_mC_m,\eta_mC_m)\Big\}\nu+O(\nu^2) \tag{17}$$

where the component of the solution denoted by the subscript f is driven by the zero depth-average fully developed component of the secondary flow while the remaining part is required to account for the effect of longitudinal variations of curvature. The former component is easily solved for and provides some forcing for the latter component, which can be reduced to satisfy the following differential equation for V_m (or similar eqns. for U_m, D_m, η_m)

$$L_mV_m=c_{1m}\mathscr{C}''+c_{2m}\mathscr{C}'''+c_{3m}\mathscr{C}^{IV} \tag{18}$$

where L_m is identical with the homogeneous free bar operator (14) with $\partial/\partial t\equiv 0$ and the coefficients c_{im} are functions of the various parameters of the problem.

Eqn. (18) is essentially the equation of a forced damped oscillator with a damping effect which is a function of β (for given θ_o, d_s/D_o) and vanishes at $\beta=\beta_R$ and $\lambda_r=\lambda_R$.

Thus the response of bottom topography to the forcing effect of curvature can be thought of as the superposition of responses of an infinite number of forced damped oscillators, each corresponding to a degree of freedom in the transverse direction, i.e. to a free mode. Taking the latter viewpoint the recently discovered phenomena of resonance and overdeepening in curved channels are readily understood.

i) Resonance in sinuous channels.

Let us first consider the case of sinuous channels, which, since the original work of Langbein & Leopold (1966) who suggested that a 'sine generated curve' well describes the actual shape of natural meanders, has been treated in the literature by assuming that

$$\mathscr{C}(s)=\exp i\lambda_m s+c.c. \tag{19}$$

Under these conditions a solution for the forced problem can be readily obtained in closed form (see Blondeaux & Seminara (1985)) and describes the typical structure of point bars forming at convex banks with a slight phase lag with respect to channel curvature. The work of Blondeaux & Seminara (1985) has also displayed an important feature which had not arisen in the several previous attempts to determine the structure of bottom topography in sinuous channels (Engelund (1974), Gottlieb (1976), Ikeda & al. (1981), Kitanidis & al. (1984)). Within a relatively wide range of values of the width-depth ratio β and meander wavenumber, centred at (β_R, λ_R), the solution for any quantity describing the perturbation (U,V,D,η) shows a peak which becomes an infinite peak when the resonant conditions (β_R, λ_R) are exactly met. Eqn. (18) indeed shows that, under the latter conditions, one of the infinite number of oscillators which describe the behaviour of the system, namely the oscillator corresponding to the first mode (m=1), resonates. In principle resonance may occur for any mode, but in practice the first mode is the one which resonates for reasonably low values of β_R such to be commonly encountered in actual rivers. Fig. 6 shows a typical response of the channel as a function of λ_m for typical values of β and given values of θ_o and d_s/D_o. Figs. 7 and 8 respectively show the dependence of β_R and λ_R on θ_o for various modes for given values of d_s/D_o.

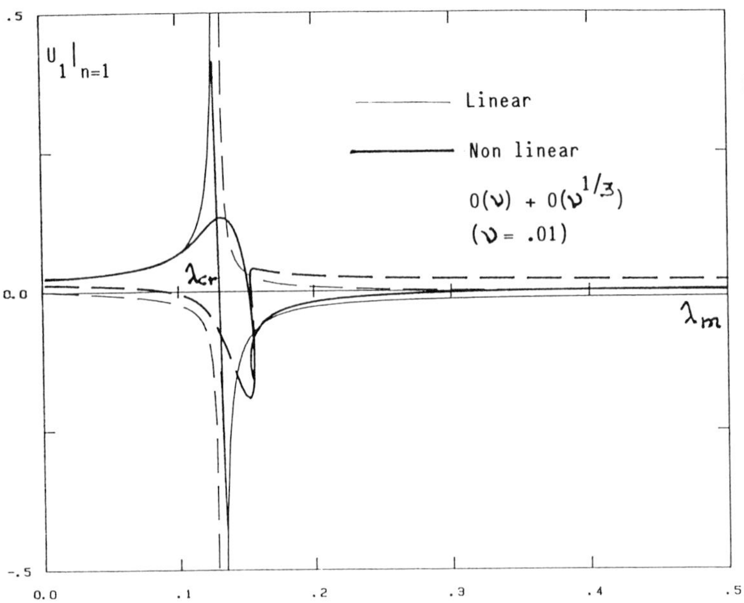

Fig. 6
Linear and non linear forced response in sinuous channels. (Solid line = $Re(U_1)|_{n=1}$; (Dotted line = $Im(U_1)|_{n=1}$)

$\theta_o = .1 \quad d_s/D_o = .01$

Deficiencies of previous models unable to predict resonance have been recently analyzed in detail by Parker & Johanneson (1989).

The linear approach discussed herein obviously breaks down close to resonance when non linear effects are required to provide the damping needed to keep the response finite. A non linear theory of resonant meanders can be formulated close to resonance (Seminara & Tubino (1990)). A brief presentation of it is deferred to sect. 3.3 after the non linear behaviour of free bars will have been analyzed.

ii) Overdeepening at the entrance region of curved channels.

The natural response of the system is also excited by the forcing effect of curvature at the entrance region of curved channels as observed by Struiksma & al. (1985). This can be readily understood in the case of uniformly curved channels through our formulation in terms of damped linear oscillators. In fact in this case we can write

$$\mathcal{C} = cost \tag{20}$$

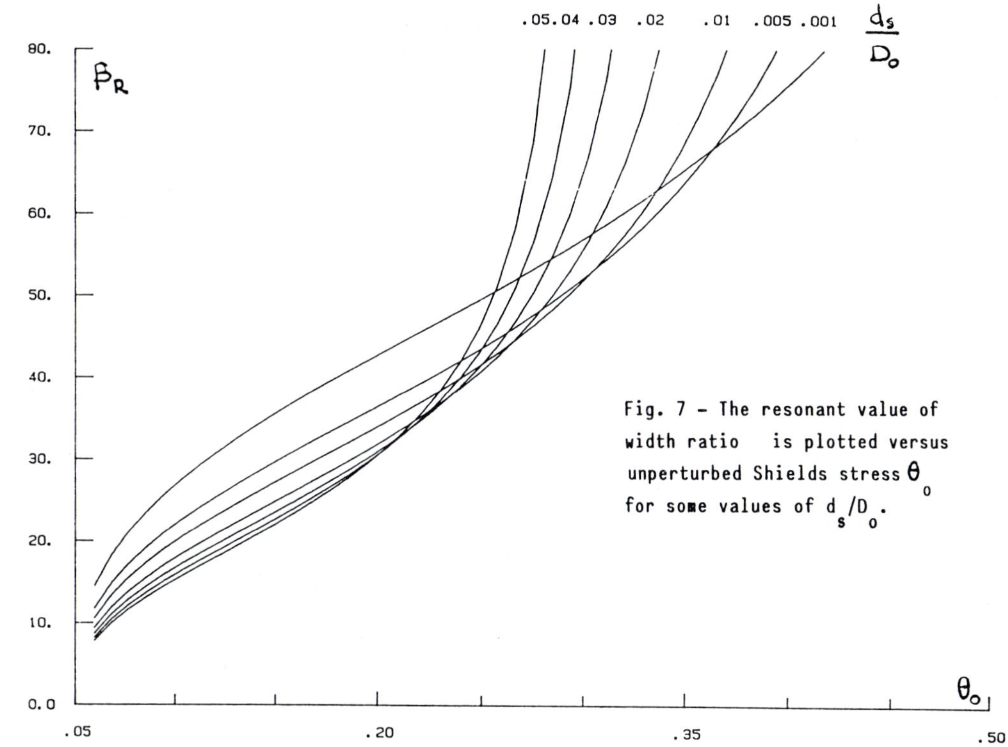

Fig. 7 - The resonant value of width ratio is plotted versus unperturbed Shields stress θ_0 for some values of d_s/D_0.

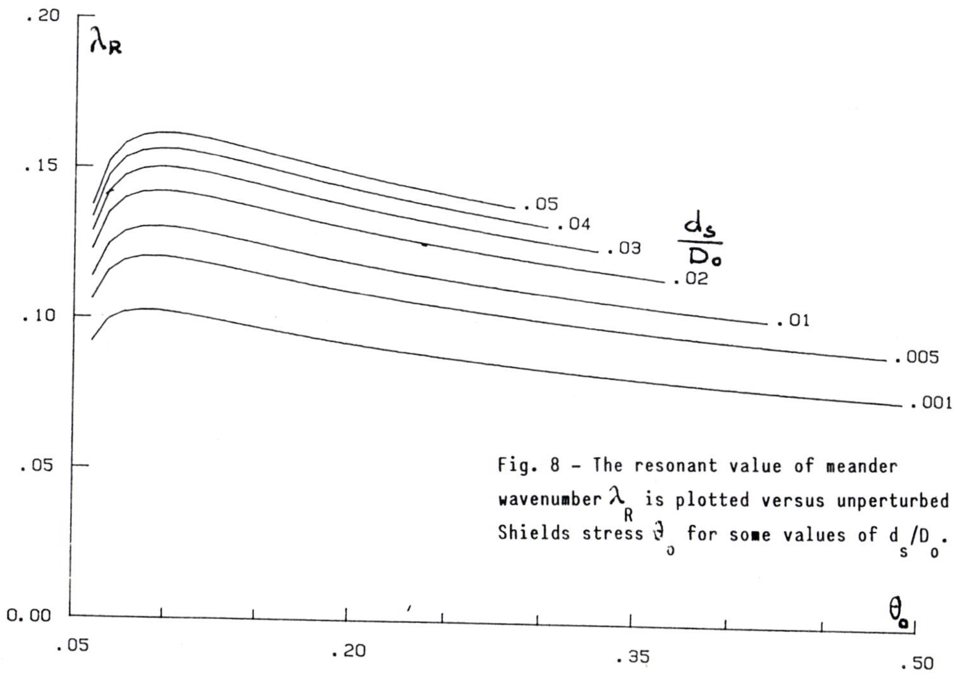

Fig. 8 - The resonant value of meander wavenumber λ_R is plotted versus unperturbed Shields stress ϑ_0 for some values of d_s/D_0.

whence eqn. (18) becomes homogenous and the free modes can only be excited by the initial conditions. Indeed under these circumstances the fully developed solution takes the form (17) with U_f, D_f, η_f constants and $U_m = V_m = D_m = \eta_m = 0$ (m = 1,2,3...) corresponding to a bottom topography exhibiting a uniform pool on the concave side and a uniform bar on the convex side. Moreover, since the fully developed solution is linear in n, and we can write:

$$n = \sum_{m_{odd}} \frac{8(-1)^m}{(\pi m)^2} S_m \tag{21}$$

it follows that all the odd free modes are excited in the entrance region of the bend in order for flow and bottom topography to adjust from the uniform straight configuration imposed at the entrance, to the fully developed solution described above. The adjustment process occurs through damped spatial oscillations which involve all the (odd) modes, with decreasing amplitudes and increasing spatial damping as m increases. This is one of the features which makes the overdeepening effect distinct from the resonance effect where only one free mode is linearly excited by the forcing. The contribution of higher harmonics is evident in Struiksma & al. (1985) results. The latter authors also developed an analytic (linear) - numerical (non linear) interpretation of the process which explains the major features of the observations leaving only some aspects unresolved (presence of higher harmonics, some discrepancy in the rate of damping for some of the experiments). More recently Johannesson & Parker (1989) have reconsidered the problem in connection with that of resonance in sinuous channels.

2.4 A still debated problem : meandering initiation

When coming to the fundamental question regarding the mechanism of meandering initiation the picture, as it emerges from the literature, is still rather speculative. Indeed, in spite of the several attempts to reproduce the process in the laboratory even in very large flumes like the one available at Disaster Prevention Research Institute (Kyoto), one major shortcoming has invariably prevented the development of a well developed sequence of coherent meanders in the laboratory. Indeed whereas in the field the time scale for bank erosion is generally orders of magnitude larger than that associated with bottom erosion, in the laboratory the two time scales are comparable. This implies that the process, as it is observed in the laboratory, invariably leads to a braided or quasi-braided channel. The sequence of phenomena observed in some very detailed experiments reported by Fujita (1989) has been recently analyzed by Seminara & Tubino (1989b). The key observation is the formation of free migrating bars in the originally straight channel. Free bars lenghten and slow down as the channel widens. Eventually bank erosion occurs apparently in the form of a width instability rather than a bend instability. The details of the latter process are not yet understood, but the presence of free bars appears to play an important role since in the experiments they are able to induce bank erosion on a relatively fast time scale.

The phenomenon occurring in nature, at least in the case of well-developed flood plains, though influenced by a variety of other factors such as flow unsteadiness, grain sorting, end effects, appears to be characterized by two sharply different time scales that of bank erosion (of the order of metres/year) and that of free bar migration (of the order of metres/day). This observation suggests that in nature migrating free bars should not be primarily responsible for meander formation which should be associated with some mechanism acting at a larger time scale. The idea of 'bend instability' put forward by Ikeda, Parker & Sawai (1981) provides a possible such mechanism. Indeed if we assume the original straight channel to be 'in equilibrium' (no bank erosion or deposition) any sinusoidal perturbation of the channel axis will induce some flow perturbation in the near-bank region such to alter bank equilibrium. Ikeda, Parker & Sawai (1981) proposed a bank erosion equation which can be written in the form:

$$\cos\vartheta Y_{,t} = E(U|_{n=1} - U|_{n=-1}) \tag{22}$$

where Y is the transverse coordinate of the channel axis relative to a cartesian frame with the longitudinal axis lying in the downvalley direction, ϑ is the angle that the channel axis forms with the latter direction (see fig. 2) and E is some erosion coefficient the value of which is of the order of metres/year.

Using a linear model to evaluate the near bank velocity one can readily evaluate the ability of a small sinusoidal perturbation of the channel axis to grow in time: the real part of $U_1|_{n=1}$ (see fig. 6) is found to be proportional to the bend growth rate, while the opposite of its imaginary part gives the bend wavespeed. Fig. 6 shows that meanders grow and migrate downstream for $\lambda < \lambda_{cr}$ and their growth rate peaks at the resonant wavenumber. When compared with experimental observations the above theory seems to pick up the general trend satisfactorily. (Blondeaux & Seminara (1985), Johannesson & Parker (1989)).

Does this imply that the process is understood? Too many features are still ignored in the above picture to allow for a conclusive answer. In particular what is the role of finite amplitude migrating bars present in the originally straight channel? In order to investigate this role a non linear theory of free bars is required. This is discussed in the next chapter.

3 . NON LINEAR THEORIES

3.1 Finite amplitude free bars : a Hopf supercritical bifurcation

The linear theories of free bars outlined in sect. 2.2 have been fairly successful in predicting the conditions required for bar formation and the wavelength and wavespeed of migrating bars.

On the other hand the prediction of bar amplitude is outside the range of capabilities of any linear theory: in fact the mechanism which may inhibit the indefinite exponential growth predicted by linear theory is the damping effect of non linearities. This is very well known in the context of hydrodynamic stability (see Stuart (1971)). An extension of the techniques of non linear stability theory to the problem of finite amplitude development of free bars is quite instructive. Let us examine the case of alternating bars and that of multiple row bars separately.

i) Alternating bars.

This problem was analyzed theoretically for steady uniform basic flows by Colombini, Seminara & Tubino (1986, 1987). An independent theoretical attempt to solve the same problem is due to Fukuoka & Yamasaka (1985): the limits of their approach are discussed in ST. The problem has also been treated numerically by Shimizu & Itakura (1985) and Nelson & Smith (1989). The main idea behind the theoretical analysis of the problem is that, within a relatively small neighbourhood of the critical conditions for alternate bar formation conveniently defined in terms of some small parameter ϵ (see below), two main features arise:

- the linear growth rate of perturbations is small;
- the non linear character of perturbations may be assumed to be 'weak'.

The first consequence is that the amplitude A of the fundamental perturbation excited linearly, which is exponential in the linear regime, will evolve in the non linear regime on a 'slow' time scale defined by the 'slow' time variable

$$T = \epsilon t \tag{23}$$

The second consequence is that higher harmonics produced by non linear interactions have increasingly smaller amplitudes in terms of ϵ.

More precisely Colombini, Seminara & Tubino (1987) showed that, if ϵ is defined as

$$\epsilon = \frac{\beta - \beta_c}{\beta_c} \ll 1 \tag{24}$$

the solution for the flow depth D can be expressed in the form:

$$\begin{aligned} D = 1 &+ [\epsilon^{1/2} A(T) S_1(n) d_1 E_1(s,t) + c.c.] + \\ &+ \epsilon\{A^2 E_2 [C_2(n) d_{22} + d_{02} + c.c.] + A\bar{A}[C_2(n) d_{20} + d_{00}]\} \\ &+ \epsilon^{3/2}\{E_1 S_1(n) d_{31}(T)\} + higher \quad harmonics \end{aligned} \tag{25}$$

where

$$E_k = \exp ki(\lambda_c s - \omega_c t) \qquad k = 1, 2, 3 \ldots) \tag{26}$$

Similar expansions can be set up for U, V, η. Furthermore the theory shows that, in order for 3rd order ($O(\epsilon^{3/2})$) effects to counteract the linear growth in a neighbourhood of the critical conditions the amplitude A(T) must satisfy the following non linear ordinary differential equation:

$$\frac{dA}{dT} = \alpha_1 A + \alpha_2 A^2 \bar{A} \tag{27}$$

where α_1 and α_2 are complex coefficients expressed in terms of the $O(\epsilon^{1/2})$ and $O(\epsilon)$ solutions, and an overbar denotes complex conjugate.

Eqn. (27) essentially shows that bar instability leads to a Hopf bifurcation, which can be of two different types (see figures 9 and 10):

- Supercritical bifurcation occurring when $sgn[R_e(\alpha_1)] \neq sgn[R_e(\alpha_2)]$.

  In this case the solution tends to the exponentially growing behaviour as $T \to -\infty$ and to an equilibrium amplitude as $T \to \infty$.

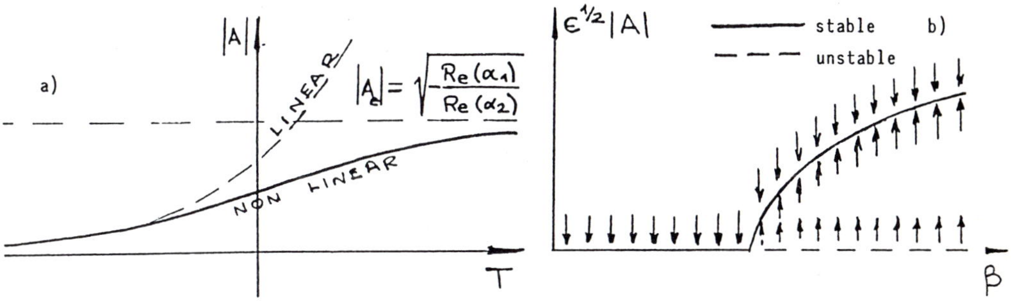

Fig. 9 - Hopf supercritical bifurcation
a) time development of the amplitude modulus $|A|$
b) bifurcation diagram

- Subcritical bifurcation, occurring when $sgn[R_e(\alpha_1)] = sgn[R_e(\alpha_2)]$.

  In this case any finite amplitude solution starting from initial values below the threshold value $\sqrt{R_e(\alpha_1)/R_e(\alpha_2)}$ decays to zero as $T \to \infty$, whereas larger initial values lead to an exponentially growing solution.

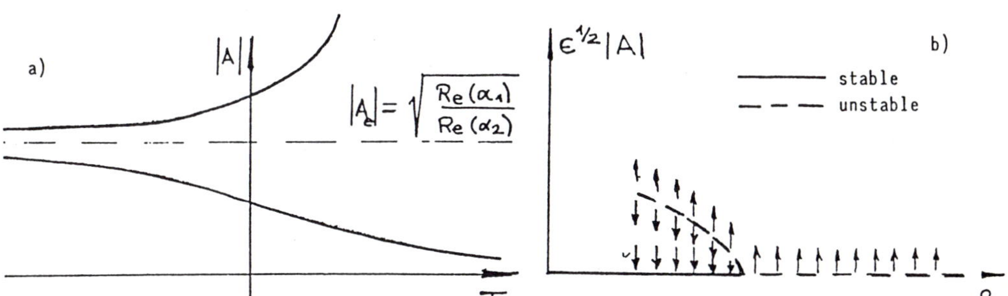

Fig. 10 - Hopf subcritical bifurcation
a) time development of the amplitude modulus $|A|$
b) bifurcation diagram

Colombini, Seminara & Tubino (1987) have shown that the bifurcation of free migrating bars is always supercritical, at least when the unperturbed bed is plane. This allowed the above authors to evaluate the equilibrium height of alternating bars which was found to be proportional, at leading order, to $(\beta-\beta_c)^{1/2}$ as depicted in fig. 9.

Further features revealed by the theoretical results are the experimentally observed tendency of finite amplitude bars to develop diagonal fronts and the reduction of bar wave speed as bar amplitude increases. The latter is a typical feature of non linear dispersive waves.

Comparison of theoretical results regarding several features of finite amplitude free bars with many experimental observations is quite satisfactory and can be found in Colombini & al. (1987) and ST.

Recently Tubino (1989) has extended the above results to the case of unsteady basic flows which is particularly relevant to the context of gravel rivers (see also Tubino's contribution at the present meeting).

ii) Multiple row bars.

The problem of finite amplitude development of multiple row bars is of relevance to those braided rivers which are characterized by ordinary floods able to let the entire bed be submerged and thus modelled by sediment transport. Some beautiful examples of such rivers are reported in the Japanese literature. In particular the migration of multiple row bars has been observed by aerial photographs and measured in the Ohi river by Kinoshita (1980) during two floods occurred in 1977 and 1979.

Another fascinating field observation of multiple row bars is reported by Ashida (1982).

An extension of the finite amplitude weakly non linear theory developed for alternating bars to cover the case of multiple row bars is straightforward. Indeed a fairly simple similarity solution can be obtained (Colombini & Tubino (1990)) such that the order of the mode is scaled out of the problem. However fig. 4 shows that higher order modes are linearly the most unstable ones for values of β much larger than the critical value associated with the respective mode. This makes a weakly non linear approach probably inadequate to tackle this problem.

Furthermore fig. 4 shows that for large enough values of β, high order modes exhibit linear amplification rates which are very close to each other. This suggests that in this range of values of β there are several competing attractors and the system may possibly choose one or the other depending on the initial conditions. Hystereses phenomena are thus quite likely, as observed in the laboratory by Fuijta (1989).

Finally the highly non linear character of the system suggests that along with periodic attractors of the type suggested by weakly non linear theories some chaotic attractors might also exist. The discovery of such attractors would pose a problem of predictability of bed topography in such systems. We are currently investigating along these lines.

3.2 Free-forced bar interactions in meanders and suppression of free bars

Having set up a theory which enables us to treat the non linear development of free bars we can now proceed to investigate the problem of coexistence of free and forced bars in sinuous channels. Kinoshita & Miwa (1974) showed experimentally that free migrating bars coexist with forced steady bars in channels which are not too sinuous, the threshold sinuosity for free bar suppression being a function of meander wavelength. A theoretical treatment of this problem is both of conceptual and of practical importance: indeed prediction of bed topography in sinuous channels cannot be in general pursued ignoring free-forced interactions. Furthermore suppression of free bars may possibly be achieved by the designer of artificial channels by conveniently shaping the channel itself.

Free-forced bar interactions have been analyzed theoretically by Tubino & Seminara (1989). Numerical reproductions of free-bar suppression has also been achieved by Shimizu, Itakura & Yamaguchi (1987) (see Fukuoka (1989)).

In order to tackle this problem the fully non linear 2-D de Saint-Venant equations formulated such to account for transverse dispersion of longitudinal momentum by secondary flow are needed. The reader is referred to Tubino & Seminara (1989) for details of the formulation. It suffices here to point out the crucial mechanism which controls free bar suppression in sinuous channels. This is depicted in fig. 11 where it appears that if free and forced bars are allowed to coexist and interact between each other in channels of typical curvature ratio ν mixed interactions reproduce the free fundamental bar at $O(\nu^2\ \epsilon^{1/2})$, where $\epsilon^{1/2}$ is again the order of magnitude of the free fundamental. Since the latter is also reproduced by free interactions at $O(\epsilon^{3/2})$, it follows that mixed interactions are likely

Free	Mixed	Forced
$O(\epsilon^{1/2})$		$O(\nu)$
$Ad_1 S_1 \exp i(\lambda_c s - \omega_c t)$		$D_1(n)\exp(i\lambda_m s)$
$O(\epsilon)$	$O(\nu\epsilon^{1/2})$	$O(\nu^2)$
$A^2(d_{22}C_2 + d_{02})\exp 2i\lambda_c(s-\omega_c t)$	$AD_{11}\exp i(\lambda_c+\lambda_m)s-\omega_c t$	$D_2(n)\exp 2i\lambda_m s$
$A\bar{A}(d_{20}C_2 + d_{00})$	$\bar{A}D_{11}\exp i(\lambda_m-\lambda_c)s-\omega t$	$D_o(n)$
	$O(\epsilon^{3/2})\quad O(\nu^2\epsilon^{1/2})$	

Free fundamental $\propto \exp i(\lambda_c s - \omega_c t)$
reproduced by free and mixed interactions.

Fig. 11 - Conceptual sketch of mixed interactions.

to affect free bar development as long as : $\nu^2 \epsilon^{1/2} \sim \epsilon^{3/2}$ or

$$\nu \sim \epsilon^{1/2} \tag{28}$$

i.e. provided the orders of magnitude of free and forced bars are comparable. Thus, after setting

$$\nu = k_i \epsilon^{1/2} \qquad k_i \sim O(1) \tag{29}$$

Tubino & Seminara (1989) were able to derive a modified amplitude equation for free bar amplitude A(T) where the effect of mixed interactions is accounted for. It reads:

$$\frac{dA}{dT} = (\alpha_1 + k_i^2 \alpha_{11}) A + \alpha_2 A^2 \overline{A} \tag{30}$$

where α_{11} is a function of θ_o, d_s/D_o arising from the effect of mixed interactions, while α_1, α_2 are identical with the coefficients of (27).

Eqn. (30) shows that:

- as $k_i \to 0$ the case of finite amplitude free bars in straight channels is recovered;
- a threshold value of k_i for free bar suppression exists such that

$$k_{thr} = \sqrt{-\frac{R_e(\alpha_1)}{R_e(\alpha_{11})}} \tag{31}$$

provided

$$sgn R_e(\alpha_{11}) \neq sgn R_e(\alpha_1) \tag{32}$$

Tubino & Seminara (1989) found that conditions (32) is satisfied and gave plots for k_{thr} as a function of θ_o, d_s/D_o. Thus the main outcome of the theory is a relationship for the threshold curvature ratio ν_{thr} for free bar suppression, in the form

$$\nu_{thr} = k_{thr}\left(\theta_o, \frac{d_s}{D_o}\right)\left(\frac{\beta - \beta_c}{\beta_c}\right)^{1/2} \tag{33}$$

The latter relationship is of immediate practical use. Comparison with Kinoshita & Miwa's (1974) experimental observations is quite satisfactory (fig. 12 a,b) and reveals an important feature: maximum free bar suppression occurs at resonance. This was not unexpected.

However it must be pointed out that the above expansion breaks down close to resonance where linear theory of forced bars becomes inadequate. The non linear response close to resonant conditions has been recently investigated by Seminara & Tubino (1990) and exhibits a variety of quite interesting features. This is outlined in the next section.

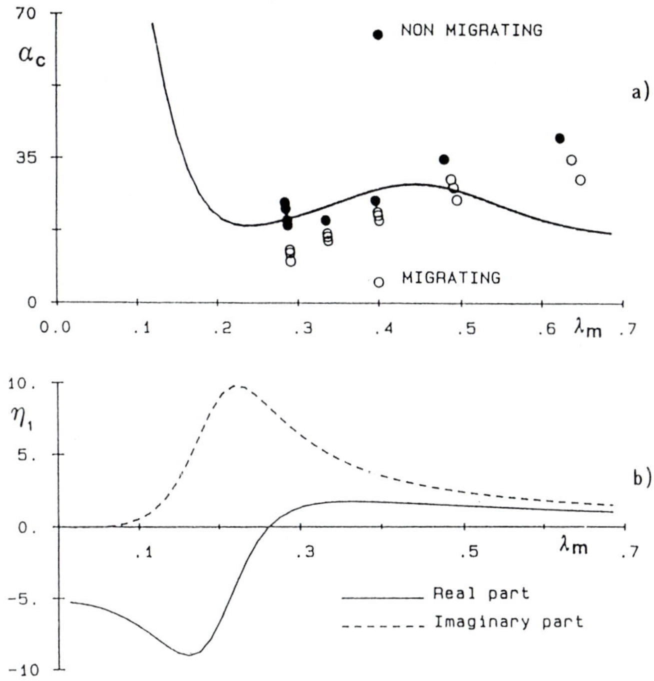

Fig. 12 – a) The critical value of Kinoshita & M. wa's (1974) measure of channel sinuosity α_c predicted by the present theory is compared with observations

b) The fundamental component of bottom elevation plotted versus λ_m shows that the resonant peak corresponds to maximum free bar suppression.

3.3 Non linear damping of resonant forced bars and widening of the resonant range

The structure of the non linear solution for forced bars close to resonance must be such that the singular behaviour occurring at the linear level ($O(\nu)$) be suppressed. Furthermore the fundamental component of the solution cannot be of the order of the forcing effect of curvature ($O(\nu)$) as in the linear theory but rather of some order ν^x with $x < 1$ such that, at the lowest approximation, the forcing effect of curvature does not appear and the 'natural' resonant solution of the homogeneous linear problem is excited. This is possible because resonance coincides with the marginal conditions for the special class of free bars characterized by vanishing migration speed. Thus, exactly at resonance, and at the lowest order of approximation, the solution (say for D) reads ($d_1 S_1 \exp i\lambda_R S + c.c.$) being a particular case of Colombini, Seminara & Tubino's (1987) solution evaluated at $\beta=\beta_R$, $\lambda=\lambda_R$.

On the other hand non linear interactions reproduce this fundamental solution at $O(\nu^{3x})$. Thus provided 3x = 1 non linear effects act on the fundamental at the same order as the forcing effect of curvature, so that non linear damping may suppress the singular behaviour predicted at linear level.

Following the above ideas Seminara & Tubino (1990) expand the solution (say for flow depth) in the form

$$D = 1 + \nu^{1/3}[A S_1(n) d_1^R E_{1R} + c.c.]$$
$$+ \nu^{2/3}\{A^2 E_{2R}[C_2(n) d_{22}^R + d_{02}^R + c.c.] + A\bar{A}[C_2(n) d_{20}^R + d_{00}^R]\} +$$
$$+ \nu[D_1(n) E_{1R} + c.c.] + O(\nu^{4/3}) \tag{34}$$

with

$$\lambda = \lambda_R + \nu^{2/3}\lambda_1 \tag{35}$$

$$\beta = \beta_R + \nu^{2/3}\beta_1 \tag{36}$$

and

$$E_{mR} = \exp m i \lambda_R s \tag{37}$$

The coefficients $d_1^R, d_{22}^R, d_{02}^R, d_{20}^R, d_{00}^R$ are identical with those of expansion (25) evaluated at the resonant conditions. Conditions (35) and (36) define an appropriate window close to the resonant conditions, within which resonance effects are felt. Suppression of the singular behaviour at $O(\nu)$ leads to the following amplitude equation for A:

$$A^2\bar{A} + (c_1\lambda_1 + c_2\beta_1)A + c_3 = 0 \tag{38}$$

where the complex coefficients c_1, c_2, c_3 are functions of θ_0 and (d_s/D_o). The solution of (38) allows one to determine the amplitude of forced bars as a function of λ_1 and β_1, i.e. in the appropriate neighbourhood of the resonant condition defined by (35) and (36).

Some important features are revealed by the solution of the amplitude equation (38) and by the consequent structure of the complete solution (34):

- The effect of resonance is felt within a range of values of λ and β whose width increases with ν as suggested by (35) and (36). In particular for very small values of ν (fig. 6) non linear effects suppress the infinite peak at resonance as expected but are negligible for values of λ very close to λ_R where the linear response is recovered. However for higher (though still relatively small), values of ν (fig. 13) the non linear response is sharply different from the linear one within a fairly wide range of values of λ.

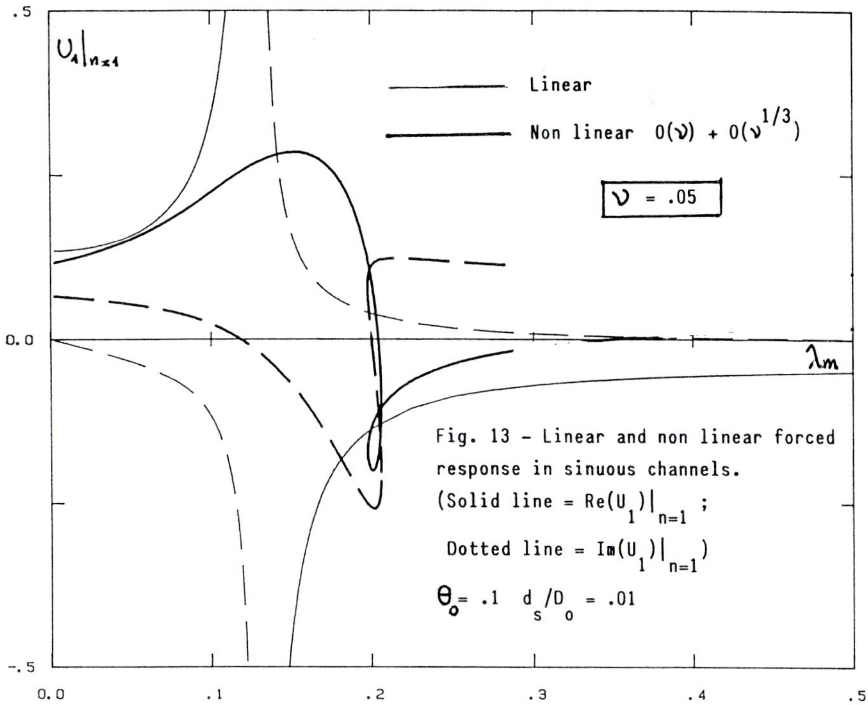

Fig. 13 - Linear and non linear forced response in sinuous channels. (Solid line = $Re(U_1)|_{n=1}$; Dotted line = $Im(U_1)|_{n=1}$) θ_o = .1 d_s/D_o = .01

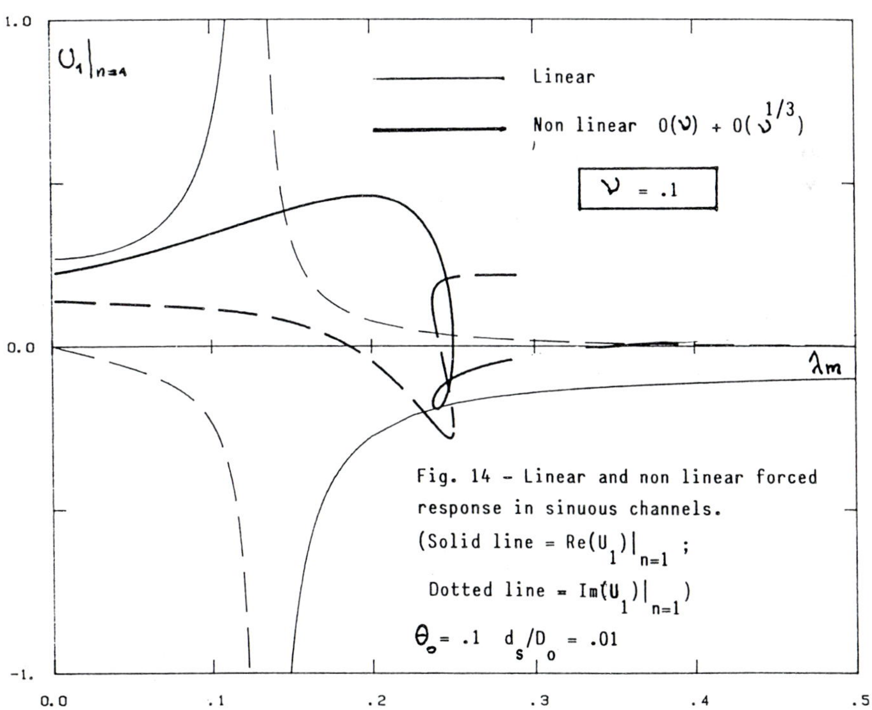

Fig. 14 - Linear and non linear forced response in sinuous channels. (Solid line = $Re(U_1)|_{n=1}$; Dotted line = $Im(U_1)|_{n=1}$) θ_o = .1 d_s/D_o = .01

- Multivalued responses appear in the solution within some range of values of λ_m, a typical feature of non linear resonant oscillators. Some of the solutions may be unstable, as it occurs for example in the case of Duffing's equation (see Thompson & Stewart (1986), pg. 72). We cannot state whether this is the case in our problem, since the stability of the present solution has not been investigated yet.

The above findings, if confirmed by fully non linear solutions of the problem [1], appear to lead to significant changes in the way meander development can be treated theoretically: indeed within a fairly wide range of values of wavenumbers λ_m and width-depth ratios β, covering the range usually found in nature, the forced response of the system is dominated by non linear resonance and cannot be described by linear models unless curvature be very small. Further implications for the mechanism controlling the development of finite amplitude meanders are discussed in the next section.

4 . A FEW CHALLENGES FOR FUTURE RESEARCH

The above findings set up a quite exciting scenario for future research. Let us discuss the case of meandering rivers and that of braiding rivers separately.

The ideas outlined in sect. 3.3 suggest that in a fairly wide near resonant range of values of λ_m and β flow non linearity is likely to control finite amplitude meander development. The geometric non linearities considered by Parker & al. (1983) and Parker and Andrews (1986) may possibly play some equally important role in the very latest stage of meander growth. The seemingly important feature of the results discussed in sect. 3.3 is that they may open the way to an analytic treatment of finite amplitude meanders which preserves the main non linear mechanism operating even for relatively large curvatures. In particular preliminary work seems to suggest the possibility to predict the shape of finite amplitude equilibrium configurations of meanders of permanent form (though probably unstable) of the type discussed by Parker & al. (1983), the controlling mechanism being flow rather than geometric non linearities. We are currently investigating along these lines.

A second important step to pursue is the establishment of a bank erosion equation to apply to conditions (laboratory experiments, braiding rivers) where the time scale od bank erosion is of the same order of magnitude as the time scale of bottom erosion. The availability of such a model would probably allow some further insight into the mechanism of meandering initiation both in the laboratory and under field conditions. Understanding of the latter process would also benefit from a riformulation of bend stability theory such to allow for the presence of free migrating bars in

1 The weakly non linear solution has been compared with the fully non linear solution for a model equation of the Duffing type. Comparison suggests that the weakly non linear approach is fairly accurate for values of λ_1 and $\beta_1 \sim O(1)$, up to relatively large values of ν.

the initially straight channel. The latter, interacting with forced bars produce a steady effect able to affect significantly the dependence of bend growth rate on meander wavenumber close to resonance.

The observed presence of more than one steady bar in tortuous meanders also needs be investigated. One may envisage different possible explanations of these observations like for instance some ultra-harmonic Mathieu type response of the system, occurring for very long meanders. The solution of this problem may also help understanding the formation of secondary loops in large amplitude meanders.

Finally all the theoretical premises now exist to dictate the importance of performing a detailed experiment on flow and bottom topography in sinuous channels aimed at clearly detecting resonance and ascertaining its role both in the linear and non linear regime starting from conditions such that free migrating bars coexist with forced bars and then increasing sinuosity till free bars are suppressed.

As for braiding rivers, it may be useful to distinguish between two different types of braiding:

i) Multiple bars braiding, characteristic of rivers which are periodically subject to formative floods which do not leave any bars above the free surface;

ii) Multiple channel braiding, such that ordinary floods do not lead to the formation of a unique channel.

For the former type of braiding the development of strongly non linear numerical models of free bars may help understanding features like mode selection and hysteres phenomena of the type mentioned in sect. 3.1. This study may possibly lead to the discovery of chaotic behaviour, with interesting engineering implications. The second type of braiding does not seem to be liable to theoretical investigations till some sound model for bank erosion of the type suggested above will be available. The latter would allow the investigation of channel shift and should be possibly coupled with investigations of the bar structure at channel bifurcations.

The above ideas appear to already pose some formidable problems even though the actual phenomena occurring in rivers exhibit some further complicating features which have not been included yet in our models, namely unsteadiness of the basic flow, grain sorting and bed armouring. However notice that some recent contributions (Tubino (1989), Parker & Andrews (1985), Ikeda (1989)) suggest that the main mechanisms controlling some of the above effects are essentially understood: unsteadiness of the basic flow is likely to play an important role when its time scale is comparable with that of bar growth, grain sorting in transversely sloping beds occurs due to the transverse component of gravity acting with different intensity on the direction of motion of sediment particles of different size.

Let us come to a general concluding remark. The above brief overwiev of recent achievements in the area of river bars, suggests that the mechanical behaviour of bars displays most features of highly non linear systems, whence their understading is greatly enhanced by cross-fertilization achieved through the knowledge of the general behaviour of non linear dynamical systems. Forecasting that the latter approach will soon lead to further important discoveries does not seem too hazardous. And our beautiful tale will continue.

ACKNOWLEDGEMENTS

The author wishes to thank dr. Colombini and dr. Tubino for their help in the preparation of this paper and for several interesting discussions. Part of the paper was written during the author's visit of St. Anthony Falls Hydraulics Laboratory, University of Minnesota. Grateful thanks in this respect are due to prof. Gary Parker for supporting the visit and for several enlightening discussions. Finally dr. Sekine's translation from Japanese of part of Kinoshita's (1980) paper is gratefully acknowledged.

REFERENCES

ASCE Committee on Sedimentation (1966). "Nomenclature for bed forms in alluvial channels", J. Hydr. Div., ASCE, 92, HY3, p. 51.

Ashida, K. (1982). Report of JSCE Committee on "Study on 3-D flow pattern and river configuration during a flood" (in Japanese).

Blondeaux, P. & Seminara, G. (1983). "Bed topography and instabilities in sinuous channels". Proc. ASCE Rivers '83, New Orleans, 747-758.

Blondeaux, P. & Seminara, G. (1985). "A unified bar-bend theory of river meanders". J. Fluid Mech. 157, 449-470.

Callander, R.A. (1968). "Instability and river meanders". Ph.D Thesis, Univ. Auckland.

Church, M. & Jones, D. (1982). "Channel bars in gravel-bed rivers" in Gravel-bed rivers, Hey R.D., Bathurst J.C., Thorne C.R. (eds), Wiley & Chichester, UK, (291-338).

Colombini, M., Seminara, G. & Tubino, M. (1986). "Equilibrium amplitude of alternate bars". Proc. 3rd Int. Symp. on River Sedimentation, Jackson, March 31 - April 4.

Colombini, M., Seminara, G. & Tubino, M. (1987). "Finite - amplitude alternate bars". J. Fluid Mech. 181, 213-232.

Colombini, M. & Tubino, M. (1990). "Non linear development of multiple row bars". (In preparation).

Engelund, F. (1974). "Flow and bed topography in channel bends". J. Hydraul. Div., ASCE, 100 (HY11), 1631-1648.

Fredsoe, J. (1978). "Meandering and braiding of rivers". J. Fluid Mech. 84, 609-624.

Fuijta, Y. (1989). "Bar and channel formation in braided streams". In 'River Meandering' eds. S.Ikeda & G. Parker, AGU Water Resources Monograph, n. 12.

Fukuoka, S. (1989). "Finite amplitude development of alternate bars". In 'River Meandering' eds. S.Ikeda & G. Parker, AGU Water Resources Monograph, n. 12.

Gaster, M. (1962). "A note on the relation between temporally-increasing and spatially increasing disturbances in hydrodynamic stability", J. Fluid Mech., 14, p. 222-224.

Gottlieb, L. (1976). "Three-dimensional flow pattern and bed topography in meandering channels". Tech. Univ. Denmark ISVA Series paper 11.

Ikeda, H. & Ohta, A. (1986). "On the formation of stationary bars in a straight flume", Ann. Rep. Inst. Geosci., Univ. Tsukuba, n. 12, p. 42-46.

Ikeda, S. (1989). "Sediment transport and sorting at bends". In 'River Meandering' eds. S.Ikeda & G. Parker, AGU Water Resources Monograph, n. 12.

Ikeda, S., Parker, G. & Sawai, K. (1981). "Bend theory of river meanders. Part 1 - Linear development". J. Fluid Mech. 112, 363-377.

Johannesson, H. & Parker, G. (1989). "Linear theory of river meanders". In 'River Meandering' eds. S.Ikeda & G. Parker, AGU Water Resources Monograph, n. 12.

Kalkwijk, J.P. Th. & De Vriend, H.J. (1980). "Computation of the flow in shallow river bends". J. Hydr. Res. 18(4), 327-342.

Kinoshita, R. (1980). "A model experiment using bedform similarity to study the consequences of removing the Ushio constriction on the Ohi river". Report (in Japanese).

Kinoshita, R. & Miwa, H. (1974). "River channel formation which prevents downstream translation of transverse bars". Shinsabo 94, 12-17 (in Japanese).

Kishi, T. (1980). "Bed forms and hydraulic relations for alluvial streams". In 'Application of stochastic Processes in Sediment Transport', Water Resources publ., Littleton, Colorado, Chap. 5.

Kitanidis, P.K. & Kennedy, J.F. (1984). "Secondary current and river-meander formation". J. Fluid Mech. 144, 217-229.

Langbein, W.B. & Leopold, L.B. (1966). U.S. Geol. Survey Prof. Paper 422-H, H1-H15.

Nelson, J. M. & Smith, J. D. (1989). "Evolution and stability of erodible channel beds". In 'River Meandering' eds. S.Ikeda & G. Parker, AGU Water Resources Monograph, n. 12.

Olesen, K.W. (1983). "Alternate bars and meandering of alluvial rivers". Commun. Hydraul., Delft Univ. of Technology, Rept. 7-83.

Parker, G. & Andrews, E.D. (1985). "Sorting of bed load sediment by flow in meander bends", Water Resources Res., 21(9), 1361-1373.

Parker, G. & Andrews, E. D. (1986). "On the time development of meander bends". J. Fluid Mech., 162, p. 139-156.

Parker, G. & Johannesson H. (1989). "Observations of several recent theories of resonance and overdeepening in meandering channels". In 'River Meandering' eds. S.Ikeda & G. Parker, AGU Water Resources Monograph, n. 12.

Parker, G., Diplas, P. & Akiyama, J. (1983). "Meander bends of high amplitude". J. Hydraul. Div. ASCE, 109.

Seminara, G. & Tubino, M. (1987). "Curvature driven evolution of alternate bars into point bars". Euromech 215, S. Margherita Ligure (Genoa), September 15-19.

Seminara, G. & Tubino, M. (1989a). "Alternate bars and meandering: free, forced and mixed interactions". In 'River meandering' eds. Ikeda & Parker, AGU Water Resources Monograph, n. 12.

Seminara, G. & Tubino, M. (1989b). "On the process of meander formation". IV Conference on River Sedimentation, Beijng.

Seminara, G. & Tubino, M. (1990). "Non linear theory of resonant meanders". (In preparation).

Struiksma, N., Olesen, K.W., Flokstra, C. & De Vriend, H.J. (1986). "Bed deformation in curved alluvial channels". J. Hydr. Res. 23 (1), 57-79.

Thompson, J.M.T. & Stewart, H.B. (1986). "Non linear dynamics and chaos". Ed. Wiley & Sons.

Tubino, M. & Seminara, G. (1989). "Free-forced interactions in developing meanders and suppression of free bars". Accepted for publication in the J. Fluid Mech.

Tubino, M. (1989). "Unsteady alternate bars: theory and experiments". (Submitted for publication).

Shimizu, Y., Itakura, T. & Yamaguchi, H. (1987). "Numerical simulation of bed topography of river channels using two-dimensional model". Proc. of 31st Japanese Conference on Hydraulics (in Japanese).

Sediment Movement in Mountain Streams:

Erosion-Deposition Models
Deposition in Reservoirs
Ordinary and Catastrophic Events;
Debris Flow
Sediment Control Devices

SEDIMENT YIELD AND RIVER BED CHANGE IN MOUNTAIN RIVERS

Takahisa Mizuyama

Pubblic Works Research Instutute
Ministry of construction, Japanese Government
Tsukuba, Japan

ABSTRACT

Sediment discharge in mountain rivers is given not only by sediment transport ability of flow, but also by sediment supply from headwater areas, torrent beds and hill slopes. This is the most different point from rivers in plains. It is a key to estimating sediment yield accurately for river bed change in mountain areas. Major modes of sediment yield in mountains are shallow landslides and debris flows. The volume of sediment yield has been estimated empirically. Not only the volume, but also the times and places of their occurrences are necessary to discuss sediment discharge and the effect of sediment control structures. A model of the stability of slopes is studied. Though the strength and the depth of soil layers cannot be estimated well at present, it becomes possible to estimate the times, places and the volume of shallow landslides.

Sediment discharge and river bed change are estimated mainly by bed load equations. Some field data concerning large sediment discharges are introduced, and they are interpreted theoretically.

1. A Model of Landslides Which Considers Unsaturated Seepage

1.1 Outline of the Drainage Basin and Conditions of Sediment Yield

Among the various forms of sediment yield, shallow landslides are a phenomenon triggered by heavy rains. During rainy periods, these occur frequently in all areas; and along with erosion of river banks, these landslides--including those which develop into debris flows--are the major source of sediment yield in drainage basins. The current paper presents a model of landslides which considers the permeation and runoff of rainwater in the surface soil layer of hillside slopes.

The Oguro River is a torrent which has its sources in the Irikasadake, Mt. Hodokubo and Mt. Kamanashi regions of the Southern Alps. The area of the drainage basin is 39.95km^2, and the main stream has a length of 13.6km and an average river bed slope of 1/13.5. The topography of this river is characterized by a gentle slope in the upper reaches, while the middle and lower reaches are quite steep. This basin has the geological feature of being divided by the Todai Tectonic Line which extends north and south along the side of the right bank, with the region west of this line belonging to the Sanhagawa zone and the eastern area belonging to the Chichibu zone. Many landslides occurred in the Oguro River drainage basin as a result of heavy rains induced by typhoon No.10 in 1982. The landslides taken for analysis here covered 1.33% of the total area of the drainage basin. (Figure 1)

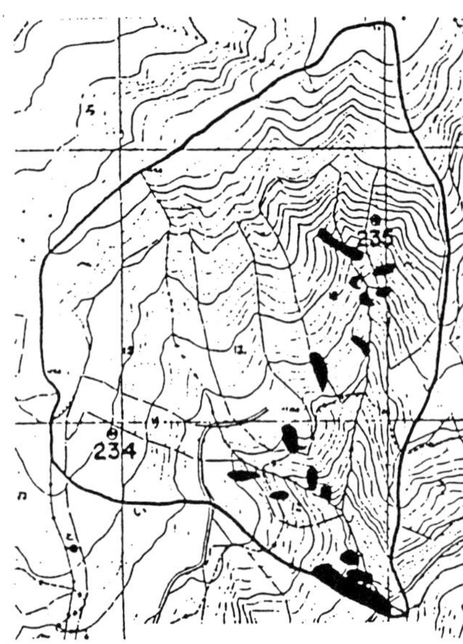

Figure 1. Distribution of Landslides

At this time, the amount of continuous precipitation reached 403.0mm at the Oguro observation station, and the maximum hourly precipitation of 45.0mm occurred 31 hours after the start of the rain.

1.2 Basic Structure of the Model

Rainfall, or rain water, is the the chief cause of shallow landslides. For this reason, assuming specific amounts of rainfall, it is necessary to evaluate permeation and runoff in actual conditions in order to be able to predict the position and scale of landslides as well as the degree of danger in the drainage basin. As shown in figure 2, the model put forth in this paper considers shallow landslides to occur as a result of decreasing stability through rainwater which permeates the unsaturated sloped surface layer, creating a saturated zone (ground water surface) on the impermeable layer, with the final result of saturated lateral flow. The internal condition of hillsides affected by the permeation and runoff of rainwater is thought to actually be a complicated three-dimensional saturated/unsaturated flow. Considering applicability to large drainage basins, however, the present model utilizes a theory of one-dimensional vertical unsaturated permeation for modifying the hyetograph, and uses Darcy's law for saturation to trace the amount of rainwater on the bedrock as the amount of lateral flow.

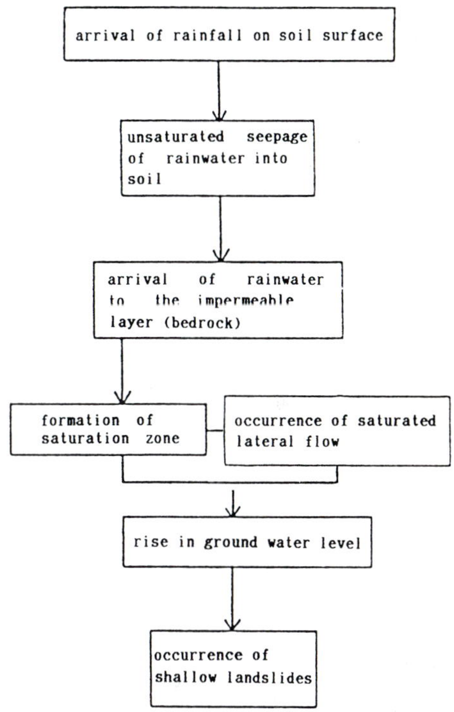

Figure 2.
Conceptual Diagram of the Landslide Model

The basic structure of the model, as shown in figure 3, is such that the drainage basin is partitioned according to a horizontal mesh, with each partition treated as a vertical column of earth according to the depth of the surface soil. These partitions serve as the basic elements for analysis. Here, the rainwater supplied to each of these basic elements permeates into the soil, forming a saturated zone where it reaches bedrock, and then moves between elements as a saturated lateral flow.

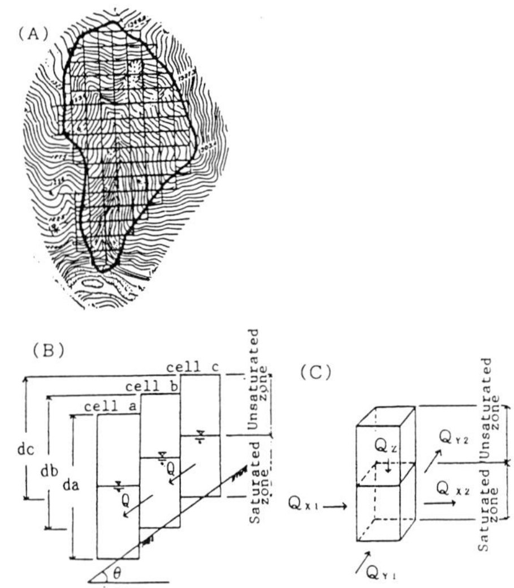

Figure 3. Basic Structure of the Model

1.3 Analysis of Filteration Flow

1.3.1 One-dimensional vertical Unsaturated Seepage

A numerical solution was sought for one-dimensional vertical seepage of rainwater in unsaturated zones by combining continuous equations with Darcy's law enlarged for unsaturated areas, and then finding the differential expansion of equation (1),

$$\frac{\partial \theta}{\partial t} = \frac{\partial}{\partial z}\left(D(\theta)\frac{\partial \theta}{\partial z}\right) - \frac{\partial k(\theta)}{\partial z} \quad \cdots\cdots\cdots\cdots (1)$$

$$q = -k(\theta)\cdot\frac{\partial \psi}{\partial z} + k(\theta), \ \frac{\partial \theta}{\partial t} = -\frac{\partial q}{\partial z} \quad \cdots\cdots\cdots\cdots (2)$$

where ψ stands for soil-moisture tension head, t for time, Z for the vertical coordinate (positive in the downward direction), $k(\theta)$ for the coefficient of unsaturated permeability, θ for the amount of soil-moisture, $D(\theta)$ for the coefficient of moisture diffusion, and q for the amount of soil-moisture movement. As boundary conditions, it was assumed that flux in the amount of flow from the surface into the soil layer was the amount of precipitation, and that the base of the soil layer is always maintained in a condition of saturation. As an initial condition, it was assumed that there is no movement of moisture immediately prior to precipitation, as shown in the following equation.

$$\psi = H - Z \quad \cdots\cdots\cdots\cdots\cdots\cdots\cdots\cdots (3)$$

In this equation, H stands for soil-moisture tension head, and Z for the distance from bedrock.

1.3.2 Saturated Lateral Flow

Saturated lateral flow was calculated, assuming that rainwater permeates to bedrock, forming a saturated layer there which runs off. The continuous equation for this process of saturated lateral runoff is expressed as follows:

$$\lambda\frac{\partial h}{\partial t} + \frac{\partial q_x}{\partial x} + \frac{\partial q_y}{\partial y} = q_z \quad \cdots\cdots(4)$$

The equation for motion, based on Darcy's law, is:

$$q_x = h \cdot k_s \cdot I_x, \quad q_y = h \cdot k_s \cdot I_y \quad \cdots\cdots(5)$$

Here, λ stands for effective porosity, h for the depth of ground water in meters, q_x and q_y stand for the flux in flow along the X and Y axes respectively, q_z for the amount of water supplied from unsaturated regions, t for time, x and y for distance along the X and Y axes, k_s for the coefficient of saturated permeability, and I_x and I_y for the hydraulic gradient along the X and Y axes.

1.3.3 Analysis of Slope Stability

Concerning the morphology of landslides, their length is long in comparison with depth, and cases where the sliding surface forms a plane are common. Hence, analysis of slope stability was performed by the method of stability analysis for infinite slope,

$$F_s = \frac{C + (\sigma_o - h\gamma_w)\cos^2\beta\tan\phi}{\sigma_o\cos\beta\sin\beta} \quad \cdots\cdots(6)$$

$$\sigma_o = \int_0^z \{\theta(i)\gamma_w + (1-n)G_s\gamma_w\}\,\Delta z \quad \cdots\cdots(7)$$

where C stands for cohesion of soil, Δz for the thickness of vertical strata divisions, h for the depth of ground water, γ_w for the unit volume per weight of water, β for the inclination of the unit slope, ϕ for the angle of internal friction, G_s for the specific gravity of soil particles, and $\theta(Z)$for moisture content per volume for minute spaces in the vertical direction.

1.4 Application of the Landslide Model

1.4.1 Conditions of the Analysis

The landslide model was applied to a drainage basin in which landslides actually occurred. The object for analysis was the small drainage basin shown in figure 1 (area = 0.52 km^2). Information concerning the topography was taken from a basic forest map having a scale of 1:5000. The area was partitioned according to a mesh of 25m, and the ground levels were measured for each unit slope. The depth of surface soil was set for unit slope by means of a surface soil depth contour map (landslide depth contour map, figure 4) which utilizes the depth of slides on slopes where they occurred. Considering the permeation of rainwater in soil strata, equation (1) cannot be solved in isolation; the relations of ψ to ϕ and k to ψ must be examined. Here, the results of a physical soil test (depth of 40cm, figures 5 and 6) were used as input conditions. Also, results of the same test were used for values of soil characteristics (G_s = 2.66, C = 0.7tf/m^2, ϕ = 20.44°).

1.4.2 Underground Water Level Response Characteristics Due to Differences in Shapes of Slopes

Representatives were chosen from among valley type, parallel type and ridge type slopes in drainage basins. These were then used to examine the effects which topographic features have on underground water level response characteristics. (See figure 7.) In the case of every type of slope, the height of the underground water level increased sharply 25 hours

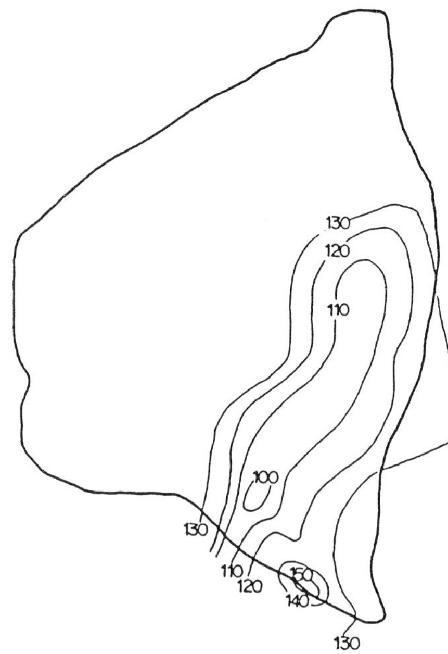

Figure 4. Surface Soil Depth Contour Map (scale 1:10000)

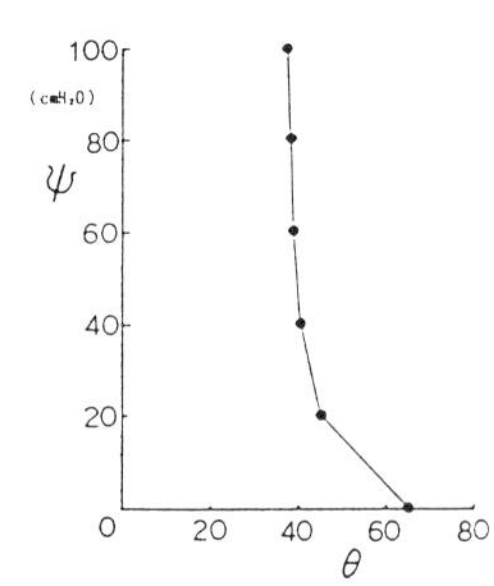

Figure 5. Soil-moisture Tension Head

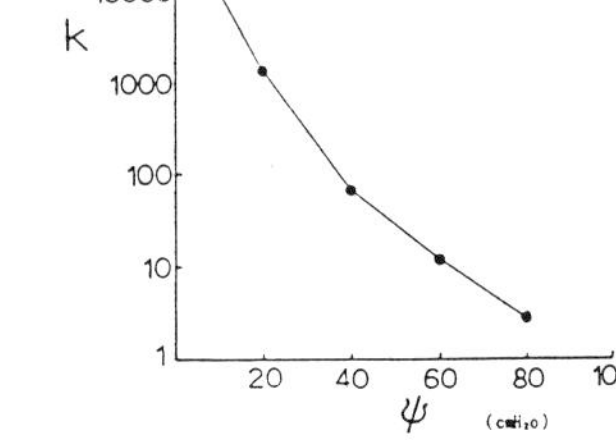

Figure 6. Coefficient of Unsaturated Permeability

after the start of rainfall. The rate of increase began to slow upon the cease of precipitation (40 hours after commencement). This tendency was particularly marked in the case of ridge type slopes. This is because the supply of rainwater into the soil from the surface stops upon the cease of precipitation; and along with a reduction in the amount of vertical unsaturated permeation, the water stored in the ridge type slope runs off as saturated lateral flow. This then becomes inflow for valley type slopes.

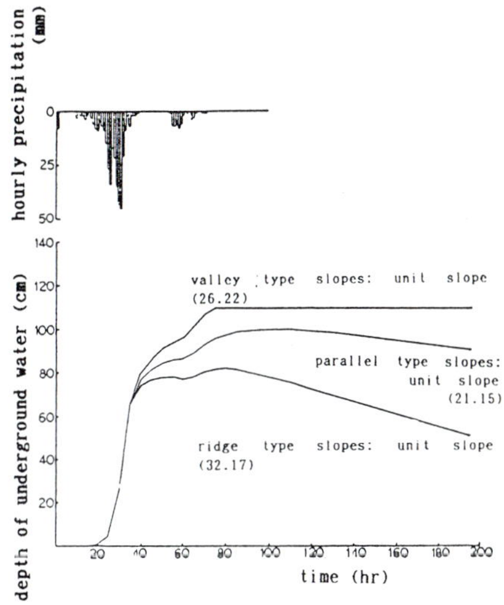

Figure 7. Underground Water Level Response Characteristics Due to Differences in Shapes of Slopes

1.4.3 Locations of Occurrences of Landslides

Figure 8 shows the results of the analysis. Amounts of precipitation at the times landslides occurred were used, and the size of the mesh used was 25m. The start time for the analysis was set to 6:00 p.m. on 31 July 1982, the time at which precipitation began. It can be seen from this figure that there is a strong correspondence between slopes with a safety factor below 1.0 and slopes on which slides actually occurred, and that there is a tendency for the recurrence of landslides in the left upper reaches, right lower reaches, and lowermost reaches of the drainage basin. However, there were actual cases of slides in areas with a safety factor above 1.0, as well as many slopes with a safety factor less than 1.0 but where slides did not occur. This is thought to be a result of the randomness in the values set earlier for precipitation, depth of surface soil and soil strength, as well as the degree of accuracy in the reproduction of the topography.

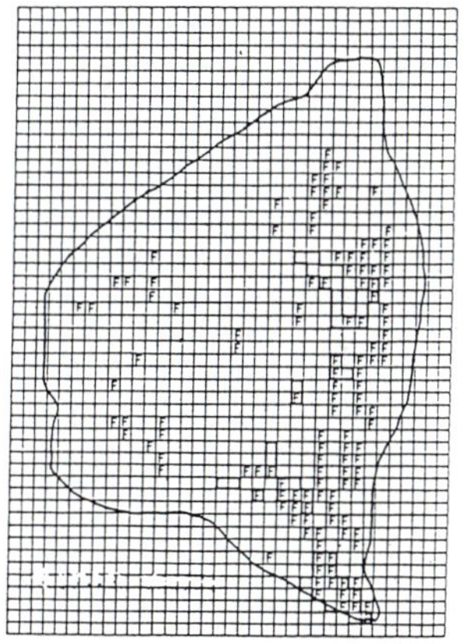

Figure 8. Distribution of Slopes Where Landslides Occurred

2. River Bed Change in Mountainous Areas

2.1 Characteristics of River Bed Change in Mountainous Areas

In the case of mountain rivers, river beds sometimes change their height by several meters over a short time. In most cases, this rapid change is brought about by large-scale landslides or debris flows induced by heavy rains or earthquakes, or when the continuity of flow has been interrupted due to the construction of reservoirs or the excavation of gravel. Among these, change induced by the former of these is most rapid.

Thinking about these problems requires the following basic consideration. Excepting special cases, sediment transport ability of flowing water is determined by gradient of the river bed, amount of water, and material composition of the river bed. On the one hand, the gradient of mountain rivers is steep in comparison with rivers in plain areas, and the amount of sediment supplied to the river during normal times is small. Hence, little sediment is supplied to the river bed during times of normal water level, and the sediment transport ability is correspondingly large, resulting in a tendency to erode the river bed. However, when the amount of supplied sediment increases quickly due to new landslides or debris flows brought about by heavy rains, the sediment transport ability of the water flow is unable to carry all of this, and the river bed necessarily rises. Figure 9 expresses this situation conceptually.

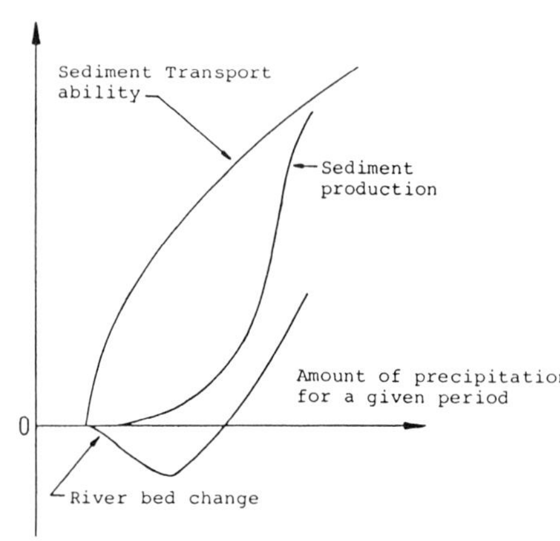

Figure 9.
Schematic diagram showing the relation of rainfall to sediment production, sediment transport ability and river bed change.

However, examining river beds after floods have abated shows that the river bed has not necessarily risen over the course of the whole river. The distributions of time and location, in which sediment produced by landslides or debris flows is supplied to the river, determine the conditions which result following a flood. For example, considering the temporal distribution of sediment supplied to rivers, in the case where large amounts of sediment are supplied during the first half of a flood while the amount decreases afterward, the river bed near the source of the sediment yield rises during this early period. However, during the last half of the flood, deposits are quickly eroded, and the final result may be a net decrease in the height of the river bed. In the lower reaches of the river, however, sediment is gradually transported downstream, with the possible result of a net rise in the river bed.

Accordingly, changes in the river beds of mountain rivers is complex, and while one can guess what changes might occur before and after floods, it is impossible to remove uncertainty from the discussion. Of course, a careful discussion of the manner of sediment transport and of the supply and loss of sediment is possible based on facts gathered from the traces left by floods and and from the deposits found there; but consideration of soil hydraulics is necessary in a treatment of such changes over time.

In the section below, using data collected from the Inamata Valley, a tributary of the Haya River that is also a tributary of the Fuji River, in 1982 as an example, an explanation in terms of soil hydraulics for the process of river bed change in mountain rivers is attempted.

2.2 Conditions of Sediment Yield in 1982

In August of 1982, typhoon No.10 caused heavy rains with a maximum hourly rate of 80mm in Nagahata near the Inamata Valley. Total rainfall was 830mm, well above the 500mm figure recorded during the previous 1959 disaster. Also, in September of the same year, typhoon No.18 produced a large total rainfall of 470mm, though the maximum hourly rainfall was about 40mm.[1]

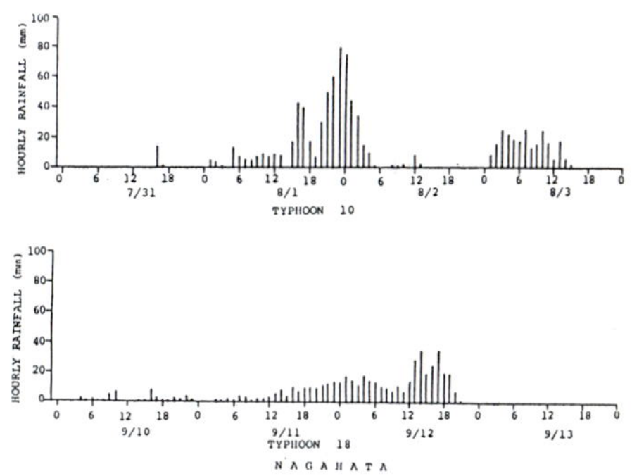

Figure 10. Hourly Rainfall of Typhoons No.10 and No.18 in 1982

Aerial photographs taken after these two large rainfalls were compared with pictures taken in October of 1981 to determine the areas in which new landslides (including debris flows) had developed. The distribution of these is shown in figure 11.

The Inamata Valley merges into the Amahata River. The catchment area at this confluence is 24.5km^2, and in 1982 there were 579,600m^2 of new landslides. This drainage basin contains a large-scale landslide area called Yashio, though it showed almost no change in 1982.

The geological features of the Inamata Valley drainage basin are Paleozoic and Paleogene sandstone, slate, and a small grain size of less than a few centimeters for material in the river bed. Figures 12 and 13 show the distribution of grain size for river bed materials at fourteen areas according to surveys conducted in November of 1978 and February of 1983. Generally, due to the effects of armoring phenomenon, river beds in mountainous areas are rather course, and become finer when large-scale sediment production and flows occur. River bed materials of the Inamata Valley drainage basin are small even during normal times, and grain size distribution does not vary greatly before or after flooding. However, the randomness of grain size distribution lessens, with 65% of grain diameters shrinking by about one half.

Generally, the changes in mountain rivers are largest. Hence, in explaining large-scale sediment flows based on grain size distribution taken from surveys of river beds before flooding, or later of locations which have been strong affected by flows, calculations may produce results which greatly underestimate the actual amounts of

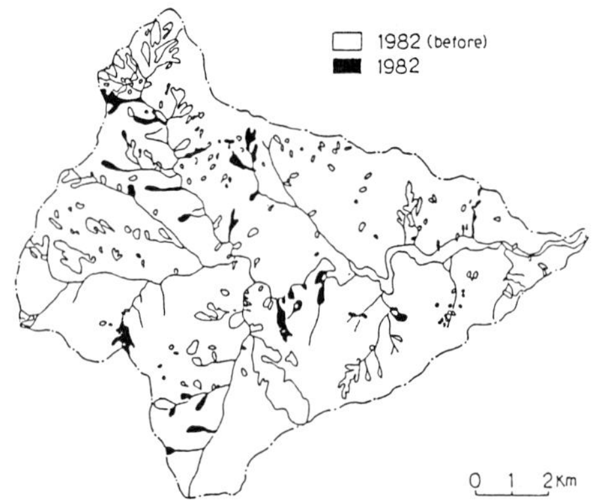

Figure 11. Distribution of New Landslide Areas

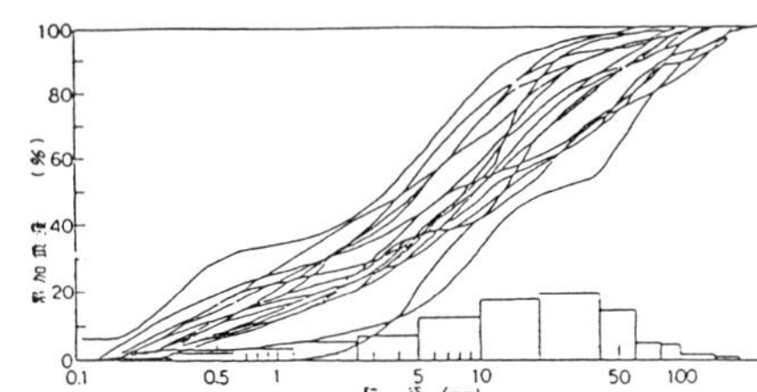

Figure 12. Grain Size Distribution of River Bed Materials in November of 1978

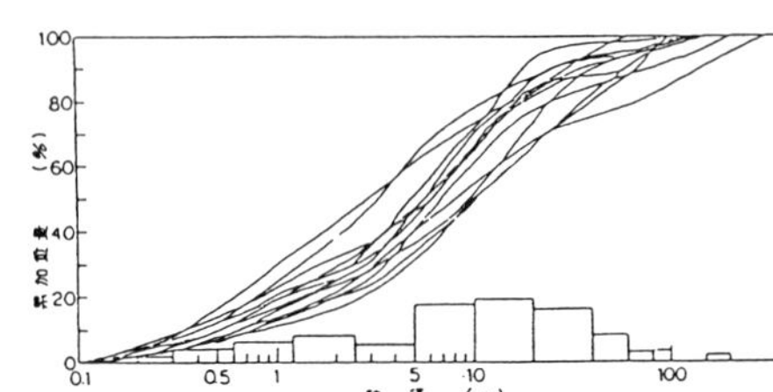

Figure 13. Grain Size Distribution of River Bed Materials in February of 1983

sediment flow.

2.3 Conditions of River Bed Change Immediately Following Floods in 1982

As stated in the previous section, large river bed changes are caused by the supply of large amounts of sediment to the river. Figure 14 shows river bed changes of the lower reaches of the Inamata Valley from the point of its confluence with the Amahata River. Prior to the floodings, river bed height was almost unchanged since 1978; but as a result of two floods in 1982, the river bed rose as much as 20m, burying the 13m high Inamata Dam and the 17m high Inamata 2nd Dam. The gradient of the river bed increased rapidly to about what it had been before the construction of the control dams. Generally, even when there has been a large sediment flow and rise in the river bed during flooding, there is still usually a net lowering of the river bed once the flooding has ceased. In this case, however, because the amount of sediment yield was large and the grain size was small, sediment yield seems to have continued until the end of the flooding, and river bed height in the area around the Inamata 2nd Dam only decreased by about 1.5m. Photograph 1 shows the situation around the Inamata 2nd Dam in July 1983.

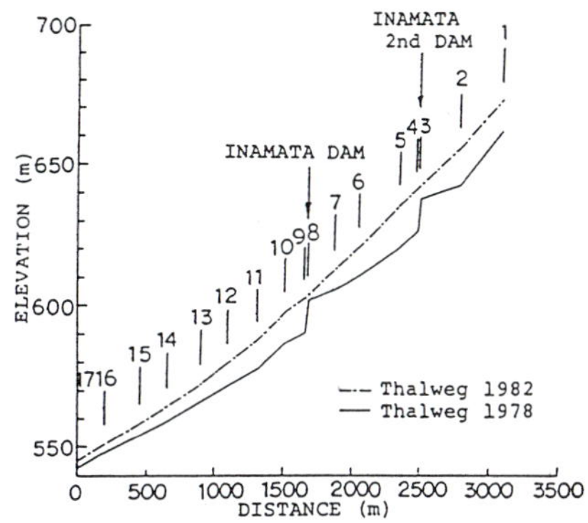

Figure 14. Changes in River Beds of Lower Reaches of the Inamata Valley (November 1982)

Photograph 1. River Bed around the Inamata 2nd Dam (July 1983)

As a further note it might be pointed out that there is an intake dam used for electric power generation at about the 3.2km point on the Inamata Valley, with an normal maximum intake of 1.5m³/sec. This is one of the causes for the fact that the river bed did not go down after flooding.

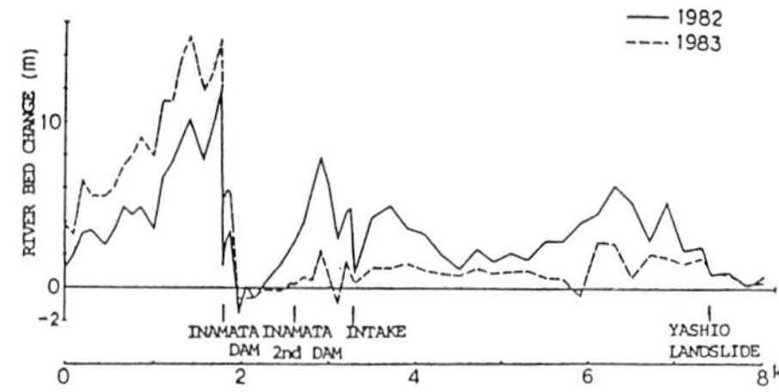

Figure 15. Comparison of River Bed Heights in 1982 and 1983 (Using 1978 as a Standard)

2.4 River Bed Change after Large-scale Sediment Discharge

The results shown of a comparative survey of the 8km interval above the junction with the Amahata River are shown in figure 15. A large amount of the sediment deposited in the river bed above the Inamata 2nd Dam was discharged by the end of 1983, and the river bed below the Inamata Dam was still rising. However, the amount of sediment which exists in the stream bed is finite, and the river bed in the lower reaches should also begin to decline.

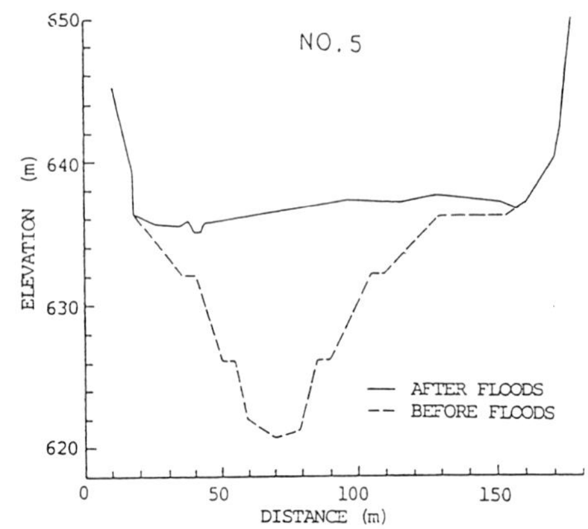

Figure 16. Example of Change in Shape of Cross Section (Point 5 of Figure 14)

Photograph 2 was taken in November of 1985 of the same area shown in photograph 1. It can be seen that the river bed had by then lowered over 2m. The river bed had not lowered by the end of 1984, and this is thought to largely be the result of heavy rains from June 30-31 in 1985 when 190mm fell in four hours. Figure 16 shows change in the shape of a cross section of point 5.

Photograph 2. River Bed around the Inamata 2nd Dam (November 1985)

2.5 Explanation of 1982 Floods by a One-dimensional Analysis of River Bed Change

An attempt is made to explain river bed changes in the lower reaches during the 1982 flooding in terms of one-dimensional river bed change analysis. From the fact that the area used in the calculation has a grade of 1/23, it is treated as bed load. There is no water level observation station in the Inamata Valley; so since data cannot be obtained regarding the amount of flow, this was calculated by means of a characteristic curve method. Results from analyzing the discharge of the Amahata Dam were used to compute the effective rainfall. The following bed load equation from Ashida, Takahashi, and Mizuyama was used:

$$\frac{q_B}{\sqrt{(\sigma/\rho-1)gd_m^3}} = 12\,\tau_{*m}^{1.5}\left(1-0.85\frac{\tau_{*cm}}{\tau_{*m}}\right)\times\left(1-0.92\sqrt{\frac{\tau_{*cm}}{\tau_{*m}}}\right) \cdots\cdots\cdots (8)$$

where q_B stands for the water discharge per unit width and unit time, g for gravitational acceleration, d_m for average grain size, σ and ρ for the density of gravel and water respectively, τ_{*m} for the tractive force in dimensionless quantity, and τ_{*cm} for the critical tractive force.

where q_B stands for the water discharge per unit width and unit time, g for gravitational acceleration, d_m for average grain size, σ and ρ for the density of gravel and water respectively, τ_{*m} for the tractive force in dimensionless quantity, and τ_{*cm} for the critical tractive force.

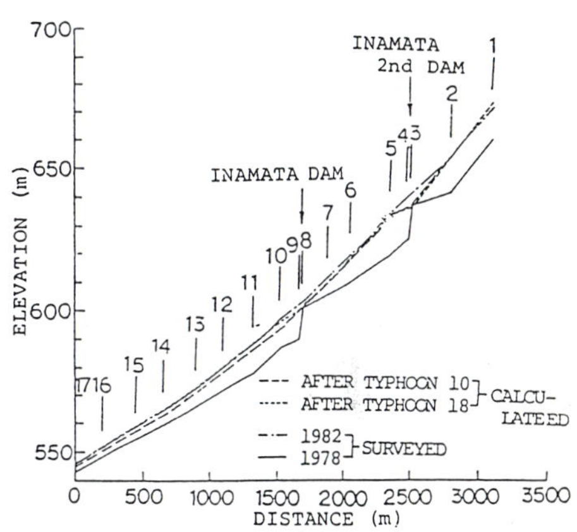

Figure 17. Computational Results

Figure 17 shows the computational result. In the case of typhoon No.10, the river bed rose for most of the whole region. In the case of typhoon No.18, the river bed in the upper stretches began to go down, explaining well the tendency in the lower stream for a rise of about 2m. It was assumed that their was no supply of sediment from the side slopes in the region computed, and that 65% of the supposed sediment from landslides in the upper regions, or about 565,110m^3, flowed into the area in question.

It was also assumed that at times when the river bed rises, it rises across the whole width of the valley. At times when the river bed goes down, the actual width of the valley and the width of the water course were compared by means of the regime theory, and it was assumed that river bed change occurs across the width of the smaller of these.

$$B = \alpha \sqrt{Q} \qquad \cdots\cdots (9)$$

Here, B stands for the width of the course of flow, and Q for the water discharge. α is a coefficient given the value of 5.0 when B is given in meters and Q in m^3/sec.

2.6 Form of Sediment Transport

The calculation of river bed change given in the previous section treated the form of sediment transportation as bed load. As a matter of fact, and as photograph 3 shows, deposits in the upper reaches following the 1982 flooding were beautifully stratified, indicative of their bed load transportation character. As shown in photograph 4, however, in the case of the stretch below the Inamata 2nd Dam in November 1985, deposits containing round cobbles with a grain size of about 20cm were found below stratified deposits with a thickness of about 1.5m. It is thought that these are the deposits of a hyperconcentrated flow.

Photograph 3. Structure of Sedimentation in the Upper Level Deposits

Photograph 4. Structure of Sedimentation below the Inamata 2nd Dam

3. Topics for Further Research

Because of an insufficiency of information concerning the characteristics of slopes, it is not yet possible to predict sediment production for mountain rivers. In assessing river bed change, for which fundamental studies are needed, it is necessary to evaluate, as in the case of old river bed deposits and river banks, the resistance to erosion of sediment possessing cohesion. Though various formulas have been proposed for the amount of lateral erosion, these are as yet inadequate for the matter. Mountain rivers rarely form armor coats of uniform gravel or round cobbles as in the case of lower rivers, and stabilize more along the form of a step-pool system. Further research is necessary concerning the formation process of stable river beds in which huge rocks lie scattered.

Bibliography

1) Takahisa Mizuyama, "Catastrophic Changes of Mountain River and its Hydraulic Explanation," Japanese Geomorphological Union 5-3, 1984, pp.195-203

2) Kyoichi Matsumura, "A Study on the Effects which the Topography of Mountain Rivers has on Sediment Movement," M.A. Thesis at Tsukuba University, March 1985

3) Kazuo Ashida, Tamotsu Takahashi and Takahisa Mizuyama, "Study on Bed Load Equations for Mountain Streams," Journal of the Japan Soc. of Erosion Control Eng., 107, pp.9-17

A GENERAL MODEL FOR INTENSE SEDIMENT TRANSPORT OF PLANE BED

Alberto Lamberti [(1)] and Luigi Montefusco [(2)]

[(1)] Hydraulics Institute, University of Bologna, Italy
[(2)] Department of Civil Engineering, University of Florence, Italy

ABSTRACT

A mathematical model representing sediment transport by a free surface stream is proposed, hypothesizing two-dimensional and uniform flow, the bed of monogranular incoherent material and a sufficiently intense transport.

The global results appear to confirm the validity of the schematization, while the analysis of the detailed results points out its considerable potential also as an instrument for better comprehension of the actual physical phenomenon.

1. Introduction

Among the studies on sediment transport by free surface streams, the analysis of the phenomenon in very intense transport conditions occupies a particularly prominent position, both due to its importance in natural physical situations (especially in mountain areas), and to its possible application in specific instances such as, for example, the overflowing of a dam in loose materials.

In recent times Wilson [1][2][3] has been particularly engaged in this problem, and has also conducted experiments with high bed shear stresses; he interpreted the results making use of Bagnold's [4] dispersive stress theory. This theory has been recently re-adopted, examined in depth and used by various other authors [5][6][7][8][9][10].

In this note, the authors present a mathematical model which allows both detailed and overall simulation of the phenomenon of solid transport by a free surface flow, with particular reference to intense transport conditions: as far as the authors are aware, this is the first time that the problem is approached with such a complete physical scheme.

The results achieved, presented here in part, appear interesting.

2. Definition of the problem

Although it is very general and complete, the model refers, at least in this phase, to a simplified schematic situation, generally accepted in this type of study. More specifically, it is assumed that the flow is, on average, steady and uniform, i.e. its characteristics (time averages) are identical in each normal section, as well as independent from time: the flow will also be considered two-dimensional.

In addition, the bed is assumed to be plane and composed of incoherent single-size sediments; it is also hypothesized that it remains flat during the transport phenomenon, thus assuming that no bed forms are present.

Finally, it is assumed that both the liquid and solid phases may be treated as continuous media, characterized by specific constitutive relations.

With reference to figure 1, an x-y orthogonal cartesian reference system was adopted, with the x-axis at the summit of the first still sediment layer and oriented in the direction of motion; the y-axis, perpendicular to the x-axis, is oriented upward. The angle α measures the horizontal inclination of the bed.

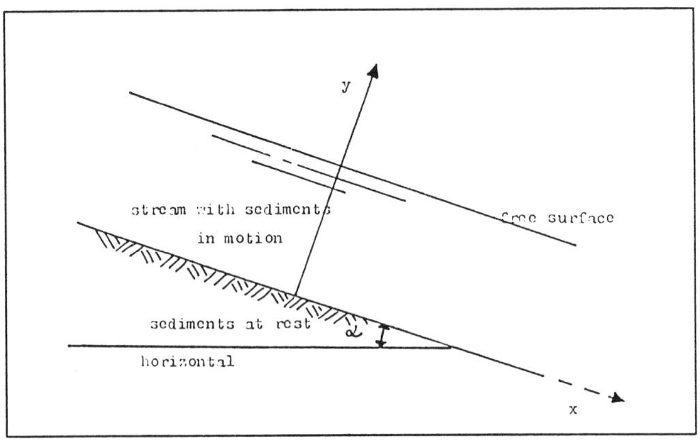

Fig. 1

The field of motion studied lies between the immobile bed and the free surface; the interpenetrating liquid and solid phases are assumed to be two continuous media, both simultaneously present in the space field.

In the same field, the following state variables are considered, with reference to their values averaged over time:
- $\tau_f(y)$, shearing stress in the liquid phase
- $\sigma_f(y)$, pressure in the liquid phase
- $U_f(y)$, x-component of the liquid phase velocity
- $V_f(y)$, y-component of the liquid phase velocity
- $\tau_s(y)$, shearing stress in the solid phase (intergranular stress)
- $\sigma_s(y)$, y-normal stress in the solid phase (intergranular stress)
- $U_s(y)$, x-component of the solid phase velocity
- $V_s(y)$, y-component of the solid phase velocity
- $C(y)$, volumetric concentration of the solid phase.

In the proposed scheme some more variables also appear; it is worth making explicit reference to the following here:
- f_x x-component of the force per unit volume applied from the liquid phase to the solid phase (net of buoyancy),
- f_y y-component of the force per unit volume applied from the liquid phase to the solid phase (net of buoyancy),
- ϵ turbulent diffusion coefficient.

With reference to the above, the solid-liquid interaction, created by the surface forces on the individual particles, is schematized, in coherence with the hypothesis of two continuous media, through equal and contrary forces of volume applied to the two phases.

Among the nine state variables listed, the following nine independent scalar relations hold:

- 4 relations which express the momentum balance along direction x and direction y, both for the liquid phase and the solid phase,
- 2 relations which express the mass balance for each of the two phases,
- 1 relation between shear stress and shear rate in the liquid phase, assumed to be in turbulent motion,
- 2 relations expressing the constitutive law of the solid phase, referring to the shear and y-normal stresses.

The expressions necessary for f_x, f_y, ϵ and the appropriate boundary conditions must be added to these.

3. Equations used

The equations which express the balance of mass and momentum for a continuum are the classic equations:

$$\frac{\partial \rho}{\partial t} + \text{div}\,(\rho\, V) = 0 \tag{1}$$

$$\text{div}\,\mathbb{T} + \rho F = \rho \frac{DV}{Dt} \tag{2}$$

with:

- $\mathbb{T}$ stress tensor
- F resultant of the mass force per unit mass
- V velocity vector
- ρ density of the continuum.

In permanent and uniform motion, local density derivative, the longitudinal terms of divergence and acceleration are zero. Including Reynolds stresses in $\mathbb{T}$, the previous equations can be reduced to the balance of vertical mass fluxes and the balance of vertical stress gradients and mass forces.

The x-components of (2) are the scalar relations

$$\frac{d\tau_f}{dy} + \rho_f\, g \sin\alpha - f_x = 0 \tag{3}$$

for the liquid phase, and

$$\frac{d\tau_s}{dy} + (\rho_s - \rho_f)\, g\, C \sin\alpha + f_x = 0 \tag{4}$$

for the solid phase, where:

- ρ_f is the liquid proper density
- ρ_s is the sediment proper density
- g is the gravity acceleration,

having chosen conventionally the positive directions for τ_f and τ_s so that their values result positive in ordinary flow conditions (the upper layers drag the lower layers).

Again the y-components of eqs. (2) produce the scalar relations

$$\frac{d\sigma_f}{dy} + \rho_f\, g \cos\alpha + f_y = 0 \tag{5}$$

for the liquid phase, and

$$\frac{d\sigma_s}{dy} + (\rho_s - \rho_f)\, g\, C \cos\alpha - f_y = 0 \tag{6}$$

for the solid phase.

Note how in the previous eqs. (3) to (6) σ and τ are forces transmitted through the considered phase per unit of apparent surface, and how the effect of buoyancy on the sediments ($\pm\rho_f gC$) is already included in all the second terms, together with the pure effect of gravity on the phase considered.

The mass balances for the liquid and solid phases become respectively:

(7) $$(1\text{-}C)\,V_f - \langle c'v'_f\rangle = 0$$

(8) $$C\,V_s + \langle c'v'_s\rangle = 0$$

where $\langle.\rangle$ represents the time average over a sufficiently long period, and c', v'_f, v'_s represent, as usual, the differences between the instantaneous and the average values for the concentration and the vertical components of velocity in the liquid and the solid phase.

Eq. (7) was used in a different form derived as follows: the sum of eqs. (7) and (8), which expresses the mass conservation for the merging of the two phases, is:

(9) $$(1\text{-}C)\,V_f + C\,V_s - \langle c'(v'_f\text{-}v'_s)\rangle = 0$$

where it seems appropriate to neglect the correlation between c' and $(v'_f\text{-}v'_s)$, thus approximating eq. (9) with

(10) $$(1\text{-}C)\,V_f + C\,V_s = 0\,.$$

Beyond the relations expressing the basic principles of mechanics, the assumed constitutive equations are to be specified.

It was assumed that the flow is completely turbulent, and that the term $\langle c'v'_s\rangle$ in eq. (8) may be expressed, in the entire domain, by the classic diffusive scheme:

(11) $$\langle c'v'_s\rangle = -\,\epsilon\,\frac{dC}{dy}$$

where the turbulent diffusion coefficient ϵ is assumed to be the same for sediments and momentum diffusion:

(12) $$\epsilon = \frac{\tau_f}{\rho_f(1\text{-}C)\,dU_f/dy}$$

For the liquid phase, assumed in fully developed turbulent motion, reference was simply made to the well-known Prandtl relation

(13) $$\tau_f = \rho_f(1\text{-}C)L^2\left(\frac{dU_f}{dy}\right)^2$$

with L (mixing length) expressed by

(14) $$L = k\,y.$$

In the question whether the value of k should be assumed as constant in the presence of transported sediments, it was preferred to conform to that which appears to be the most recent and careful study on the subject [11], opting for k to be dependent on the C value.
An experimental expression to supply this dependency as a function of the local time average concentration value was also not immediately available; the only usable experimental result

seemed to be that presented by Yalin [12], graphically and partially dubious; this, translated by the authors into a computable expression, may be stated as:

$$k = \frac{.4}{2 - 10^{-333.C}} \tag{15}$$

According to the previous expressions, for values of C greater than approximately $5 \cdot 10^{-3}$, k assumes the value 0.2. It is also opportune to specify that the minimum dimension of the mixing length L has always been assumed to be equal to the diameter of the bed sediments. In fact, combining the two indications: L=d for y<5d .

For the constitutive equations relative to the solid phase, reference was made to Bagnold's experimental relations [4], which may be written:

$$\tau_s = 0.013\, \rho_s\, \lambda^2\, d^2 \left(\frac{dU_s}{dy} \right)^2 \tag{16}$$

and

$$\sigma_s \,\mathrm{tg}\, \phi_d = \tau_s \tag{17}$$

where d is the diameter of the bed sediments, ϕ_d is the dynamic friction angle, determined experimentally by Bagnold as being approximately 18° , and λ is the so-called linear concentration, related to the volumetric concentration by the relation

$$\lambda = \frac{1}{(C_*/C)^{1/3} - 1} \tag{18}$$

with C_* the concentration of maximum grain packing (obviously at rest). C_* was set in the calculations as equal to .62 .

It should be remembered that the expressions (16) and (17), although still used today, have been partially modified (in particular the value of the numerical coefficient) by later authors [6][8], and their validity has been circumscribed [10]. In this instance, as the present task has a basically methodological character, it was preferred to refer to Bagnold's original relations: the results were rather comforting, as shown below.

The equations used to calculate the components f_x and f_y of the volume force exerted by the liquid on the grains are still to be specified.

Having otherwise taken buoyancy into account, each dynamic interaction between liquid and sediments is attributed to a relative local motion between liquid and the grains. No account is given for forces related to velocity gradients.

Having set the intensity of said relative motion

$$q = \sqrt{(U_f - U_s)^2 + (V_f - V_s)^2} \tag{19}$$

it is assumed that the following force is exerted on an isolated grain, in the direction of the relative velocity:

$$F = c_D \frac{\pi d^2}{4} \rho_f \frac{q^2}{2} \tag{20}$$

As the number of grains (in this case assumed to be spherical) per unit volume is $6C/\pi d^3$, the expressions

$$f_x = \frac{3}{4} c_{D\text{-eff}} \frac{1}{d} \rho_f\, C\, q\, (U_f - U_s) \tag{21}$$

and

(22) $$f_y = \frac{3}{4} c_{D\text{-eff}} \frac{1}{d} \rho_f \, C \, q \, (V_f - V_s)$$

are thus obtained.

In the expressions (21) and (22) $c_{D\text{-eff}}$ is used instead of c_D in order to take into account the presence of the surrounding sediments.

The classical Rubey's [13] expression appears appropriate for isolated sand grains

(23) $$c_D = \frac{24}{Re} + 2$$

with $Re = qd/\nu$ (ν kinematic viscosity coefficient of the liquid). The correction to take into account the presence of the adjacent particles follows the proposal by Maude & Withmore [14].

(24) $$c_{D\text{-eff}} = c_D \, (1 - C)^{-4.5}$$

Finally, the ϵ expression may be put into the form

(25) $$\epsilon = L \sqrt{\frac{\tau_f}{\rho_f (1-C)}}$$

which was used with the lower limitation for L, as noted above.

4. Some elaborations

Combining equations (4), (6) and (17) aiming to eliminate σ_s and τ_s

(4) $$\frac{d\tau_s}{dy} + (\rho_s - \rho_f) \, g \, C \sin\alpha + f_x = 0$$

(6) $$\frac{d\sigma_s}{dy} + (\rho_s - \rho_f) \, g \, C \cos\alpha - f_y = 0$$

(17) $$\frac{\tau_s}{\sigma_s} = \mathrm{tg}\, \phi_d = \frac{\sin\phi_d}{\cos\phi_d} \quad ,$$

the following relation can be obtained

(26) $$(\rho_s - \rho_f) \, g \, C \sin(\alpha - \phi_d) + f_x \cos\phi_d + f_y \sin\phi_d = 0 \quad .$$

This expression shows that (according to the present scheme) solid transport in current terms (in the sense that the water pulls the solid masses) is possible only as long as the value of α is lower than that of ϕ_d, as the second and third terms of eq. (26) are intrinsically positive in this case.

For steeper channels, the transport phenomenon should be considered in another light (debris flows and/or landslides), in the sense that water, although it may have had an effect on the breaking of the static equilibrium, is no more dynamically essential during the motion.

The equation (26) also permits a modest simplification of the model, as it makes it possible to uncouple the normal stresses from the other variables in the solution of the problem.

The variables τ_f , τ_s , U_f , U_s , C (the proper state variables) and V_f , V_s may in fact be determined separately from σ_f and σ_s by using the following seven equations (five differential and two algebraic)

(3) $$\frac{d\tau_f}{dy} + \rho_f \; g \; \sin\alpha - f_x = 0$$

(4) $$\frac{d\tau_s}{dy} + (\rho_s - \rho_f) \; g \; C \; \sin\alpha + f_x = 0$$

(8') $$\frac{dC}{dy} - \frac{V_s}{\epsilon} C = 0$$

(13') $$\frac{dU_f}{dy} - \frac{\sqrt{\tau_f/\rho_f(1-C)}}{L} = 0$$

(16') $$\frac{dU_s}{dy} - \frac{\sqrt{\tau_s/(.013\;\rho_s)}}{\lambda\; d} = 0$$

(10) $$(1-C)\; V_f + C\; V_s = 0$$

(26) $$(\rho_s - \rho_f)\; g\; C\; \sin\,(\alpha - \phi_d) + f_x \cos\phi_d + f_y \sin\phi_d = 0$$

together with the expressions shown above for f_x, f_y, ϵ , L and λ.

5. Boundary conditions and solution

In order to get a solution of the proposed model, it is necessary to specify the five boundary conditions to be added to the five differential equations above.

Obvious physical reasons suggest that the values of τ_f and τ_s must be set at zero at the free surface (y=h).

As for y<0 the sediments are at rest and the water is seeping parallel to the channel bed, both $V_f(0)=0$ and $V_s(0)=0$ must be true as a consequence of the mass balance near y=0. It is thus possible to calculate at y=0 the value U_{fo} of U_f-U_s which verifies eq. (26) with V_f-V_s null: this condition replaces the continuity condition near the bed, which cannot be directly expressed in terms of the proper state variables.

Two additional conditions representing the dynamics of the mobile-immobile sediment interface must be established.

One is $U_s(0)=0$, upheld by the fact that Bagnold's scheme for inter-granular stresses appears valid down to the immediate vicinity of the immobile sediments, and that a finite velocity would cause infinite stresses (σ and τ) not compatible with the finite mass of sediments in motion. The condition $U_f(0)=U_{fo}$ follows from the above continuity condition. This value is generally different from the actual seepage velocity; this lack of "adherence" of the liquid phase near the bed seems to the authors to be acceptable due to the presence near the bed of a thin layer where the effective viscosity is lower than represented by the above equations and for the small value of the difference.

The remaining boundary condition must represent the compatibility of the stresses transmitted to the bed by the static equilibrium of the sediments present in it.

With reference to figure 2, and considering the first layer of sediments at rest, it is observed that both normal and tangential stresses

$$\sigma_s(0) \quad \text{and} \quad \tau_f(0) + \tau_s(0)$$

act on its upper face while both the normal and tangential stresses:

$\sigma_s(-d)$ and $\tau_s(-d)$

act on the lower face.

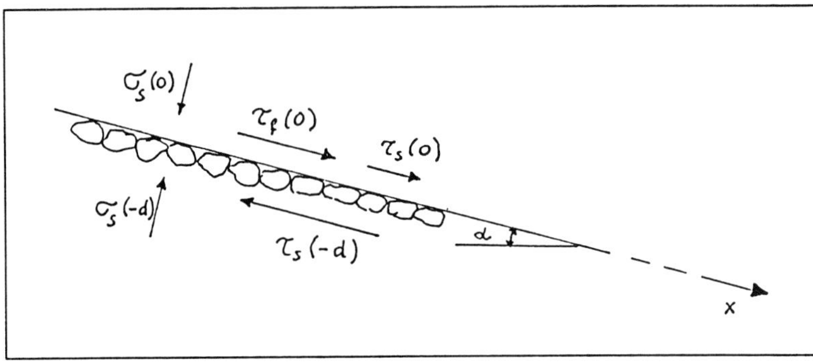

Fig. 2

Considering the equilibrium of this layer, assumed to have the maximum packing concentration C_*, the following relations hold:

$$\sigma_s(0) - \sigma_s(-d) + (\rho_s-\rho_f)C_* \, gd \cos\alpha = 0 \tag{27}$$

$$\tau_f(0) + \tau_s(0) - \tau_s(-d) + [(\rho_s-\rho_f)C_*+\rho_f] \, gd \sin\alpha = 0 \tag{28}$$

Between $\sigma_s(0)$ and $\tau_s(0)$ the relation

$$\frac{\tau_s(0)}{\sigma_s(0)} = \mathrm{tg}\, \phi_d = \mathrm{tg}\, 18^\circ = 0.32 \tag{29}$$

continues to be valid, and between the quantities $\sigma_s(-d)$ and $\tau_s(-d)$ the corresponding static relation must be valid:

$$\frac{\tau_s(-d)}{\sigma_s(-d)} = \mathrm{tg}\, \phi_s = \mathrm{tg}\, 31^\circ = 0.60 \tag{30}$$

The static friction angle ϕ_s for immersed sediments is assumed to be about 31°.

The combination of the eqs. (27), (28), (29) and (30) makes it possible to derive the desired boundary condition:

$$\tau_s(0) = \frac{\mathrm{tg}\phi_d}{\mathrm{tg}\phi_s} \left[\tau_f(0) + \tau_s(0) - g\, d\, (\rho_s-\rho_f)\, C_* \frac{\sin(\phi_s - \alpha)}{\cos \phi_s} + \rho_f \, g\, d \sin \alpha \right] \tag{31}$$

which shows how only a portion of the total stress is transferred to the bed by the sediments in motion.

It must however be observed that even the numerical integration of the set of five first order ordinary differential equations (strongly non-linear) with the boundary conditions

- $\tau_f(h) = 0$
- $\tau_s(h) = 0$
- $\mathcal{F}(\tau_f(0),\tau_s(0))=0$, eq. (31)
- $U_f(0) = U_{fo}$, derived as described above
- $U_s(0) = 0$

is not trivial.

Nonetheless, the solution was conducted in a series of cases referred to below, treating the problem as an initial values problem rather than a boundary one.

More specifically, rather than setting the zero values of τ_f and τ_s at h, an arbitrary (large enough) value of the total shear stress was set at the bed, using eq. (31) for determining the separate values of $\tau_f(0)$ and $\tau_s(0)$.
Then a value of the bed concentration C_o was determined, by trial and error, in order to have a solution which makes τ_f and τ_s equal to zero for the same value of y, which was then assumed as h.
This value C_o was used as the last necessary initial condition.

It must be added that Bagnold's equation (16') is abandoned at that distance from the bed for which the longitudinal component of the relative liquid-sediment velocity becomes zero, keeping $U_s=U_f$ in the above layer.
This was generally the case for concentration values which were in good agreement with the limits recognized to Bagnold's scheme.

Integration was alway performed with a step of d/10.

6. Overall results.

Integrations were performed according to the procedure described above for the cases listed in table 1 along with the used parameters and the overall results achieved.

The cases considered were chosen in order to verify the model over a wide range of situations, although not all are completely realistic, and in some cases they go beyond the hypoththesis assumed.

Figure 3 illustrates the representative points of the cases considered on the plane U_*d/ν and U_*/w (with w indicating the sedimentation velocity of isolated grains): it appears evident that no bed forms of the lower regime could be present.

Table 1

	Data			Results				
Case	α	d	τ_o/τ_{cr}	C_o	h	q_{liq}	q_{sol}	C(Chezy)
	rad	m	_	m^3/m^3	m	$m^3/m/s$	$m^3/m/s$	_
A-1	.100	.100	10.	.272	0.84	2.65	.118	3.22
A-2	.100	.100	20.	.334	1.61	10.84	.806	4.85
A-3	.100	.100	30.	.352	2.30	22.93	2.37	5.87
B-1	.010	.010	10.	.188	0.89	2.85	.00263	10.79
B-2	.010	.010	20.	.302	1.76	10.51	.0287	14.19
B-3	.010	.010	30.	.332	2.62	22.04	.106	16.34
C-2	.001	.001	20.	.239	1.15	2.24	.000236	18.42
C-3	.001	.001	30.	.295	1.70	4.54	.00091	20.62
C-4	.001	.001	40.	.319	2.22	7.45	.0023	22.30

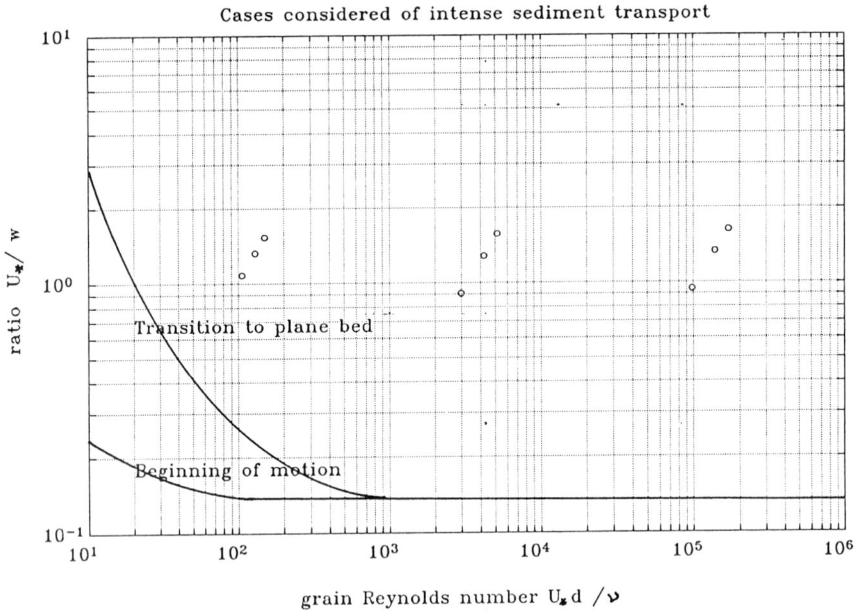

Fig. 3

Figure 4 shows the value of bed concentration C_o versus the Shields parameter $\tau_o/(\rho_s-\rho_f)gd$: a qualitative agreement with the theory of Engelund & Fredsoe [15] is evident.

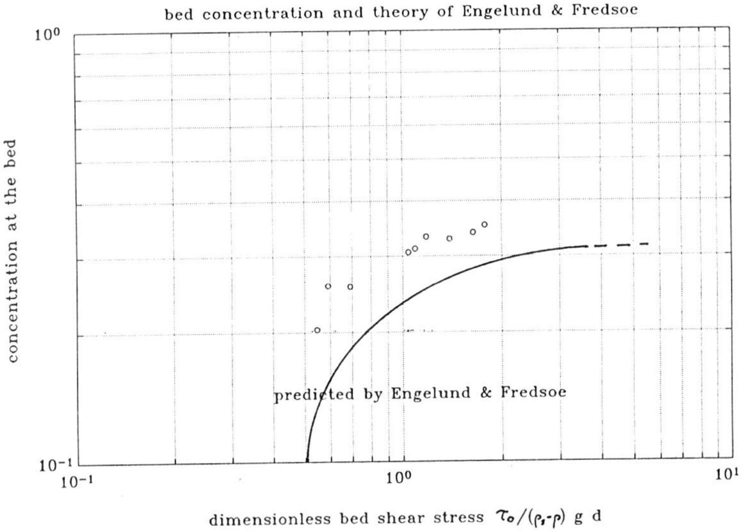

Fig. 4

Figures 5 (a and b) show the comparison between the values of solid transport obtained here and those supplied by the classical formulas of Meyer-Peter & Müller and

Engelund & Hansen in the same conditions (α, h, d).

A satisfying agreement with both formulas is obtained, as the most significant differences occur where these formulas are less reliable, that is for larger solid transport in the case of the Meyer-Peter & Müller formula, and for the lower values in the Engelund & Hansen's formula.

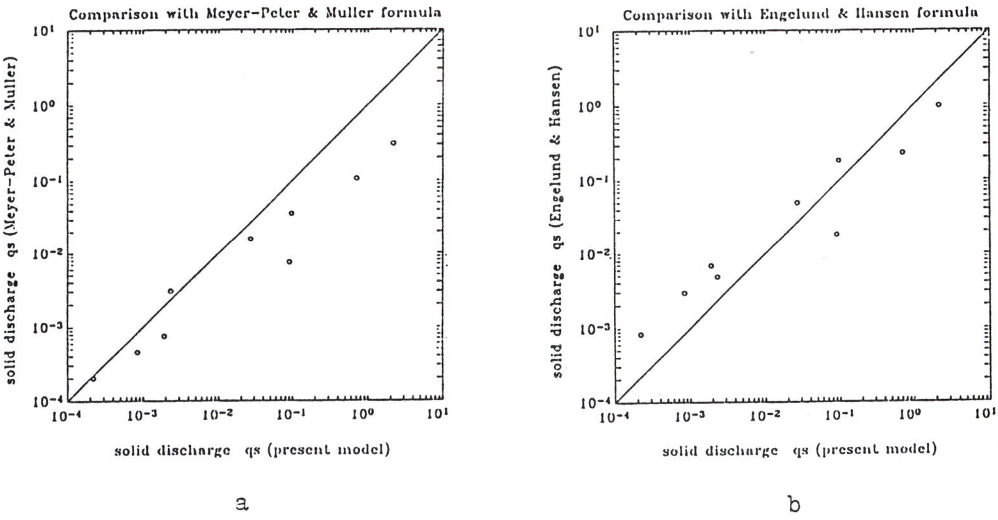

Fig. 5

Figure 6 (a and b) illustrates the comparison between the values here obtained for Chezy's nondimensional coefficient (x coordinate) and those obtainable from the following formulas (y coordinate):

- a) the formula suggested by Graf for mountain beds [16],
- b) the formula proposed by Wilson [3].

The expression $C = 3.35 \ln[h/(3.25d)]$ represents the best interpolation of the calculation results with a formula of the type used by Graf.

As interesting as they may be, the results require further examination.

It does not seem appropriate to dwell any further on the analysis of the overall results, as it is sufficient here to point out that they show a good internal coherence and a reasonable agreement with empirical formulas, legitimizing, at least in its general set-up, the model.

7. Detail results

The detail results which the model provides appear to be of considerable interest, especially near the bottom, where experiences have shown to be of prohibitive difficulty.

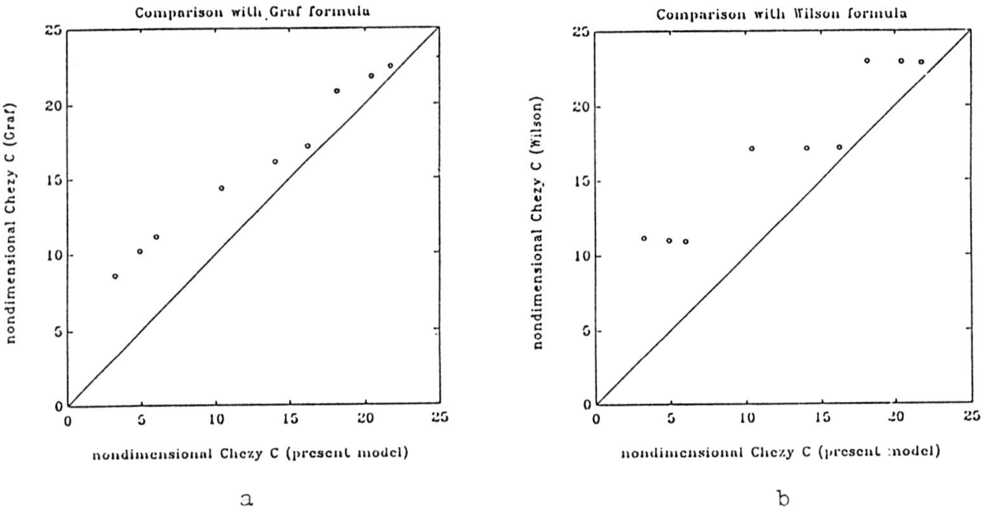

Fig. 6

Figure 7 illustrates, for an intermediate case (B-2), the distribution of the shear stresses in the liquid phase and solid phase as they result from the model: it is evident that part of the total stress is supported by the sediments (and transmitted to the immobile bed) in the form of inter-granular stresses for a layer very close to the bottom (only a few grains thick).

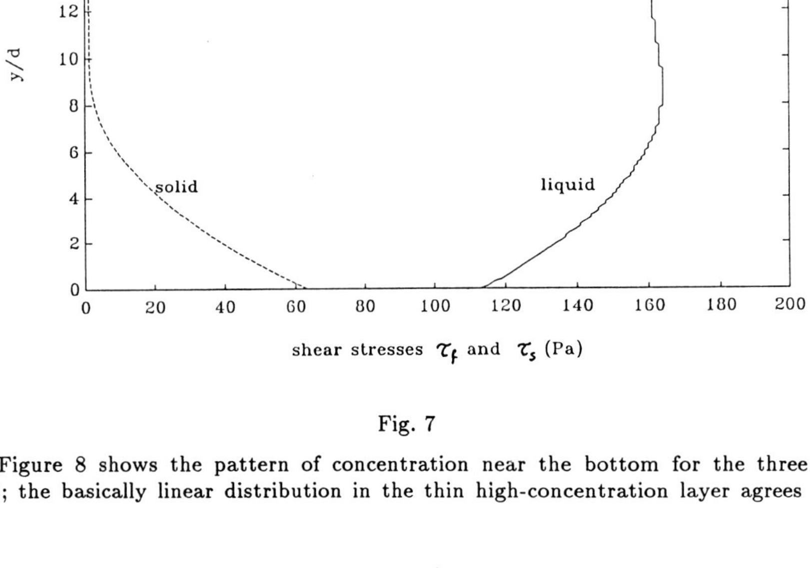

Fig. 7

Figure 8 shows the pattern of concentration near the bottom for the three cases in series B; the basically linear distribution in the thin high-concentration layer agrees with the experimental results cited by Wilson [2].

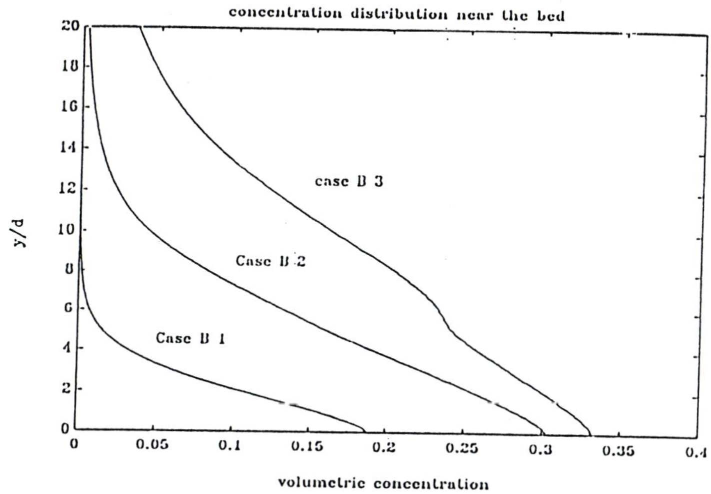

Fig. 8

Figure 9 illustrates the pattern of concentration over the entire depth of the stream for case C-4; the classical profile of all known experimental results is obtained.

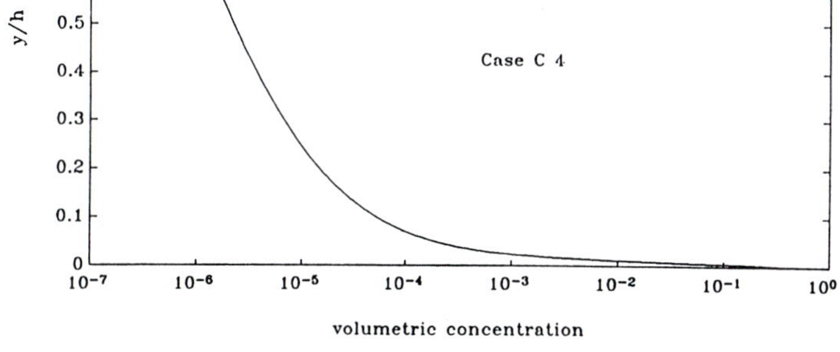

Fig. 9

Figure 10 illustrates the pattern, again for case B-2 and near the bottom, of the longitudinal components U_f and U_s of the velocity in the liquid and solid phases, while figure 11 represents the pattern of the corresponding components V_f and V_s: the latter appears obviously negative and tends toward the value of free sedimentation, when the quota increases, and simultaneously the concentration decreases.

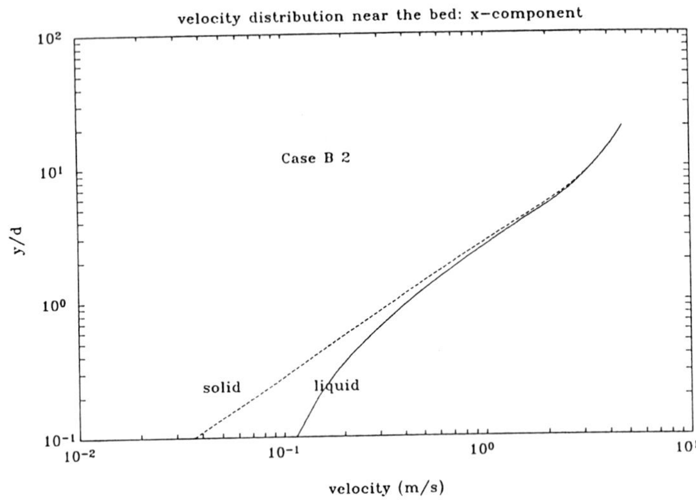

Fig. 10

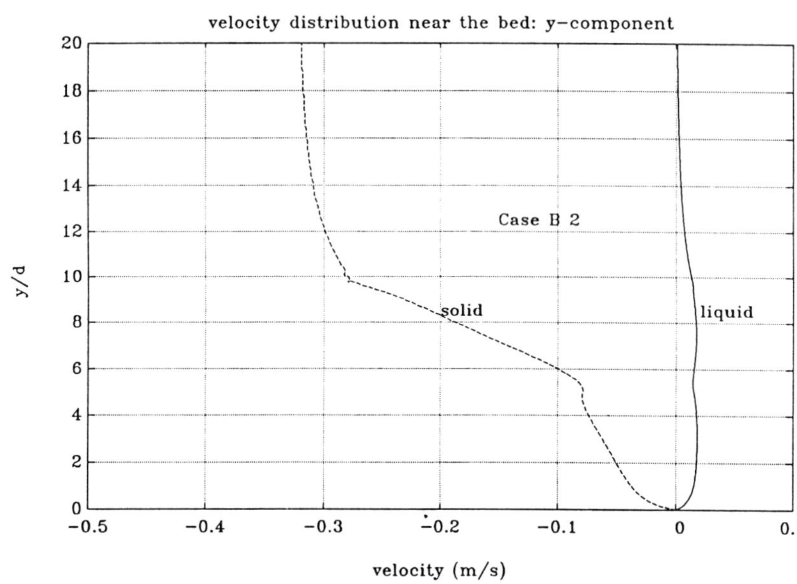

Fig. 11

8. Conclusions

A possible, rather general mathematical model to represent the phenomenon of solid transport by a free surface current in uniform motion has been presented.

The model is currently descriptive-interpretive, but it seems that it could also become predictive, once the necessary comparisons with the experimental data have been performed.

The applications provided as examples have given sufficiently credible results, and have shown some probable new aspects of the phenomenon.

Conscious that the proposed model is susceptible to improvements in all aspects, the authors hope to have contributed to the development of physical comprehension and mathematical modeling of the solid transport phenomenon.

References

[1] Wilson, K. C. (1966). "Bed-load transport at high shear stress" J. Hydr. Div., ASCE, 92 (6).
[2] Wilson, K. C. (1987). "Analysis of bed-load motion at high shear stress" J. Hydr. Engrg., ASCE, 113 (1).
[3] Wilson, K. C. (1989). "Mobile-bed friction at high shear stress" J. Hydr. Engrg., ASCE, 115 (6).
[4] Bagnold, R. A. (1956). "The flow of cohesionless grains in fluids" Phil. Trans. Roy. Soc. London, A 249.
[5] Jenkins, J. T. & Savage, S. B. (1983). "A theory for the rapid flow of identical, smooth, nearly elastic, spherical particles" J. Fluid Mech., 130.
[6] Savage, S. B. & Sayed, M. (1984). "Stresses developed by dry cohesionless granular materials sheared in an annular shear cell" J. Fluid Mech., 142.
[7] Hanes, D. M. & Bowen, A. J. (1985). "A granular-fluid model for steady intense bed-load transport" J. Geoph. Res., 90 (C5).
[8] Lun, C. K. K. & Savage, S. B. (1987). "A simple kinetic theory for granular flow of rough, inelastic, spherical particles" J. Appl. Mech., 54 (3).
[9] Campbell, C. S. & Gong, A. (1986). "The stress tensor in a two-dimensional granular shear flow" J. Fluid Mech., 164.
[10] Pasquarell, G. C. , Ackermann, N. L. , Shen, H. H. & Hopkins, M. A. (1988). "Collisional stress in granular flows: Bagnold revisited" J. Engrg. Mech., ASCE, 114 (1).
[11] Valiani, A. (1988) "An open question regarding shear flow with suspended sediments" Meccanica, 23 (1).
[12] Yalin, M. S. & Finlayson, G. D. (1972). "On the velocity distribution of the flow carrying sediment in suspension" in "Sedimentation, Symposium to honor Prof. H.A.Einstein" , Shen Editor and Publisher.
[13] Rubey, W. (1933). "Settling velocities of gravel, sand and silt particles" Amer. J. Sci. , 225.
[14] Maude, A. D. & Whitmore, R. L. (1958). "A generalized theory of sedimentation" Brit. J. Appl. Phys., 7 (12).
[15] Engelund, F. & Fredsoe, J. (1982). "Hydraulic theory of alluvial rivers" Advances in Hydroscience, 13.
[16] Graf, W. H., Cao, H. H. & Suszka, L. (1983). "Hydraulics of steep mobile bed channels" XX IAHR Congress, Moscow, vol. VII.

MODEL INVESTIGATIONS ON THE SEDIMENT TRANSPORT OF A LOWER ALPINE RIVER

W.Bechteler, G.Vogel and H.J.Vollmers

Laboratory for Hydraulics and Hydraulic Structure
University of the Armed Forces, Munich, Germany

ABSTRACT

In this paper the difficulties will be demonstrated in estimating the sediment transport in mountain rivers which marked hydrological events. In such cases model investigations with movable bed are often the only way to find physical explanations. Firstly the situation of the Weissach river is described and than after a short similarity consideration the most important model devices are presented.

The model could not be calibrated because there were no field measurements available. Therefore a discharge/transport relation was developed. The main result of the investigations was the very stable behaviour of the model against changing of the slope and the constructional measures realized in the course of the renaturalization of the Weissach river.

1. Introduction

In recent years it has been tried to renaturalize rivers that had been previously regulated in a schematic or monotonous way. The constructional steps that are necessary to achieve the goal of "renaturalization" may influence both water levels and sediment transport behaviour of a river. In the case of mountain rivers the assessment of the transport behaviour not only depends on the relatively steep slopes but also on the discharge characteristics. The example presented in this paper is meant to show the difficulties in realizing a model investigation as well as in interpreting the result when sufficient field measurements are not available.

2. Situation

The Weissach is the main tributary of lake Tegernsee. In the western part of the village of Rottach-Egern it pours its waters into the lake after a course of about 20 km length. The whole reach downstream of the weir at km 2.841 (Fig. 1) can be regarded as the inflow section. At the weir a channel diverges up to 2.5 m^3/s of discharge for the supply of a small hydropower plant. The tailrace channel of the plant flows directly into lake Tegernsee.

The river inflow section lies between dikes that are overgrown with trees and bushes and is rather straight in order to carry floods fast and without problems towards the lake. At the beginning of the century a trapezoidal cross section had been chosen for the river. Its dike crests are above the mostly built-on adjacent area.

The simplified longitudinal section (Fig. 1) shows that there is a change of slopes at km 1.695 as well as at km 0.407. Moreover, there are numerous ground weirs which react as drop structures for small discharges. This means that the slopes given are averaged values. Changes of slopes may cause changes of the transport behaviour. Therefore in both cases a reduction of the transport capacity i.e. sedimentations can be expected. As a matter of fact most of the material remains downstream of km 0.407 from where it is dredged for commercial use.

In the course of the constructional works for renaturalization of the Weissach river ground weirs had been repaired and the monotony of the river bed interrupted by groups of boulders. In August 1985 shortly after the conclusion of these works a flood wave caused considerable sedimentation in the reach of the first change of slopes at km 1.695. As a possible future flood wave could not be expected to carry away the settled material, but rather on the contrary a raising of water levels and overflowing of dikes had to be feared, the material was removed pretty soon. Suspicions saw the sedimentations as a consequence of the just finished constructional works for renaturalization.

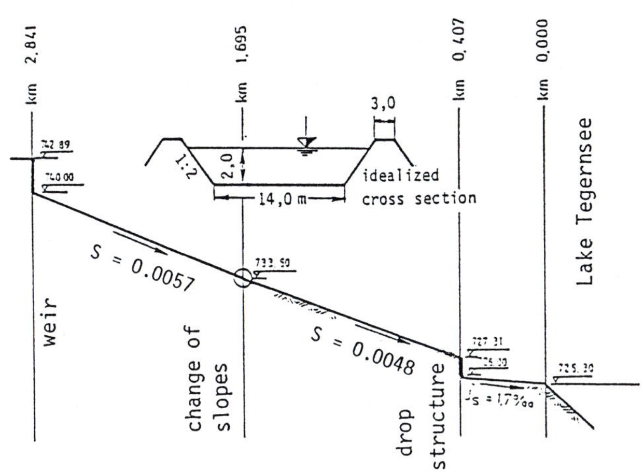

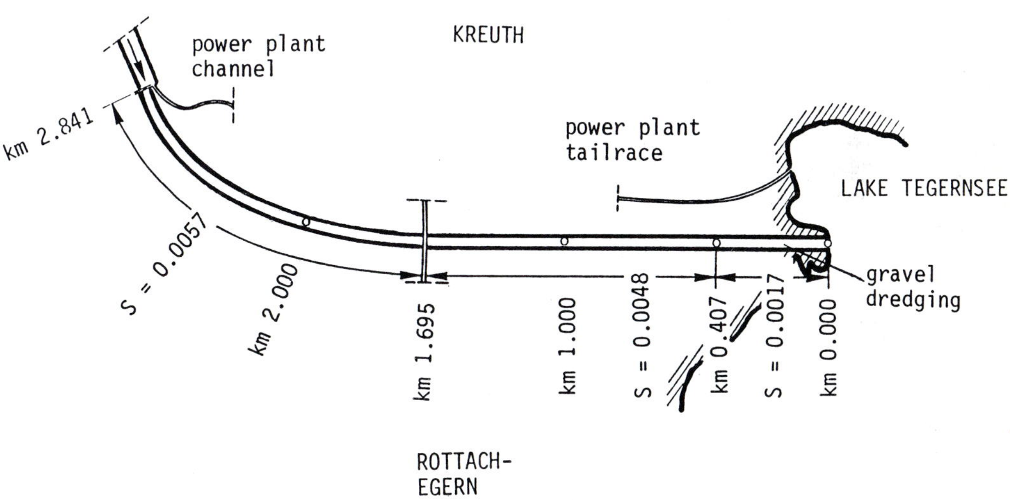

Fig. 1: Simplified plan and longitudinal and cross sections of the Weissach river

3. Hydrology and Morphology

The river catchment area at the Oberach gauging station just upstream of the considered reach is 92 km^2. The discharge behaviour is mainly influenced by the mountains and exceptional meteorological events. The ratio between medium and peak discharges is about 1:25. The maximum discharge ever was observed on September 8, 1978. For rivers like the Weissach it is important to consider the pattern of the flood wave, which is responsible for the transport balance; i.e. from 80 up to 90% of the annual bed load may be transported in the course of a single flood wave. Two representative waves are given in Fig. 2. Fig. 3 shows the discharge duration curve for the years 1958 through 1986.
Numerous samples had been taken to analyse the bed material. Using the averaged distribution curve (Fig. 4) an averaged grain diameter of 1.43 cm could be obtained.

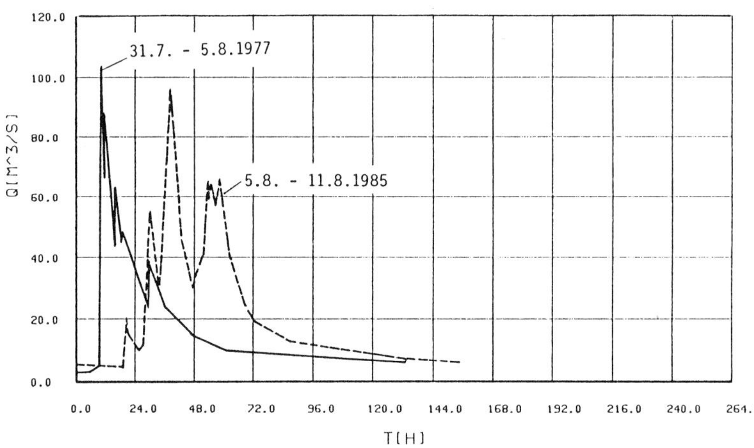

Fig. 2: Two flood events of the Weissach river

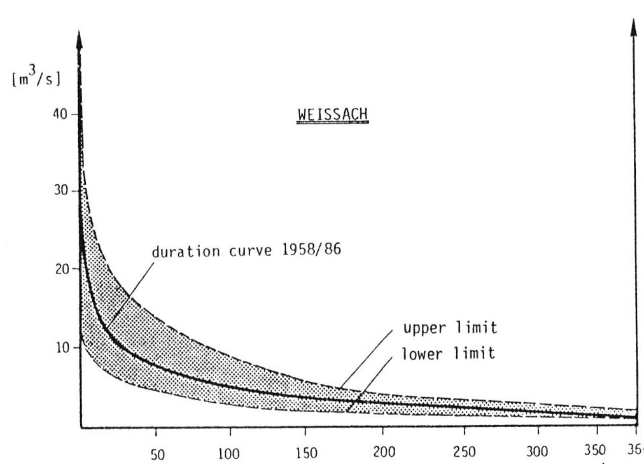

Fig. 3: Duration curve for the years 1958 through 1986

4. Similarity Consideration

In the case of great grain diameters of the bed material so-called "sand models" can be realized, where the natural material is reduced geometrically according to the model scale. The prerequisite for this is that for both natural and model discharges a hydrodynamically rough flow regime prevails. By this the range of applications for such models is automatically limited, because they need small scale numbers and in consequence become rather spacious. It is important that the initial conditions for the sediment transport show the same Froude numbers in the model as in nature. The model scale chosen was 1:20. The model sand is rather uniform to avoid the transport of suspended load which in reality does not take place either (Fig. 4).

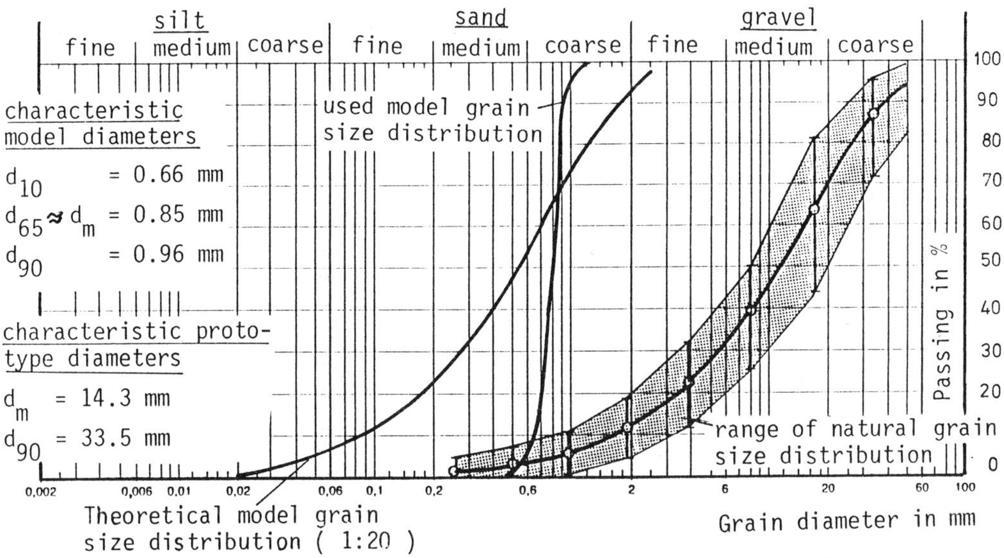

Fig. 4: Bed material distribution curves

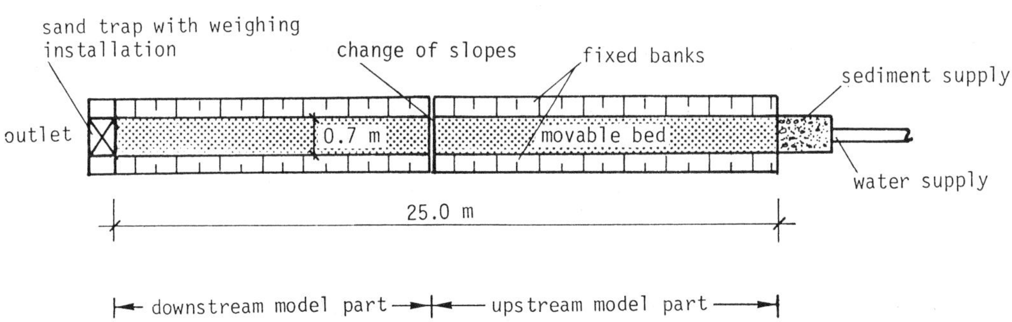

Fig. 5: Schematic plan of the model

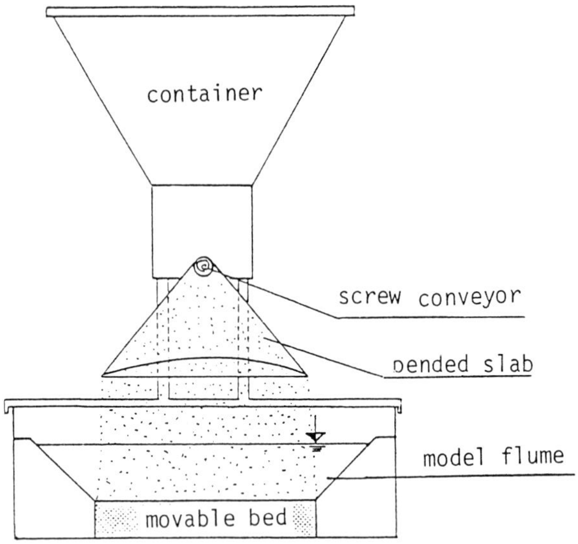

Fig. 6: Sand supply at the model inflow

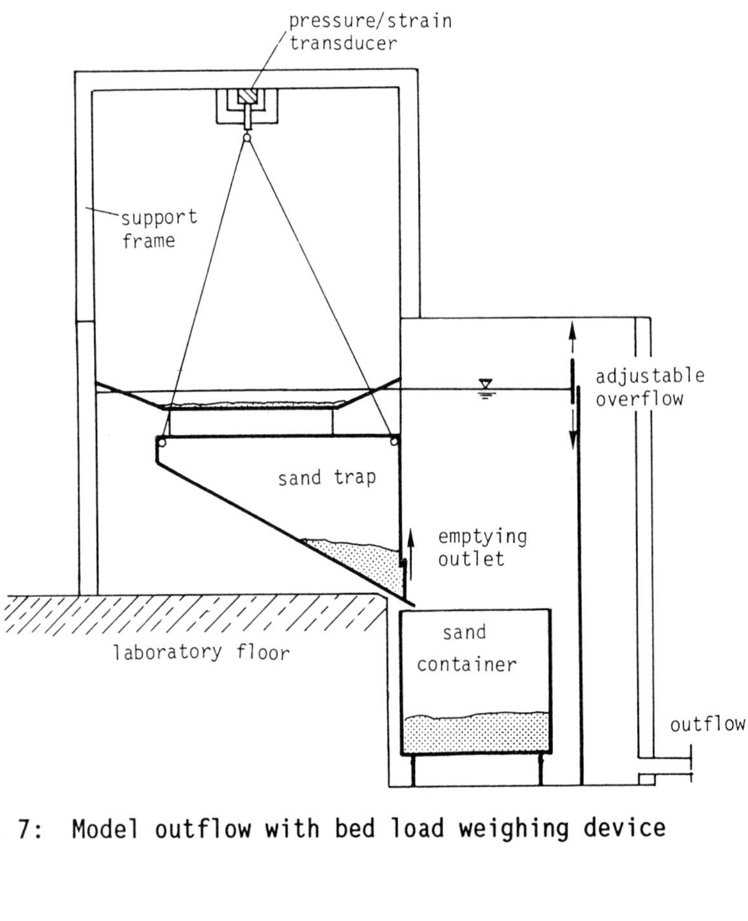

Fig. 7: Model outflow with bed load weighing device

5. Model Description

Special devices had to be developped for the sand supply at the model inflow and the sand trap at the outflow to provide a mass balancing as precise as possible. The input material is stored in a container holding some 150 kg. At its bottom a screw conveyor with varying velocity transports the dry sand onto a bended slab from where it drops almost uniformly into the model (Fig. 6). The maximum input rate is 138 g/s. Special attention was given to the construction of the "sand trap" at the outflow end of the model. Free outflow without any backwater influence and continuous registration of the transported bed load material had to be guaranteed. Fig. 7 shows the important features of the whole construction. The sand trap is a balance pan suspended from a pressure/strain transducer which on its turn is fixed to a stable support frame. The filled sand trap is emptied into another container which can be removed by the laboratory crane. The adjustable overflow regulates the model outflow water level.
The rest of the measuring installation consisted of a flowmeter before the inflow, four water level gauges alongside the model and a Delft bottom profile follower for the registration of bed profiles. All measured data are transferred directly to a PC for storing and evaluation purposes.

6. Model Investigations

The model could not be calibrated, because there were no field measurements available. Thus it was only possible to find a discharge/-transport relationship for the model representing an approximate state of balance. For this purpose the model was used with a continuous slope of S = 0.0048 without any constructional elements in its course. In order to get an idea about the expected discharge/transport dependence the bed load transport was calculated using the MEYER-PETER/MÜLLER formula.
Measuring results for steady discharges showed the usual characteristic deviations for such transport processes: the transport is significantly influenced by turbulent oscillations. Fig. 8 shows a comparison of registered data (5-minute averages) and values calculated from different bed load or total load formulas. Especially for greater water depths it can be seen that the calculation results differ from each other by factors up to 10.
A further general problem are the transport bodies that appear in any discharge event in a movable bed flume. Due to the lack of physical knowledge about their causes we still have to rely on empirical - even though mostly dimensionless - criteria to at least make predictions on the basic form of the bed configurations. According to Fig. 9 both in reality and in the model "dunes" had to be expected which proved to be correct. As especially the height of dunes depends on the water depth, the observed bottom configuration at the end of a flood wave always showed a flattened form of the dunes. Moreover, lateral reflexions

which are typical for straight flumes played an important role in the build-up of bed forms. Corresponding configurations in the model are shown in Fig. 10.

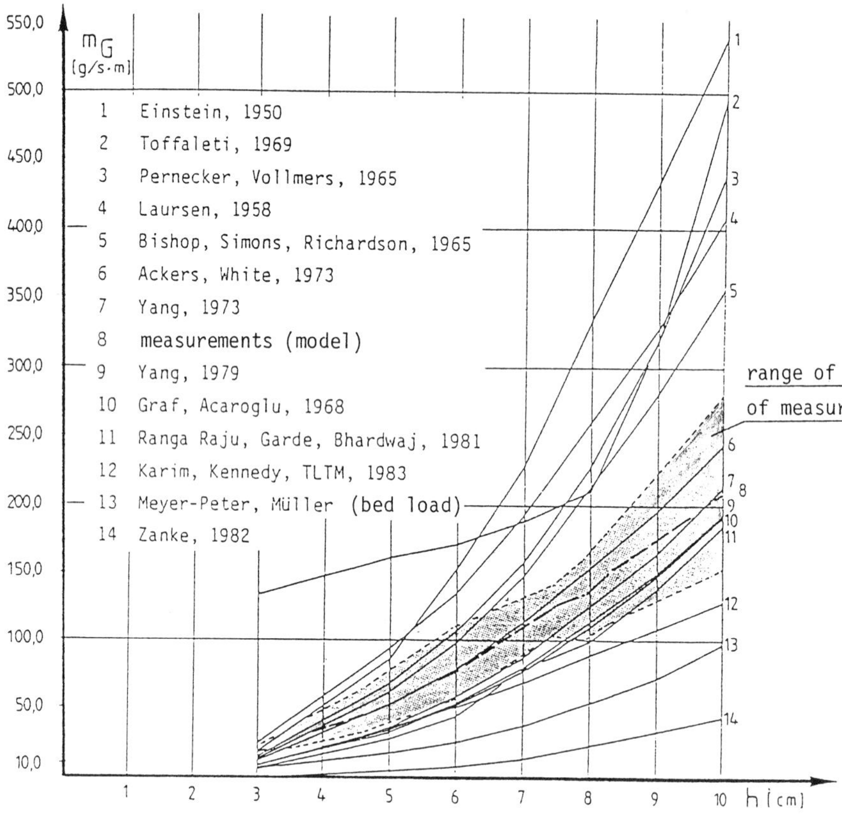

Fig. 8: Sediment transport rate versus water depth for the model. Comparison of registered and calculated values.

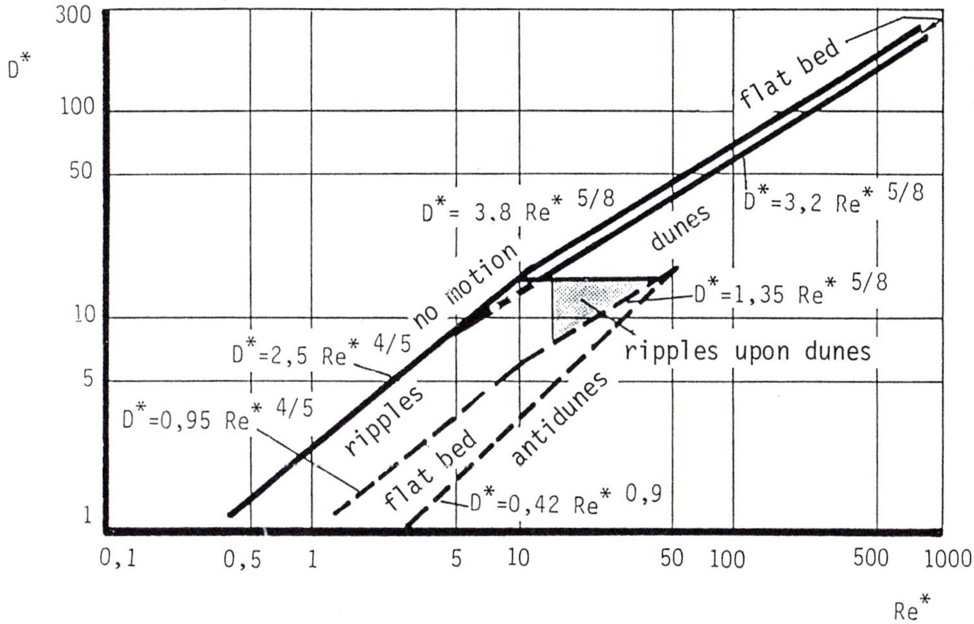

Fig. 9: D^* versus Re^* transport body criterion after BONNEFILLE/PERNECKER (VOLLMERS/GIESE 1970)

Fig. 10: Typical bottom configurations at the end of a test

6.1 Experiments with continuous slope (S = 0.0048)

The first series of tests produced a rather stable discharge/transport relationship. A change of the sand material input without the corresponding change of the water inflow resulted in erosion or deposition. The reaction to constructional elements near the banks as well as across the flume (ground weirs) was almost negligible. The comparison of three experiments with different initial constructional situations and identical water and sand inputs proves this statement. The following conditions were investigated:

a) No constructional elements in the flume
b) Ground weirs at bottom level of the flume
c) Ground weirs at bottom level plus subsequent drops of 3 cm (Fig.11)

Water and sand input curves were symmetrical histograms (cf Table 1). For the increasing first half transport rates are even to or above the input rates, whereas for the declining part they are obviously smaller. Even though case a (no constructional elements) shows the greatest total transport no clear tendency can be derived from this fact (Table 1). Due to the fixing of the bottom level by the ground weirs transport bodies are not so well marked for cases b and c. Periodical scouring and deposition occured at the weirs because of the batchwise bed load transport.

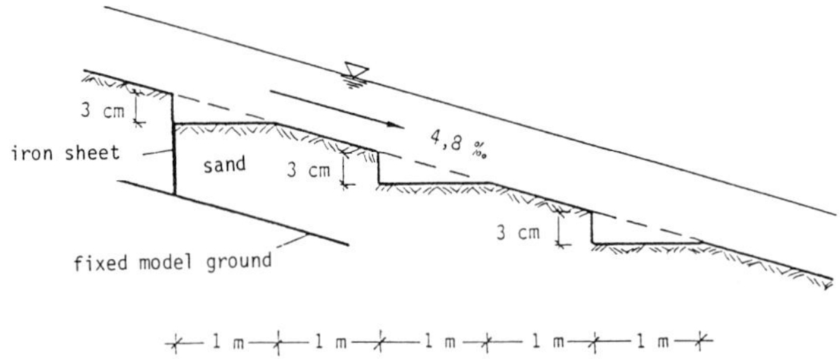

Fig. 11: Bottom modification by drops (case a: initial situation)

duration	Q	input	bed load transport		
			case a	case b	case c
min	l/s	g/s	g/s	g/s	g/s
10	12	15,0	16,4	15,2	10,9
10	20	29,9	34,6	32,5	32,8
20	28	48,5	50,3	50,9	53,6
20	41	80,4	74,3	80,1	70,3
20	28	48,5	42,8	35,8	35,1
10	20	29,9	21,2	19,2	19,7
10	12	15,0	6,7	9,3	6,2
100 min		267 kg	248 kg	246 kg	233 kg

Table 1: Comparison of three experiments with identical water and sand inputs and different initial constructional situations

6.2 Experiments with change of slopes (S = 0.0048 and 0.0057)

Again a discharge/transport relationship was established (Fig. 12). It practically coincides with the one for continuous slope. On this basis now significant changes of both discharge and sand input were investigated. The initial test I used a symmetrical flood wave with a step length of 30 min and the corresponding sand input rate, which is verified by the actually registered transport rate. Duration time was 5.5 hours (Fig. 13 and Table 2).

Test II left the discharge characteristic unaltered, whereas the variable sand supply was replaced by a constant value (60.4 g/s) representing nearly the average value of test I (63.3 g/s). Registered transport rates differ from the previous ones, but are basically discharge dependent (Fig. 13 and Table 2).

During test III the discharge was kept constant (= average value of test I) and the sand supply (due technical reason) slightly increased in comparison to test I. Transport rates are nearly constant for the first half and increase moderately towards the end of the test.

The interesting thing is, that the total loads and in consequence the average transport rates are the same in all tests. As the input varies there is a 60 kg erosion (1196 - 1256 = -60) in test II and a 84 kg deposition (1358 - 1274 = +84) in test III.

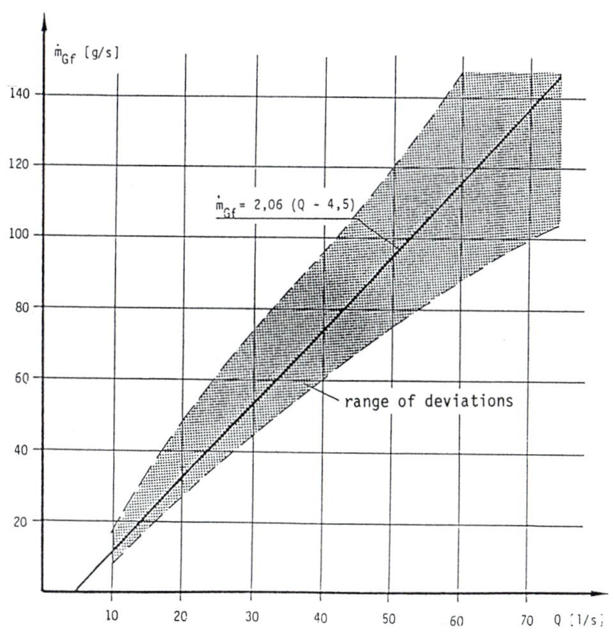

Fig. 12: Sediment transport rate versus discharge for the model

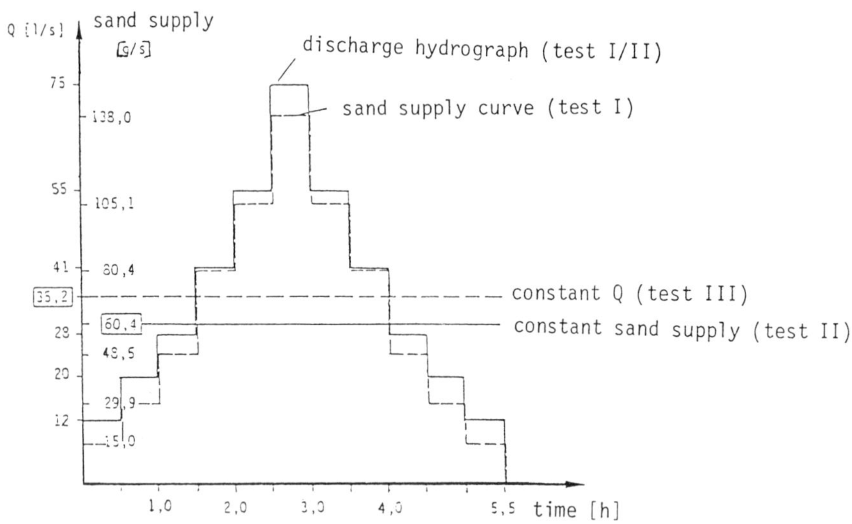

Fig. 13: Discharge hydrographs and sand supply curves

	I initial test			II constant sand supply			III constant discharge		
duration	Q	input rate	trans-port	Q	input rate	trans-port	Q	input rate	trans-port
min	l/s	g/s	g/s	l/s	g/s	g/s	l/s	g/s	g/s
30	12	15,0	6,8	12	60,4	15,0	35	16,2	60,7
	20	29,9	30,8	20		36,0		31,9	61,4
	28	48,5	49,2	28		58,9		53,5	60,2
	41	80,4	83,9	41		88,9		88,6	59,8
	55	105,1	110,5	55		127,4		118,0	59,3
	75	138,0	150,0	75		152,0		138,0	60,0
	55	105,1	103,2	55		96,8		118,0	59,3
	41	80,4	71,4	41		59,4		88,6	65,8
	28	48,5	43,2	28		37,1		53,5	66,2
	20	29,9	30,0	20		18,8		31,9	76,5
30	12	15,0	15,9	12	60,4	7,4	35	16,2	78,3
averages	35	63,3	63,2	35	60,4	63,4	35	68,6	64,3
Σ 5,5 h		1252 kg	1251 kg		1196 kg	1256 kg		1358 kg	1274 kg

Table 2: Compilation of test results

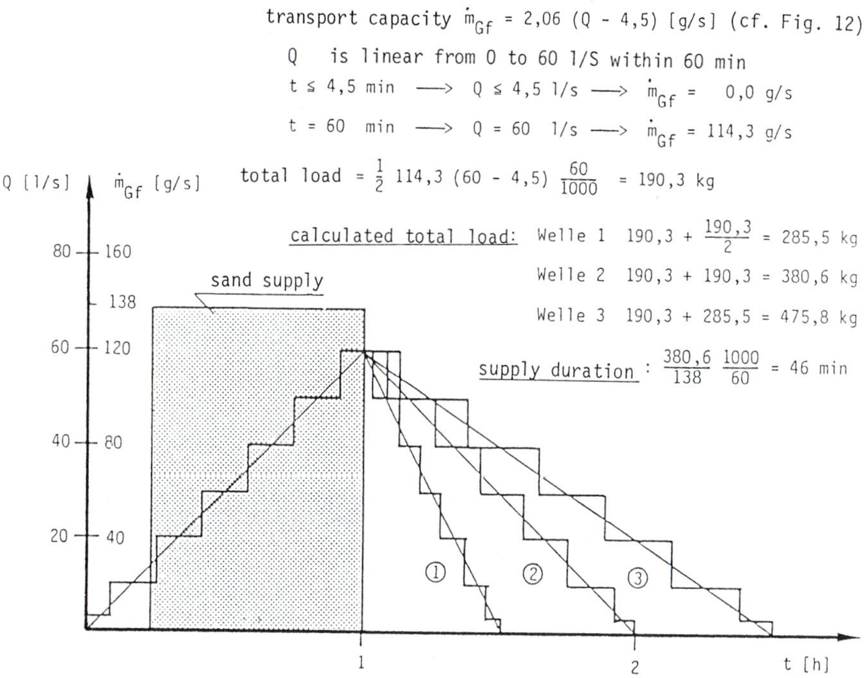

Fig. 14: Three discharge hydrographs and sand supply curve

6.3 Natural River Situation

For the concluding test series the model was prepared according to the natural situation in the river. A first experiment simulating an actual flood wave event did not produce significant results with respect to sedimentation near the change of slopes. Therefore it was investigated if the decreasing part of a flood wave shows any special influence on sedimentation processes after a very concentrated sand supply during the increasing part of the wave. Idealized triangular flood events, again approximated by histograms, were used to allow systematic comparisons.

Fig. 14 presents both the three discharge hydrographs and the time-dependent sand supply. The hydrographs are identical for the first hour. After having reach the peak they decline differently. The sand input happens for 46 min during the increasing part of the waves at its maximum rate of 138 g/s. This corresponds to an even transport balance of wave 2 calculated according to Fig. 12.

A summary of the registered mass values is given in Table 3. First of all the numbers show the good reproducibility of the tests, as in all three cases during the increasing part of the hydrograph identical masses have been transported (187, 192, 181 kg). This proves that re-

sults registered in physical movable bed models are not random, but rather stable. In our case they, moreover, even coincide with the precalculated value of 190,3 kg (cf Fig. 14).
For the declining part the rapid decrease of wave 1 shows a significant influence. Due to the unbalance between the sand supply and the actual transport capacity of the wave a total deposition of 143 kg is registered. Wave 2 shows an almost even balance and for wave 3 the balance is negative, which means some erosion took place, as its total transport capacity went beyond the sum of the sand material provided at the inflow.

Besides the overall weighing balance additional bottom registrations were performed before and after these experiments in order to find out if and where transport material had been deposited around the change of slopes (16 m registration area upstream and downstream). The calculated volume differences are converted into masses by means of the deposition density 1,5 t/m^3 and listed as well in Table 3.
The bottom registrations inform most of all about the local distribution of deposition and erosion. They were based on a 4 cm lattice in the model, which is 0.8 m in natural size. For the graphical representation of Fig. 15 the 4 cm stripes across the bottom width (70 cm) were lumped by tens and the differences of their volumes before and after the tests calculated.
Now an interpretation of the longitudinal development of the transport process is possible taking into account the sand input characteristics and the duration of the hydrograph decline.

For <u>wave 1</u> it becomes obvious that the sediment material added massively for 46 minutes could not be transported any further, because of the rapid recession of the flood wave, and had been deposited as far downstream as ground weir II. Sedimentation heights reach slightly above 40 cm.

For <u>wave 2</u> the situation looks somewhat more balanced due to the longer lasting transport process. The bottom is not raised as much as before; maximum deposition is about 30 cm.
The transport capacity of wave 2 had been planned to balance the sand input, which did not work out completely. As the total balance results in a 44 kg deposition and the bottom registration shows 101 kg, the difference of 57 kg must have been eroded from the not registered parts of the model.

For <u>wave 3</u> the recession time after having reached the peak is so lang that the river almost finds back to a balanced bottom level. There is still a minor elevation around the change of slopes; the slight erosion at the beginning, however, indicates the lack of sand supply, which is demonstrated as well by the weighing balance of -36 kg (Table 3).

flood wave	input [kg]	transport [kg]			deposition [kg]	
		in-creasing	de-clining	total	weighing	bottom registration
1	381	187	51	238	143	148
2	381	192	145	337	44	101
3	381	181	236	417	-36	29

Table 3: Transport balance of the three discharge hydrographs

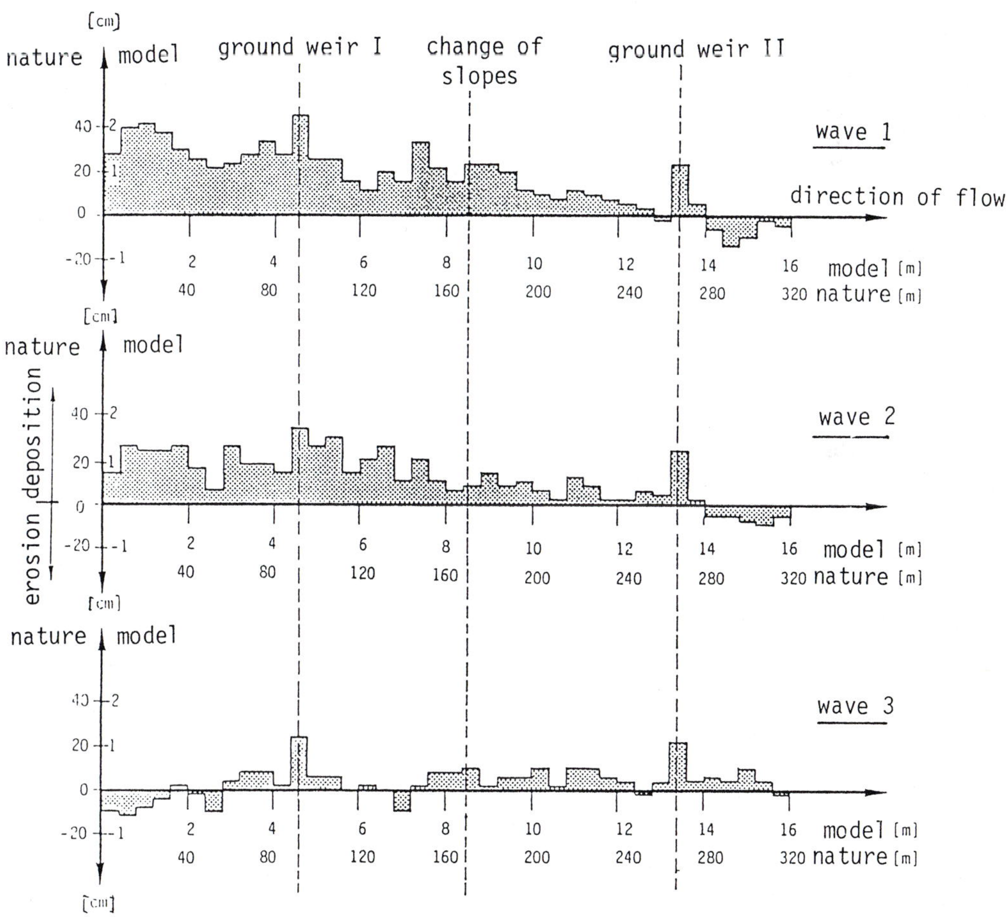

Fig. 15: Bottom changes due to the three flood waves

7. Concluding Remarks

The model investigations dealt with basic and practical aspects of bed load problems that arose after the renaturalization of a river which has a relatively steep slope and is hydrologically influenced by the mountains. Special devices had been developped for running the model and registering data. The results may be summarized in the following way.

The discharge/transport relation was basically the same for the flume having a continuous slope and having an increased slope in its upper half. This relation proved to be stable even though the sand supply did not correspond directly to the transport capacity, and only the averaged values over a sufficiently long experiment time corresponded.

The final application of a very concentrated sand input for the simulation of the natural river situation did not change either the basic tendency of the flume, but showed a certain hydrograph dependent sedimentation in the area of the change of slopes. It may, however, be excluded that there is any original context between the depositions and the constructional measures realized in the course of the renaturalization of the Weissach river.

8. References

BECHTELER, W., VOLLMERS, H.-J. 1988, Modelluntersuchungen über die Auswirkungen des naturnahen Gewässerausbaues auf den Geschiebetrieb, Interpraevent 1988, Graz

VOLLMERS, H.-J., GIESE, E. 1970, Instability of flat bed in alluvial channels (Hill, Srinavasan, Unny) Discussion, Proc. ASCE Vol. 96, HY 6

\- 1988, Feststofftransport in Fließgewässern - Berechnungsverfahren für die Ingenieurpraxis, DVWK Schriften Nr. 87, Verlag Paul Parey, Hamburg/Berlin

SEDIMENT MOVEMENT ON THE KUROBE ALLUVIAL FAN

Tadaharu Ishikawa

Department of Civil Engineering
Tokyo Institute of Technology, Tokyo, Japan

ABSTRACT

Annual volume of bed material load brought forth by the Kurobe river, a Japanese steep river, is estimated by considering the process of the formation of an alluvial fan which has been developing between the mountain region and the Japan sea for these 10,000 years. The distribution of flood disasters along the Kurobe river on the alluvial fan is related to the distribution of long-term sediment accumulation which was estimated from the difference of topography between the present alluvial fan and the newest diluvial fan.

1.Introduction

The annual volume of bed material load is one of basic factors to be estimated for the management of steep rivers. However, we usually do not have enough hydraulic data for the estimation, and are compelled to estimate or sirmise it from some indirect data such as geographical, geomorphological or geological data. Accordingly, it is necessary for us to accumulate the indirect estimation techniques for practical use. This paper presents a geomorphological consideration on the sediment load brought forth by the Kurobe river which is one of Japanese typical steep rivers.

The Kurobe river has its headwater atop in Japan Alps 3,000m above the sea level and its length of 70km. The river carries a considerable amount of sediment forming a wide alluvial fan between the mountain rigion and the Japan sea. The radius of the alluvial fan is approximately 13km, and the averaged slope is 1/100. Flooding and sedimentation on the alluvial fan is a clue to the estimation of the annual load of sediment discharged by the Kurobe river.

2.Flood disaster distribution and sediment accumulation distribution

Fig.1 shows the topographical map of the Kurobe alluvial fan. The river flows down meandering in the slightly western side of the center of the alluvial fan. This channel was fixed about 300 years ago. Before that time, the river had changed its location every time a big flood took place.

The dots in the figure indicate the locations where the embankments were severely damaged in these 50 years.(1) Although the dots are considerably scattered, it is seemed to the author that two groups of the dots enclosed by dotted lines are formed. This study has started with a conception that the formation of the two groups has something meaningful; the river wants to change its course at the two locations.

An alluvial fan is a topography developed by swing of a channel at the top the fan where the river comes out of the mountain region. Therefore, it is a matter of necessity that the river tends to change its course there. Here arises a question: Why it wants to change in the mid-slope too? When looking at Fig.1 anew with such scrutinizing eyes, it is noticed that the axis of the channel obliquely intersects the contour lines at the location in question as if it evaded an invisible obstacle. Then, what is the obstacle?

Fig.2 shows a longitudinal distribution of sediment accumulation in the channel during a big flood in 1969 whose return period was estimated as more than 100 years.(2) This flood brought so much sediment that even regularly practiced survey revealed its distribution with enough accuracy. The abscissa denotes the distance from the mouth of the river.(Hereafter, the distance from the mouth of the river is indicated by "K" which means km.) In the figure, the rigions where embankment damages concentrate(I,II) are noted. They are located at the upstream side of the piles of sediment accumulation. It seems that the invisible obstacle which turns the flow aside is a pile of sediment accumulating between 5-K and 0-K. If so, the next question is whether this tendency of sediment accumulation is general or was temporary.

Fig.3 shows the longitudinal profile of the river bed. There is an inflection point in the mid-slope at the point of 5K. The downstream segment is concave and the upstream segment is linear or slightly convex. Since the width of the Kurobe river is almost constant, it is considered that the second derivative of the bed profile stands for the long-term tendency of the channel bed variation : Sediment accumulation takes place on the downstream segment while sediment movement on the upstream segment is almost in equilibrium.

The inflection point corresponds to the upstream end of a pile of sediment made by the flood(1969) shown in Fig.2. Therefore, it is seemed that the sediment accumulation between 5-K and 0-K has lasted for a long time.

It can be deduced from the above circumstantial evidence that there exists a singular point of sediment movement in the mid-slope of the alluvial fan, where the river has a tendency to change its course.

3.The origin of the singular point

There are some remains of old (diluvial) fans on both the sides of the present alluvial fan as illustrated in Fig.1. The old fans are named the Junikanno Plain, the Maesawa Plain, and the Funamino Plain. Their longitudinal profiles are shown in Fig.4. The whole ground in this region has gradually been inclined with ages because of the upheaval of the Japan Alps and the subsidence of the Toyama Bay. Accordingly, The older the fan is, the steeper the slope is.[3] The Funamino Plain (hereafter abbreviationarily shortend to the F-plain) had been expanded before the present fan (hereafter called G-plain) was formed. Since the remain of the F-plain goes underneath the G-plain approximately 40m above the sea, the total configuration of the F-plain was not known until a recent date.

Geographical Survey Institute carried out echo soundings in the sea around the Kurobe alluvial fan in order to map the marine bottom surface, and they clarified that there exists an unconformable strata under the sea bottom enclosing the Kurobe alluvial fan, which is supposedly a surface of diluvial deposit.[4] Fig.5 shows an example of geological maps published in a GSI report. The locations of the strata are plotted with (+) in Fig.4 and they are on the extended line of the F-plain surface. Supposing that the strata is a part of the F-plain, it is concluded that the F-plain once formed a conic shape over the whole area which is covered by the G-plain now with an edge of the cone 90m below the present sea level. Therefore, in the course of the formation of the G-plain, the area above the intersection point has been eroded and sediments have accumulated on the area below the point. It is noted that the intersection point of the F-plain and the G-plain coincides with the inflection point of the present river bed profile shown in Fig.3.

This fact implies that the sediment movement of the present days are related to the formation process of the alluvial fan. Accordingly, by acquiring a knowledge of the formation process, we might be able to get more information on the sedimentation on the Kurobe alluvial fan.

4.The period of the formation of the present alluvial fan

Some geomorphologists have studied the formation process of the Kurobe alluvial fan.[3] Among them, Fujii[5] roughly estimates the rate of sediment accumulation by assuming that the F-plain and the M-plain are an identical one and that the formation of the G-plain started 20,000 years ago when the regression of the sea took the maximum during the Wurm Gracier Epoch. However, there are two questions on Fujii's assumptions : Firstly, his first assumption is inconsistent with the above mentioned GSI data which were obtained after his study. Secondly, he did not take into account the movement of ground. The followings are author's considerations on the formation process of the G-plain.

Since the edge of the F-plain locates 90m lower than the present sea level, the F-plain is considered to have existed when the sea level relative to the ground stagnated around -90m. The variation of the relative sea level is caused by the movement of the ground and the movement of the sea surface.

Fig.6 shows the movement of the ground obtained from the level survey along a highway near the coast. It is noticed that the ground near the coast is subsiding with a rate of 3mm per year. Although it is not known how long this subsidence has continued, simple extraporation is adopted here because this subsidence is considered to be related to the formation of a techtonic basin of Toyama Bay which has continued for considerably long time.

On the other hand, the movement of the sea surface during the Holocene has been studied by many researchers.[6] The curve in Fig.7 shows the result presented by Gohara.[7] On the upper side of the figure, the extraporated line of the ground movement is drawn. Looking for the age when the difference between
the ground and the sea surface stagnated around 90m, it was 10,000 years ago that the difference stagnated at 80m. Therefore, the formation of the G-plain is assumed to have started 10,000 years ago.

5.Annual rate of sediment accumulation

For the simplicity, the sea level is assumed to have raised abruptly from -55m to 0m 10,000 years ago, and the ground subsidence is assumed to have continued with a constant rate of 3.5mm per year for these 10,000 years. (* The error of the estimation (90m-80m=10m) is alloted evenly to the sea movement and the ground movement.) The sea level relative to the ground raised from -90m to -35m 10,000 years ago and has raised from -35m to 0m in these 10,000 years. Fig.8 illustrates the formation process of the G-plain. It is similar to the formation of a sand delta at the upstream end of a reservoir. The edge of the plain moves with the sea surface, and sediments partly deposit in the sea in front of the edge and partly settles on the land.

Fig.9 shows the volume of the space between the F-plain and the G-plain. It is devided into three sub-rigions. The upper rigion (*A*) was eroded during the formation of G-plain, while sediments accumulated on the lower rigion (*B*,*C*). (*B*) is a rigion of deposit on the ground and (*C*) is a region of deposit in the sea.

6.Some corrections for the volume of sediment accumulation

According to the geological survey carried out by Yamagata,[8] there are sandstone stratas of the Tertiary Period 20~40m below the surfaces of the F-plain and the G-plain in the eroded rigion (*A*). This fact means that an amount of sandstone was once scraped at the early stage of the formation. By estimating the volume of scraped sandstone and by subtracting it from the volume of rigion (*A*), a result is obtained that sediment accumulation of $0.06 \times 10^9 m^3$ has been produced in the rigion (*A*). This amount of accumulation is much smaller than that in the rigion (*B*) and (*C*). Therefore, the sediment movement in rigion (*A*) seems to be in equilibrium as was inferred from the longitudinal profile of the present river bed shown in Fig.3.

A strong drift current flows from north-east to south-west along the shore of the Kurobe alluvial fan. Sediments are transported by the current and fall into submarine canyons locating around the alluvial fan. The rate of sediment loss is estimated from the following considerations. The coast east of the river is eroded year by year because no sediment has been supplied after the river channel was fixed at the present location. Yamada et al.[9] estimated the movement of coastal line based upon old maps and photographs. Fig.10 shows the spacially averged distance of past coastal lines from the present one. The mean recession rate is 1.3m per year. Fig.11 shows beach profiles along several radial lines shown in Fig.1. Coastal bed erosion by the recession of the coastal line is limited to the extent of 40m depth, and the average erosion depth is 20m. Accordingly, The rate of sediment loss per 1m of the coastal line may be roughly estimated as 26m^3 per year. The total rate of sediment loss is 44,200m^3 per year by assuming that the rate is constant along the whole coastal length of 17km. By dividing the volumes in Fig.9 by 10,000 years and by rectifying them as mentioned above, the averaged annual sediment movement on the Kurobe alluvial fan is obtained as illustrated in Fig.12. It should be noted that all considerations here do not include wash load which may be several times as much as bed material load.

7. A correlation between the flow rate and the sediment accumulation

It is desired to examine the above mentioned result from a hydraulic or hydrological point of view. The annual sediment accumulation on the land is compared with the annual maximum flow rate which is the only hydraulic factor available for decades. Fig.13 shows a probability distribution of the annual maximum flow rate Qp which is expressed as follows.

$$f(\xi)=\frac{1}{\sqrt{2\pi}}\sigma \exp\left\{-\frac{(\xi-\bar{\xi})^2}{2\sigma^2}\right\} \quad (1)$$

where, $\xi=\log(Qp)$, $\bar{\xi}=3.1$, $\sigma=0.27$ The following form is assumed as a relation between Qp and the annual sediment accumulation on the land, S.

$$S=\alpha\cdot Q_p^{\beta} \quad (2)$$

Then, the averaged value of S is obtained by the composition of Eqs.(1) and (2).

$$\bar{S}=\frac{\alpha}{\sqrt{2\pi}}\sigma\int_{-\infty}^{\infty}\exp\left\{\beta\cdot\ln(10)\cdot\xi-\frac{(\xi-\bar{\xi})^2}{2\sigma^2}\right\}d\xi \ =1.7\times10^5\mathrm{m}^3 \quad (3)$$

Total volume of sediment accumulation in 1969 caused by the flood(1969) was about 800,000m^3 as shown in Fig.2. The maximum flow rate was 5,700m^3/s.

$$800{,}000=\alpha\ 5{,}700^{\beta} \quad (4)$$

Coefficients α and β are determined by Eqs.(3) and (4).

$$S=21Q_p^{1.22} \quad (5)$$

It should be noted that Eq.(5) does not express the one by one correspondence of S and Qp, because S varies even with the same Qp by the change of flood frequency and their hydrographs. Eq.(5) is expected to be a long-term statistical correlation.

According to Eq.(5), The sediment accumulation is proportional to the 1.22th power of the flow rate. The value of power is not very different from the one obtained from bed load formulas for large tractive force. Accordingly, the numerical values obtained in the previous chapter is considered to be acceptable at least in the scale of magnitude.

8.Conclusion

Geomorphological considerations have been made to estimate the sediment movement on the Kurobe alluvial fan. Characteristics of flood disasters as well as the sediment movement are found to be considerably affected by a long-term geomorphological phenomenon which has continued since 10,000 years ago. This fact suggests that the long-term history of topography around a river as well as the river itself must be taken into account for plannings of river improvement works on steep rivers because rigions of steep topography are often under the long-term crustal movements.

Reference

(1) Kurobe Construction Office (1977), History of the Kurobe river (in Japanese), Ministry of Construction, p.87

(2) Suga,S., Ishikawa,T.and Kasai,T.(1978): Hydraulic model test of the Kurobe river (in Japanese), Tech.Memo., No.1942, PWRI, Min. Const., p.44

(3) Fukai,S. (1966): On the small-scale landforms on the Kurobe alluvial fan (in Japanese), in Kurobegawa editted by Toyama Univ. Science Research Organization, pp.121-138.

(4) Geological Survey Institute (1982): Report of the basic survey of the coastal area - on the eastern part of the Toyama Bay - (in Japanese), Tech.Rep.of GSI, D.3-No.40, pp.131-138

(5) Fujii,S.(1965): The development of the Kurobe fan and the submarine forests around the Toyama Bay (in Japanese), Earth Science, No.78, pp.11-20

(6) Ota,Y., Matsushima,Y. and Moriwaki,H.(1982), Notes on the Holocene sea-level study in Japan (in Japanese), The Quaternary Research, vol.21, No.3, pp.133-143

(7) Gohara,Y.(1976): Climatic fluctuation and sea level change during the late Pleistorene and early Holocene, Pacific geology, No.11, pp.88

(8) Yamagata,D.(1966): On the geological structure and ground water of the Kurobe alluvial fan (in Japanese), in Kurobegawa editted by Toyama Univ. Science Research Organization, pp.61-84

(9) Yamada,T., Tanii,F. and Takemura,T.(1978): Coastal erosion around the Kurobe alluvial fan (in Japanese), Bulletin No.2, Research Institute of the Kurobe Alluvial Fan, pp.3-9

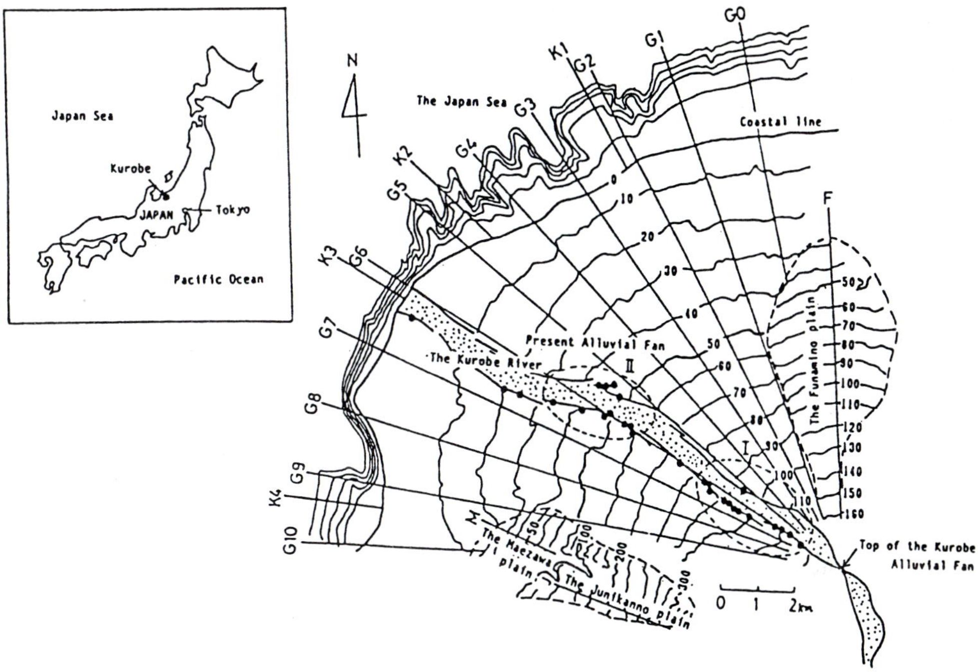

Fig.1 Topography of the Kurobe alluvial fan

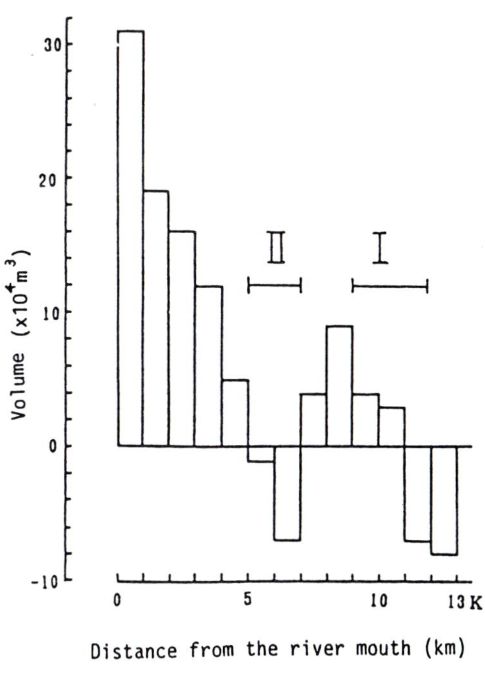

Fig.2 Distribution of sediment accumulation

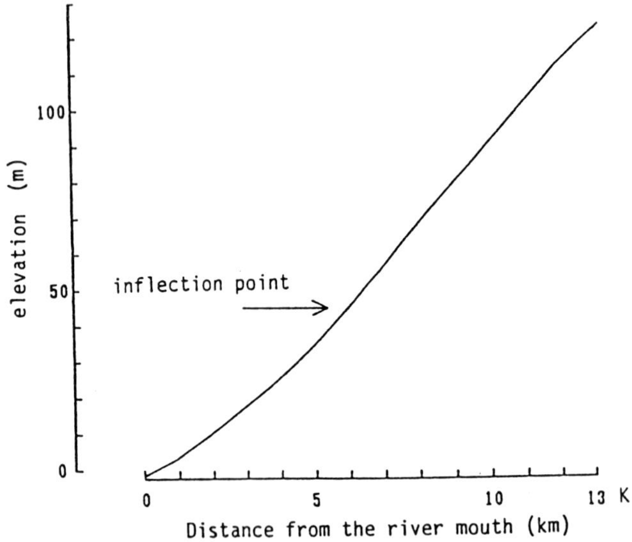

Fig.3 Lengitudinal profiles of the river bed

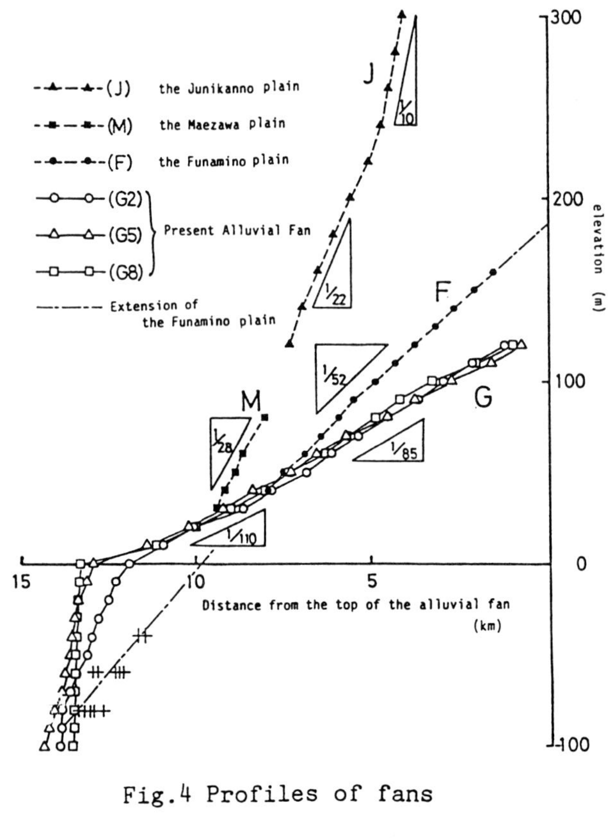

Fig.4 Profiles of fans

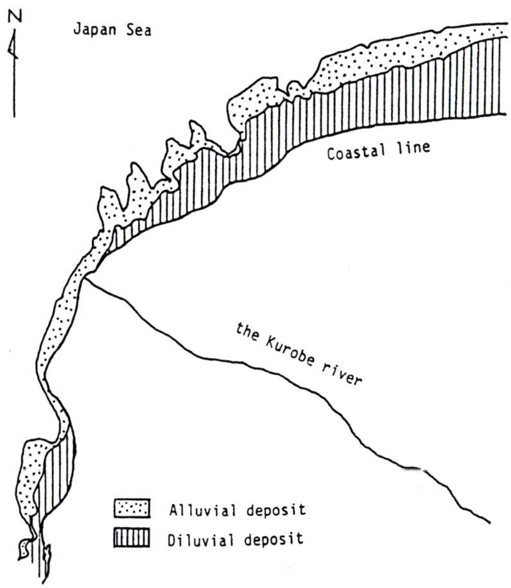

Fig.5 Geological map around the Kurobe alluvial fan

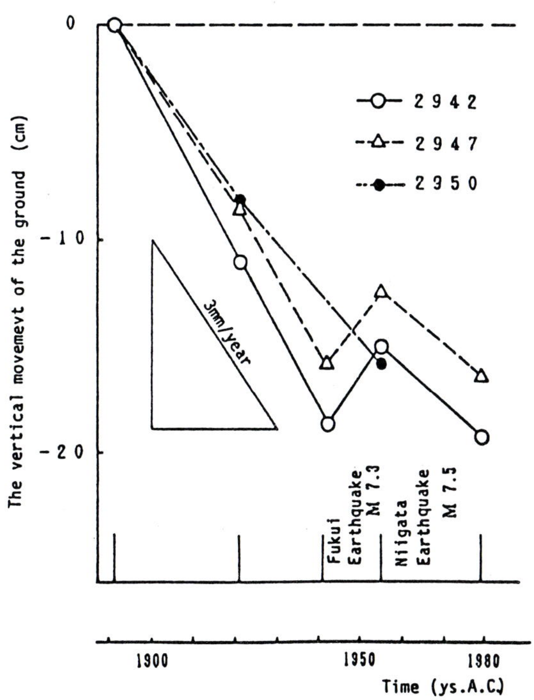

Fig.6 Movement of the ground

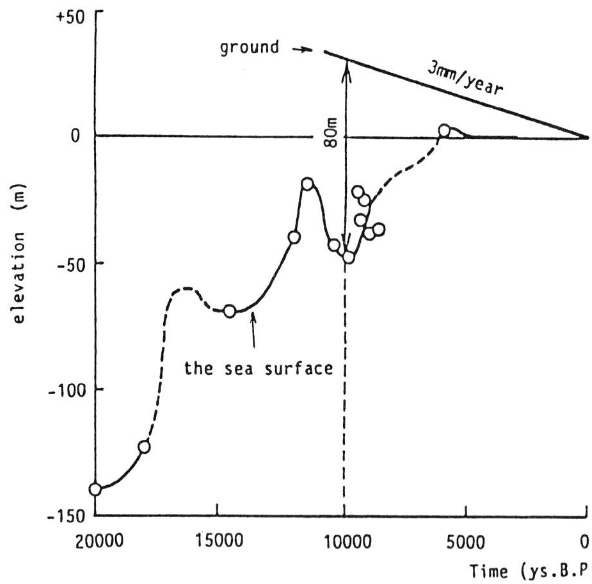

Fig.7 Movement of the sea surface

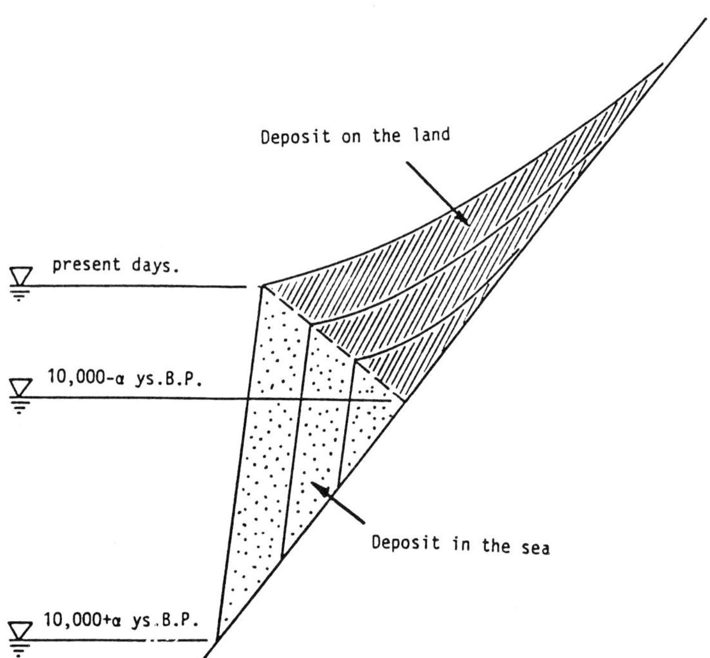

Fig.8 Illustration of the formation process of the Kurobe alluvial fan

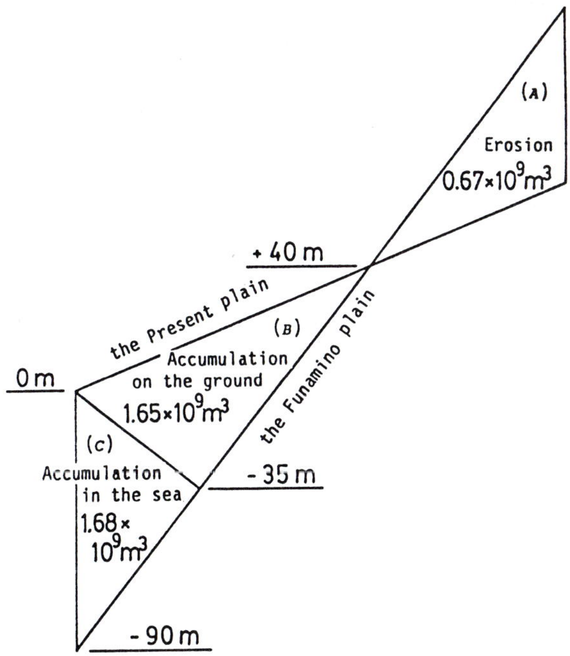

Fig.9 Volumes of sediment accumulation and evosion

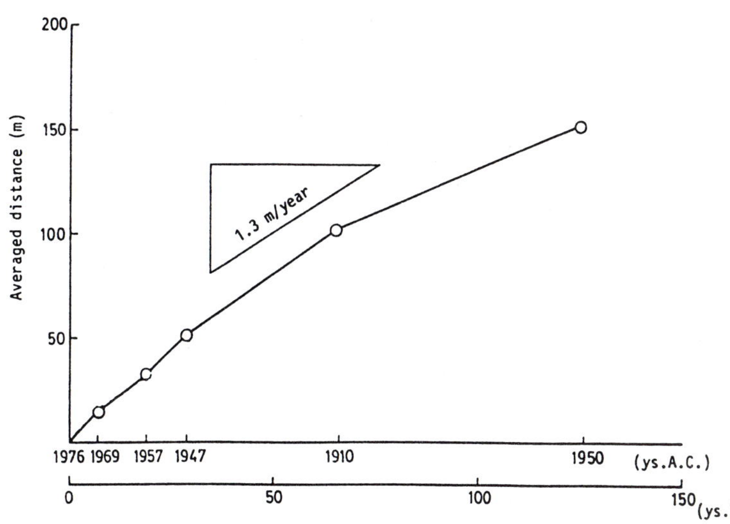

Fig.10 Recession the coastal line

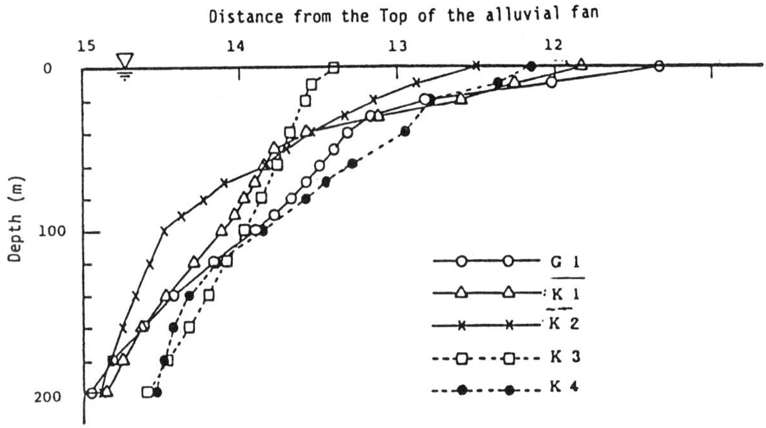

Fig.11 Beach profiles

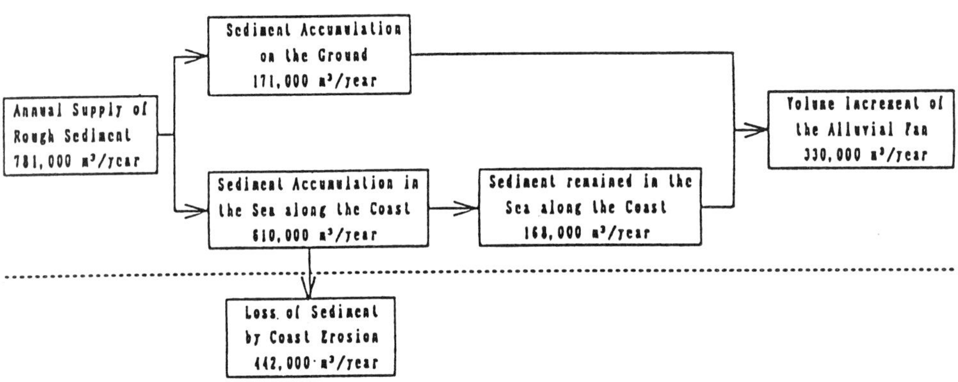

Fig.12 Sediment movement on the Kurobe alluvial fan

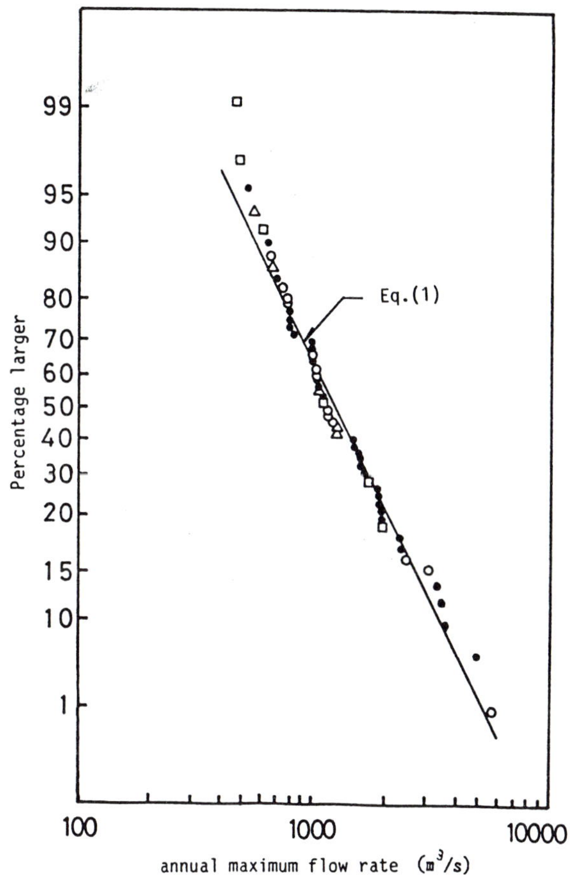

Fig.13 Probability distribution of annual maximum flow rate

SIMULATION OF RESERVOIR SEDIMENTATION IN MOUNTAIN REGIONS

Masaharu Fujita[(1)], Masanori Michiue[(1)] and Kazuo Ashida [(2)]

[(1)]Department of Civil Engineering
Tottori University, Tottori, Japan
[(2)] Disaster Prevention Research Institute
Kyoto University, Uji, Japan

ABSTRACT

If a catastrophic landslide occurs in mountain regions, the influence upon sediment transportation remains so great for a long time that the reservoir sedimentation will become a more serious problem. In such a case, the sediment discharge into the reservoir will change remarkably with erosion of the sediment deposited in the valleys. So, the way to determine the boundary condition on sediment discharge is discussed by using the field data in the case of "Ontake landslide(1984)". Then a method to predict the reservoir sedimentation in one dimension is developed on the basis of consideration of the changes of the boundary condition on sediment discharge into the reservoir. This method is applied to the reservoir sedimentation in a field and its applicability is very good.

Mt. Ontake what is called "Ontake landslide", whose rock fall volume is estimated to be 36 million m^3 as shown in Photo 1. Just after the event, the sediment yielded by the landslide run in Nigori river (Fig.1) in the mode of debris flow and a large amount of sediment was deposited in the mild slope areas and in the reaches upstream of two gorges (Gakiganodo Gorge and Kohrigase Gorge shown in Fig.1). The longitudinal profile of the deposition in the channel upstream of these gorges is shown in Fig.2. [1] The maximum thickness of the deposition was over 50 m and the volume of deposition is estimated to be 26 millions m^3. A large amount of sediment eroded in these regions was transported into Ontake reservoir, whose dam site is located about 10 km downstream from Kohrigase Gorge. After the occurrence of the landslide, the volume of sedimentation in the reservoir reached 2.3 millions m^3 in only 3 months, which is 13 times of the annual sedimentation volume in the ordinary year, and it reached 4.5 millions m^3 in one year. So, the front of sedimentation moved forward rapidly as shown in Fig.3. After that, the annual volume of sedimentation has been decreasing, but in 1988 it is still 3 or 4 times of the ordinary one.

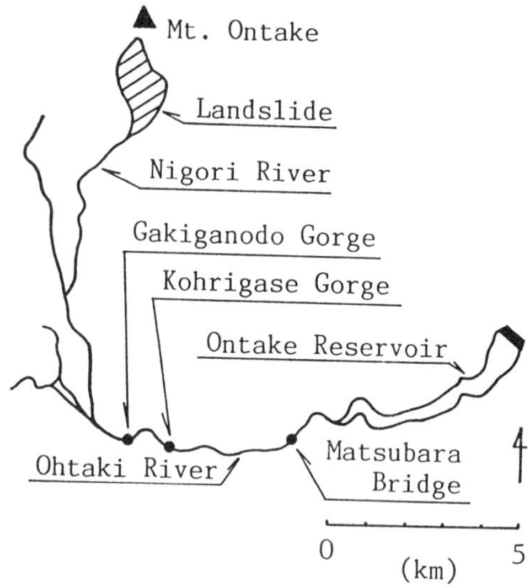

Fig.1 Basin of Ontake reservoir

Photo 1 Ontake landslide

1. INTRODUCTION

If a catastrophic event such as a landslide or a volcanic eruption occurs in mountain regions, the sediment transport rate will increase remarkably in the river of the basin so that it must immediately have an great influence on the reservoir sedimentation, i.e. the reduction of reservoir life. In such a situation, it is very important to predict how long this influence will remain and how much sediment will deposit in the reservoir in that period.

In 1984, a catastrophic landslide occurred due to an earthquake in western Nagano of Japan. A large amount of sediment yielded by the landslide was transported in the mode of debris flow and deposited in the mountain basin, and then transported in the modes of bed load, suspended load and wash load. As a result the annual volume of the sedimentation of the reservoir in this basin reached about 26 times of the ordinary one just after the occurrence of the landslide. Sediment transport rate from the mountain basin, however, has been decreasing gradually with erosion and armoring of the deposited material in the valleys and by the constructions of the check dams for some years. Thus, it is very significant to consider the process of decreasing of the sediment transport rate in the case that a catastrophic landslide occurred in prediction of the reservoir sedimentation.

Mathematical models to simulate the reservoir sedimentation in one dimension have already been developed and finite difference methods are usually used in calculation. By these methods, the bed profile and the grain size distribution of sedimentation are calculated under the appropriate boundary conditions on sediment discharge. These boundary conditions must be given concerning both wash load and bed material load. The empirical relationship between water discharge and wash load transport rate is generally used for the former condition and the grain size distribution of bed material is usually set up for the latter condition. It is, however, difficult to determine these conditions, because the situation of the valley where a large amount of sediment is deposited changes with time due to armoring and erosion of the bed material after the occurrence of a large landslide. The information on such movement of the deposited material will be required in order to estimate these conditions precisely.

In this paper, a method to simulate the reservoir sedimentation in one dimension in the case that a catastrophic landslide occurred is discussed in consideration of the changes of the boundary condition on sediment discharge. In order to estimate the changes appropriately the case of Ontake landslide (1984) is taken and it is investigated. Then, using the results, the present method is applied to the simulation of the reservoir sedimentation.

2. RESERVOIR SEDIMENTATION AFTER ONTAKE LANDSLIDE

The river basin of Ontake reservoir is shown in Fig.1. In September, 1984, an earthquake with a magnitude of 6.8 occurred in this basin. This earthquake gave rise to a catastrophic landslide in

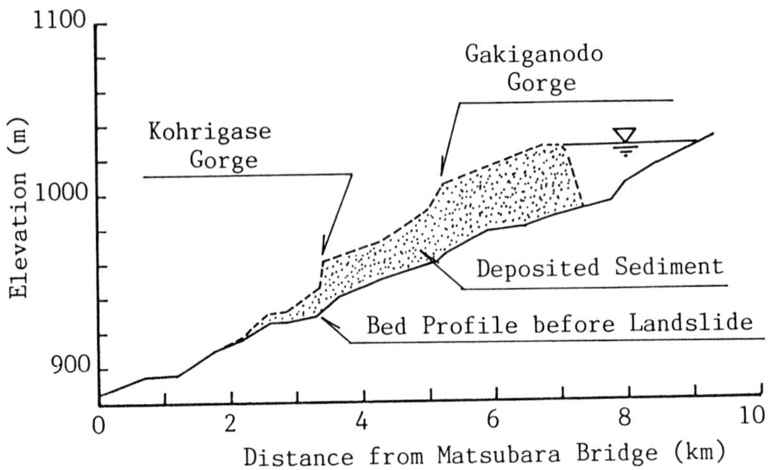

Fig.2 Longitudinal bed profile in the reaches upstream of two gorges just after the occurrence of Ontake landslide [1]

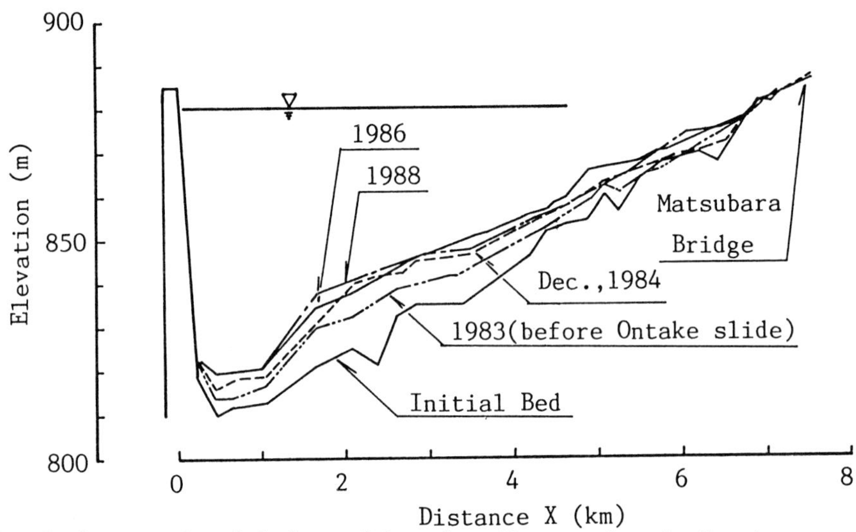

Fig.3 Longitudinal bed profile of sedimentation in Ontake reservoir

According to the investigation of the sediment movement by the field data of bed profile, almost of the sediment transported into the reservoir was eroded in Nigori river and in the reaches upstream of two gorges.

The grain size distributions of the material deposited in the reservoir were investigated for two periods, namely, the period between Sep., 1984 (the occurrence of the landslide) and Dec., 1984 and the period between Jan., 1985 and Aug., 1985 as shown in Fig.4. This figure indicates that the deposited material becomes slightly coarser with time.

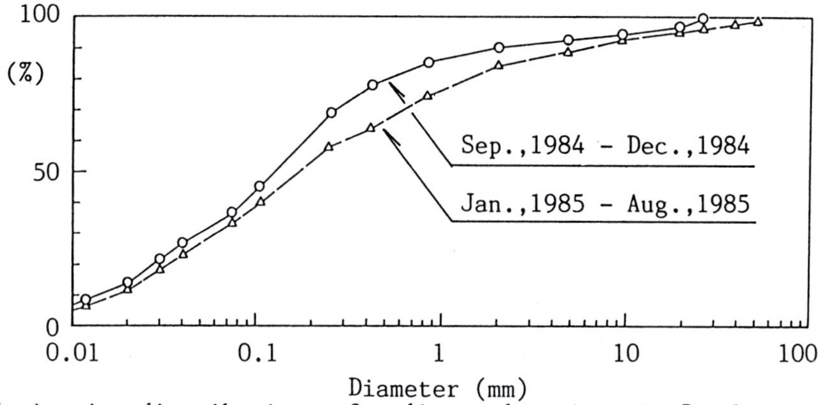

Fig.4 Grain size distributions of sediment deposited in Ontake reservoir

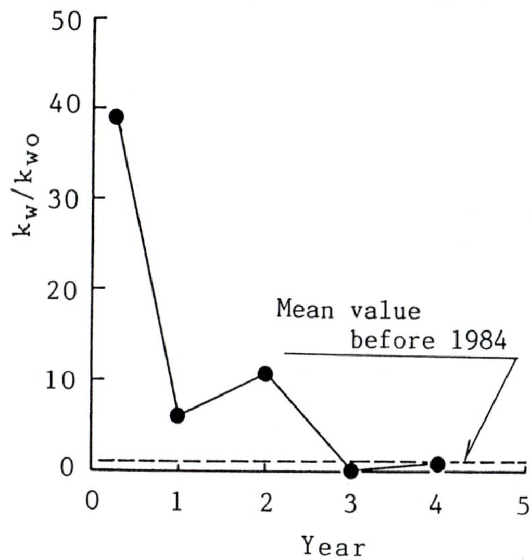

Fig.5 Variation of the coefficient on wash load transport rate with time

Assuming that the particle diameter of wash load is less than 0.1 mm, wash load forms about 40 % of total deposited sediment after the occurrence of the landslide. If wash load transport rate can be expressed as $Q_w = k_w Q^2$, the influence of the landslide on wash load is investigated by the variation of k_w, where Q = water discharge; and k_w = a constant. Assuming that the deposition rate of wash load is equal to the product of the concentration Q_w/Q by the fall velocity of the particle, k_w can be estimated from the deposition rate at the bottom set bed, where only wash load deposits. In Fig.5, the variation of k_w/k_{wo} with time is shown, where k_{wo} = mean value of k_w before the occurrence of the landslide. This figure indicates that the deposition rate of wash load is extremely large during 3 years after the occurrence of the landslide, but recently is decreasing to approach to the ordinary one due to erosion control works such as construction of check dams and turfing works.

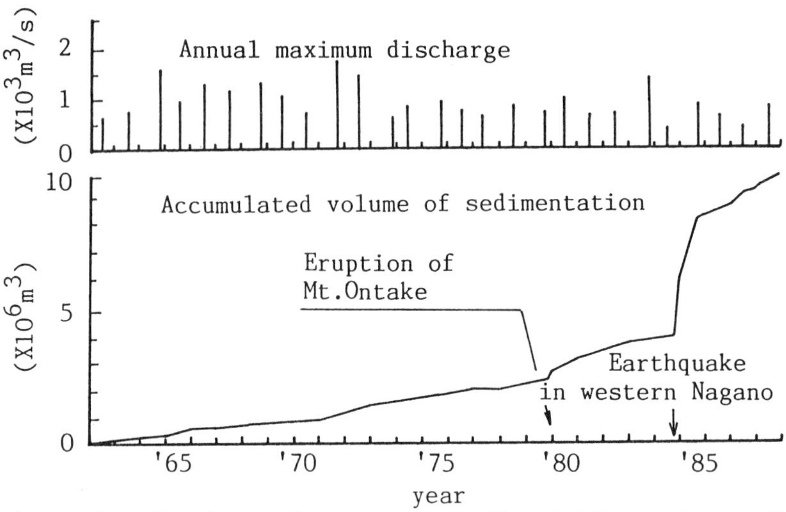

Fig.6 Accumulated volume of reservoir sedimentation and annual maximum discharge

Accumulated volume of the sediment deposited in the reservoir is shown together with the annual maximum water discharge in Fig.6. This figure shows that two strong events occurred to cause the sedimentation volume increase rapidly. One is the eruption of Mt. Ontake (1979), the other is the earthquake in western Nagano(1984). The influence of the eruption on the reservoir sedimentation continued for 2 or 3 years. The periods during which the influence remains great have a close relation with the scales of eruption and landslide. The landslide in 1984 is so great that the influence may continue for a long time.

Transport rate of bed material load into the reservoir is mainly dependent to the geometric conditions such as slope and width, the water discharge and the grain size distribution of the bed material in a section just upstream of the back water region of the reservoir, which can be regarded as "a control section of sediment discharge". Fig.6 shows that the sediment transport rate into the reservoir increased rapidly just after the occurrence of the landslide though the maximum water discharge was not so large. This means that the channel bed became much steeper or the bed material became much finer than the previous one in the control section of sediment discharge. As shown in Fig.3, the slope did not change so much. It seems, therefore, that the bed material became much finer during a flood after the occurrence of the landslide.

The volume of sedimentation can be calculated by the formulae of sediment transport rate, the daily water discharge and the grain size distribution of the bed material in the control section assuming that 40 % of total load was wash load. Thus, this grain size distribution can be estimated so that the calculated volume of sedimentation agrees with the observational one. In Fig.7 the estimated grain size distributions in the control section (in Matsubara bridge section shown in Fig.1) are shown for three periods, namely, the period between Sep., 1984 (the occurrence of the landslide) and Dec., 1984 (I-1), the period between Jan., 1985 and Aug., 1985 (I-2) and the period between Sep., 1985 and Nov., 1988 (I-3). The previous grain size distribution which was evaluated by the data of sedimentation in 1975, I-4, is also shown in this figure. This figure indicates that the bed material became much finer just after the occurrence of the landslide and has gradually been returning to the previous ordinary one (I-4).

If the bed material in the control section was very fine just after the occurrence of the landslide as shown in Fig.7, the grain size distribution of sedimentation calculated by this distribution (I-1 in Fig.7) is much finer than the observed one even if the calculated result on the volume of sedimentation agrees correctly the observational one. It seems that the sediment was transported in the mode of hyper-concentrated flow, whose mechanism is far different from that of bed material load, just after the occurrence of the landslide. This difference of transport mechanism seems to be one of the reasons why the grain size distribution of sedimentation can not be simulated at all. In Fig.8 the grain size distribution of sedimentation between Jan., 1985 and Aug., 1985 calculated by using the distribution I-2 in Fig.7 is compared with the observed one for the same period except wash load. As the agreement between both curves is very good, the distribution I-2 seems to be valid in calculation of both quantity and quality of the transported sediment.

In Fig.9 the grain size distribution of the bed material in the control section estimated for the period between Jan., 1985 and Aug., 1985, that is, the distribution I-2 shown in Fig.7 is compared with the observed grain size distributions of the material deposited in Kohrigase Gorge and Gakiganodo Gorge. Both distributions are very similar. This gives the result that the bed material in the control section just after the occurrence of the landslide was replaced by the sediment yielded in the mountain region. Thus, it seems that the bed material in the control section varies from the material yielded in the mountain region to the previous ordinary one after the occurrence of a catastrophic landslide.

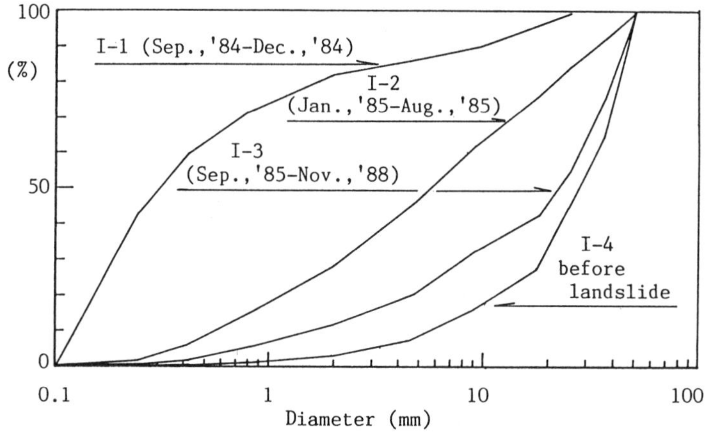

Fig.7 Grain size distributions of bed material in boundary section (in Matsubara Bridge section)

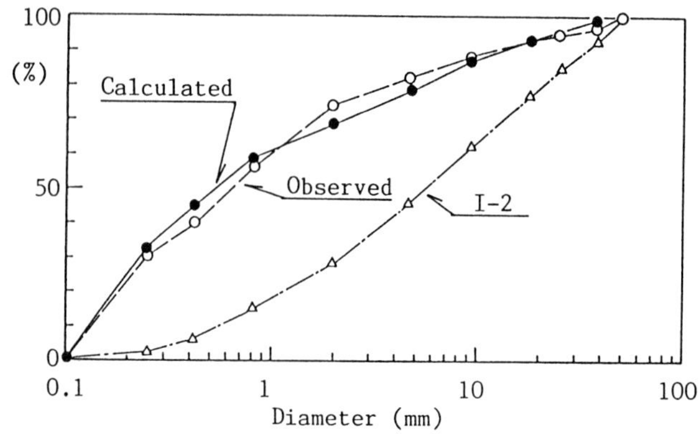

Fig.8 Comparison between predicted grain size distribution of sediment transported into reservoir and observed one

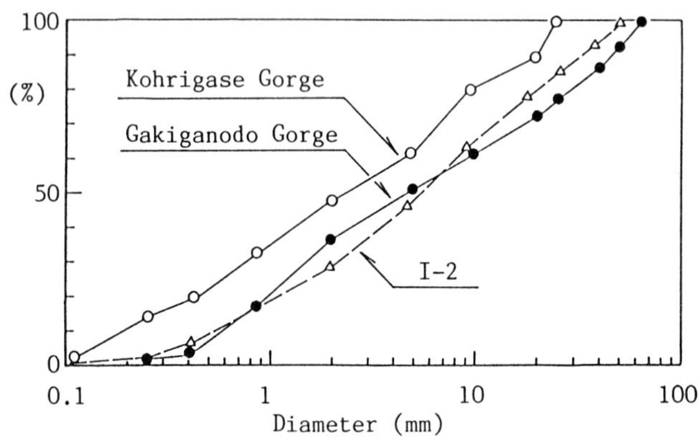

Fig.9 Comparison between predicted grain size distribution of bed material in boundary section and observed ones of sediment deposited in mountain region

3. A METHOD TO SIMULATE RESERVOIR SEDIMENTATION

(1) Basic equations

Coordinate system is shown in Fig.10, in which x, y and z axes express longitudinal, lateral and vertical directions, respectively. Reservoir sedimentation is simulated by the following equations:

Conservation equations of mass and momentum for water are written as

$$\partial Q/\partial x = 0 \tag{1}$$

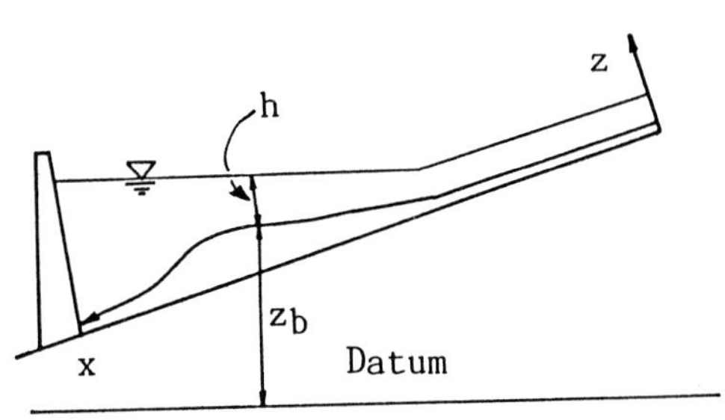

Fig.10 Coordinate system

$$(v/g)(\partial v/\partial x) + \partial h/\partial x = I_o - I_e \tag{2}$$

where Q = water discharge; v = mean velocity ; h = water depth ; I_o = channel slope; I_e = energy slope; and g = acceleration of gravity. Applying Manning's relation to resistance law,

$$v = (1/n)R^{2/3}I_e^{1/2} \tag{3}$$

where n = coefficient of roughness; and R = radius.

Continuity equation of transported sediment is written as

$$\frac{\partial z_b}{\partial t} + \frac{1}{(1-\lambda)B}\frac{\partial[(q_b+q_s)B+Q_w]}{\partial x} = 0 \tag{4}$$

where z_b = bed elevation; B = channel width, q_b = bed load transport rate per unit width; q_s = suspended load transport rate per unit width; Q_w = wash load transport rate; and λ = porosity.

According to Ashida and Michiue's relation [2], the bed load transport rate of the grain with a diameter of d_i is expressed as

$$q_{bi}/(p_i u_{*e} d_i) = 17\tau_{*ei}(1-\tau_{*ci}/\tau_{*i})(1-\sqrt{\tau_{*ci}/\tau_{*i}}) \tag{5}$$

where q_{bi} = bed load transport rate pertaining to d_i; $\tau_{*i} = u_*^2/sgd_i$; $\tau_{*ei} = u_{*e}^2/sgd_i$; $\tau_{*ci} = u_{*ci}^2/sgd_i$; u_* = shear velocity ($\sqrt{gRI_e}$); u_{*ci} = critical shear velocity for the fraction with a diameter of d_i; u_{*e} = effective shear velocity; p_i = percentage of the fraction with a diameter of d_i on bed surface; and s = specific gravity of sediment in water. Effective shear velocity is calculated by Ashida and Michiue's relation [2] as

$$v/u_{*e} = 6.0 + 5.75\log[(R/d_m)/(1+2\tau_*)] \tag{6}$$

where d_m = mean diameter of the material on bed surface.

According to Egiazaroff's theory [3] and Ashida and Michiue's theory [2], critical shear velocity is expressed as

$$\frac{d_i}{d_m} > 0.4 : \quad \frac{u_{*ci}^2}{u_{*cm}^2} = \left[\frac{\log 19}{\log(19d_i/d_m)}\right]^2 \tag{7}$$

$$\frac{d_i}{d_m} < 0.4 : \quad \frac{u_{*ci}^2}{u_{*cm}^2} = 0.85 \tag{8}$$

where u_{*cm} ($=\sqrt{\tau_{*cm} sgd_m}$) = nondimensional critical shear stress for the fraction with a diameter of d_m. τ_{*cm} is usually taken to be 0.05.

Suspended load transport rate of d_i is calculated as

$$q_{si} = \int_a^h uC_i dz \tag{9}$$

where C_i = concentration distribution of suspended load pertaining to d_i; a = reference level of suspended load concentration; and u = velocity distribution. Using Rouse's distribution or Lane and Kalinske's distribution, which are most conventional, and logarithmic law for velocity distribution, suspended load transport rate can be calculated by numerical methods. If Lane and Kalinske's distribution is applied and velocity distribution is represented by depth-averaged streamwise velocity, suspended load transport rate per unit width can easily reduced to

$$q_{si} = C_{ai}[\exp(c_1 a)/c_1]v[\exp(-c_1 a)-\exp(-c_1 h)] \tag{10}$$

where $c_1 = 6w_{oi}/(\kappa u_* h)$; κ = Karman constant; w_{oi} = fall velocity of particle with a size d_i; and C_{ai} = concentration at reference level. Reference level is assumed to be 0.05h. Concentration at reference level is given by Ashida and Michiue's theory [4] as

$$C_{ai} = p_i \, K[g(\xi_o)/\xi_o - G(\xi_o)] \tag{11}$$

where

$$g(\xi_o) = (1/\sqrt{2\pi})\exp(-0.5\xi_o^{\,2}) \tag{12}$$

$$G(\xi_o) = (1/\sqrt{2\pi})\int_{\xi_o}^{\infty} \exp(-0.5\xi^2)d\xi \tag{13}$$

where K = 0.025; and $\xi_o = w_{oi}/(0.75u_*)$.

Total bed load and total suspended load rate are written as

$$q_b = \Sigma\, q_{bi} \tag{14}$$

$$q_s = \Sigma\, q_{si} \tag{15}$$

Wash load transported into reservoirs is usually assumed to be in proportion to the square of water discharge according to many observations in fields. So wash load is expressed as

$$Q_w = k_w Q^2 \tag{16}$$

where k_w = coefficient related to the characteristics of sediment yield. The deposition rate of wash load, $\partial z_{bw}/\partial t$, is calculated by $\partial Q_w/\partial x$, but wash load approximately deposits at its falling velocity in still water if shear velocity is not so larger than the falling velocity. Thus, the deposition rate of wash load is expressed as

$$u_* > w_{oi} \;:\; \partial z_{bw}/\partial t = 0 \tag{17}$$

$$u_* < w_{oi} \;:\; \partial z_{bw}/\partial t = c_w w_{oi}(Q_w/Q)/(1-\lambda) \tag{18}$$

where c_w = a constant; and w_{oi} = fall velocity of the particle with a diameter of d_i.

If bed material is composed of nonuniform sediment, the continuity equations for each fraction of bed material are required. According to Hirano's model [5], in which the concept of exchange of sediment in surface bed layer is considered, the relations are as follows:

$\partial z_b/\partial t > 0$:

$$\frac{\partial}{\partial t}(Bp_i) = -\frac{1}{\delta(1-\lambda)}\left[\frac{\partial}{\partial x}(Bq_{bi}+Bq_{si})-(1-\lambda)B\frac{\partial z_b}{\partial t}\right] - \frac{p_i}{\delta}\frac{\partial}{\partial t}(Bz_b) \quad (19)$$

$\partial z_b/\partial t < 0$:

$$\frac{\partial}{\partial t}(Bp_i) = -\frac{1}{\delta(1-\lambda)}\left[\frac{\partial}{\partial x}(Bq_{bi}+Bq_{si})-(1-\lambda)B\frac{\partial z_b}{\partial t}\right] - \frac{p_{io}}{\delta}\frac{\partial}{\partial t}(Bz_b) \quad (20)$$

where p_{io} = percentage of the fraction with a diameter of d_i under surface exchange layer; δ = thickness of exchange layer, which is taken to be a maximum diameter of bed material.

(2) Boundary conditions

Reservoir sedimentation can be simulate using the basic equations if the boundary conditions on supplies of water and sediment are known correctly. The records of water discharge and water level of reservoir can be used as the boundary conditions on water supply.

The boundary conditions on sediment supply, which is the conditions in the control section of sediment discharge, must be given concerning both sediment transport rate and its grain size distribution or only the grain size distribution of bed material. If the rate of sediment yield in the mountain region is stable, the grain size distribution may not change so much, but for example after the occurrence of a large landslide, the distribution will change remarkably as discussed in section 2. This changes relate to the erosion and armoring processes of the sediment deposited in the mountain region. Thus in general it is difficult to predict the process of changing of the distribution in detail. As shown in Fig.8 and Fig.9, the bed material is replaced by the sediment yielded in the mountain region just after the occurrence of the landslide and after that returns to the previous ordinary one. Therefore the grain size distributions between the distributions of the material deposited in the mountain region and the bed material before the occurrence of landslide can be used as the boundary conditions.

Using only the grain size distribution of the material deposited in the mountain region as the boundary condition on sediment supply, the sedimentation volume is overestimated, and using only the previous ordinary grain size distribution, the sedimentation volume is underestimated. So, roughly speaking, the grain size distribution of the material yielded in the mountain region should be used as the boundary condition on sediment supply in order to predict how the sediment deposits in the reservoir for a few years after the occurrence of a landslide. In the prediction for the next period the grain size distributions coarser than the distribution of the material deposited in the mountain region and finer than the previous ordinary distribution of the bed material in the control section should be taken as the boundary condition. In order to evaluate the boundary condition on sediment supply precisely, the sediment movement in the valleys must be investigated further.

(3) Application

Using this idea, the sedimentation in Ontake reservoir is simulated for two periods, that is, the period between Jan., 1985 and Aug., 1985 and the period between Nov., 1987. and Nov., 1987. The initial conditions and the boundary conditions are as follows: The initial bed profile and the initial grain size distributions of bed material are determined from their observational results in Dec., 1984 and Nov., 1987. The grain size distributions of bed material are given in each area of the mesh which is 100 m long and 1m thick. The daily records of water level and water discharge of the reservoir are used and the boundary conditions of sediment supply are given by the distributions of bed material shown in Fig.7 for each period.

The material, whose minimum and maximum grain sizes are 0.001 mm and 44 mm, was divided into 11 fractions. The product of coefficient on wash load k_w by c_w was taken to be $5X10^{-5}$ in the 7 month period from Jan., 1985 and $5X10^{-6}$ in the 1 year period from Nov., 1987. These values

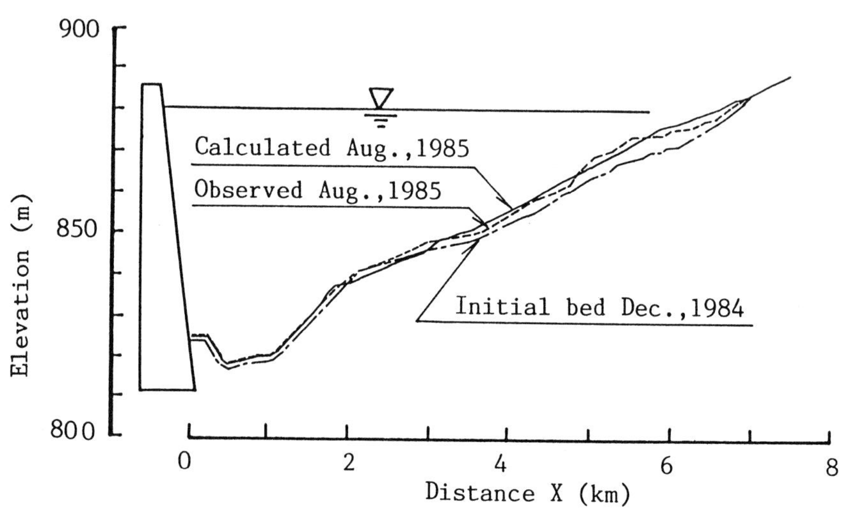

Fig.11 (a) Simulation of longitudinal bed profile in Ontake reservoir

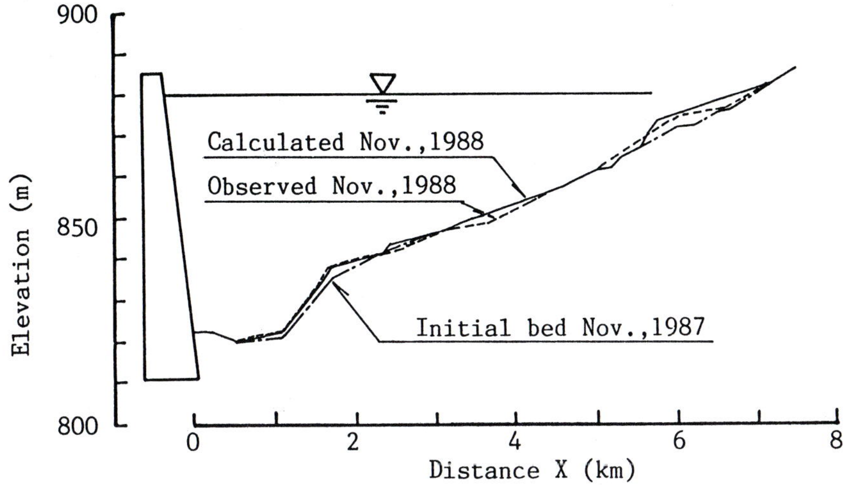

Fig.11 (b) Simulation of longitudinal bed profile in Ontake reservoir

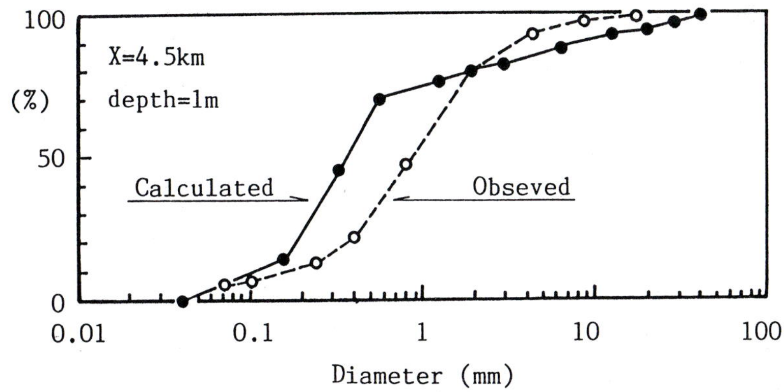

Fig.12 Simulation of grain size distribution of bed material in Ontake reservoir

were estimated by using the observed deposition rates of wash load at the bottom set bed. The basic equations were calculated by finite difference method. The longitudinal distance scheme is taken to be 100 m and the time steps between 3 sec and 60 sec are selected according to the water discharge.

In Fig.11(a) and (b), the results of the simulation of the sedimentation are shown concerning the longitudinal bed profile. The calculated results are in good agreement with the observed ones. In Fig.12 the calculated grain size distribution of bed material is compared with the observed one. The calculated distribution agrees roughly with the observational one. As the boundary condition of sediment supply is appropriately determined using the field data of bed material, the applicability of the present model is very good, but it is important to predict how the boundary condition will change after this.

4. CONCLUSIONS

In this paper a model to calculate the reservoir sedimentation in the case that a catastrophic landslide occurred is discussed. This method is developed with attention focused on the changes of the grain size distribution of the bed material in a boundary section, which is a control section of sediment discharge into the reservoir, with time. The results are mainly as follows:

(1) After the occurrence of a catastrophic landslide the bed material in such a control section of sediment discharge into the reservoir as a section just upstream of the back water region of the reservoir immediately replaces the finer sediment that is almost same as the sediment yielded in the mountain region. But, it is gradually coarser to return to the previous ordinary one.

(2) The grain size distribution of the bed material in a control section of sediment discharge into the reservoir is required as the boundary condition. It is difficult to determine the condition in detail, but the grain size distribution of the material yielded in the mountain region can be used to predict how the sedimentation move forward for some years just after the occurrence of a landslide. When the sedimentation is predicted for the far future, the grain size distribution must be properly taken to be the distribution between the material yielded in the mountain region and the previous ordinary one. The condition can be estimated more correctly if the information on the movement of the material deposited in the valley is obtained further.

(3) A method to simulate the reservoir sedimentation in one dimension was developed. If the condition on grain size distribution in a control section of sediment discharge into the reservoir is given appropriately, this method is applicable to the prediction of reservoir sedimentation.

5. REFERENCES

[1] Komura,S. : Synthetic Investigation of Earthquake and Damage of the Western Nagano Prefecture Earthquake, 1984, pp.217-223 (in Japanese).

[2] Ashida,K. and M.Michiue : Study on hydraulic resistance and bed load transport rate in alluvial stream, Proc. JSCE, pp59-69,1972(in Japanese).

[3] Egiazaroff,I.V. : Calculation of nonuniform sediment concentration, Proc. ASCE, Journal of Hydraulics Division, 91, HY.4, pp.225-246, 1965.

[4] Ashida,K. and M.Michiue : Study on the suspended sediment (1), Annuals, D.P.R.I., No.13B, pp.233-242, 1970 (in Japanese).

[5] Hirano,M. : River-bed degradation with armoring, Proc. JSCE, pp.55-64, 1971 (in Japanese).

SEDIMENT SLUICING IN MOUNTAIN RESERVOIRS

Helmut Scheuerlein

Obernach Hydraulics Laboratory
Technical University Munich, Munich, Germany

ABSTRACT

Reservoir sedimentation generally follows a similar pattern. Material carried as bed load is deposited close to the reservoir entrance, material transported in suspension settles down uniformly all over the reservoir. In mountain reservoirs, it usually is the deposition of coarse material that causes problems.

Removal of the gravel deposits at the reservoir entrance by means of flushing can be a suitable remedy to solve the problem, particularly when it is combined with water level drawdown. Sluicing of coarse sediment through the reservoir can also be supported by creating a defined channel at the bottom of the reservoir.

Analytical treatment of reservoir sedimentation and sediment sluicing is difficult due to the many parameters involved and due to the uncertainties concerning the evaluation of reliable data.

A simplified method to estimate the efficiency of sediment sluicing is presented. Its practical application is demonstrated by means of an example.

INTRODUCTION

Sedimentation of man-made reservoirs is one of the major problems hydraulic engineers will have to face in future. In spite of the various measures that can be taken to minimize sediment yield from the watersheds, the intrusion of sediment into reservoirs can never be avoided completely. Sediment enters the reservoir either as bedload or as suspended load. Usually the amount of sediment carried in suspension exceeds the bed load transport by a factor of 5 to 10. The sediment deposits in a reservoir are composed accordingly. As it is well-known, the distribution of sediment deposits in a reservoir is not uniform. It usually develops forming a pattern shown in Fig. 1.

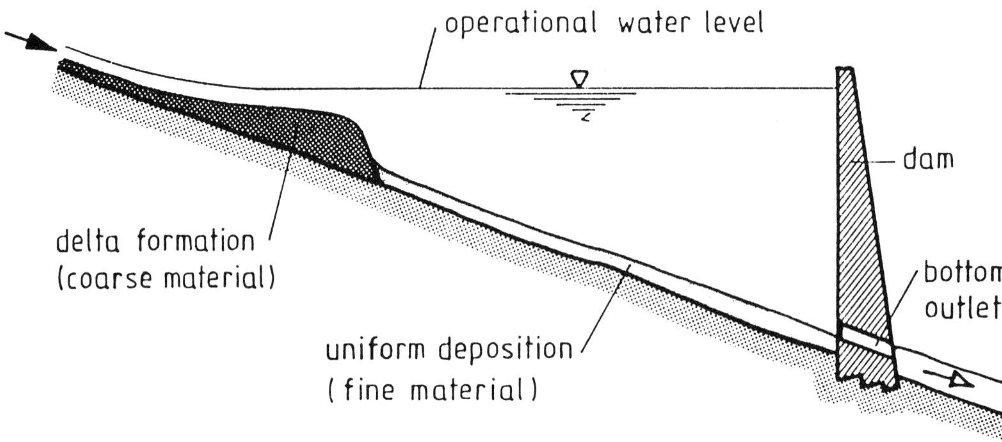

Fig. 1: Distribution of sediment deposits in reservoirs

As soon as the flow enters the reservoir the hydrodynamic conditions change considerably. The flow velocity and in turn the sediment transport ability are reduced. First the coarse material transported close to the bed settles down, forming a delta at the reservoir entrance. Material transported in suspension is carried further into the reservoir and deposited more or less uniformly in the whole reservoir.

In principle, the mechanism described above takes place similarly in any reservoir, independent from size or location. However, the magnitude of the depositions, the pace of the deposition process, and the importance of the depositions for the operation of the reservoir may vary considerably.

In mountain reservoirs, very often the uniformly distributed fine deposits are of minor importance, although totalling perhaps 5 times the amount of the coarse material. It is the gravel and sand settling right at the entrance of the reservoir that causes problems. Firstly, the coarse deposits affect directly the most active part of the reservoir as they cover the reach close to the operational level. Secondly, the deposits at the reservoir entrance are apt to cause a significant raise of the water level in the river upstream of the reservoir during flood season due to backwater effect. Thirdly, the coarse material trapped in the reservoir is missing downstream of the dam to balance the tractive forces of the flow. As a consequence, erosion of the bed takes place causing continuous degradation until equilibrium conditions are reached.

With respect to the reasons mentioned above, removal respectively avoidance of coarse deposits at the upstream end of mountain reservoirs can be much more important than excavation of the uniformly distributed fine deposits. A rather straightforward method to deal with the problem would be the mechanical excavation of the gravel and sand deposits by means of scraping equipment or backshovels. The method is simple and its application merely a question of economics. However, it rather helps to regain some storage capacity and to avoid backwater effects upstream, it does not contribute to fight degradation downstream of the dam, unless the excavated material is dumped at the toe of the dam which rarely may be the case.

Another - from the viewpoint of the hydraulic engineer more presumptuous - method would be to use the hydraulic energy of the flow to clear the delta region by sluicing the deposits through the reservoir and pass it on downstream by means of the bottom outlets of the dam. In the following, the paper is dealing mainly with this method.

PRINCIPLE OF SEDIMENT SLUICING

Deposition of sediment occurs when the flow velocity drops below a certain threshold value. With respect to sediment sluicing it would be desirable to keep the flow velocity always above the threshold value. This, however, usually conflicts with the operation rules dictated by the purpose of the reservoir. In the most cases, deposition of sediment during the year cannot be avoided. Sometimes, only the flood season can be used to get rid of some of the deposited material, sometimes there is also the possibility of planning and implementing regular flushing activities.

The efficiency of flushing or sediment sluicing depends significantly on the water level in the reservoir during the measure (ACKERS et al, SCHEUERLEIN, WHITE et al). Other important factors are the river flow and the duration of the measure.

The best results are achieved when at high river flow the water level in the reservoir can be kept low for some time. As this, on the other hand, means a substantial loss of water over a considerable period of time, effective sediment sluicing must be oriented towards minimization of drawdown and sluicing time.

Sediment sluicing efficiency can be supported also by auxilliary measures in the reservoir as f.i. the creation of a defined channel at the bottom of the reservoir to concentrate the flow where the tractive forces need to be as high as possible.

PROBLEMS CONCERNING ANALYTICAL APPROACH TO SEDIMENT SLUICING

Theoretical treatment of reservoir sedimentation and sediment sluicing is difficult for two reasons. Firstly, the mechanism is complex and hardly comprehensively describable. Extensive simplifications cannot be avoided. Additional problems arise from the considerable variation of parameters such as

- flow conditions in the approaching river,
- water and sediment discharge,
- grain size distribution and sorting of sediment,
- flow conditions in the reservoir (3 D flow configuration).

Secondly, reliable verification of most of the above mentioned parameters is difficult if not impossible. Lack of information frequently forces to work with inadequate data and to fill remaining gaps with rough guesses.

With respect to the inconvenient data situation any sophisticated theoretical analysis of the complicated mechanism seems hardly justified. Elaborate computer work becomes questionable when the variation of the basic data used for the calculations can be of the magnitude of ± 50 % which is easily possible.

SIMPLIFIED METHOD TO ESTIMATE SLUICING EFFICIENCY

The following analysis purposely uses extreme simplifications in order to keep the computational work as straightforward as possible. The calculation is not intended to be more than a first rough estimate concerning the possibility of sediment sluicing at reasonable expenses.

Consider the situation as given in Fig. 2. The river flow enters the reservoir at point A. Upstream of this point we assume steady flow conditions, i.e. energy gradient, water level and river bed are parallel.

Using the Darcy-Weisbach friction factor the flow velocity at A can be written as

$$v_A = \left(\frac{8g}{f} h_A \cdot S\right)^{1/2} \qquad (1)$$

with

S = slope of the river bed

h_A = water depth at A

f = Darcy-Weisbach friction factor

g = acceleration due to gravity

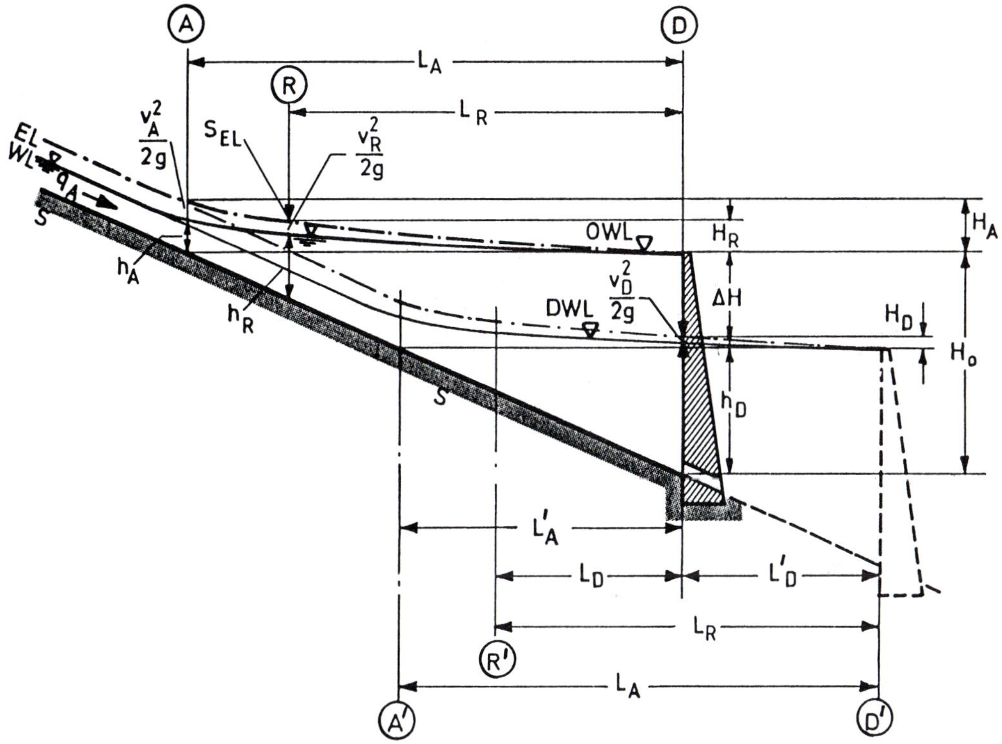

Fig. 2: Definition sketch of the parameters used in the calculation

The energy head at A is

$$H_A = h_A + \frac{v_A^2}{2g} \tag{2}$$

According to the continuity equation the river flow per unit width at A is

$$q_A = v_A \cdot h_A \tag{3}$$

Combination of equations (1), (2) and (3) leads to

$$H_A = q_A^{2/3} \, \frac{1 + 4\,\frac{S}{f}}{\left(8g\,\frac{S}{f}\right)^{1/3}} \tag{4}$$

As soon as the flow enters the reservoir the slope of the energy gradient and the slope of the bed can no longer be considered to be parallel. Assuming that the original river bed slope is also valid in the reservoir, equation (1) changes to

$$v_R = \left(\frac{8g}{f} h_R S_{EL}\right)^{1/2} \tag{5}$$

with

v_R = flow velocity in the reservoir

h_R = flow depth in the reservoir

S_{EL} = energy gradient in the reservoir.

The discharge per unit width in the reservoir (which still needs to be defined) reads

$$q_R = v_R \cdot h_R \tag{6}$$

From (5) and (6) one arrives at

$$q_R = \frac{1}{8g} \frac{f}{S_{EL}} \cdot v_R^3 \tag{7}$$

The behaviour of the sediment in the reservoir is directly related to the flow velocity. Using the Hjulstrom diagram as reference one can define distinct critical velocities for any given grain size (Fig. 3)

Introducing v_{Rc} instead of v_R in equation (7) the threshold condition for sediment motion in the reservoir can be formulated as

$$S_{ELc} = \frac{f}{8g} \frac{v_{Rc}^3}{q_R} \tag{8}$$

Here, S_{ELc} stands for the minimum energy gradient that is necessary to produce the critical velocity v_{Rc} (d). Assuming parabolic shape of the energy gradient between A and the dam of the form

$$H_R = p\, L_R^{\alpha} \tag{9}$$

one can verify the parameters p and α by means of the two constraints:

1. $\left.\begin{matrix} H_R = H_A \\ L_R = L_A \end{matrix}\right\} \rightarrow H_A = p\, L_A^{\alpha}$

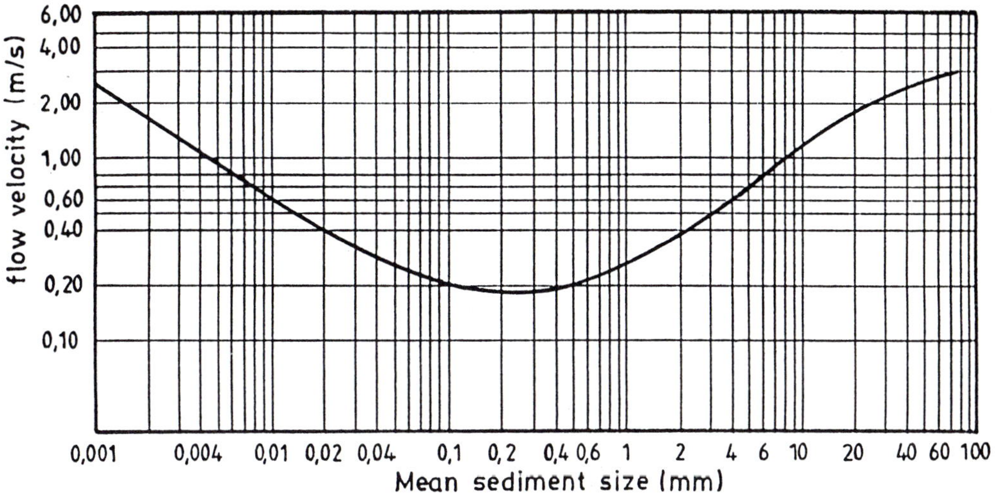

Fig. 3: Critical velocities as function of mean grain size after Hjulstrom (from ASCE)

2. The gradient of function (9) at A must be S, i.e.

$$\alpha \cdot p \cdot L_A^{\alpha-1} = S$$

Bearing also in mind that

$$L_A \cdot S = H_O$$

one arrives at

$$\alpha = \frac{H_O}{H_A}$$

$$p = \frac{H_A}{L_A^{\left(\frac{H_O}{H_A}\right)}}$$

The gradient of the parabolic function (9) can now be linked with equation (8)

$$\alpha \cdot p \cdot L_{Rc}^{(\alpha-1)} = S_{ELc} = \frac{f}{8g} \cdot \frac{v_{Rc}^3}{q_R} \tag{10}$$

In equation (10) L_{Rc} describes the distance from the dam to which a given grain size d will be transported by a given flow q_R respectively q_A at operational water level (OWL).

Strictly, q_R is also a function of L_R due to lateral mixing of the flow on its way through the reservoir. However, in mountain reservoirs which usually have limited length and width the assumption $q_R \approx$ const. $\approx q_A$ can be considered an acceptable approximation. Particularly, when sediment sluicing is carried out with water level drawdown, and even more so when a defined sluicing channel can be presupposed.

The approach presented above can also be used to calculate the effect of a certain water level drawdown (DWL) on the transport of the grain size in question. According to Fig. 2 the distance from the dam to which a given grain size d will be transported at a water level drawdown Δ H can be calculated as follows:

Water depth at the dam at drawdown

$$h_D = H_O - \Delta H \tag{11}$$

Flow velocity corresponding to h_D

$$v_D = \frac{q_D}{h_D} \tag{12}$$

Energy gradient according to equation (10)

$$S_{EL_D} = \frac{f}{8g} \cdot \frac{v_D^3}{q_D} \tag{13}$$

Virtuel distance L'_D according to equation (10)

$$L'_D = \left(\frac{S_{EL_D}}{\alpha \cdot p}\right)^{\frac{1}{\alpha-1}} \tag{14}$$

Real distance to which grain size d will be transported

$$L_D = L_R - L'_D \tag{15}$$

with L_R according to equation (10).

When the drawdown Δ H can be chosen freely, f.i. when the measure has to guarantee the transport and sluicing of a certain grain size through the whole reservoir then the calculation becomes even more straightforward. In this case the drawdown has to be adjusted in such a way that the cross-section which corresponds to the critical velocity v_{Rc} coincides with the dam itself. In this case the necessary drawdown would be

$$\Delta H = H_0 - \frac{q_R}{v_{Rc}} . \tag{16}$$

EXAMPLE

In the upper region of one of the mountain rivers in Bavaria a small reservoir has been built some time ago. Up to now the reservoir has been run only on operational water level. The yearly floods are able to clear the reservoir partly from sediment deposits which have accumulated during the year. However, the coarse deposits at the reservoir entrance have not been affected very much so far. Downstream of the dam the deficiency of coarse material causes considerable degradation of the river bed.

It is planned to sluice also the coarse material (gravel) through the reservoir by drawing down the water level at the end of the yearly flood. For environmental reasons the drawdown cannot exceed a certain maximum value. It is also planned to support the sluicing measure by creating a defined channel at the reservoir bottom (with a width corresponding to that one of the river at the entrance of the reservoir).

The questions which have to be answered before proper planning can start are as follows:

a) What drawdown would be needed to guarantee that coarse gravel (d = 30 mm) will be sluiced downstream?

b) Up to which grain size will sediment be sluiced downstream at a drawdown of 50 % of the operational water level?

c) How far is coarse gravel (d = 30 mm) transported into the reservoir at operational water level?

d) How far is coarse gravel (d = 30 mm) transported into the reservoir at 50 % drawdown?

All four questions can be answered by means of the above presented analysis.

Given:

Height of the dam $H_0 = 7$ m

Length of the reservoir $L_A = 2800$ m

Slope of the river	$S = 0{,}0025$
River flow available for sluicing	$Q_A = 400\ m^3/s$
Width of the river bed	$B_A = 60\ m$
Width of the channel at reservoir bottom	$B_R = 60\ m$
Friction factor (Strickler)	$k_s = 35\ m^{1/3}/s$

Calculation:

$k_s = 35\ m^{1/3}/s$ corresponds to $f = 0{,}05$

$$q_A = q_R = \frac{400}{60} = 6{,}67\ m^3/s/m$$

coarse gravel $d = 30$ mm $\rightarrow$ $v_{Rc} = 2.1$ m/s (from Fig. 3)

a) According to (16):

$$\Delta H = 7 - \frac{6{,}67}{2{,}1} = 3{,}82 \quad (m)$$

Sluicing of coarse gravel would require a drawdown of 3,82 m.

b) $\Delta H = 0{,}5 \cdot 7 = 3{,}5 \quad (m)$

From (16):

$$v_{Rc} = \frac{6{,}67}{7-3{,}5} = 1{,}91 \quad (m/s)$$

From Fig. 3:

$d = 20$ mm

Drawdown of 50 % will faciliate the sluicing of grains up to 20 mm.

c) According to (4):

$$H_A = 6{,}67^{2/3} \frac{(1+4\,\frac{0{,}0025}{0{,}05})}{(8\cdot 9{,}81\cdot\frac{0{,}0025}{0{,}05})^{1/3}} = 2{,}70 \quad (m)$$

From (8):

$$S_{ELc} = \frac{0{,}05}{8\cdot 9{,}81} \cdot \frac{2{,}1^3}{6{,}67} = 0{,}00088$$

From (10):

$$\alpha = \frac{7}{2{,}7} = 2.593$$

$$p = \frac{2,7}{2800^{2,593}} = 3,111 \cdot 10^{-9}$$

$$L_{Rc} = \left(\frac{0,00088}{2,593 \cdot 3,111 \cdot 10^{-9}}\right)^{\frac{1}{2,593-1}} = 1458 \quad (m)$$

At operational water level coarse gravel of 30 mm will be transported up to 1458 m upstream of the dam.

d) According to equation (11) to (16):

$$h_D = 7 - 3,5 = 3,5 \quad (m)$$

$$v_D = \frac{6,67}{3,5} = 1,91 \quad (m/s)$$

$$S_{EL_D} = \frac{0,05}{8 \cdot 9,81} \cdot \frac{1,91^3}{6,67} = 0,00067$$

$$L_{D'} = \left(\frac{0,00067}{2.593 \cdot 3,111 \cdot 10^{-9}}\right)^{\frac{1}{2,593-1}} = 1251 \quad (m)$$

$$L_D = 1458 - 1251 = 207 \quad (m)$$

At 50 % water level drawdown coarse gravel of 30 mm will be transported up to 207 m upstream of the dam.

REFERENCES

ACKERS, P., THOMPSON, G.,
Reservoir sedimentation and influence of flushing,
Sediment Transport in Gravel-bed Rivers, J. Wiley & Sons,
London, 1987.

ASCE,
Sedimentation Engineering,
ASCE Manuals and Reports on Engineering Practice No 54,
New York, 1977.

SCHEUERLEIN, H.,
Sedimentation of reservoirs - methods of prevention, techniques of rehabilitation,
First Iranian Symposium on Dam Engineering,
Tehran, Iran, June 1987.

WHITE, W., BETTESS, R.,
The feasibility of flushing sediments through reservoirs,
Challenges in African Hydrology and Water Resources (Proceedings
of the Harare Symposium, Juli 1984), IAHS Publ. No. 144.

REVIEW OF DISASTROUS TORRENT FLOOD ON THE VLASINA RIVER ON JUNE 26, 1988 - INCLUDING ANALYSIS OF FLOOD AND THE OBTAINED RESULTS

Zoran Gavrilovic and Zivorad Matovic

Institute for the Development of Water Resources
"Jaroslav Cerni", Jaraslava Cernog, Beograd, Jugoslavia

ABSTRACT

A disastrous torrent flood on the ***Vlasina River*** of **26 June 1988** was of unprecedented intensity in Yugoslavia. The catchment area affected by the catastrophe covered more than **1000 km²** and the damage was estimated to one billion dollars. It is interesting to note that this area was treated with erosion control works and that percentage of eroded land (soils) was negligible. According to the experts opinion, or should we say delusion, torrent flood could not possibly occur in thus protected catchment. However, we have to admit that impossible did happen. The analysis of the flood yielded significant results that revealed that experts were mistaken in many points. Those results indicated a need to find different analytical approach because, besides visible traces, the actual records of flood were sparse. Rain gauging stations in the zone affected most seriously were completely destroyed as well as water level gauges. Since the torrent flood computations on the base of flood traces are not reliable, the causes of flood occurrence and development of flood itself had to be reconstructed. In that respect, meteorology - radar surveys were used for reconstruction of rainfalls. Using these data, as well as those obtained by torrent bed survey, by grain size analysis and by determination of peak flood on some tributaries and along the river itself, the flood could be analysed. In addition, the efficiency level of erosion control works was analysed and also the importance of torrent control as basic element of complex erosion and torrent control works.

1. INTRODUCTION

The *Vlasina River* catchment area is **985 km²** large, at **1300-1700 m a.s.l.** The confluence of the ***Vlasina*** and ***Južna Morava River*** is near **Leskovac town**, at **250 m a.s.l.** This shows that the ***Vlasina*** and its tributaries flow at a steep slope which is one of the basic prerequisites of torrent flow. The catchment is of fan - like geometry which is characteristic for torrent flows.

Torrent-like properties of the ***Vlasina*** and its tributaries were noticed a long time ago, so that as early as before II World War this area was treated with torrent control works, although not so intensively as in adjacent Grdelica Gorge. The catchment was also subjected to erosion control works, mostly due to legal obligation to treat ones own estate(s) with soil conservation works, involving afforestation and other measures.

These details are necessary for full comprehension of the complex situation on **26 June 1988**, i.e. of torrent flood on the ***Vlasina river***, and for enlightenment of possible events leading to its occurrence.

2. REVIEW OF DISASTROUS TORRENT FLOOD ON THE VLASINA R. OF 26 JUNE 1988

In the afternoon of that day, the upstream part of the ***Vlasina*** catchment covering **300 km²**, was stricken by an **intensive 3 hour rainfall**. Excessive precipitation led to development of torrent flood which was of disastrous impact even in the upper catchment area. The same happened with all tributaries affected by heavy rainfall.

The movement of cloudburst was convenient for formation of numerous torrent floods on tributaries, which jointly formed disastrous torrent flood, first on large tributaries and then, on the ***Vlasina*** itself.

The upper part of the ***Vlasina*** catchment is rather craggy, with not so much of agricultural land so that, in this respect, flood was not so harmful. However, the flood destroyed all the bridges on its way and most of the road leading through the gorge of the ***Vlasina river***. The flood also caused many landslides in the area, which means additional layer of sediment in the river.

Since torrent lava consists of water and sediments of different grain size, its disastrous energy was great. All structures on its way were completely demolished or severely damaged.

The mountain villages were the first ones attacked: all bridges and buildings that were exposed to lava were destroyed. Only ruins and river full of rocks were left behind. The sediment layer was even several meters thick at some places, **Fig.1**.

Fig.1. Before flood, torrent bed was 3 m underneath the road. The flood destroyed the bridge leaving visible traces of sediment

The torrent lava overflowed cultivated soils in lowlands, depositing sediments. The total area permanently destroyed in this way was 500 ha, which is 10 % of the whole region. In many villages, the best quality land was flooded.

The picture of demolished and damaged settlements, bridges and roads, as well as of arable land, was the same all the way up to the confluence of the ***Vlasina*** and ***Južna Morava rivers***, Fig.2.

Vlasotince is the biggest settlement attacked by torrent flood. It is situated on the ***Vlasina river***, about 10 km from its confluence. There was a dam here, but it could not withstand the burst of water which caused dam breach on the right lateral side and then run wildly through the town, crushing everything on its way. Fortunately, it was still daytime so that people could desert the endangered area and avoid sure death.

In the upstream reaches of the ***Vlasina river*** the huge rocks (maximum grain size) were left after the flood. The biggest sample found in Yugoslavia up to this flood was **68 m³ (in 1968)**, while in the ***Vlasina***, many rocks were twice as big as that one, **Fig.4. and Fig.5.**

Fig.2. Ruins of a badly damaged bridge (Tegošnica))

Fig.3. After dam breach, the torrent flood destroyed part of the town

Fig.4. Torrent flood was intensive enough to move and even carry sediment of *"maximum grain size"* often exceeding 100 m^3. The actual magnitude of such a grain can be seen in comparison to a height of a man.

Fig 5. The fact that maximum sediment grain size of diameter over 5 m, weighing more than 200 tons, was common in that area could be best comprehended from perspective view.

3. ANALYSIS OF PRECIPITATIONS ACCOUNTED FOR TORRENT FLOOD

The disastrous impact of the torrent flood can not be seen clearly from this summarized description, without knowing the causes of its occurrence.

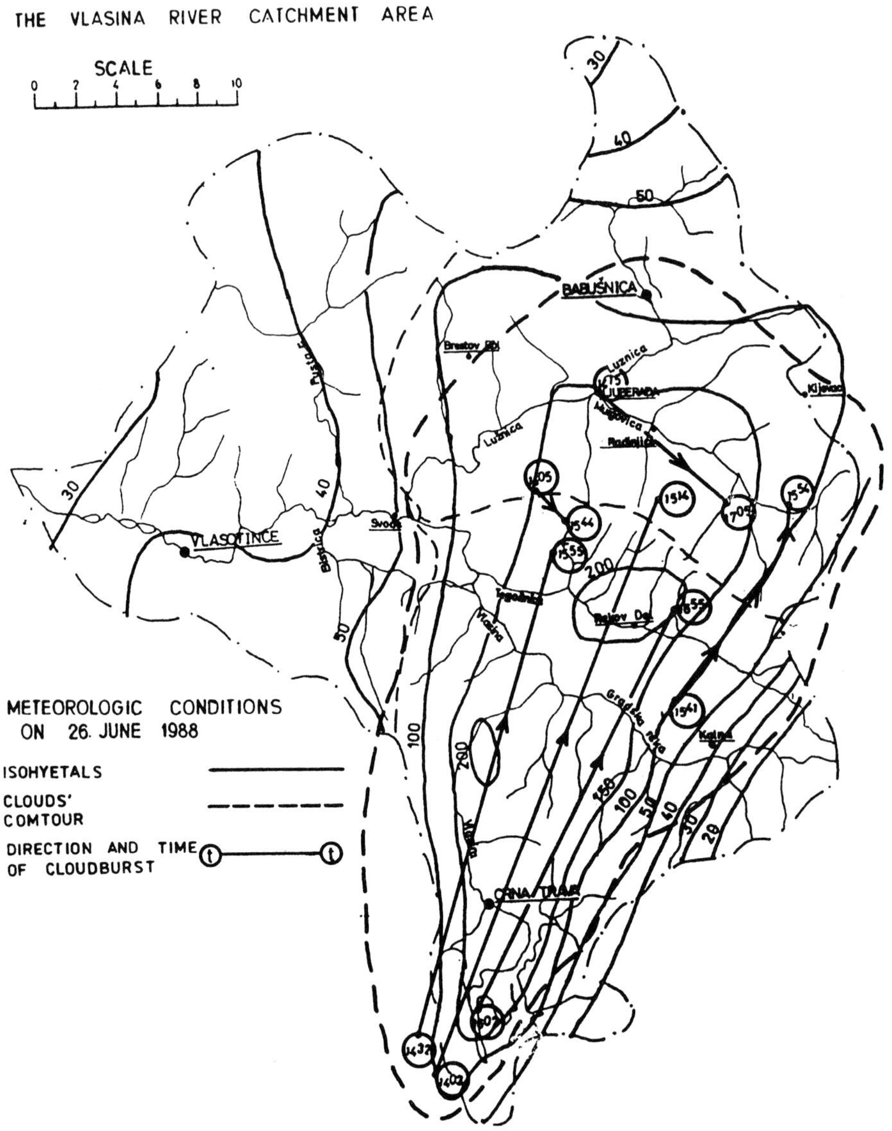

Fig.6 Meteorologic situation over the Vlasina river catchment

At the very beginning of the field investigations, it was found that only one rain gauging station existed in the centre of the zone affected by cloudburst, which recorded the precipitation rate of **220 mm** for two hours of intensive rainfall. The other stations were destroyed. There were also several adjacent stations but their records of flood were not satisfactory so that radar data on cloudburst movement had to be analysed in order to get real picture of its development in space and time and to make isohyetal map of the flood. Due to limited scope of this paper, it is presented in the most convenient graphic form, **Fig.6**.

Obviously, the rainfall was of short duration, **3 hours** at most, but of very high intensity. The mean annual precipitation in the ***Vlasina*** catchment is **830 mm**. Using results of long - term observations of torrent floods, as well as statistically data, the storm rain depth was determined ***(rainfall intensity was about* 1 mm/min)**, for different rainfall probabilities. In the case of the ***Vlasina*** and its tributaries, the calculated rainfall depth was **90 mm** which is characteristic of a rain causing catastrophic flood. The registered values could not have been foreseen at all. **Ref.1.**

4. RESULTS AND METHODS OF TORRENT LAVA FLOW COMPUTATIONS

To compute torrent lava flow is hardly an easy task. Most often it has to be determined on the basis of flood traces. In the case of phenomenon discussed, it was the only possible way since all gauging devices were completely destroyed. But even if they had been found intact, they could have supplied no records of flood since they were not designed to operate under torrent flood conditions. **Ref.2.**

The available methods of floodplain discharge computations by flood traces are based on different compilations of ***Bassin's*** and ***Manning's*** formulas. However, although convenient for flood flow computations on rivers, they do not yield reliable results in case of torrent floods, due to large amount of sediment and, as in this case, lava as well, carried by the torrent. The solid proof of torrent lava density is a rock sample found in the ***Gradska reka***, tributary of the ***Vlasina river***. It was stuck between branches of a tree, at the height of **1 m. Fig.7.**

The situation in the downstream reaches of the river was similar as it can be seen from a close look at the photograph of a river bank **(8 m *above river bed)*** and tree which was completely stripped due to gravel fractions sediment, **Fig.8.**

Every expert is well - aware of the difficulties encountered in computing on the basis of torrent flood traces, even when the flood is much less intensive than the one described here.

The torrential flows discharge velocity could be measured only on the surface (using buoys, under the condition of really being present at the moment of passage of the flood water peak)

In this case, although there were no experts present during the flood, the torrent water movement velocity could be calculated.

Namely, the river Vlasina flood captured a large watershed area and passed along the total length of 80 km. The torrental water velocities were not high, the flood occured at day time so that the data could be collected from the inhabitants.

The detailed poll conducted among the inhabitants confirmed the assuption. There were many cloks which stopped at the moment of the flood and many people remembered the hour. In many cases the hourly wages of television and radio programmes was of help.

The detailed reconstruction of the torrential flood wave travel time obtained this way, enabled the accuracy which is difficult to achieve in such situations. The greatest error could be of the 15 *minutes* order of magnitude which is negligible if the flood waves travel time of 9 *(eight)* hours is considered.

Fig.7. Rock carried by dense torrent lava (weight ab. one ton)

According to the Fig.6. the duration time of the catastrophal rain (**14.^{00}h** to 17.^{00}h with the flood waves starting at **14.^{30}h** to reach the estuary around **23.^{30}h**) could be calculated.

We already said that standard methods of computation using ***Bassin's*** and ***Manning's*** formulas, as well as their compilations, did not yield satisfactory results.

The solution to the problem was derived from the results of laboratory investigations carried out 19 years ago, concerning water and sediment transport under conditions similar to torrent flow. **Ref.3.**

The results were rather interesting, particularly those regarding lava flow. In analysing the results, it was concluded that the principles of torrent lava flow and of fluid flow in hydraulic systems are different.

Fig.8. Gravel sediment completely stripped off a tree at height of 6 m

This was the reason to cease the tests, since there was no concordance between the test results and those already known and accepted.

Nevertheless, the results were sufficient to derive several formulas for computing the mean velocity of torrent and lava flow, which were later improved and modified. The flood on the Vlasina river was the real opportunity to check the validity of all relevant methods and formulas, including those improved on the basis of already mentioned investigation results.

5. FORMULA FOR COMPUTATION OF TORRENT LAVA FLOW VELOCITY

The basic formula for computation of the torrent lava flow velocity is:

$$V = R_v \times \frac{(g \times f \times I)^{1/2}}{w \times (b)^{1/2}} \ (\mathrm{m/s})$$

where

V - mean velocity of torrent lava flow expressed in **m/s**

g - gravitational acceleration **g=9.81 m/s²**

f - cross-sectional area up to height of flood traces expressed in **m²**

I - torrent bed slope (m/m)

b - mean width of torrent bed in meters

w is experimentally determined roughness coefficient of river bed and banks for natural and treated torrent beds, the values of which are the following:

I. Values of coefficient w for untreated torrent bed (**w**)

I.1. Torrent bed in vary bad condition, covered with coarse - grained sediment, overgrown with weed and bushes....................0.100

I.2. Torrent bed in bad condition, covered with coarse - grained sediment, overgrown with weed..............................0.090

I.3. Common torrent bed in soil, with gravel and schist bottom or pebbles, covered with weed..................................0.085

I.4. Torrent bed in sand deposits or loess, rather regular, covered with weed..0.080

II. Values of coefficient w for treated torrent bed (**w**)

II.1. Bottom and banks of excavated torrent bed in natural state; composed of earth fill or schist; meanders are regular; banks under grass......................................0.065

II.2. Bottom and banks of excavated torrent bed composed of earth fill or schist, but at certain distances masonry walls or concrete consolidation belts are raised; banks under grass..........0.060

II.3. Earth fill bottom; banks are protected with walls of crushed stones in cement grout or with coarse cement lining.........0.055

II.4. Bottom and banks completely covered with linings of crushed stone in cement grout....................................0.045

II.5. Bottom and banks completely covered with normally coarse concrete linings...0.040

R_v is a coefficient of torrent lava velocity reduction due to sediment content. The similar coefficients were used in other compilations of well-known formulas, but they were computed in different way.

The basic reason for introduction of R_v coefficient into computations is the considerable content of sediment in torrent lava which increases its density and, therefore, can not be neglected. Density of torrent lava will be most successfully determined using method of sampling, but it is usually inapplicable since requires being at the gauging point at a right moment. The investigations regarding torrent lava density as a function of bed slope and sediment grain size distribution have not been completed yet, so at this moment, we can only give some approximate values of torrent lava density:

Description of a torrent bed as an indicator of magnitude of torrent lava density

1. Steep slope of torrent bed, with visible traces of disastrous torrent flood; maximum sediment grain size exceeding 1 m. Figs. 3, 4 and 7 .... $\gamma > 1.5$

2. Steep slope of torrent bed, with visible traces of disastrous torrent flood; maximum sediment grain size ranking from 0.5 to 1.0 m .... $1.4 < \gamma > 1.5$
3. Steep slope of torrent bed, with visible traces of disastrous torrent flood; maximum sediment grain size ranking from 0.3 to 0.5 .... $1.3 < \gamma > 1.4$
4. Moderate slope of torrent bed with visible traces of torrent flood; maximum sediment grain size ranking from 0.1 to 0.3 m .... $1.2 < \gamma > 1.3$
5. Minor slope of torrent bed; visible traces of torrent flood, but there are none of destruction; maximum sediment grain size ranking from 8 to 15 cm ....$1.1 < \gamma > 1.2$

6. Minor slope of torrent bed; traces of torrent flood are hardly visible and there are none of destruction; maximum sediment grain size is under 8 cm .... $\gamma < 1.1$

Torrent lava density, if cannot be measured directly, should be determined on the basis of the flood traces and mentioned criteria.

Grain size analysis of bed load should be done as well. In that respect, only the sediment deposited by flood in question should be analysed and not layers formed during earlier floods. The sediment layer resulting from one flood may be even several meters thick.

Fig.9. Usual view of torrent bed

Grading curve should be plotted only for large fractions since the percentage of small ones in torrent flood flows is small, **Fig.9., Fig.10.**

Coefficient R_v is derived on the basis of the formula:

$R_v = K_g \times (1/\gamma)$ where

is measured or estimated density of torrent flow

K_g is a coefficient of grading curve, which is computed by pondering the mean value of percentage offractions, as a function of defined coefficients. These coefficients are computed using abundance of measurement data on very strong torrent flows. The example showing computation of coefficients for some fractions is:

FRACTION	coeff. K_g	%
rocks with diameter over 50 cm	0.84	6
large pebbles 30-50 cm	0.87	66
pebbles 6-30 cm	0.90	18
large-size gravel3-6 cm	0.93	9
medium-size gravel 10-30 mm	0.96	1
small-size gravel 2-10 mm	0.99	0
the mean pondered value.....................	**K_g=0.89**	

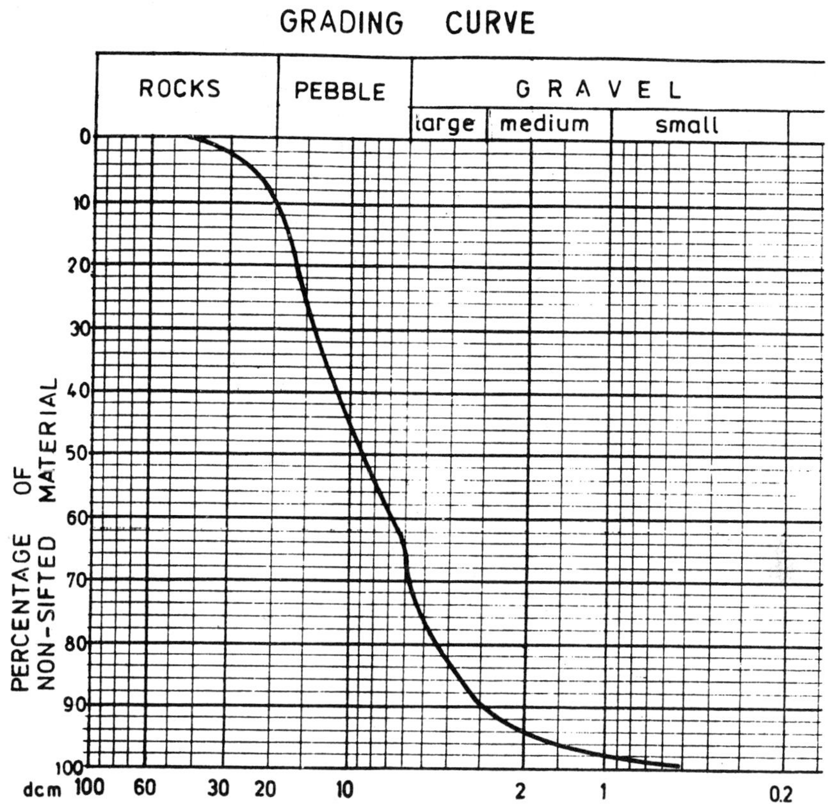

Fig.10. Grading curve for bed load

The formula used here proved to be applicable and at certain levels, showed some advantages over other methods in terms of accuracy. Computations by this formula were performed immediately after the surveying of flood traces. Later, meteorological analysis was completed and its results used for computation of possible peak discharge. These values were approximate, so there was no need for additional calibration. Since there is not enough space to present all results in detail, we will keep to main indicator of magnitude of torrent flood on the ***Vlasina river***, and that is the computed value of flood plain discharge which destroyed a dam and part of the **Vlasotince town**; it is Q_{max}=1200 m^3. One should have in mind that the flood wave was a result of a previously described heavy rain which affected **300 km^2** vast area.

6. EFFECTS OF EROSION CONTROL AND TORRENT FLOW CONTROL WORKS

The ***Vlasina river*** catchment area was sufficiently treated with various erosion control works and measures which nowadays ensure complete protection. But there is still an open question: how the catastrophic torrent flood of **26 June 1988** could have occurred in spite of all precaution?

Fig.11. The wiew of one part of the Vlasina river catchment (visible protection works against surface erosion)

The answer is quite simple: the catchment was protected only against surface erosion, while more than insufficiently against torrent flood **Fig.11.** The proof of need and benefits from complete torrent flow and erosion control works should not have, however, come from so painful experience such as the disastrous torrent flood causing damage which far exceeds the cost of torrent control works not only in the ***Vlasina*** catchment, but in others too. It was a mistake which showed the importance of torrent control works, beside erosion control, and which also revealed many mistakes in the experts point of view.

The only part of the ***Vlasina river*** catchment affected, but not destroyed by disastrous impact of the flood is a river stretch at **Crna Trava** which was previously treated with torrent control works. The works were carried out in accordance to the prescribed rules and, at the same time, they are the only ones completed in the Vlasina catchment, **Fig.12.**

As the torrent flood of such a magnitude ***(catchment area of* 1000 km²*)*** is unprecedented in Yugoslavia and, as much as we know, worldwide as well, the erosion intensity was analysed, yielding sufficient level of protection of erosion control works.

In analysis of this kind, the complex method of erosion intensity has been used in Yugoslavia, i.e. ***"The method of erosion potential"*** *(developed in the* ***"Jaroslav Černi"*** Institute, in the period 1954-1987), Ref.4.

Fig.12. The only section of a river affected by torrent lava which remained intact

The current state of erosion and torrents has not been analysed separately since the investigation results 10 years ago had shown tendency of decreasing soil erosion intensity by some 30 %, as compared to the conditions back in the time when erosion control works were started. Field observations confirmed this statement. To represent of erosion intensity, according to the modified methodology, the erosion map is used, but it is not necessary here.

The efficiency levels of erosion control works and of torrent control works do not show concordance, but are intermingled. Torrent-like characteristics of a flow result from intensity of erosion, but also from the level of torrent control. Namely, torrent control works and structures, beside reducing sediment transport, exercise another important function, and that is, consolidation of river bed. **Ref.5.**

The violent cloudburst which caused torrent flood on the **Vlasina** had variable intensity over the affected area which enabled analysis of the efficiency level of the erosion control works. It was concluded that the erosion control works ensure high level of protection, and noted that before they were executed, torrent flood was often encountered, resulting when the storm rainfall depth was over **40 mm**. Nowadays, the torrent floods occur when the rainfall depth exceeds **75 mm**. For example, the catastrophic rainfall depth for the ***Vlasina*** catchment is only **90 mm**.

It means that the erosion control works ensure protection even in case of some storm rains of longer return period. But the real level of their efficiency has been still investigated.

7. CONCLUSION

Torrent floods cannot be compared with floods on big rivers which cause serious damage but are of smaller intensity. The structures which happen to be on the way of the torrent flood are very alike ruins after strong earthquake, if can be found at all.

The disastrous impact of a torrent flood is a high content of sediment in a water. This mixture was so dense in some cases that it was called ***"muddy and stony"*** or just ***"stony flow"***. However, most often maximum grain size of sediment varies from 30 to 50 cm in diameter, but they are not the hugest samples found.

In such cases, standard torrent flood control measures and structures are not suitable due to disastrous impact of torrent floods, so that special torrent and erosion control systems must be designed.

The catastrophic torrent flood of **26 June 1988** on the ***Vlasina river*** is the most obvious proof of disastrous impact of torrent floods and of need for torrent control works even when the catchment is completely protected against erosion.

REFERENCES:

1. Hydrometeorological Report on the Torrent Flood on the Vlasina River of 1988 (State Hydrometeorological Service)

2. M. Holy: Erosion and Environment, PERGAMON PRESS . OXFORD . NEW YORK. TORONTO. SYDNEY. PARIS . FRANKFURT. 1985.

3. S. Gavrilović: Torrents and Erosion (Bujični tokovi i erozija), Beograd, Građevinski kalendar 1976.

4. Z. Gavrilović: THE USE OF AN EMPIRICAL METHOD (EROSION POTENTIAL METHOD) FOR CALCULATING SEDIMENT PRODUCTION AND TRANSPORTING I UNSTUDIED OR TORRENTIAL STREAMS, International Conference on RIVER REGIME, Wallingford 1988

5. Z. Gavrilović: Natural disasters in Atmosphere and on Rivers - Erosion and Torrents (Atmosferske nepogode i nepogode na rekama - Bujčni tokovi i erozija), Budva 1986.

RESEARCH OF FLUVIAL PROCESSES IN MOUNTAINS
A CHANGE OF EMPHASIS

Tim R.H. Davies

Natural Resources Engineering Department
Lincoln College, University of Canterbury, New Zealand

ABSTRACT

The impact of fluvial events on inhabitants and facilities in mountain valleys is increasing. In order to better protect inhabitants and developments, knowledge of the possible magnitudes of hazard events is vital, and knowledge of the conditions that allow their occurrence should form the basis of a forecasting and warning strategy. Too much current and past research focusses on theoretical aspects of mountain fluviology that cannot contribute to these aims in the forseeable future. There is an urgent need for directly applicable research on possible event magnitudes and preconditions; of necessity this is likely to be empirical rather than theoretical or statistical, in order to yield realistic and useful results.

INTRODUCTION

In recent years there has appeared a wealth of publications on the hydrology and fluviology of mountain regions. There are two reasons for this. One is the increasing consciousness of the risks to human life and facilities in mountain regions, in spite of the almost universal reduction in their population; the use of the areas by humans is however probably increasing through sport, communications, hydro-power and forestry, so the impact of fluvial events is probably increasing also. Thus, an obvious reason for carrying out research on mountain areas is to better predict these hazards and mitigate their effects.

The other reason is that hydrologists and fluviologists have recognised in the mountain environment a hitherto almost unexplored arena for their scientific skills, one furthermore in which processes are clearly much different to those in lowland regions. The complexity of this environment is clear, and the scene is thus ripe for a myriad of publications describing the processes observed, and for hypotheses and models to represent them which have little chance of being unequivocally tested because of the difficulty of accounting for the (random?) effect of unconsidered variables. A tempting field indeed.

Cursory scrutiny of many of these publications, including mine, reveals that most of them fit the second reason better than the first. That is, they are 'basic science' in a subject which is justifiable to a funding body because the knowledge obtained could perhaps in due course be useful in understanding or even mitigating hazards. In most cases, though, the number of extra steps needed to make these contributions practically useful is large - often very large.

It is the thesis of this essay that, while lives are being lost due to mountain hazards, this ivory-tower game of Monopoly is rather immoral. Clearly there is a need to reduce death and destruction in mountain areas due to snow avalanches, floods, debris flows, rockfalls, etc.; this is moreover an urgent need, better methods are needed now, and much more research is needed that is immediately and directly useful to a strategy for hazard mitigation.

Herein, the conventional methods of assessing and predicting hazard events (theoretical-simulation and empirical-statistical) are briefly outlined, and a scrutiny of the highly variable fluviology of mountain catchments allows the usefulness of these methods to be evaluated. Based on what can realistically be known about mountain fluvial hazards, and their future occurrence, a strategy involving possible event magnitudes, acceptance that the probability of occurrence of an event is unknown but low, and a combination of land-use planning with real-time monitoring-warning-evacuation procedures is put forward as a realistic goal for hazard management. Finally, some research areas are suggested that could contribute directly to the development of such a strategy.

FLUVIAL HAZARD ASSESSMENT

Mountain rivers become a hazard only when they threaten human life or property, by inundation, erosion, sediment deposition or destruction. By contrast, lowland rivers often pose only the threat of inundation, and the simple extension of hazard assessment methods from a lowland to a mountain situation is likely to be inadequate. Similarly, while inundation in a lowland valley results primarily from a high water level which is essentially horizontal across the valley, this is much less likely to be the case in the much steeper and rougher terrain of the mountains.

The theoretical-simulation approach to fluvial events is based on some fundamental understanding of the flow processes involved, for example, turbulent sediment-laden flow over a rough boundary, or high-concentration slurry flow in a confined channel. This understanding usually takes the form of a theory or model which is related to more or less realistic boundary conditions and time by a computer simulation. To be useful the theory must accurately describe the process over the likely range of flow conditions, and the boundary conditions must also be able to vary realistically in response to the flow (and vice versa); this feedback between flow and boundary is an essential feature of most erosion processes and is fundamental to their prediction.

The empirical-statistical approach uses recorded field data describing the event as it has actually occurred in the past, and predicts future events by extrapolating from a statistical description of the past data. The 'model' or 'theory' in this case is usually an assumption that the field data fit some standard statistical distribution of known properties, whose parameters remain constant with time. Typically a magnitude-frequency relationship is derived from this distribution and used for an economic analysis to find the 'best' solution to the problem of hazard mitigation.

FLUVIOLOGY OF MOUNTAIN REGIONS

Here we consider the nature of mountain fluvial processes in order to assess the practicability of these two approaches to predicting hazard events.

The outstanding characteristic of fluvial and hydrologic processes in mountain regions is their extreme variability in both time and space. Science has classically treated such variability by calculating averaged parameters, assuming these to be steady and/or uniform and analysing them on the basis of theory. Mountain hazards, however, usually result from the occurrence of just these localised, extremely intense events that are concealed by such averaging, and hence it is imperative to try to quantify the variability. Precipitation is a good example of the difficulty of doing this. Rain- and snowfall in mountains can be very localised because of orographic effects, and events such as debris flows and flash floods result from the occurrence of very intense and localised 'bursts' of rain (Church and Miles, 1987; Okuda et. al, 1980). It is very difficult to obtain data to describe the spatial and temporal distribution of these rainbursts, and their intensity, simply because it is hard to put sufficient raingauges in a catchment for long enough. Hence, although clearly vital for predicting the event, this information is simply too difficult (expensive) to obtain.

Other processes contributing to fluvial hazard events show a similar degree of variability. Runoff, for example, can be extremely rapid or non-existent, depending on antecedent conditions, vegetation, soil depth and infiltration capacity, type of bedrock, local slope, the presence or otherwise of a snow pack, and so on; slope erosion can range from very severe to zero, local to very extensive, depending on all these factor together with the occurrence or otherwise of slope oversteepening by an undercutting river, wind load on vegetation, earthquakes, etc. The entry of water and sediment to streams (the independent variables which dictate the stream behaviour) are therefore very variable in time and space, and quite similar storms can give rise to very different responses of flow rate, bed and bank erosion, sediment load, water level and damage, especially in low-order streams (Hayward, 1980). The morphology of mountain streams is itself very variable (Newson, 1981), channel width being about the only consistent parameter. A very wide range of grain-sizes is often present; step-pools, riffles, boulder accumulations and (sometimes vegetated) channel bars are common as are bedrock steps and gorges (Hayward, 1980) and log-jams in forested regions. Wider rivers are often braided in planform, with the number of channels varying strongly along the valley (e.g., from one to ten to one again within 5 km in the 1 km wide Rakaia River, New Zealand), and with channel patterns that vary rapidly with time at high flows. Even under steady flow, the bedload transport rate in a channel of a gravel-bed mountain stream can vary by an order of magnitude within a few minutes or a few metres (Hubbell, 1987).

During high-intensity events the nature of the flow in a mountain river can change in a very short time from a turbid flood transporting gravel as bedload to a very unsteady, pulsing flow of an extremely dense slurry of boulders, gravel, sand, silt and clay in water - a debris flow. The behaviour of debris flows is poorly understood (Davies, 1988a) and their infrequent occurrence makes good description difficult; obtaining field data on their composition and motion is clearly not an easy task (Pierson, 1985). The conditions under which they occur are known qualitatively at least, but the degree of variability intrinsic in their behaviour is again very high. Figure 1, for example, shows how shear stress varies with shear rate when debris flow material is sheared at a <u>steady</u> rate in a cone-and-plate viscometer (Phillips and Davies, 1989) - the data scatter is real, not the result of measurement error, and results from the interaction of large stones in the flow.

Where a mountain stream has a forested catchment, dead trees will enter the stream from time to time and might wedge across the channel, forming a barrier against which other trees can accumulate, forming a log jam. This will in turn accumulate a large volume of sediment behind it. In due course the logs forming the jam will rot or will be removed by a very large flow, releasing a large quantity of water, sediment and woody debris in a wave that might well destroy other log jams downstream. Such an event can form an extremely destructive flash flood (Kellerhals, 1988).

The influence of events external to the stream is potentially very much greater on a mountain stream than on a larger lowland river. Snow avalanches and rock falls, as well as log jams, can temporarily block the stream and later release a destructive flood wave. Mountain streams rarely have self-formed 'alluvial' channels; boulders emplaced by glacial or gravity action might well be too big to be transported by the stream, but can have a major effect on its behaviour. Similarly rock steps (waterfalls) often result from external factors such as earthquake fault lines.

All the above factors make it extremely difficult to apply fundamental models for sediment transporting flow processes - for example the Meyer-Peter and Müller bedload transport equation or the Manning flow resistance equation - to streams in mountain catchments. How is mean flow depth to be measured in a braided river with ten individual channels or in a single channel in which boulders protrude from the flow (Bathurst, 1978)? Where is the mean bed level in a channel that has boulders protruding from the flow (MacMurray, 1985)? How is sediment concentration to be measured in a debris flow? How often are the common assumptions of steady, uniform flow met? Even a model using gross flow parameters, such as Chang's (1980) suggestion that the product of flow rate and slope is minimised in a self-formed channel, is of little use when the channel is <u>not</u> self-formed, slope cannot be measured accurately because the water surface is so rough, and the flow rate varies widely within a few minutes. How accurately can the continuity equation be applied when the flow is aerated to an unknown degree? The prospect of practicable theoretical simulation models thus seems to be extremely remote, particularly when the complexity of the boundary conditions, the importance of extreme values of the variables, and the possibility of fundamental changes of behaviour (e.g. from water flood to debris flow) is acknowledged.

The alternative approach, of gathering empirical field data on hazard events to put into a statistical model, has the advantage that it starts from practical reality rather than conceptual theory, and the parameters as measured incorporate all the complications listed above. The measured data, however, are <u>extremely</u> difficult to obtain. Hazard events in the mountains usually occur in the early hours of Sunday morning, in winter, during severe rain- or snowstorms. Access to the hazard area is often difficult and sometimes dangerous, and it is very expensive to install sufficiently robust equipment for automatic data collection. Events occur relatively rarely at a given place (for example, very few people have seen debris flows except in experimental catchments). Hence, even with well-funded data collection programmes it is difficult to foresee large quantities of data soon becoming available to describe, for, example, flash flood wave velocities or debris flow surge heights (Pierson, 1985; Costa, 1987). Given the sparseness and inevitably approximate nature of data on extreme events, it seems unlikely that an empirical-statistical model will be able to yield useful information about the future occurrence of mountain hazards. In effect, this means that the frequency of occurrence of a hazard is not able to be predicted in this way; indeed, as will be shown later, even a well-defined magnitude-frequency relationship is of little practical value for infrequent events, since there is always a very large degree of uncertainty involved in predicting when an event of a prescribed magnitude will occur.

TOWARDS A STRATEGY OF HAZARD MITIGATION

The urgent need to find a way of reducing death and destruction in mountain valleys requires that we be honest about how much we <u>can</u> know about hazards in such an environment. Since lives are at stake, we cannot allow pride or scientific rigour to affect our judgement of what should or should not be done.

The data available on hazard events span at most a few hundred years, usually much less; in this time there might be approximate estimates of the magnitudes of several large events and better estimates of recent moderate ones. The magnitude of the former events can probably be quantified to within, say, ± 30%, even for catastrophes recorded in folklore or old newspapers. The statistical frequency of occurrence of major events is much more difficult to establish, simply because so few data exist. Even with sophisticated modern analysis, future predictions cannot be made with any confidence from sparse past data. The most appropriate method of analysis seems to be the simple distribution of exceedences, which allows the risk, that an event equal to or greater than a recorded event will occur within a chosen number of years, to be specified; it requires no assumptions other than that the future (unknown) distribution is the same as the (unknown) past distribution, and that events are independent of each other (Young and Davies, 1989). Even with a complete knowledge of the magnitude-frequency relationship, however, the future <u>risk</u> of an event occurring tells one nothing about whether or not it will occur within the next, say, 100 years; the risk will only match the occurrences on average over a very long period of time, say, 10,000 years in the case of an event with an annual exceedence probability of 0.01. During the next one hundred years, this event will be exceeded perhaps 0 times, or once, or twice, or 3 times; it certainly will not be exceeded 0.63 times, which is the theoretical risk of exceedence. Since we are concerned with hazard events, which are by definition relatively rare, the number of times they will occur in the next 100 years (a realistic planning time-frame) is essentially unknown, but low. Any attempt to calculate a risk or a frequency thus seems to be a waste of time. This is equivalent to admitting that the probability of occurrence is also unknown, but low. The two important characteristics of a hazard event, knowledge of which would allow its effects to be mitigated, are its magnitude and time of occurrence; the latter cannot be predicted.

Accepting that the risk of an event is unknown is not, in fact, a retrograde step; again, if the risk were accurately known, it would be almost impossible to decide whether this risk was acceptable. In principle an economic decision could be made to do this, but the end result is likely to be largely spurious (Young and Davies, 1989). Such a procedure also ignores unquantifiable costs and benefits due to environmental, social and cultural issues. The acceptable degree of risk is thus essentially a subjective decision, and little is lost by admitting the actual risk to be unknown.

Event Magnitudes

What remains, then, as useful information, is knowledge of the magnitudes of past events. This shows that events up to a certain size have occurred at this place; unless circumstances have changed significantly one must assume that an event of this size can and will occur again at any time. The record can perhaps be extended by paleoflood data using old sediment deposits (Baker, 1983; Costa, 1983), ancient debris-flow lobes (Costa and Jarrett, 1981), or vegetation patterns (Veblen and Ashton, 1978). The imprint on the landscape of the largest past event is likely to have survived its smaller successors, and gives an indication of the size of the largest possible event. Such an indication is also able to be calculated, for flood events, using the worst possible combination of meteorological and catchment circumstances to give the 'probable maximum flood' (PMF). Here again it is important not to assign a frequency to this, and to be realistic about the precision of the estimate.

In a somewhat similar way it seems possible to establish a physical limit to the possible size of, for example, debris flow surges. Davies (1988) deduced from laboratory work that there could well be a calculable limit to the height of a steady-state surge, and even if only very approximate this could be better than no estimate at all. The experimental data of Fig.1 illustrate this point well; no regression relationship is credible, but the upper and lower bounds of the data are quite well defined. The same is often true of sediment transport data, and in general it seems likely that more useful information on the possible magnitude of events in mountain streams is to be found in envelope curves than in regression lines.

If the maximum likely event magnitude can be estimated, the extent of the area threatened by this event can also be estimated, knowing the local topography. In the case of debris flows one needs to know where the maximum event can go, and where it might stop, and rules of thumb exist to predict these factors (Ikeya, 1981; Takahashi and Tsujimoto, 1987); the use of small-scale physical models can also give useful indications, but since the processes of the event differ between model and prototype, the model should be viewed rather as an analogy and its predictions treated accordingly (Mizuyama, et. al, 1984).

Event Thresholds

Where the probability of a major event is unknown but low, as will be the case for the largest likely event, the best strategy for damage limitation is difficult to determine. In many cases structural countermeasures will be inappropriate because the cost of coping with a large event will simply be too high, and the economic return cannot even be estimated without any idea of a frequency of occurrence. In hitherto unsettled areas, the location of future developments can be restricted by zoning if a hazard map can be produced; alluvial fans are prime sites for development, and a recent model investigation (Zarn, 1989) has shown that flood hazard-prone areas might be able to be delineated knowing the fan topography, soil age distribution and present position of the stream. There might be resistance to such zoning by landowners who hope to make large profits by selling to developers.

In areas already settled, particularly by recent developments which are more likely than traditional ones to be in hazard zones, the prospect of permanent removal of facilities is remote, given the distinct possibility that a major event might not occur within the next 100 years.

The most practicable hazard mitigation strategy seems to be a hierarchy of 'soft' (i.e., non-structural) countermeasures (Takahashi, 1981; Zarn, 1989), which requires knowledge of the weather, ground and flow conditions likely to precipitate an event.

Because fluvial events intense enough to cause hazards in mountain regions usually require large amounts of sediment to be available, the likelihood of such an event, given the appropriate meteorological conditions, could be assessed by conducting a survey of potential sediment sources in the catchment, including undercut slopes, potential slope failures, sediment in channel storage, etc. If the catchment were thus monitored at regular intervals, perhaps using remotely-sensed data with some ground control, a worrying build-up of sediment could be reflected by an increased state of readiness or awareness in the threatened areas. If the type of sediment in the catchment suggests that debris-flows could occur (Davies, 1986), or if surveys indicate that log-jams might collapse, the extent of this state of readiness could be varied accordingly.

Given such a state of preparedness, some knowledge of the meteorological threshold conditions likely to trigger the event is clearly vital - for example the antecedent conditions and rainfall intensities that commonly cause debris flows to occur (Church and Miles, 1987). If weather forecasts suggest that these intensities are likely to be exceeded (in the next two or three days, perhaps), a preparatory warning can be issued to the susceptible areas. Telemetered raingauge data and/or weather radar can be used in real time to update the severity of the warning with, ultimately, a decision to evacuate requiring a good judgement by someone responsible. A sensible adjunct would be an automatic warning device triggered by the occurrence of the hazard event, though whether this would give sufficient time for evacuation is questionable; an example is the lahar detection device that gives about ten minutes' warning to skiers on Mt Ruapehu, a volcano in New Zealand. The decision to evacuate requires very serious judgement; too soon, and people might think a mistake has been made and return to their dwellings, too late is too late. Clearly, the people in danger must be very clear about the nature of the threat and the state of knowledge of the Civil Defence authorities - to ensure credibility, an honest and thorough programme of public education is vital.

The establishment of threshold conditions for debris flow, flash flood and other hazard events needs a great deal of good research, and must be pursued with great vigour. Conditions suitable for the formation and collapse of log-jams are particularly important, and little is known about this topic at present (Kellerhals, 1987; Keller and Swanson, 1979).

Hazard Prediction

Data describing past event magnitudes are only relevant insofar as they are applicable to future events. Changes of climate and catchment land use can clearly affect the occurrence of hazard events, and a little is known about the direction of these changes, but even if future climate and land use can be predicted there is little prospect of predicting the change in hazard behaviour quantitatively. Indeed, knowledge of past events is usually so sparse that any change of hazard behaviour would need to be gross to be detectable as a trend given the inevitable wide scatter of data; for example, Davies (1988b) estimated that for a recognisable change in debris-flow activity in the Gangbach catchment, Switzerland, due to deforestation, the frequency or magnitude of events would need to change by about three-fold. If such a change were to occur, it could not be maintained for long (in a region of competent bedrock, at least) because in the longer term the rate at which sediment leaves a catchment must be equal to or less than the rate at which it is produced by weathering on hillslopes. This remark clearly does not necessarily apply in regions of very weak bedrock where enhanced debris-flow activity can cause severe bed degradation and hill-slope undercutting, making very large volumes of sediment available.

The problem of predicting catchment response to a change of climate or land use seems to be so difficult, given the paucity of past data, that the best one can do is assume that major events will continue to occur in the future as they have in the past, and that climate and land use changes will not have detectable effects in the timescale considered.

CONCLUSION

This brief discussion leads inevitably to the conclusion that fluvial hazard assessment, prediction and mitigation in mountain regions cannot be accomplished with anything like the precision that many scientific publications would imply. This is because the physical situations involved are extremely variable and unpredictable, and complicated by many independent random external factors. Furthermore, many of the investigations undertaken in the name of mountain fluvial hydraulics have little prospect of contributing useful results to the search for ways of reducing mountain fluvial hazards (e.g., Whittaker and Davies, 1982; Phillips and Davies, 1989). It is obvious, however, that many of these studies serve the very useful purpose of training scientists to be logical, to solve problems, and sometimes even to test hypotheses. It is important to see them clearly in the context of real-world human problems, however, and the stark reality is that all we know about mountain fluvial hazards are the rough magnitudes of some past events.

Given this, the suggested approach to hazard mitigation is to plan for the occurrence at any time of events such as have occurred in the past and can occur in the future, and a strategy of monitoring/readiness/warning/evacuation is suggested as being the most practicable; further development of hazardous areas can perhaps be limited also.

To allow progress to be made in implementing such a strategy, the following research areas need urgent attention:

(i) Estimation of past event magnitudes from sediment deposits, erosion scars, and vegetation and soil age distributions and other techniques of paleoflood hydrology (Baker, 1983).

(ii) Estimation of antecedent and threshold meteorological and sediment availability conditions for debris flows, flash floods, etc., probably on a local basis.

(iii) Study of conditions for the formation and collapse of log-jams.

(iv) Estimate physical limits for the height and velocity of flash-flood waves and debris-flow surges.

(v) Develop methods for rapid monitoring of catchment and sediment source conditions from remotely-sensed data.

(vi) A serious attempt to educate the local inhabitants about the nature of the hazards that threaten them, and about the limitations of science in predicting their occurrence.

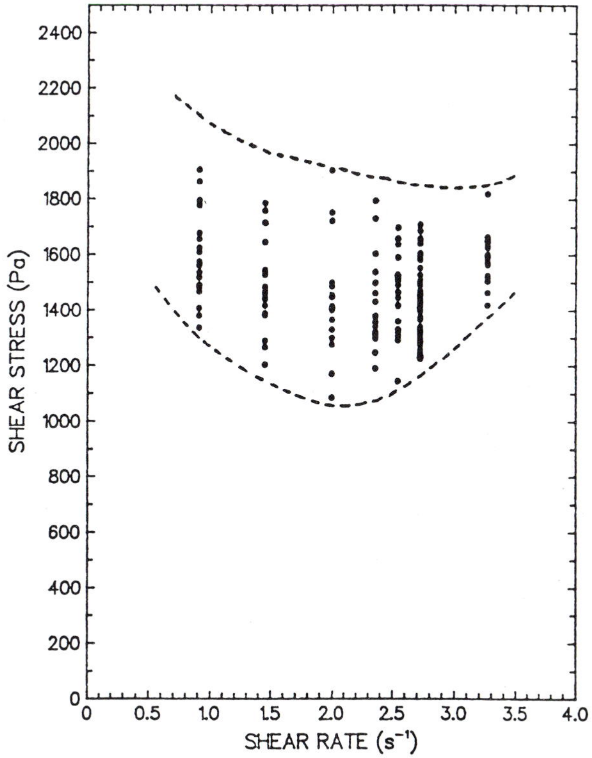

Figure 1. Laboratory data on variation of shear stress with shear rate in debris-flow material. From Phillips (1989).

REFERENCES

Baker, V.R. (1983). Large-scale fluvial paleohydrology. In. K.J. Gregory (ed.), Background to Paleohydrology. Wiley, New York, N.Y., 4543-478.

Bathurst, J.C. (1978). Flow resistance of large-scale rough-ness. Journal of the Hydraulics Division, A.S.C.E., 104, HY12, 1587-1603.

Chang, H.H. (1980). Minimum stream power and river channel patterns. J. Hydrol., 41, 303-327.

Church, M., and Miles, M.J. (1987). Meteorological antecedents to debris flow in southwestern British Columbia; some case studies. In. Reviews in Engineering Geology, III, Geology Society of America, 63-79.

Costa, J.E. (1983). Paleohydraulic reconstruction of flash-flood peaks from boulder deposits in the Colorado Front Range. Bull. Geol. Soc. Amer., 94, 986-1004.

Costa, J.E. (1987). Hydraulics and basin morphometry of the largest flash floods in the conterminous United States. Journal of Hydrology, 93, 313-338.

Costa, J.E., and Jarrett, R.D. (1981). Debris flows in small mountain stream channels of Colorado, and their hydrologic implications. Bull. Assoc. Eng. Geol., 18, 309-322.

Davies, T.R.H. (1986). Occurrence and characteristics of debris flows following Waldsterben. Unpubl. report, Department of Agricultural Engineering, Lincoln College, Canterbury, N.Z., 52 p.

Davies, T.R.H. (1987). Problems of bedload transport in braided gravel-bed rivers. In. Sediment Transport in Gravel-bed Rivers. (C.R. Thorne, J.C. Bathurst, R.D. Hey, eds.), Wiley, 793-828.

Davies, T.R.H. (1988a). Debris flow surges - a laboratory investigation. Mitteilung No.96, V.A.W., ETH-Zurich, 122 p.

Davies, T.R.H. (1988b). Die Auswirkung des Waldsterbens auf Murgänge im Gangbach. In. Folgen der Waldschäden auf die Gebirgswässer in der Schweiz. V.A.W., ETH-Zurich, 63-69.

Hayward, J.A. (1980). Hydrology and stream sediments in a mountain catchment. T.G.M.L.I. Special Publ. 17, Lincoln College, N.Z.

Hubbell, D.W. (1987). Bed load sampling and analysis. In. Sediment Transport in Gravel-Bed Rivers, (C.R. Thorne, J.C. Bathurst, R.D. Hey, eds.), Wiley, 89-120.

Ikeya, H. (1981). A method of designation of area in danger of debris-flow. IAHR Publ. No.132, 576-588.

Keller, E.A., and Swanson, F.J. (1979). Effects of large organic material on channel form and alluvial processes. Earth Surf. Proc. Vol.4, pp.361-380.

Kellerhals, R. (1988). Prozesse in den Bachgerinnen. in. Folgen der Waldschäden auf die Gebirgswässer in der Scheiz, V.A.W., ETH-Zurich, 70-77.

MacMurray, H.L. (1985). The use of the salt-velocity method for the precise measurement of resistance to flow in rough-boundary channels. Unpub. Ph.D. Thesis, Univ. of Canterbury (Lincoln College), New Zealand, 274 p.

Mizuyama, T.; Ikeya, H.; and Imamuna, R. (1984). Methods for zoning debris flow vulnerable areas. Proc. Symp. on Effects of Forest Land Use on Erosion and Slope Stability, Hawaii, 281-287.

Newson, M.D. (1981). Mountain streams. In. British Rivers (J. Lewin, ed.). Geo. Allen and Unwin, 58-89.

Okuda, S., et al (1980). Observations on the motion of a debris flow and its geomorphological effects. Zeitschrift für Geomorphologie, Suppl. Bd 35, 142-163.

Phillips, C.J. (1989). Rheological investigations of debris-flow materials. Unpub. Ph.D. Thesis, Univ. of Canterbury (Lincoln College), Canterbury, New Zealand, 211 p.

Phillips, C.J., and Davies, T.R.H. (1989). Debris flow rheology-direct measurement. Proc. Int. Symp. on Erosion and Volcanic Debris Flow Technology, Djakarta.

Pierson, T.C. (1980). Erosion and deposition by debris flows at Mt Thomas, North Canterbury, New Zealand. Earth Surf. Proc., 5, 227-247.

Pierson, T.C. (1985). Field techniques for measuring debris-flow dynamics. Proceedings, International Symposium on Erosion, Debris Flow and Disaster Prevention, Tsukuba, Japan, 203-208.

Takahashi, T., (1981). Estimation of potential debris flows and their hazard zones; soft countermeasures for a disaster. Natural Disaster Science.

Takahashi, T., and Tsujimoto, H. (1985). Delineation of the debris flow hazard zone by a numerical simulation method. Proc.. Int. Symp. on Erosion, Debris Flow and Disaster Prevention, Tsukuba, 457-462.

Veblen, T.T., and Ashton, D.H. (1978). Catastrophic influences on the vegetation of the Valdivian Andes, Chile. Vegetatio, 36, 3, 149-167.

Whittaker, J.G., and Davies, T.R.H. (1982). Erosion and sediment transport processes in step-pool streams. In. Recent Developments in the Explanation and Prediction of Erosion and Sediment Yield, IAHS Publ. No. 137, 99-104.

Young, J.R., and Davies, T.R.H. (1989). Realistic criteria for flood control design. Proc. Australasian Conf. on Hydrology, Christchurch, N.Z.

Zarn, B. (1989). Hazard zone delineation on alluvial fans. Unpub. M.E. Thesis, Univ. of Canterbury (Lincoln College), Canterbury, New Zealand.

MECHANICS AND THE EXISTENCE CRITERIA OF VARIOUS TYPES OF FLOWS DURING MASSIVE SEDIMENT TRANSPORT

Tamotsu Takahashi

Disaster Prevention Research Institute
Kyoto University, Kyoto, Japan

ABSTRACT

Massive sediment transport in which solid fraction and water are altogether drived by gravity as virtually a continuum appears in several aspects depending on the channel condition, discharge and the properties of solid and fluid. In this paper, first the characteristics of the macro-viscous flow are discussed and a formula to estimate viscosity in highly sediment laden flow is given. The modified Bagnold's number to divide the flow ranges into macro-viscous and inertial is introduced.

Then, the flow formulae in inertial range corresponding to the stony debris flow, turbulent mud flow, immature debris flow and a hybrid of stony and muddy debris flow are presented focusing on the characteristics of the velocity profile and the extent of solid particle dispersion in the flow.

Finally, the existence domains of those various type flows in inertial range are given on a plane whose two orthogonal axes represents slope and relative depth respectively.

1. Introduction

Large scale topographical change in steep channels distributing in very steep mountain basin is mainly induced by the massive sediment transport which takes place occasionally with an outbreak of severe rainstorm. A steep loosely packed sediment layer will become unstable with appearance of surface water stream and the particles that start moving will form a moving particle layer being mixed up with water. All particles as well as the interstitial fluid in the particle moving layer, regardless it occupies the whole flow depth or only a lower part of the flow, are moved by gravity, so that the relative velocity between solid phase and surrounding fluid would be small and if any it would play only a minor role in transporting particles(Takahashi 1978). By contrast, in fluid flow, lift and drag forces produced by the relative velocity are essential for individually transporting particles.

The sediment gravity flows which massively transport sediments can appear in several types, even though they belong to one category in the sense in those flows sediments are moved by gravity together with water, depending on topographical conditions of the channel, properties of the transported materials and water supply conditions. Definement of existing criterion and the characteristics of each type of flow is essential for assessing, understanding, predicting and controling the fluvial processes in mountain region.

2. Momentum Conservation Equations and the Constitutive Relations

Considering the momentum conservation law in steady uniform two phase open channel flow composed of solid grains and fluid, we obtain

For x direction(positive to downstream):

$$\{\sigma c+\rho(1-c)\}g\sin\theta+\frac{\partial T}{\partial z}+\frac{\partial T'}{\partial z}=0 \tag{1}$$

For z direction(positive to upward):

$$-\frac{\partial P}{\partial z}+\{\sigma c+\rho(1-c)\}g\cos\theta+\frac{\partial p}{\partial z}=0 \tag{2}$$

$$p=\rho g\cos\theta(h-z) \tag{3}$$

where T,P : x and z components of the inter-particle stress, c: volume concentration of solid in the flow, θ: slope gradient of the channel, g: acceleration due to gravity, p: pressure in the interstitial fluid, T': x component of the stress in the interstitial fluid due to turbulence and viscosity, h: flow depth, x: coordinate axis to mean flow direction, z: coordinate axis to the direction perpendicular to mean flow(0 at the bottom of flow), σ: density of particles, ρ: density of fluid.

Integrating (1) from z to h and noticing $T^h=0$ and $T'^h=0$, we obtain

$$T^z+T'^z=g\sin\theta\int_z^h\{(\sigma-\rho)c+\rho\}dz \tag{4}$$

Substituting (3) into (2) and integrating this from z to h, we obtain

$$P^z=-(\sigma-\rho)g\cos\theta\int_z^h c\,dz \tag{5}$$

where T^z, T'^z and P^z means T, T' and P values at height z.

Many constitutive relations describing T^z and P^z have been proposed since the pioneering work of Bagnold(1954). Majority of them claims that in rapid flow range(inertial range) both T^z and P^z should be proportional to $(du/dz)^2$, whereas in slow flow range(macro-viscous range) they should be proportional to (du/dz), in which u is velocity of the mixture.

3. Flow in Macro-Viscous Range

A series of open channel experiments was carried out by Takahashi and Fujii(1988) to obtain the flow equations for debris flow(in which particles are moving dispersing all over the depth of flow) generated by fluid whose viscosity is far larger than that of plain water.

Debris flows were produced by abruptly supplying fluid at a constant discharge onto the erodible sand bed. The velocity profile and the flow depth were measured at the rigid bed part existing downstream of the erodible bed part. The mean solid concentration within the flow was obtained by sampling the flow. High viscosity in the experimental fluid was attained by mixing the high absorbent rosin powder with water. This powder can absorb water as much as a thousand times of its weight and the viscosity of the fluid depends on the content of this powder.

Assuming solid grains are uniformly dispersing in the whole depth and considering the mixture as a continuum, the left hand side of (4) can be described as following:

$$\tau(=T^z+T'^z)=(\mu_s+\mu_f)\frac{du}{dz}\left(\equiv\mu_T\frac{du}{dz}\right) \tag{6}$$

where μ_s denotes increment of viscosity due to effects of dispersing granules, μ_f the viscosity of the interstitial fluid and μ_T is the apparent viscosity of the mixture. It should be noted here that the flow we are considering does not contain such very fine materials as cohesion plays an important role in the mechanics of flow. Eq.(6) means that the mixture is nothing but the normal Newtonian fluid and from (4) we can obtain

$$\frac{u}{u_*^2}=\frac{\rho_T z(2h-z)}{2\mu_T h} \tag{7}$$

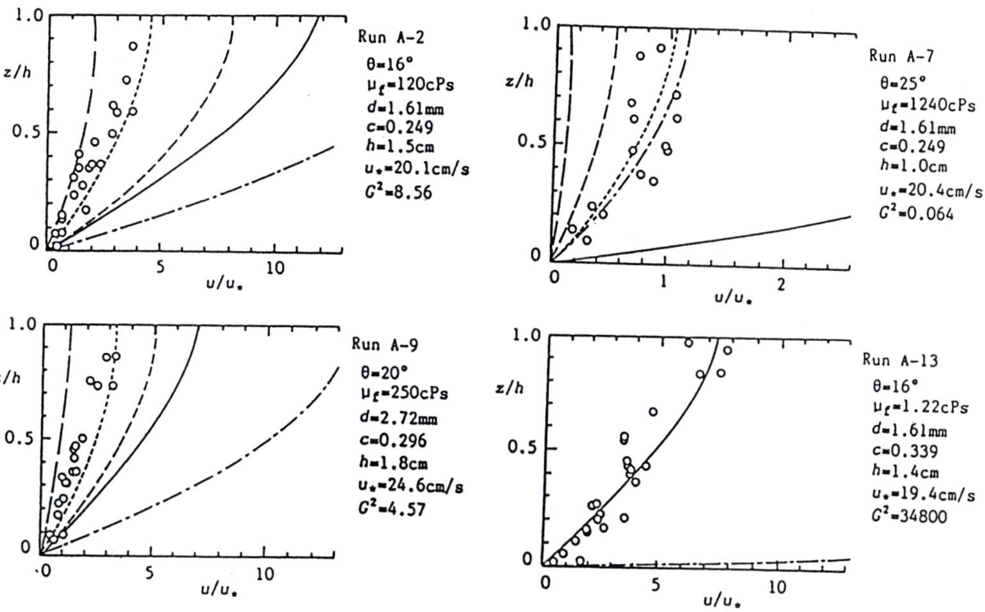

Bagnold(viscous): $\mu_T=(1+\lambda)(1+\lambda/2)\mu_f$

Laminar(fluid viscosity): $\mu_T=\mu_f$

Mori & Ototake: $\mu_T=\{1+3/(1/c-1/c_*)\}\mu_f$

Takahashi & Fujii: $\mu_T=\mu_f+\mu_c/(1/c-1/c_*)$

Inertial flow

Fig.1 Various theoretical velocity profiles and the experimental data

where $u_*(=\sqrt{gh})$ is the shear velocity and $\rho_T\{=(\sigma-\rho)c+\rho\}$ is the apparent density of the mixture.

So far presented μ_T (or μ_s) evaluating formulae are, however, not applicable as is clear in Fig.1. Remark Run A-13 is the case for inertial range.

According to authors' experiments the ratio μ_s/μ_f became smaller as μ_f became larger, suggesting μ_s was independent of μ_f by contrast to many other investigation. Therefore, herein, the author proposes a new formula which well predicts the velocity and velocity profile in the experiments:

$$\mu_s = \frac{\mu_c}{(1/c - 1/c_*)} \tag{8}$$

where c_* is the volume fraction of solid when packed.

The dotted curves in Fig.1 were depicted by substituting $\mu_c = 1$ Pa·s (10 Poise) into (7). Those curves and other experimental cases affirm (8).

Because μ_T obtained by the author is different from that obtained by Bagnold(1954), Bagnold's number, N, which represents the ratio of the inertia stress to the viscous stress should be modified as following:

$$N_m = \frac{\lambda^2 \sigma d^2}{\mu_T}\left(\frac{du}{dz}\right) = \frac{N}{\{1+(\mu_c/\mu_f)c_*/(c_*/c-1)\}\{(c_*/c)^{1/3}-1\}^{3/2}} \tag{9}$$

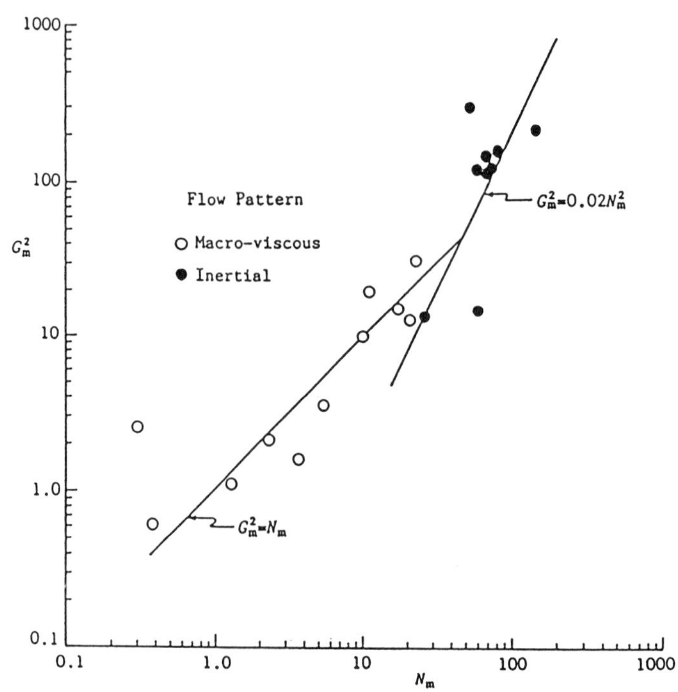

Fig.2 G_m^2 against N_m in the open channel experiment

Similarly, G^2 which represents the nondimensional τ may be modified:

$$G_m^2 = \frac{\lambda^2 \sigma d^2 \tau}{\mu_T^2} = \frac{G^2}{\{1+(\mu_c/\mu_f)c_*/(c_*/c-1)\}^2\{(c_*/c)^{1/3}-1\}^3} \tag{10}$$

where d is the grain diameter and $\lambda = \{(c_*/c)^{1/3}-1\}^{-1}$ is the linear concentration.

G_m^2 for the open channel experiments are plotted in Fig.2 against corresponding N_m. In the calculation of N_m the nominal value of du/dz was used. Namely, for the case of viscous range $2(u_s/h)$ and for the case of inertial range $(3/2)(u_s/h)$ were used, in which u_s is surface velocity and the range each flow belonged was judged from pattern of velocity profile. The tendency of plotted data suggests viscous range flow appears when N_m is larger than about 30 while inertial range flow appears when N_m is larger than about 50. For case of very slow flow, i.e., quasi-static range, by analogy of usual soil mechanics tests, the mixture should have a certain yielding stress, meaning G_m^2 can not approach to infinitely small value, meaning flow can not exist under very small G_m by contrast to usual Newtonian fluid.

4. Flow in Inertial Range

1) Stony debris flow

When $G_m^2 > 30$ and depth is less than a few ten times of grain diameter, the experimental results approximately obeys the following equation:

$$G_m^2 = 0.02 N_m^2 \tag{11}$$

This means in inertial range the total shearing stress is proportional to $(du/dz)^2$ as following:

$$\tau = \frac{0.02\sigma d^2}{\{(c_*/c)^{1/3}-1\}^2}\left(\frac{du}{dz}\right)^2 \tag{12}$$

Substituting (12) into (4) and assuming c is uniformly distributed in the whole depth, one obtains the following velocity distribution formula:

$$\frac{u_s - u}{u_s} = \left(\frac{h-z}{h}\right)^{3/2} \tag{13}$$

The solid line in Fig.1 corresponds to this formula and this curve fits the experimental velocity profile for Run A-13 which is the case of inertial range flow. Many other experiments on rigid bed revealed (13) well explained the velocity profile in inertial range.

On erodible bed, however, although the cross-sectional mean velocity can be as satisfactorily estimated as on rigid bed, the formula can not explain the whole velocity distribution in the flow because the velocity profile in this case has an inflection point near the bottom. To revise this discrepancy effects of change in $\tan\alpha$ which appears in Bagnold's constitutive relation

$$T^z = -P^z \tan\alpha \tag{14}$$

are taken into account by assuming

$$\tan\alpha = (c_*/c)^{1/3} \tan\alpha_o \tag{15}$$

where α_o is a constant equivalent to kinetic friction angle which is a little smaller than the internal friction angle, ϕ, for a mass of irregularly shaped particles.

Substitution of (14) and (15) into (4) and (5) and neglection of smaller terms which appear in the process of algebra lead to the following velocity distribution formula:

$$\frac{u}{u_s} = 1 - \frac{1}{2}\left(1 - \frac{z}{h}\right)^{3/2}\left(2 + 3\frac{z}{h}\right) \tag{16}$$

Fig.3 compares (16) to the experimental results which include different slope and different discharge cases. The theoretical velocity distribution fits to experimental ones very well. This theory predicts solid concentration distribution as well.

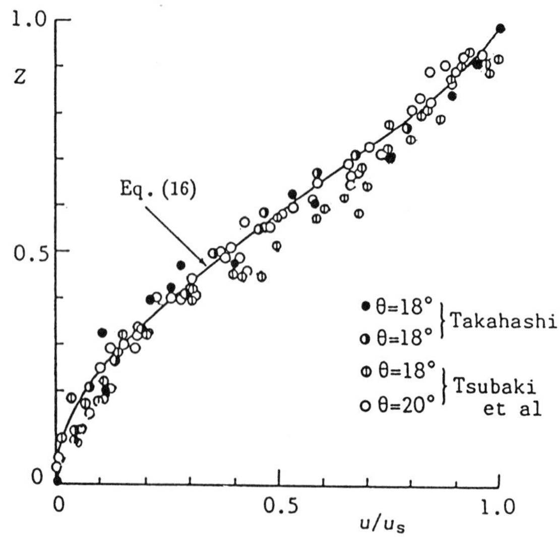

Fig.3 Theoretical velocity distribution and the experimental data

The condition in which the erodible non-cohesive bed is stable against massive slippage is written as following(Takahashi 1978):

$$c \geq \frac{\rho \tan\theta}{(\sigma - \rho)(\tan\phi - \tan\theta)} (\equiv c_{s\infty}) \tag{17}$$

On the other hand, substituting Bagnold's constitutive relation (14) into (4) and (5) and considering T'^0 should have some infinite value, one obtains

$$c \leq \frac{\rho \tan\theta}{(\sigma - \rho)(\tan\alpha - \tan\theta)} (\equiv c_{d\infty}) \tag{18}$$

The experiments prove an equilibrium flow with neither erosion nor deposition can take place under the condition $c \approx c_{s\infty}$ as depicted in Fig.4. This in turn means T'^0 is positive in an equilibrium flow.

For a very steep channel which satisfies

$$\tan\theta \geq \frac{(\sigma - \rho)c_*}{(\sigma - \rho)c_* + \rho} \tan\phi \tag{19}$$

(17) calculates c larger than c_*. But flow would not be able to exist in such high concentration and according to the experiments c is always less than about $0.9c_*$.

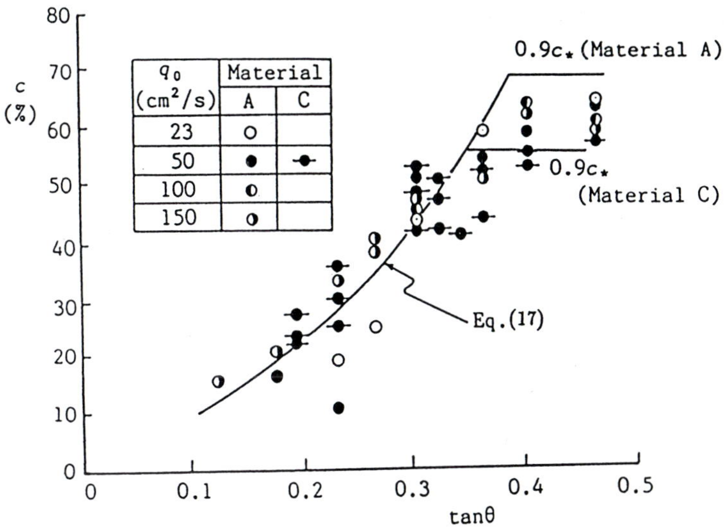

Fig.4 Equilibrium solid concentration in stony debris flow

2) Immature debris flow

The $c_{s\infty}$ defined in (17) calculates very small values on a flat slope. Actually, if the mean concentration becomes less than $0.4c_*$, particles can not disperse throughout the entire depth of flow and a fairly clear separation between the upper clearer and lower particle mixture layers is observed. This is the immature debris flow as the author defines. If the mean concentration of solid and thickness of the particle mixture layer are written c_l and h_l respectively, the following equilibrium condition of the stresses would be satisfied in the mixture layer

$$0.02\sigma\lambda^2 d^2\left(\frac{du}{dz}\right)^2 + \rho l^2\left(\frac{du}{dz}\right)^2 = \{(\sigma-\rho)c_l(h_l - z) + \rho(h-z)\}g\sin\theta \qquad (20)$$

where l is the mixing length in the interstitial fluid and may have value corresponding the average pore distance between particles ($l = \xi d/\lambda$).

In the upper water layer, the equilibrium condition should be similar to that of normal turbulent flow, i.e.:

$$\rho l^2\left(\frac{du}{dz}\right)^2 = \rho g(h-z)\sin\theta \qquad (21)$$

where

$$l = \xi\frac{d}{\lambda} + \kappa(z - h_l) \qquad (22)$$

Integration of (20) and (21) by noticing u at $h = h_l$ deduced from (20) must have the same value to that deduced from (21) gives the velocity profiles as shown by the solid lines in Fig.5 in which $\xi = 3$ and $\kappa = 0.4$ are assumed. The full circles in the figure indicate velocities of the representative particles arbitrarily selected in the mixture layer and the open circles indicate velocities of the neutrally buoyant polystyrene beads poured into the water layer to measure velocity within this layer.

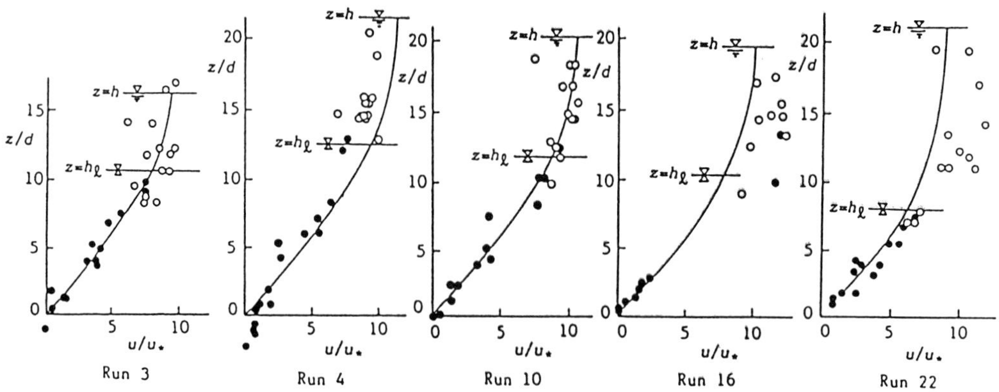

Fig.5 Velocity distributions in immature debris flows

3) Muddy debris flow/turbulent mud flow

As stated earlier in inertial debris flow the pore fluid is more or less turbulent($T'^z > 0$), and therefore, if that turbulent intensity is strong enough to suspend all the contributing particles, large eddies containing dense populations of particles move randomly and mix forming large scale turbulence. The total shearling stress in this type of flow would be the sum of the stresses produced by large scale turbulence and particle collision, i.e.,

$$\tau = \rho_T l_T^2 \left(\frac{du}{dz}\right)^2 + \frac{0.02\sigma d^2}{\{(c_*/c)^{1/3} - 1\}^2}\left(\frac{du}{dz}\right)^2 \tag{23}$$

where l_T is the mixing length and ρ_T is the apparent density of the fluid mass, $\rho_T = (\sigma - \rho)c + \rho$.

Substituting (23) into (4) and integrating it by assuming c is constant, one obtains

$$\frac{u}{u_*} = \frac{1}{\kappa}\ln\left|\frac{z/h + \sqrt{(z/h)^2 + \lambda^2(0.02/\kappa^2)(\sigma/\rho_T)(d/h)^2}}{H + \sqrt{H^2 + \lambda^2(0.02/\kappa^2)(\sigma/\rho_T)(d/h)^2}}\right| \tag{24}$$

where H is equal to a_o/R_* for a hydraulically smooth bed and to bk/h for a rough bed, $R_* = u_*h/\nu, a_o = 1/9.025, b = 1/30, k$ is the roughness height, ν the kinematic viscosity of water and κ the Kármán constant.

Equation (24) means, if the other conditions are not changed, contribution of the turbulent mixing term becomes large with increase in h/d or decrease in solid concentration and the velocity profile approaches that of turbulent plain water flow.

Fig.6 plots the experimental velocities and compares them with the theoretical profiles obtained from (24) and those in stony debris flow model(dilatant). From the tendency of plotted experimetal velocity in Fig.6 one might be impressed to be possible to adopt the Bingham fluid model(e.g. Johnson 1970) having a plug part near the surface, but there is no plug at all in this flow.

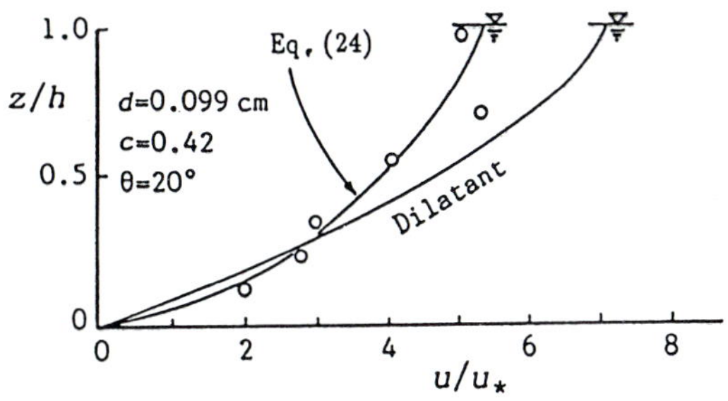

Fig.6 Velocity distributions by the turbulent mudflow model and the dilatant fluid model

4) A hybrid of stony and muddy debris flow

According to the flume experiments that use light specific weight particles, both in the case of mature(particles are flowing in entire flow depth) and immature debris flows, an approximately uniform velocity region appears in the upper part of the mixture flowing layer when the channel slope becomes steeper, the relative depth, h/d, becomes larger or solid concentration becomes thinner. The detailed observation of the flow reveals that considerably strong turbulence is present in this upper region and that particles move randomly meaning they are suspended by turbulence in this region while the dilatant flow is existing in the lower region.

Consider an immature/mature debris flow whose thickness of particle flowing layer is h_l and the constant solid concentration in that layer is c_l and assume particles located between height h_t and h_l are suspending in fluid, and those lacated between heights zero(bottom) and h_t are supported by dispersive pressure. Then, the velocity in the lower layer($0 \leq z \leq h_t$) can be obtained by integrating (19), and the velocity in the upper sediment suspending and water layer($h_t < z \leq h$) is obtainable by integrating (20), in which

$$l = (z - h_t) \tag{25}$$

would be assumed. The problem is how to determine h_t. Herein, it is assumed to be equal to the height at which the local mean velocity, u_t, satisfies the condition of Bagnold's auto-suspension(Bagnold 1962):

$$u_t = \frac{w_s}{\sin\theta} \tag{26}$$

where w_s is the settling velocity of particles in a group.

Fig.7 shows the experimentally obtained velocity profiles and the calculated ones on the rigid bed channel.

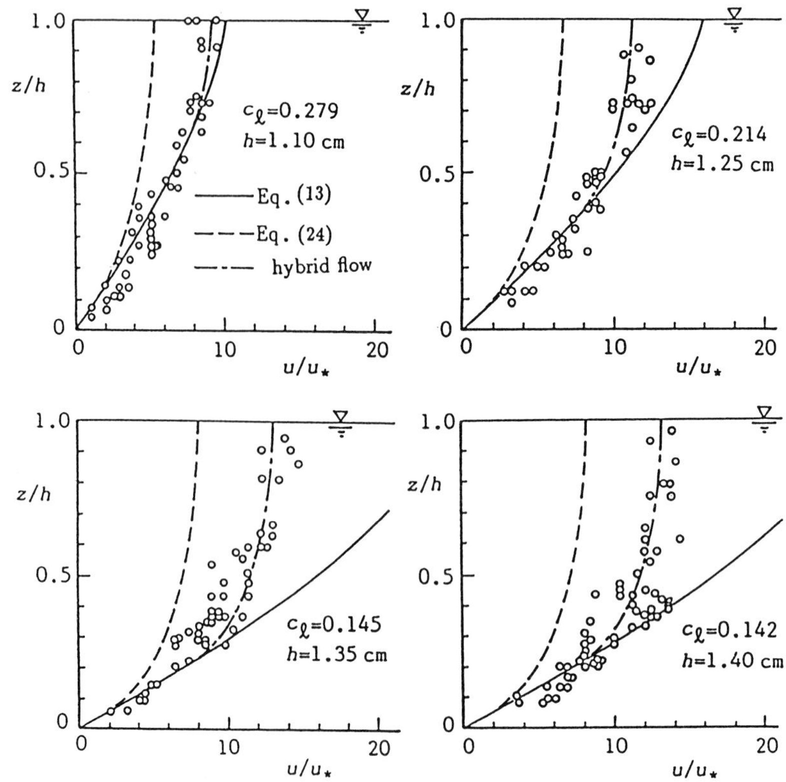

Fig.7 Velocity profiles in hybrid flow

5. Existence Criteria of Various Type Flows in Inertial Range

The critical condition to generate a massive grain transportation, i.e., the initiation of immature debris flow is defined by the critical condition of occurrence of slippage in the bed(Takahashi 1987):

$$\tau_* \geq \cos\theta\left\{\left(c_* + \frac{q_{s*}}{8.5}\right)(\tan\phi - \tan\theta) - \frac{\sigma-\rho}{\rho}\tan\theta\right\} \tag{27}$$

in which $\tau_* = u_*^2/\{(\sigma/\rho - 1)gd\}$, $q_{s*} = q_s/(u_* d)$.

The boundary between immature and mature debris flow is determined by (Takahashi 1978)

$$\tan\theta = \frac{c_*(\sigma-\rho)}{c_*(\sigma-\rho)+2.4\rho}\tan\phi \tag{28}$$

The shearing stress in the pore fluid of stony debris flow or stony immature debris flow at the bottom of flow is

$$T'^0 = \rho g h \sin\theta\left\{\left(\frac{\sigma-\rho}{\rho}\right)\left(1-\frac{\tan\alpha}{\tan\theta}\right)c+1\right\} \tag{29}$$

Therefore, the shear velocity shared by the interstitial fluid, u_{*f}, would be

$$u_{*f} = \sqrt{T^{f0}/\rho} \tag{30}$$

If the mean settling velocity of particles is w_s, the boundary between the stony debris flow or stony immature debris flow and the turbulent mud flow would be analogized from the well known theory for suspended load that

$$u_{*f} \approx w_s \tag{31}$$

The condition (31) can be rewritten by using (29) and Rubey's particle settling velocity formula(Rubey 1933):

$$\frac{h}{d} \approx \frac{(\sigma/\rho - 1)F^2}{\sin\theta\{(\sigma/\rho - 1)(1 - \tan\alpha/\tan\theta)c + 1\}} \tag{32}$$

The curves labeled A, B and C in Fig.8(a) represent (27), (28) and (32) respectively for the ordinal sand particles. Eq.(32) are depicted for various c in the figure and the experimental values are also classified by concentration. The turbulent mud flow seems to appear to the right of each corresponding boundary curve, C.

Fig.8(b) shows the curves corresponding to (32) for the case of lighter material. The curves in this case are calculated by substituting $(\sigma/\rho - 1) = 0.8$ instead of 1.6 for the former case. The hybrid flow appears around those boundary curves.

Consequently, if the relative depth is gradually increased under the constant solid concentration and the constant channel slope(in the range $I/(\sigma/\rho - 1) \geq 0.15$), at first the stony (mature) debris flow, then the hybrid flow(dispersed all over the flow depth) and finally the turbulent mud flow would appear. And if the channel slope is in the range $I/(\sigma/\rho - 1) \leq 0.15$, the immature debris flow, the hybrid flow(partly dispersed) and the turbulent mud flow would appear in this order.

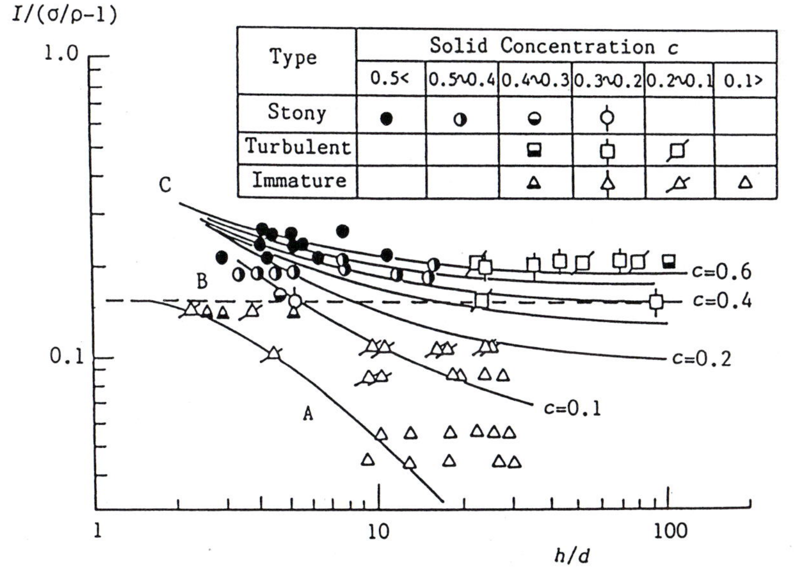

Fig.8(a) Existence domain of various flows in inertial range

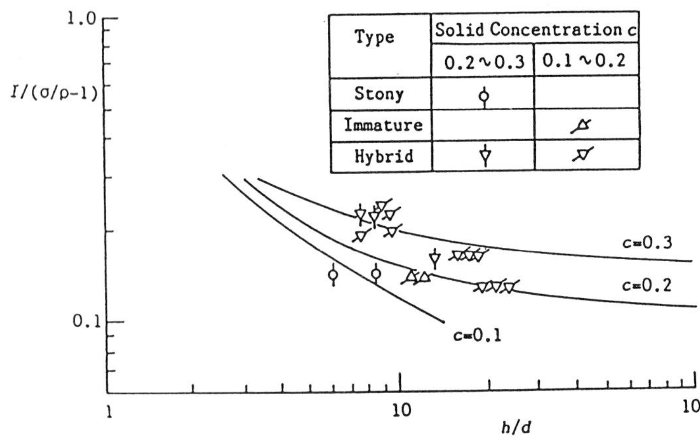

Fig.8(b) Existence domain of various flows in inertial range(lighter material)

6. Conclusion

This paper described the massive sediment transportation in equilibrium state appeared in many aspects mainly depending on channel slope, relative flow depth, solid concentration and viscosity of fluid. The mechanical significance and characteristics of each aspect of flow were discussed and the criteria for existence of those flows were presented. Thus, one can predict what kind of flow will appear by what magnitude for a given combination of hydraulic conditions. The peculiar characteristic of velocity profile in the flow depth for each type of flow was focused on. The other characteristics such as volume concentration of solid, sediment discharge and flow resistance were also disclosed in the processes of discussion. However, those were not included herein simply due to restriction of space.

References

Bagnold, R.A. (1954), Experiments on a gravity-free dispersion of large solid spheres in a Newtonian fluid under shear, Proc. Roy. Soc. London, A, Vol.225, pp.49-63.

Bagnold, R.A. (1962), Auto-suspension of transported sediment; turbidity currents, Proc. Roy. Soc. London, A, No.1322, pp.315-319.

Johnson, A.M. (1970), Physical processes in geology, Freeman, Cooper, San Francisco, Calif.

Rubey, W.W. (1933), Settling velocities of gravels, sand and silt particles, Am. Jour. of Science, Vol.25, pp.325-338.

Takahashi, T. (1978), Mechanical characteristics of debris flow, Jour. Hydraulics Div., ASCE, Vol.104, No.8, pp.1153-1169.

Takahashi, T. and Fujii, Y. (1988), Debris flow control once under way, Annuals, Disaster Prevention Research Institute, Kyoto Univ., No.31B-2, pp.633-654(in Japanese).

NUMERICAL ANALYSIS OF HILLSLOPE - CHANNEL INTERACTION IN FIRST ORDER BASINS

Paolo Ghilardi and Giovanni Menduni

Istituto di Idraulica
Politecnico di Milano, Milano, Italy

ABSTRACT

We present a numerical model whose goal is to describe the unsteady behavior of supercritical open channel flows on a loose granular bed as far as short term aggradation/degradation phenomena are concerned. Following the usual approach, the problem is formulated in terms of *De Saint Venant* equations plus a sediment continuity relationship. The presence of an unerodible bed underlying the granular layer is taken into account; to this purpose the sediment balance relationship is re-formulated using an original approach. The numerical solution is achived via a generalized Preissman scheme. A test simulation is discussed which gives rise to a solid transport mechanism essentially based on an intermittent shedding of sediment bursts; the example shows how a pulsating non periodic behavior may arise within a strictly deterministic framework this due to the strongly non linear formulation of the problem.

Introduction

Erosional processes over an hillslope are mainly the product of intense rainstorm events. Due to this phenomena overland flow may determine sheet erosion and subsequent channelization processes or, under particular conditions, debris flow and even landslides.

Recent works by different authors (see, e.g., Roth et al., 1989) show how analytical investigation of these processes can help understanding the geomorphological evolution of first-order basins, which is the preliminary step for better understanding of drainage patterns evolution. Furthermore, sediment yield from first order basins provides the upstream condition which is required to model sediment transport at the basin scale.

The present work deals with dynamic analysis of erosional and sediment transport processes as they occurr at the spatial scale which characterizes geometry of first-order basins and at a time scale which may range from the single storm event to the long term. An hydrodynamical model is developed for the purpose, so that response of the hillslope-channel system to one or more storm events can be analyzed.

This model can be roughly splitted into two components. The first one describes overland processes through the classical kinematic wave approach. The second one describes unsteady two phase flow into the channel. The present paper will mainly deal with this second part and the techniques used for the numerical solution.

The unsteady two phase flow of a loose boundary channel is a quite widely studied problem, particularly as far as the long term aggradation and degradation processes of a stream are concerned. Nevertheless, it appears that the major part of the experimental and analytical studies deal with subcritical flows, for which many analytical and numerical schemes were tested and validated (See, e.g., the reviews of Mahmood & Yevjevich, 1975, Cunge et Al., 1980, and the works of Ponce et al., 1979, Lyn, 1987).

Mathematical modelling of the hydraulic behaviour of a first order stream during a storm event, usually implies the analysis of the unsteady flow in supercritical steep-slope channel reaches, with both sediment and water lateral afflux which may be up to some order of magnitude greater then the upstream incoming discharge.

Simulations performed at the storm event time scale often require a number of *second order* phenomena inherent flow resistance and sediment transport to be taken into account. We may recall here the free surface instability due to high Froude numbers (Rouse, 1965, Montuori, 1984), the unsteady behavior of hydraulic resistance, due to the time required for the bed configuration to reach the equilibrium with the flow conditions (Lamberti & Menduni, 1987), the temporal and spatial lags between the suspended transport and the transport capacity of the stream, and the lag between the actual grain size composition of the bed and the same composition that should attain when equilibrium conditions are reached (Armanini and Di Silvio, 1988).

Furthermore, despite the usual assumption of channel bed excavated into an infinite depth sediment layer, the morphology of mountain streams is often characterized by reaches where sediments lie over an unerodible rocky bed which can be uncovered e.g. during high flows.

Due to these factors, the development of a model simulating such a complex system results to be quite difficult. Apart from the numerical problems arising from the application of usual unsteady flow numerical schemes in such extreme conditions, many problems are in fact due to a lack of knowledge in the physical mechanisms of erosion and deposition (See e.g. the review of Dawdy and Vanoni, 1986, and the works of Lyn, 1987 and Lyn and Goodwin, 1987).

The present work is mainly concerned with the mathematical one-dimensional modeling of the unsteady supercritical flow over a fixed bed with sediment transport and lateral, non uniform non stationary, afflux of both water and sediments. Furthermore, bed degradation is downward limited by the presence of an unerodible layer while aggradation can occur due to lateral or upstream sediment affluxes exceeding the stream transport capacity.

Though morphological effects due to non-stationary behavior of a sediment mixture may be of fundamental importance (Armanini and Di Silvio, 1988) in first order streams dynamics, a monogranular cohesionless sediment is considered here, being the multigranular approaches available in literature not specifically designed to take directly into account the presence of an unerodible layer; a development in this direction is presently in progress.

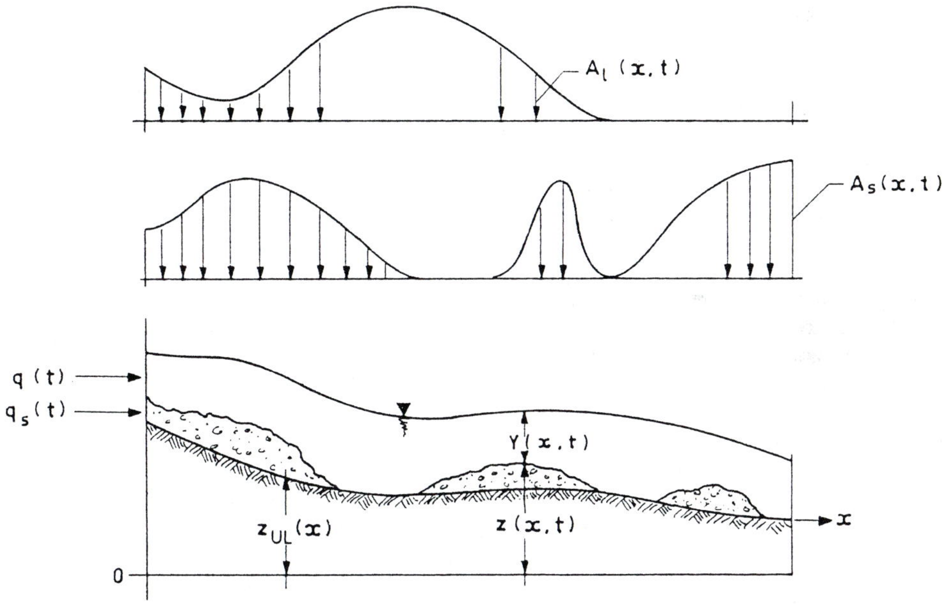

Figure 1.

Basic Equations

Following the classical approach (See, e.g., De Vries, 1965, Cunge and Perdreau, 1973, Vreugdenhil and De Vries, 1973), three balance equations, namely for water mass, sediment mass and momentum can be written. If we consider a wide prismatic rectangular channel with lateral sediment and water affluxes (see Figure 1), we can write

$$\frac{\partial}{\partial x}(UY) + \frac{\partial Y}{\partial t} = A_w \tag{1}$$

$$\frac{\partial q_s}{\partial x} + \frac{\partial}{\partial t}(C_s Y) + \frac{\partial z}{\partial t} = A_s \tag{2}$$

$$\frac{\partial Y}{\partial x} + \frac{1}{g}\left(U\frac{\partial U}{\partial x} + \frac{\partial U}{\partial t} \right) + \frac{\partial z}{\partial x} + S_f + \frac{UA_w}{gY} = 0 \tag{3}$$

where U is the mean flow velocity [m/s], Y the water depth [m], A_w and A_s the lateral non-uniform/non-steady water and sediment affluxes per unit channel length and per unit channel width [m/s], q_s the sediment volumetric transport rate of bed material under *local equilibrium* conditions [m^2/s], $C_s = q_s / UY$ the depth averaged sediment concentration [-], z the channel bed elevation above an horizontal reference [m], S_f the friction slope [-], x the abscissa along the reach [m], t the time [s] and g the acceleration due to gravity [m/s^2].

In order to solve the above mentioned system of partial differential equations two more relations are required for sediment tranport and hydraulic resistance as a function of actual flow parameters.

Sediment transport formula

Sediment transport capacity on steep slopes, under equilibrium conditions, has been recently discussed by several authors, particularly from the Swiss school (Smart & Jaeggi, 1983, Smart 1984, Graf, 1984). Nevertheless, standing on our knowledge, a certain lack of both experimental evidence and theoretical analysis on the behavior of high-slope coarse-sediment flows is still present particularly as far as non stationary conditions are concerned.

During their laboratory work, Smart and Jaeggi observed the behavior of steep-slope coarse-sediment two phase steady and uniform flows whose slope ranged from 0.4% up to 20%. The authors derived a formula for *total load* sediment transport capacity $\overline{q_s}$, valid in the above mentioned slope range and for sediments whose mean grain size is greater than 400 μ. The expression proposed by the authors is

$$\phi = 4 \left\{ \left(\frac{d_{90}}{d_{30}}\right)^{0.2} S^{0.6}\, C\, \theta^{0.5} \,(\theta - \theta_{cr}) \right\} \tag{4}$$

where $\phi = \overline{q_s} / [g(s\text{-}1) d_{50}^3]^{0.5}$ is the dimensionless sediment transport capacity [-], s is the ratio of sediment density to water density [-], the d_i is the sediment diameter for which the i per cent weight of a non uniform sample is finer [m], S the channel slope [-], C the Chezy dimensionless conductivity coefficient, θ the *slope adjusted* (Smart, 1984) *Shields* dimensionless shear stress and θ_{cr} the same non dimensional shear stress corresponding to the incipient sediment motion.

In the present approach, an instantaneous adaptation of sediment trasport to the local flow conditions and an *adaptation length* (Armanini & Di Silvio, 1986) which is neglibile with respect to the length scale of the channel were assumed. Expression (4) was thus considered correct in order to determine the sediment transport rate along the channel.

To simulate the presence of an unerodible layer in the channel bed under steady conditions, the following expression was employed:

$$q_s(x) = \begin{cases} \overline{q_s}(x) & \text{if } z(x) > z_{ul}(x) \\ \min\left(\overline{q_s}(x),\ \overline{q_s}(x_e) + \int_{x_e}^{x} A_s(x)\, dx \right) & \text{if } z(x) = z_{ul}(x) \end{cases} \tag{5}$$

where $\overline{q_s}$ is the volumetric sediment transport capacity, z_{ul} [m] is the elevation of the unerodible layer and

$$x_e = \max\left\{ x_e \,|\, z(x_e) > z_{ul}(x_e),\ x_e < x \right\} \tag{6}$$

is the first abscissa in the upstream direction where $z > z_{ul}$ (See Figure 1).

The computation of q_s under unsteady flow conditions require additional considerations which, for the sake of clarity, will be outlined in the following paragraphs.

Resistance to Flow

Resistance to flow in mountain streams has been widely discussed in the last two decades. In a recent work of Graf (1984), results of an experimental work are discussed in order to check the influence of relative roughness, Reynolds number and Froude number on flow resistance over movable beds in a slope range between 0.5% and 9%. Though sediment transport tests on higher slopes do not show the same degree of accuracy of the others, the author states that the same logarithmic formula (Keulegan, 1938), found to succesfully fit lower slopes data, can lead to correct estimates on the whole experimental slope and sediment transport range. In the present approach the following expression is assumed

$$C = 2.5 \ln \frac{13.3\ Rf}{\epsilon} \tag{7}$$

where R is the hydraulic radius [m], f is the *form factor* [-] introduced by Marchi (1961) and ϵ is the equivalent roughness [m] assumed to be a function of the sediment mean diameter d_{50} only. By this assumption no bedform effect is considered on hydraulic resistance. A development towards this direction is presently in progress though experimental data on unsteady behavior of antidunes, particularly on coarse sediments, are extremely scarce so that a quantitative model seems quite hard to be tested. Experimental evidence (Smart & Jaeggi, 1983), however, seems to confirm how bed-form effects play a minor role in hydraulic resistance on supercritical flows where coarse sediment movable bed is present.

Boundary Conditions

After substituting relations (5) and (7) in (2) and (3), equations (1), (2) and (3) constitute a set of three first-order partial differential simultaneous equations in the three unknowns U,Y and z when $z>z_{ul}$; as one may observe (De Vries, 1965), the system is hyperbolic and admits three characteristic directions. In our case, i.e. supercritical flows, straightforward calculations show that two characteristic directions are directed downstreamwards, while one upstreamwards; one should thus get two upstream boundary conditions and one downstream to ensure the well-posedness of the problem.

The choice of the proper boundary conditions for equations (1), (2) and (3) is a rather critical point and, at present, it is a widely debated subject (Dawdy and Vanoni, 1986; Lyn, 1987; Stevens, 1987; Aparicio and Berezowsky, 1987; Croad, 1987). Of course the choice is mainly determined by the statement of the physical problem. As far as the simulation of a first order stream is concerned, it seems reasonable to impose U, Y and q_s at the upstream end of the reach, e.g., they all equal to zero. These three conditions are not independent as q_s can be univocally determined from U and Y by means of equation (4); in the more general case, this dependence is broken as soon as the unerodible layer is uncovered.

The constance of bed elevation at downstream can be imposed if, for example, the same elevation can be considered independent on the morphological events of the higher order streams. In terms of mathematical modeling, this last occurrence may be handled rather easily.

Numerical Solution

Generally, z varies much slowly in time as compared with U and Y. This consideration suggested essentially two different algorithms for a so-called *decoupled* solution. By the first approach, at each time step, momentum and continuity equations for water are first simultaneously solved considering the local slope as a constant; then, bed variations are computed by means of the sediment continuity equation. By the second approach, water transients are considered so rapid with respect to the time integration step that a steady condition is *instantaneously* achieved; in this case, terms $\partial Y/\partial t$ and $\partial U/\partial t$ can be neglected in equations (1), (2) and (3), and water continuity equation can be immediately solved while the simultaneous solution of the remaining two equations can be seeked in a further step. When simulation is performed at the time scales typical of water transients the first decoupling algorithm should be recommended.

Decoupling the water flow equations from the sediment continuity equation however, may lead to uncorrect results, as was recently addressed by Lyn (1987). A simple trick to check the correctness of the method can be achieved by substituting the decoupled solution into the original system. The error can thus be directly evaluated performing this procedure in the worst conditions, i.e. for the higher values of $\partial z/\partial t$, and the solution accuracy may be assessed.

Finite Difference Scheme

Numerical solution is achieved via a generalized nonlinear implicit scheme (Preissmann, 1961). According to this method, the continuous space-time domain is discretized by means of a rectangular grid whose mesh sizes are Δx and Δt: at any point (x,t) $\mid i\Delta x < x < (i+1)\Delta x, j\Delta t < t < (j+1)\Delta t$ a function f and its derivatives are estimated as a linear combination of the four values

$f_{i,j}$, $f_{i+1,j}$, $f_{i,j+1}$, $f_{i+1,j+1}$ computed in the corner points of the mesh:

$$f \simeq (1-\Theta)(1-\Phi)\, f_{i,j} + (1-\Theta)\Phi\, f_{i,j+1} + \Theta(1-\Phi)\, f_{i+1,j} + \Theta\Phi\, f_{i+1,j+1} \tag{8}$$

$$\frac{\partial f}{\partial t} \simeq \frac{\Phi(f_{i+1,j+1} - f_{i+1,j}) + (1-\Phi)(f_{i,j+1} - f_{i,j})}{\Delta t} \tag{9}$$

$$\frac{\partial f}{\partial x} \simeq \frac{\Theta(f_{i+1,j+1} - f_{i,j+1}) + (1-\Theta)(f_{i+1,j} - f_{i,j})}{\Delta x} \tag{10}$$

where Φ and Θ are two weighting coefficients. Substituting these expressions in equations (1), (2) and (3) a system of three nonlinear difference equations is obtained.

According to the previously mentioned decoupling technique, computations are splitted into two steps for each Δt: the discretized momentum and water continuity equations are first simultaneously solved considering $\partial z/\partial x$ as a constant; the discretized sediment continuity equation is then solved so that changes in bed elevation can be computed.

The algorithm behaves differently whether the whole unerodible bed is buried underneath the loose sediment layer or not; in the first occurrence (i) the sediment transport equals the transport capacity, (ii) an unique equivalent roughness value is assumed to be valid for the whole channel length and (iii) the only constraint for the bed elevation stays in its constant value at downstream. In the first substep, the only involved boundary conditions are the upstream values of U and Y; the computation of the U and Y new values along the channel proceed from upstream to downstream, determining $U(x+\Delta x,t+\Delta t)$ and $Y(x+\Delta x,t+\Delta t)$ via a Newton-Raphson procedure. The process stops once the downstream end is reached, where the boundary condition on z is imposed; starting from this point, the sediment continuity equation is iteratively applied along x from downstream to upstream, so that the new bed elevation z values can be computed.

The opposite evenience is detected when, in any abscissa x, a value of the bed elevation $z(x,t+\Delta t)$ is computed which results to be lower than the unerodible layer level $z_{ul}(x)$. In this case the sediment discharge $q_s(x+\Delta x,t+\Delta t)$ results to be overestimated, this leading in fact to an uncorrect over-erosion of the bed; in this case the procedure is inverted: the value of $q_s(x+\Delta x,t+\Delta t)$ is corrected by imposing the constraint condition $z(x,t+\Delta t)=z_{ul}(x)$. Due to this fact, all values of q_s and z for $x>x+\Delta x$ must be checked and, if necessary, corrected in order to take in account the new situation, accordingly to Equation (5*b*). Computations proceed until the upstream end of the channel is reached where an upstream boundary condition (i.e. a given value of sediment discharge) is required to ensure the problem closure.

In this latter case an abrupt change in bed roughness may result, due to the different characteristics of the sediment layer and the rocky bed surfaces; if the unerodible layer results to be rougher then the sediments, a decrease in sediment transport capacity may occur in the *bare* reach of the channel as time increases and the unerodible bed could be covered again. This may lead to *oscillating/non-periodic* behavior of both sediment and water discharge that can be of interest as far as short time scale simulations are concerned.

TABLE 1
Channel data for prototype case simulation

Length:	400	$[m]$
Initial Average Slope:	4.9	[%]
Initial Sediment Layer Thickness:	0.2	$[m]$
Sediment Characteristic Diameter:	0.01	$[m]$
Sediment Equivalent Roughness:	0.02	$[m]$
Sediment Specific Gravity:	1.65	[-]
Unerodible Layer Equivalent Roughness:	0.2	$[m]$
Upstream Incoming Water Discharge:	1	$[m^2/s]$
Upstream Incoming Sediment Discharge:	0.018	$[m^2/s]$
Peak Lateral Water Afflux:	0.07	$[m^2/s/m]$
Peak Lateral Sediment Afflux:	0.0007	$[m^2/s/m]$
Rising Stage Time for Lateral Afflux:	400	$[s]$
Peack Stage Time for Lateral Afflux:	200	$[s]$
Decay Stage Time for Lateral Afflux:	400	$[s]$
Simulation Duration:	1500	$[s]$

Results of simulations are described in Figure 3. Paricularly in Figure 3(a) the sediment layer thickness h is shown as a function of space and time. An abrupt aggradation can be observed in the upstream reach of the channel which is almost synchronous with the rising stage of the lateral afflux and whose decay is slowly achieved with time. In fact the lateral water afflux causes a diminishing in transport capacity at the upstream end of the channel; the upstream incoming flow becomes thus *overcharged* (Di Silvio, 1979; Armanini and Di Silvio, 1986) and aggradation starts. On the other side, degradation takes place downstreamwards; an erosion front develops with time whose celerity is almost constant for the whole peak stage of the lateral afflux. As the unerodible bed is uncovered and no further erosion is then possible, a new propagation phenomenon arises whose celerity is much greater then the preceding one.

As the sediment layer is completely eroded, an abrupt change in bed roughness takes place; as a consequence, the water level increases and a diminishing in sediment transport capacity is achieved thus determining the conditions for a new sediment layer to develop. An *almost oscillatory* process starts characterized by a sort of *sand waves*, or *clusters*, traveling downstreamwards. A wide numerical investigation proved this *pseudo-unstable* behavior not to be originated by numerical noise, seeming in fact peculiar to the physical assumptions underlying this model.

The bed time evolution is, of course, reflected by water free surface behavior, whose level is shown in Figure 3(b). Particularly free surface *waves* are evident which results to be counter phased with respect to the bed *sand waves*. Water and sediment discharges are plotted in Figures 3(c) and 3(d) respectively.

Discussion

The model described in the above paragraphs proved to be a powerful tool in simulating aggradation/degradation phenomena in open channels, when the current is supercritical and short-time scale processes are involved. In order to highlight some of the more peculiar characteristics of this approach a prototype case will be discussed. Referring to the channel whose characteristics are summarized in Table 1, a series of computer runs was performed in order to simulate the effects of the sediment layer thickness on the transient evolution.

For these tests, a triangularly distributed lateral afflux of both water and sediments was imposed on a reach whose length is about 1/4 of the total channel length (See Figure 2). The maximum pulse intensity and duration were choosen in order to result so high that a granular layers could be washed away whose thickness was up to 20 times the characteristic sediment diameter. Two different values of equivalent roughness were attributed to sediment layer and unerodible bed being the latter one order of magnitude greater than the former. During the tests Froude number ranged between 1.5 and 3.

As previously mentioned, a couple of boundary conditions were imposed at upstream, namely both steady water and sediment discharges; as a downstream boundary condition, a constant bed elevation was imposed. As an initial configuration, a granular bed of uniform thickness was assumed to overlay the unerodible bed and whose profile was in equilibrium (i.e. no erosion/no deposition) with the upstream steady discharge.

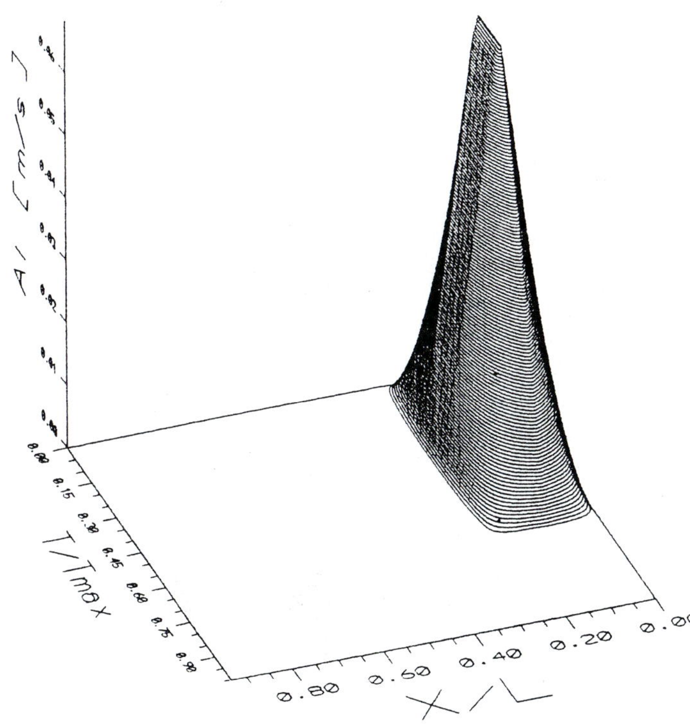

Figure 2. Lateral afflux as a function of space and time

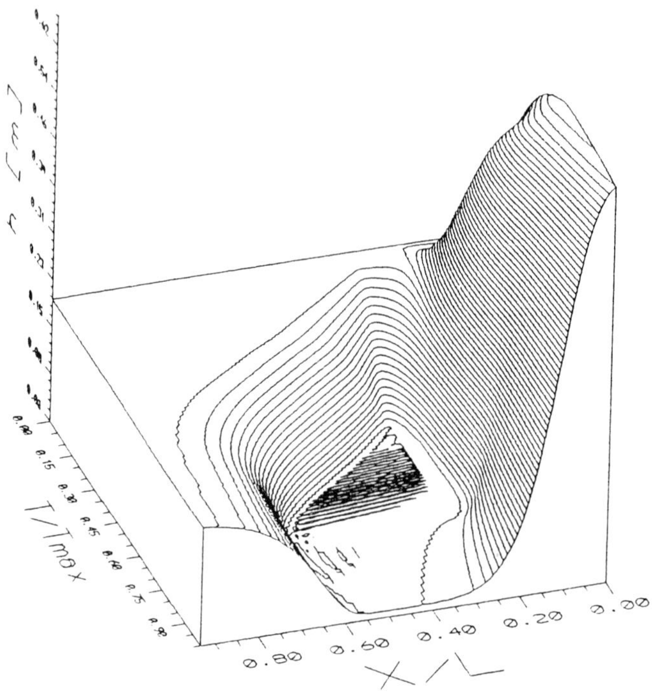

Figure 3(a). Sediment layer thickness as a function of space and time

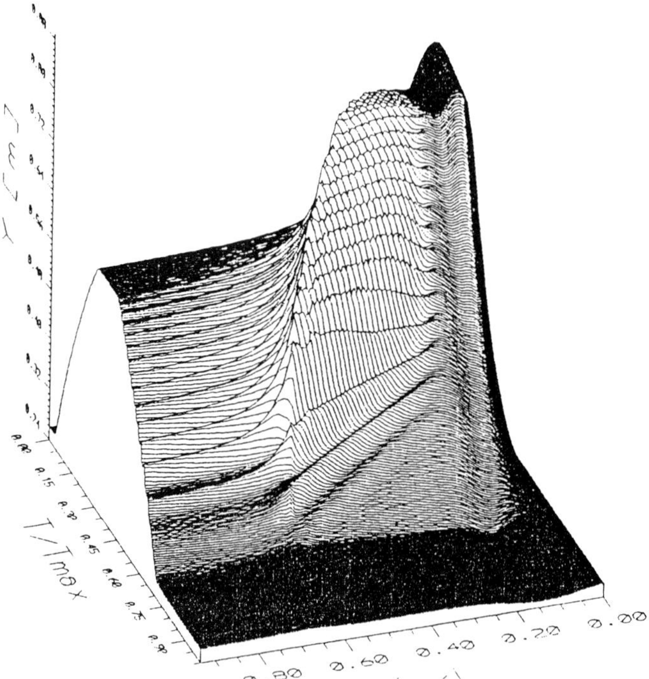

Figure 3(b). Free surface elevation as a function of space and time

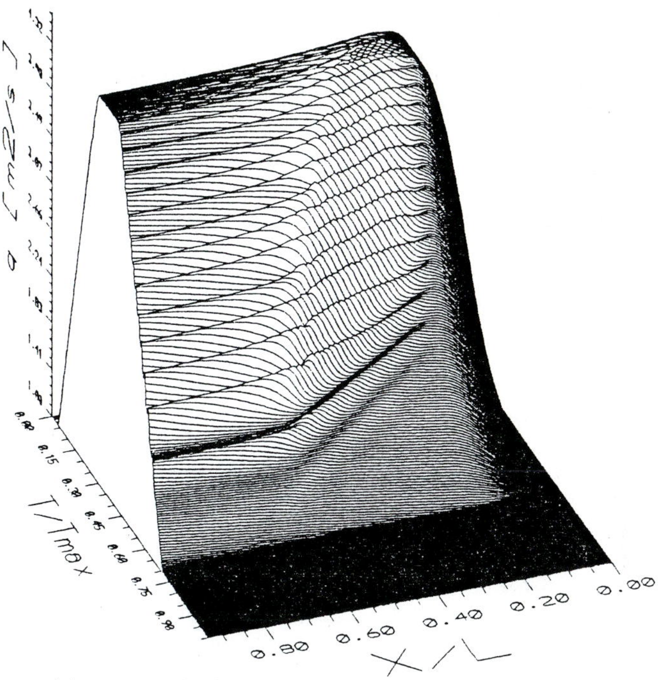

Figure 3(c). Water discharge as a function of space and time

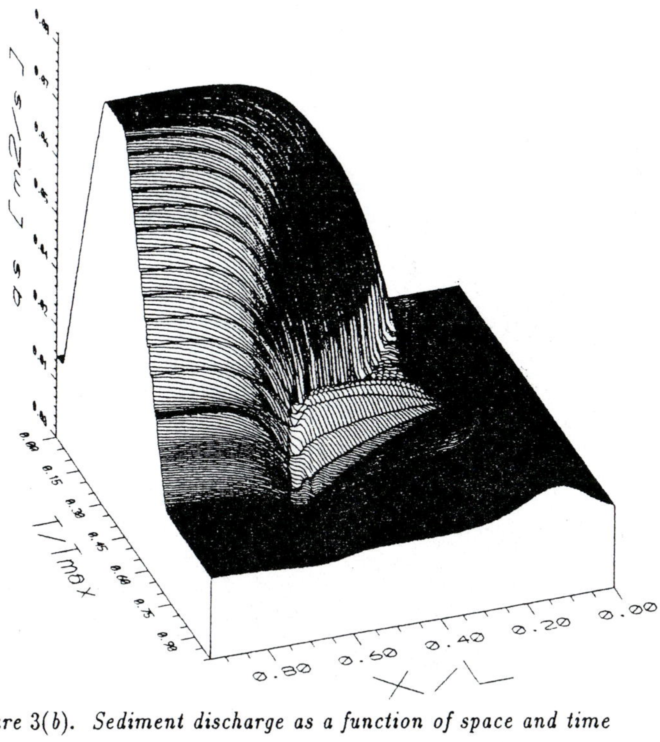

Figure 3(b). Sediment discharge as a function of space and time

Conclusion

The analysis of unsteady supercritical flow in a loose granular bed open channel is still an open problem essentially due to both a lack of knowldege of basic physical factors governing the overall phenomenon and numerical difficulties. In the present paper a numerical tool is presented whose main aim is the simulation of supercritical flow as far as short term aggradation/degradation phenomena are concerned. Following the classic De Vries approach the problem is formulated in terms of the two *De Saint Venant* equations plus a sediment continuity relationship. The presence of an unerodible bed underlying the granular layer is taken into account; to this purpose the sediment balance relationship is re-formulated using an original approach. The numerical solution is achived via a generalized Preissman scheme. The algorithm proved to be stable notwithstanding the strong discontinuities introduced by the double layer bed in both sediment transport and resistance to flow; to this purpose the simulation of a prototype case is presented where the roughness of the sediment layer is much smaller than that of the unerodible bed.

Results show how *sand clusters* may be observed traveling downstreamwards, this giving rise to a solid transport mechanism essentially based on an intermittent shedding of sediment bursts; an oscillatory though non periodic behavior is also found in both water depth and discharge. The case we discuss here was not inteded to simulate any particular real situation, being nothing else but a numerical experiment; however we stress here that intermittency is often observed in nature as a typical characteristic of sediment transport, particularly as far as supercritical flows are concerned; the present work shows how a pulsating non periodic behavior may arise within a strictly deterministic framework based on the hypotheses currently assumed in developing unsteady flow models; this, in fact, is not surprising if one observes that Equations (1), (2) and (3), together with the whole set of constraints we discussed throughout the paper, describe a strongly non-linear dynamical system (Guckenheimer and Holmes, 1983).

Acknowledgements

The authors warmly acknowledge professor Renzo Rosso for having inspired and constantly encouraged this work. Discussions with professor Alberto Lamberti were helpful at various stages of model developing.

References

Aparicio, J., Berezowsky, M., *A mathematical model for unsteady supercritical flow on a mobile sandy bed*, J. of Hydrological Sciences, Vol. 32, No. 3, 313-327, 1987

Armanini, A., Di Silvio, G., *Transport of Suspended Sediments along Channels with a Trench*, Proc. XXI IAHR Congress, Melbourne, 527-532, 1985

Armanini, A., Di Silvio, G., *Sulla non stazionarieta' del trasporto solido nei corsi d'acqua*, Proc. Seminario su Modelli dei Fenomeni Idraulico Fluviali, Bologna, 211-221, 1986

Armanini, A., Di Silvio, G., *A one-dimensional model for the transport of a sediment mixture in non-equilibrium conditions*, Journal of Hydraulic Research, IAHR, Vol. 26, No. 3, 275-291, 1988

Croad, R.N., Discussion (including authors' reply) on Aparicio and Berezowsky, 1987, J. of Hydrological Sciences, Vol. 33, No. 4, 415-418, 1987

Cunge, J.A., Holly, F.M., Verwey., A., *Practical Aspects of Computational River Hydraulics*, Pitman, 1980

Cunge, J.A., Perdreau, N., *Mobile bed fluvial mathematical models*, La Houille Blanche, n. 7, 561-580, 1973

Dawdy, D.R., Vanoni, V.A., *Modeling Alluvial Channels*, Water Res. Research, Special Issue: "Trends and Directions in Hydrology", Vol. 22, No. 9, 71S-81S, 1986

De Vries M., *Considerations about non-steady bed load transport in open channels*, Proceedings IAHR Congress, Paper 3.8, Leningrad, 1965

Di Silvio, G., *General Report on Subject D.a: Physical and Mathematical Models for Watersheds, Soil Erosion and Conservation*, Proc. XVIII IAHR Congress, Cagliari, 391-400, 1979

Lamberti, A., Menduni, G., *A Mathematical Model for Simulation of Loose Granular Bed Response to Varying Discharges*, Excerpta, Vol. 2, 63-79, 1987

Graf, W. H., *Flow Resistance for Steep, Mobile Channels*, Atti del Seminario su: Idraulica del Territorio Montano, Bressanone, 8-13 Ottobre, 341-352, 1984

Guckenheimer, J. Holmes, P., *Nonlinear Oscillations, Dynamical Systems, and Bifurcation of Vector Fields*, Springer, 1983

Keulegan, G. H., Journal of Research of the National Bureau of Standards, Vol. 21, Res. Pap. RP1151, 1938

Lyn, D.A., *Unsteady Sediment Transport Modeling*, ASCE J. of Hydr. Eng., Vol. 113, No. 1, 1-15, 1987

Lyn, D.A., Goodwin, P., *Stability of a General Preissmann Scheme*, ASCE J. of Hydr. Eng., Vol. 113, No. 1, 16-28, 1987

Mahmood, K., Yevjevich, V. (Editors), *Unsteady flow in open channels*, Water Resources Publications, Fort Collins, 1975

Marchi, E., *Il moto uniforme delle correnti liquide nei condotti chiusi e aperti (Parte I)*, L'Energia Elettrica, Vol. XXXVIII, no. 4, 289-301, 1961

Marchi, E., *Il moto uniforme delle correnti liquide nei condotti chiusi e aperti (Parte II)*, L'Energia Elettrica, Vol. XXXVIII, no. 5, 393-413, 1961

Montuori, C., *Sviluppi recenti nello studio delle correnti supercritiche*, Atti del Seminario su: Idraulica del Territorio Montano, Bressanone, 8-13 Ottobre, 205-256, 1984

Ponce, V.M., Garcia, J.L., Simons, D.B., *Modeling Alluvial Channel Bed Transients*, ASCE J. of Hydr. Eng., Vol. 105, No. HY3, 245-256, 1979

Roth, G., Rosso, R., Siccardi F., *Hydrodynamic Description of the Erosional Development of Drainage Pattern*, Water Resources Research, 25, 2, 319-332, 1989

Rouse, H., *Critical Analysis of Open-Channel Resistance*, Journal of Hydraulic Division, Proc. ASCE, Vol 91, No. HY4, 1-25, 1965

Smart, G. M., *Sediment Transport Formula for Steep Channels*, Journal of Hydraulic Engineering, Vol. 110, No. 3, March 1984.

Smart, G. M., Jaeggi M. N. R., *Sedimenttransport in steilen Gerinnen*, Mittelungen der Versuchsanstalt fur Wasserbau, Hydrologie und Glaziologie, No. 64, Zurich, 1983

Stevens, M.A., Discussion (including author's closure) on Lyn, 1987, ASCE J. of Hydr. Eng., Vol. 114, No. 8, 954-956, 1988.

Preissman A., *Propagation des intumescences dans les canaux et rivieres*, 1er Congres de l'Assoc. Francaise de Calcul, Grenoble, Pagg. 433-442, 1961

Vreugdenhil, C.B., and De Vries, M., *Analytical Approaches to Non-Steady Bed Load Transport*, Delft Hydraulic Laboratory, Res. Rep. S-78-III, 1973

MODELLING SHORT- AND LONG-TERM EVOLUTION OF MOUNTAIN RIVERS: AN APPLICATION TO THE TORRENT MALLERO (ITALY)

Giampaolo Di Silvio[(1)] and Massimo Peviani[(2)]

[(1)] Hydraulics Institute "G.Poleni", University of Padua, Padua Italy
[(2)]I.S.M.E.S. s.p.a. , Bergamo, Italy

1. Introduction

Inundations in mountain rivers are usually related to estremely large input of sediment (landslides and debris flow) and the correspondent deposit along the hydrographic network. The overaggradation phenomena showed to be strongly dependent, beside the hydrodynamic parameters, on the grain size distribution of both the bottom and the input landslide material.

The present work describes a simplified numerical model for the short time-scale study of the overaggradation phenomena, as well as for long time-scale evolution of mountain rivers. The practical aspects of the model implementation are commented and the results of a specific application of the model to the Torrent Mallero (Italy) are presented.

The model itself is in fact calibrated and verified with the data available in several places of this torrent (watershed surface 319 km^2), with special reference to the cathastrophic event of July 1987. The hypothesis adopted in the model, however, will be also checked against the results of an experimental investigation at more detailed scale presented in another contribution to this Workshop (Di Silvio and Brunelli).

2. Mathematical model for short-time scale processes

The catastrofic floods are a consequence of very intense precipitations with extremely large amounts of sediment that feed into the main stream

within a time span of few hours, with consequent rapid change of bed elevation.
An acceptable representation of the physical phenomena can be obtained by a simplified theoretical model, as described below . The model is intended for a relative coarse space-resolution (few hundred meters).
As for the time-resolution, it has been conceived for relatively short intervals (tens of minutes) in order to follow a single flood event; for stability requirements it can also be seconds.

2.1 Theoretical model

The following assumptions are made to derive the set of differential equations that describes the physical phenomena:

- The celerity of flood waves in mountain rivers is much higher than the celerity of the bottom changes. It leads to assume instantaneous adaptation in time of the discharge disturbances in each branch of the network (it means quasi-steady flow condition).

- The flow pattern in a mountain river is a sequence of subcritical and supercritical regime, passing through a critical condition and a hydraulic jump respectively, with a Froude number value very near to one. The local energy loss at the weak hydraulic jumps as well as the variation of kinetic energy are negligible comparing with the friction losses of the flow itself. Taking a relatively regular reach of river, in which many of this sequences occurs, it leads to assume that the average flow regime is quite well represented by a quasi-uniform flow.

- It is assumed that the motion equation for each sediment fraction, both as bedload and in suspension, is completely governed by the average local parameters, such as discharge, bottom slope, channel width grain size, etc. (it means, instantaneous adaptation of the vertical concentration profile).

As a consequence of the previous simplifications, the set of equations for the i-th particle size is:

$$BDi + \frac{\partial Ti}{\partial x} = 0 \qquad \text{(sediment continuity)} \quad (1)$$

$$Ti = \alpha \frac{I^n Q^m}{B^p di^q} \beta i \; ri \quad , \text{ with } ri = \left(\frac{di}{\Sigma \beta i \; di} \right)^s \quad \text{(sediment motion)} \quad (2)$$

$$\frac{\partial \delta \beta i}{\partial t} = Di - \left| \begin{matrix} \beta i \\ \\ \beta ei \end{matrix} \right. \cdot \frac{\partial z}{\partial t} \quad \text{(vertical sediment balance)} \quad (3)$$

$$\frac{\partial z}{\partial t} = \Sigma Di - \frac{\partial \delta}{\partial t} \quad \text{(bottom level change)} \quad (4)$$

$$\delta = 2 \; d90 \quad \text{(mixing layer thickness)} \quad (5)$$

$$\frac{\partial Z}{\partial x} = \left(\frac{U}{\frac{24}{d90^{1/6}} R^{2/3}} \right)^2 = I \quad \text{(water motion)} \quad (6)$$

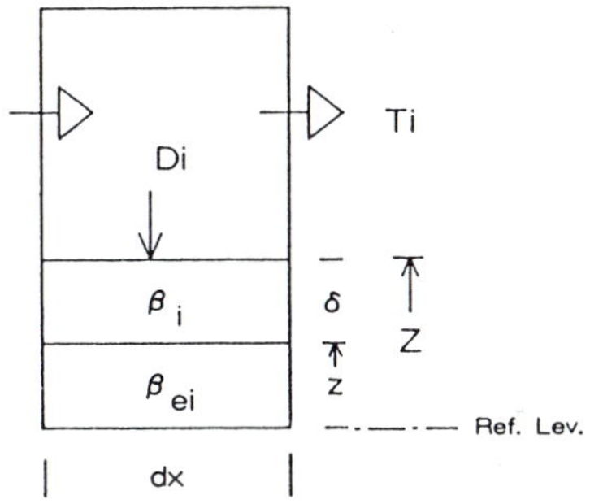

FIG 1 – Definition sketch for the sediment balance

Equation (1) expresses the balance between transport Ti and net deposition Di.

The transport equation for each fraction (2) includes an exposure-correction coefficient (ri) on accounts of the "hiding-and-exposure" effect for the non-uniform material, according to the finds of the earlier researchers (Einstein, Egiazaroff, etc.).

Q is the water discharge, B is the stream width, I is the slope and di is the grain diameter of the fraction.

The coefficient α, and the exponents m, n, p, q, and s are empirically evaluated (Ref.5,6,7). With the hypothesis commented before, the sediment transport Ti on a certain cross-section can change via Q(t), $\beta i(t)$ and I(t).

Equation (3) is the continuity expression of the vertical exchange of sediment and it leads to the computation of the mixing-layer composition changes. In this equation βei is the composition of the undisturbed material below the mixing layer, if $Di<0$ (erosion), while βi is the mixing layer material composition, if $Di>0$ (deposition).

The mixing layer thickness is assumed to be twice the representative diameter of the coarsest particles (d90), as expressed by equation (5) (Ref.8).

Equation (6) describes the water motion according to the Strickler formula.

2.2 Expression of the celerities and physical damping

The analysis of the above set of partial differential equations leads to the expression of the characteristic celerities ci and c, respectively for a disturbance of the variables βi and Z.
The expression of the physical damping, Dphi, can also be derived from the analysis (Ref.8).
For the variable βi:

$$ci = \frac{ai\ ri - s\ ai\ \beta i\ ri^{s+1}}{\delta} \qquad \text{with} \qquad ai = \alpha \frac{I^{n}\ Q^{m}}{B^{p}\ di^{q}} \tag{7}$$

$$Dphi = \frac{d\ \beta i}{dt} = - \frac{\beta i\ \Sigma Di}{\delta} \tag{8}$$

For the variable Z:

$$c = \sum_{1}^{4} \left(\frac{ai\ ri - s\ ai\ \beta i\ ri^{s+1}}{\delta} \left(\frac{Di}{\Sigma Di} - \beta i \right) \right) \tag{9}$$

Note that these expressions result to be relatively simple because of the previously mentioned hypothesis of quasi-uniform flow in each single reach, and on the further simplifying assumption that ai is a constant.

2.3 Numerical model

The set of equations is solved numerically using a finite difference approximation. A predictor-corrector method is chosen for the numerical solution of the sediment equations. The predictor step is carried out with a FTFS (Forward Time Forward Space) scheme and a Four-point scheme is used for the corrector step. The sediment continuity equation is solved, for example, with the following difference equations (superscript n = present time level, superscript ξ = predicted value, superscript n+1 = new time level):
predictor (FTFS)

$$D^{\xi}_{ij} = (T^{n}_{ij-1} - T^{n}_{ij}) / B_{j} \Delta x \tag{10}$$

```
          x     ξ
          |
          |
   x------x     n
  j-1     j
```

corrector(s)

$$D^{n+1}_{ij} = (1 - \Theta)(T^{n}_{ij-1} - T^{n}_{ij}) + \Theta(T^{\xi}_{ij-1} - T^{\xi}_{ij}) / B_{j} \Delta x \tag{11}$$

```
   x------x -     n+1
   |  -|- | -
   |      | θ
   x------x -     n
  j-1     j
```

The sequence of the numerical calculation is:
a) The motion equation for the sediment (2) is applied at every grid point at time level n, in which the channel geometric data and the sediment data are known.
b) The predicted values Di^{ξ} and βi^{ξ} are computed at every grid point by using the sediment continuity equation (1) with the predictor-scheme and the vertical balance equation (3). Then the sediment transport (Ti^{ξ}) computed again at every grid point as a function of the

predicted values βi^{ξ}.

c) By applying again equation (1) with the corrector-scheme the value of Din+1 is computed, and with equation (3) the value β in+1 is computed at every grid point.

d) The sediment transport at time level n+1 is now computed as a function of β in+1.

e) At the end, the new bottom level ($Z=z+\delta$) is computed from equation (4) with the value Din+1.

f) The new water level is computed with expression (6).

2.4 Analysis of the numerical schemes

Some preliminary tests were performed as to analyse the stability of the numerical schemes as well as to check their accuracy against the theoretical evaluations of celerity and physical damping mentioned above (par. 2.2).

All the tests were performed by studying the behaviour of a channel in uniform conditions when a sudden variation of the bottom composition is introduced in its upper part. This variation simulates the input of fine material in a mountain river, by landslides or debris flow (overfeeding);the availability of fine material in the bed generates a much higher transport (until the fine material is completely removed) and a rapid aggradation along the river. The following parameters have been used in the analysis: $Q=300\ m^3/s$, $B=67\ m$, $I=0.02$, $\alpha=0.011$, $m=1.8$, $n=2.1$, $p=0.8$, $q=1.2$, $s=0.0$.

The mixing layer thickness was kept constant and equal to 1 m , not much different from the value 2d90 reported in eq.(5).

Fig. 2a and 2b gives an example of stability analysis , showing the progressive aggradation of the bottom along the channel caused by the stepwise disturbance at the upper end.

The granulometry of the initial bed (β i0) and of the landslide material (β i*) is given in the following table:

	di(mm)	β i0	β i*
1	0.3	0.080	0.250
2	3	0.240	0.520
3	30	0.460	0.200
4	300	0.220	0.030

The distance along grid points is $\Delta x=250$ m while the time step is $\Delta t=400$ s in case a and $\Delta t=250$ s in case b.

The weight coefficent in eq.(11) was $\theta=0.6$.

The result shows an actual upper limit for the Courant number, Cr=0.65

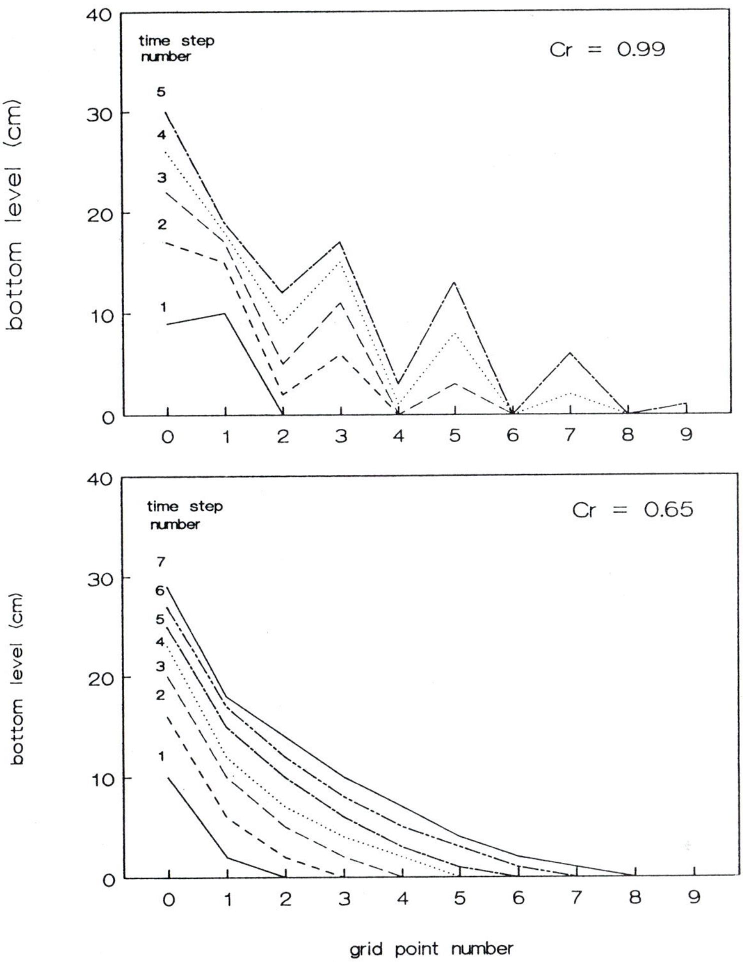

Fig. 2.a and 2.b - An example of stability analysis. Propagation of a bottom level disturbance (stepwise variation of bottom composition) with different Courant numbers.

Fig. 3a and 3b show the propagation of the disturbance of the bottom composition introduced in the upper part of the channel (only the propagation of the fraction i=1 is reported). The composition of the disturbed bottom in these tests is $\beta_{1*}=0.200$, $\beta_{2*}=0.330$, $\beta_{3*}=0.310$ and $\beta_{4*}=0.160$; but while in the case above (3a) the disturbance a $\Delta t=0$ is gradually distributed along six sections of the channel, in the case below (3b) a stepwise variation is introduced at its upstream end.
In both cases, the celerity and the damping provided by the numerical

model pratically coincide with the theoretical results obtained in par.2.2, with an accuracy of about 5% and 13% respectively.

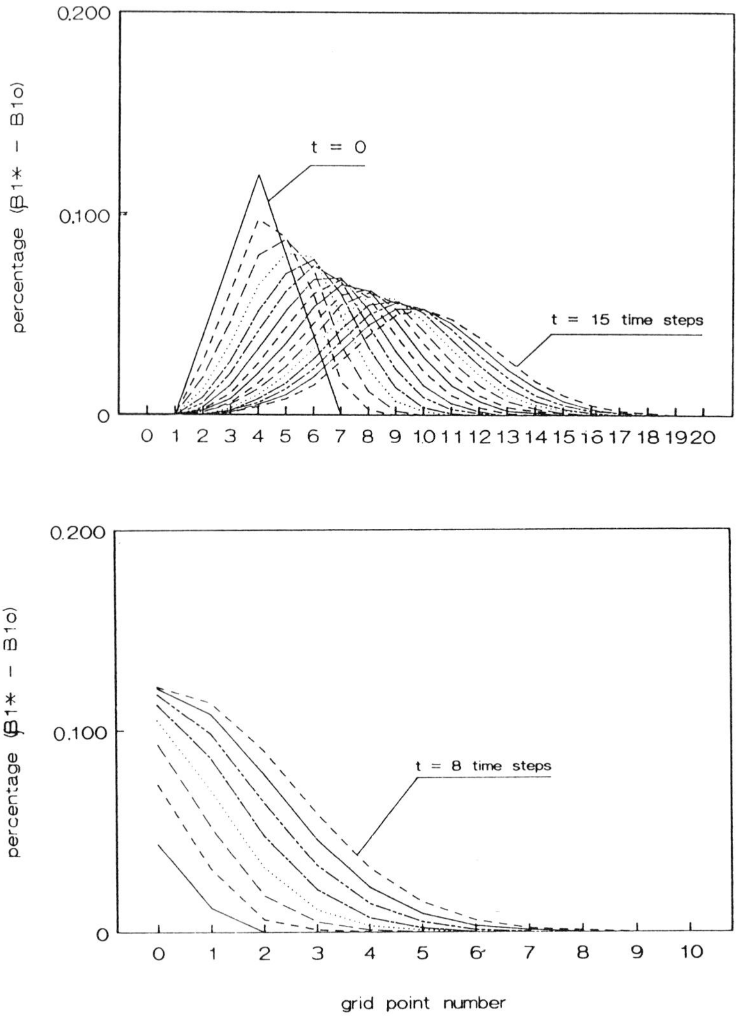

Fig 3.a and 3.b - An example of celerity analysis. Propagation of a bottom - composition disturbance: gradual variation (above) and stepwise variation (below) of the first fraction, $\beta 1$.

2.5 Sensitivity analysis of the physical parameters

Some tests were performed to see the influence of the main physical parameters on the behaviour of the disturbance propagation .
In particular, the celerity and height of the bottom aggradation was analysed for different granulometries of the initial bottom and of the landslide material.

Influence of the initial bottom composition (βio):(Fig 4.a and 4.b)
In both cases the composition of the material at the upstream boundary is $\beta_{1*}=0.250$, $\beta_{2*}=0.520$, $\beta_{3*}=0.200$, $\beta_{4*}=0.030$, with $\delta=2d90$ and s=0.6. The initial bottom composition in Case a is $\beta_{10}=0.040$, $\beta_{20}=0.20$ $\beta_{30}=0.500$, $\beta_{40}=0.260$. On the other hand, Case b has a finer material composed of $\beta_{10}=0.120$, $\beta_{20}=0.280$, $\beta_{30}=0.420$, $\beta_{40}=0.180$. It is possible to see that the disturbance wave propagates faster in case of finer initial bottom composition. As the transport at the upstream boundary is the same in both cases, the height of the wave is smaller in case of finer material.

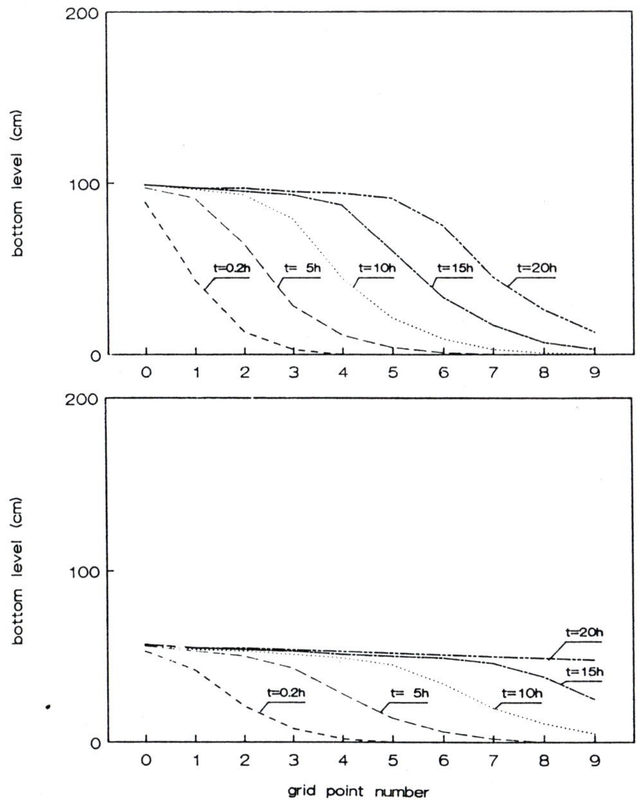

Fig 4.a and 4.b - Sensitivity analysis. Effects of the initial bottom composition (βio) on the disturbance propagation.

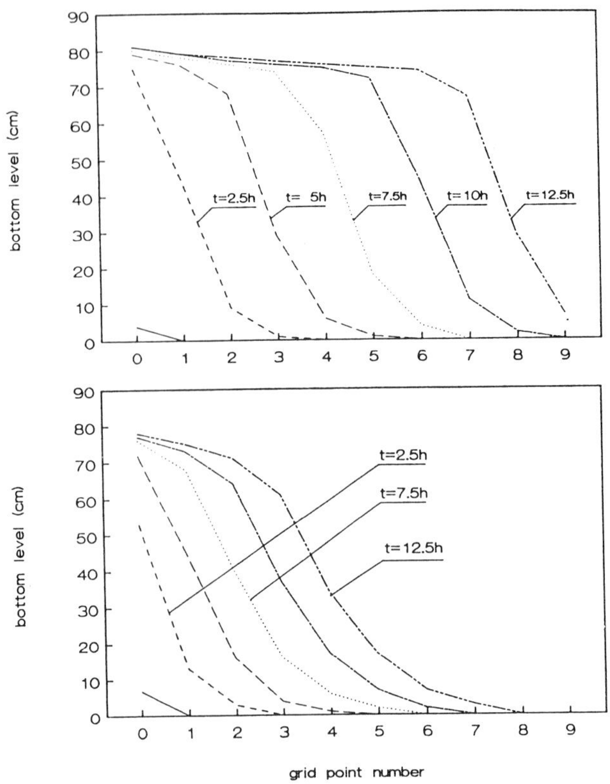

Fig. 5.a and 5.b - Sensitivity analysis. Effects of the overloading composition (βi^*) on the disturbance propagation.

<u>Landslide material composition (βi^*):</u> (Fig 5.a and 5.b)
In both cases the initial composition of the bottom is $\beta_{10}=0.080$, $\beta_{20}=0.240$, $\beta_{30}=0.460$, $\beta_{40}=0.220$, with δ =2d90 and s=0.6. The composition of the overloading material at the upstream boundary in <u>Case a</u> is $\beta_{1*}=0.463$, $\beta_{2*}=0.349$, $\beta_{3*}=0.168$, $\beta_{4*}=0.020$.
In <u>Case b</u> there is a coarser material with composition $\beta_{1*}=0.250$, $\beta_{2*}=0.520$, $\beta_{3*}=0.200$, $\beta_{4*}=0.030$. The plots show a faster propagation disturbance in case of finer material; in spite of the difference on the upstream transport value (T = 2.67 m^3/s in Case a, and T = 1.55 m^3/s in Case b) the height of the wave remains quite constant.

Further tests confirm that the <u>thickness of the mixing layer δ</u> influences the celerity of the disturbance propagation on an inverse way: with smaller values of δ the celerity increases, as in expressions (7) and (9).

The value of s on the exposure-correction coefficient influences the sediment transport values of each fraction. In this way, for higher values of s the transported material becomes coarser.

2.6 Boundary conditions

Before entering on the practical application of the short-time scale model, the lateral and internal boundaries used in the model to represent special phenomenas are described.

Type A: describes the lateral input of sediment Tt of a tributary conveying in the main stream material of a nearby landslide as sediment transport. For the computation of Tt as a function of the tributary water discharge Qt, it is assumed that the bottom composition in eq.2 is the same as the landslide material.

Type B: describes the lateral input of sediment of a tributary conveying landslide material directly in the main stream as debris flow. This is the case that usually occurs when the tributary slope is higher than 0.30. It is assumed that, at the confluence, the material composition in the main stream remains equal to the composition of the slided material till the complete volume is removed by the water flow.

Type C: describes the collapse of a check-dam. It is considered that when it occurs there is a certain volume of finer material with composition $\beta i^* = 1/2(\beta io + Ti/\Sigma Ti)$ available to be transported by the water flow.

Type D: describes special points in the stream in which the bottom is composed of rocky material. In that case the erosion can not progress any more, i.e. the transport remains constant through the rocky reach.

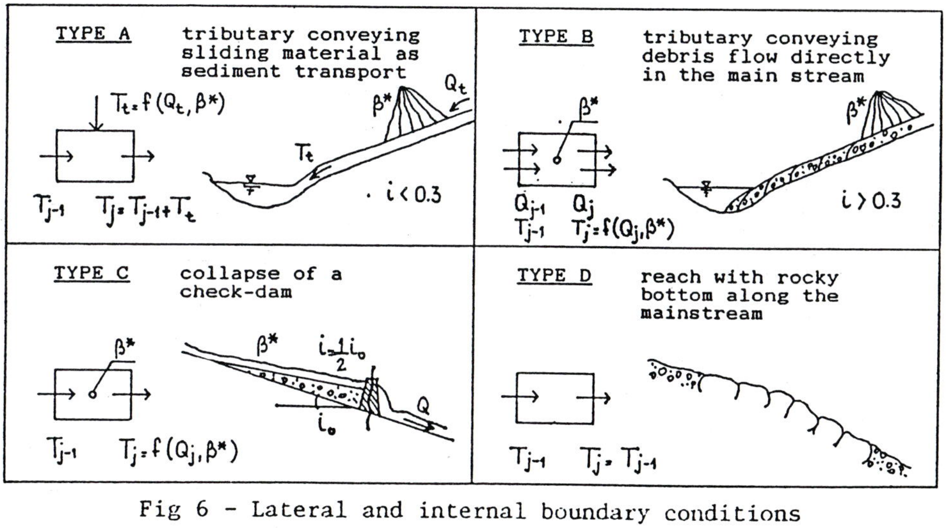

Fig 6 - Lateral and internal boundary conditions applied on the short-time scale model.

3. Mathematical model for long-time scale processes

The implementation of the numerical model described in the previous paragraph, requires the definition of the boundary conditions and the initial conditions as well.

While the topographical and the hydrological information is usually available from maps and rainfall records associated to a rainfall/runoff model, it is almost impossible to have the required required sedimentological data (grain size distribution of the bottom material previous to the catastrophic event) in all the branches of the hydrographic network. If these data are available in the upper of the network, the ordinary annual transport feeding the main streams and the corresponding bottom composition all over the network can be estimated by using a long-time scale model.

Note that the grainsize distribution in all the branches feeding the main streams should be representative of ordinary conditions; so it's to be measured previous catastrophic event or same years later, when the small watersheds have resumed their ordinary configuration.

The application of the long-time scale model during a long period (10 years or more) will reconstruct the composition of the bottom material in the main stream under ordinary flow condition, to be checked against the available data.

The theoretical model for the long-time scale process is described by the same set of equations than for the short-time scale process, except the sediment motion equation. In this case the equation of motion for each sediment fraction is

$$Ti\ year = \alpha \frac{I^{n}\ (Qo^{m-1}\ Vo)}{B^{p}\ di^{q}\ m}\ \beta i\ ri \tag{12}$$

where Qo and Vo are the annual ordinary maximum discharge (m^3/s) and the annual ordinary runoff volume ($m^3/year$), computed from the available hydrological data. Equation (12) is obtained by integrating equation (2) over a period of one year or more, and by assuming an exponetial duration curve for the water discharge (Ref.5 and 6).

Besides the reconstruction of initial condition for the short-time scale model, the long-time scale model can be used to study the equilibrium configurations of natural hydrographic networks.

4. Practical application of the model to the Torrent Mallero

4.1 General description of the basin and the event of July 87

The Torrent Mallero is a tributary of the Adda River which is the main stream of Valtellina, in the central alpine region of North Italy (Fig 7). The Torrent Mallero is 24 km long, starting at the elevation of 1636 m (on m.s.l.) from the confluence of the Vazzeda and the Ventina torrent, and ending at the elevation of 282 m (on m.s.l.) .on its confluence with the Adda River. The surface of its basin is approximatly 319 km².
There are several urban establishments along the stream and Sondrio near the confluence with the Adda river is the most important one. The Valtellina region has always been, from information coming from the Middle Age times, a place where severe storms accompanied by landslide phenomena and overaggradation have occured.
On July 1987 an exceptional event, with a discharge of around 200 years return period, has produced enormous disasters in Valtellina. Particularly in the Mallero Basin there were several landslides with a total volume of several millions cubic meters. This material reaching the .streams and transported by the water flow has produced in many places an overaggradation of the bottom with the consequent increasing of the water levels and flooding.
A detailed study of the area have been carried out after the event, with the location of the landslides and evaluation of volumes and granulometric composition of the slided material. A general description of the Mallero stream after the event, with computation of volumes of deposited and eroded material and a number of granulometric analysis of the bed have been also carried out. The rainfall during the event was registred on three pluviographs set up in the basin.

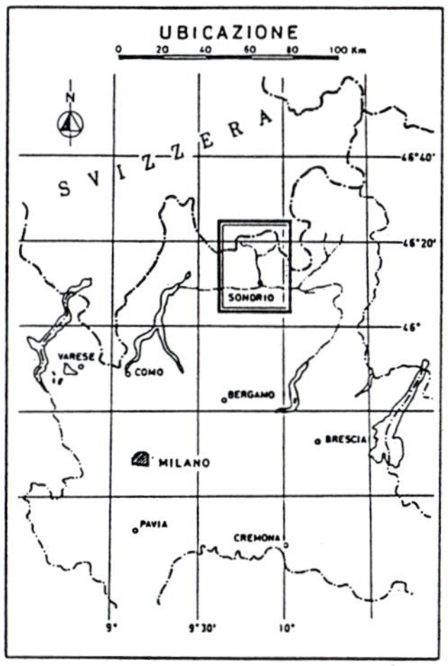

Fig.7 - Location of the torrent Mallero in Valtellina (Central Alps) in North Italy

4.2 Implementation of the model

A schematization of the model set up is showed in Fig.8.
As a general criterium, tributaries with a terminal slope steeper than 25% are not included in the model except as a lateral input of water and sediments.
Along the two branches of Mallero and Lanterna torrents, which are completely modelled, the figure shows the tributaries from where water and sediments are fed to the main network; sediments may be either fed as ordinary transport or as exceptional transport caused by land-slides, following boundary-conditions A or B described in Fig. 6 .

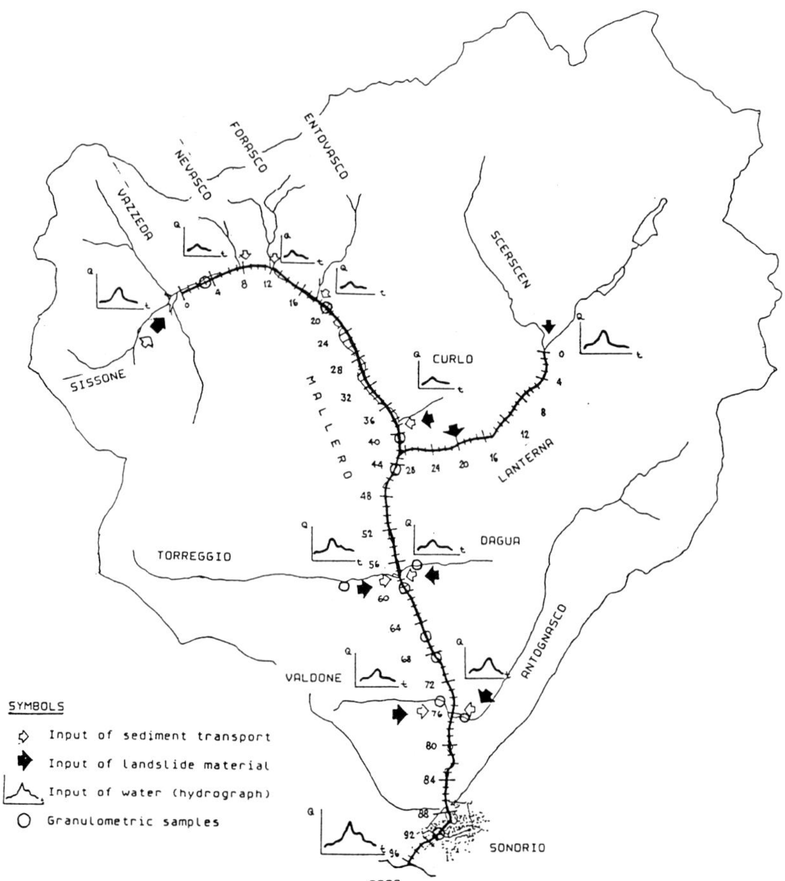

Fig.8 - Basin of the torrent Mallero and model schematization.

4.2.1 Long-time scale model

The long-time scale model is used to determine the composition of the bottom material all over the stream network under ordinary flow conditions. This result will be used as initial condition for the successive study of a short-time scale event.

The input of water from each tributary is given by the annual ordinary maximum discharge (Qo), computed at each sub-basin following the relation of the square root of the areas, and by the annual ordinary runoff volume (Vo), also computed at each sub-basin following the relation of the areas. The assumed values at Sondrio are Qo=162 m^3/s and Vo= 332*10^6/year.

The input of sediment from each tributary is computed by equation (12) with the morphological characteristics (slope and width) and the granulometric distribution (βio) at the upper reach of the confluence with the main streams. For the main streams Mallero and Lanterna the geometrical data are introduced (cross section width and longitudinal slope). As initial granulometric distribution of the bottom material in these torrents an arbitrary value (pointed line in Fig.9) is used. The long-time scale model is run until the equilibrium configuration is reached.

After 10 years of ordinary events the river arrives to a dynamic equilibrium. In Fig.9 the mean diameters of the bottom sediment composition along the stream are plotted, at different stages of computation. It shows that when the equilibrium is reached, a finer material corresponds to the flatter and wider zones of the river profile. Some granulometric samples of the bottom material at particular points of the network (indicated with circles in Fig.8) are used to verify the consistence of the model.
The values of the parameters used in the model are:
Δx=250m, θ=0.55, coefficients m=1.8, n=2.1, p=0.8, q=1.2
The value of the exponent s in the exposure coefficent has been calibrated by confronting the granulometric distribution of the bottom material computed with the model against that one from the "in situ" samples (Fig.10). In this case the value that gave the best fit was s=0.8 .
The value of α (assumed in this computation equal to 0.025, according to previous findings in alpine regions) does not affect the final granulometry of the bottom.

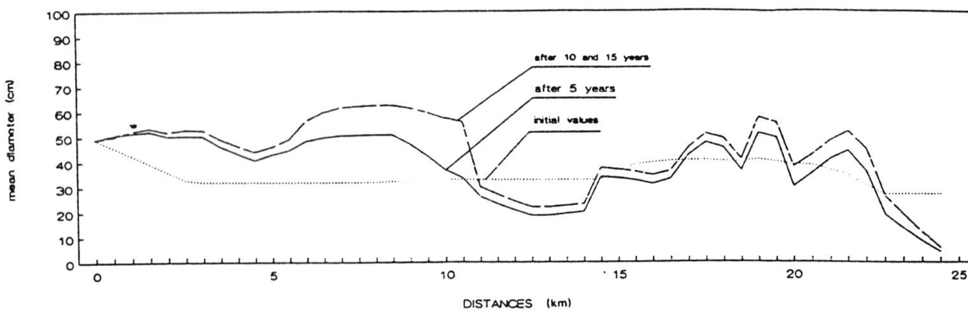

Fig.9 - Long-time scale model. Adaptation of the bottom granulometry to the equilibrium situation

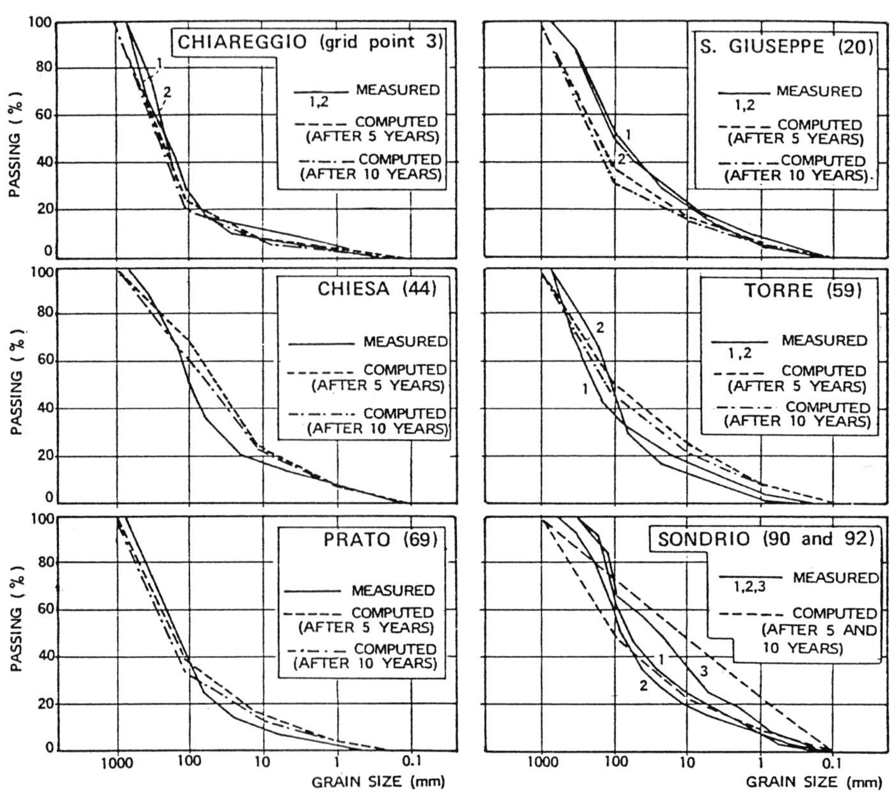

Fig.10 - Long-time scale model. Comparison between the "in situ" samples and the computation after reaching the equilibrium configuration.

4.2.2 Short-time scale model

Input of water and input of sediments as a function of time during the events are necessary boundary conditions.
The hydrographs at the end of each tributary and in some sections of the main stream as well (see examples in Fig.11), have been computed by applying a rainfall-runoff model of Rational Method type (Ref.11).

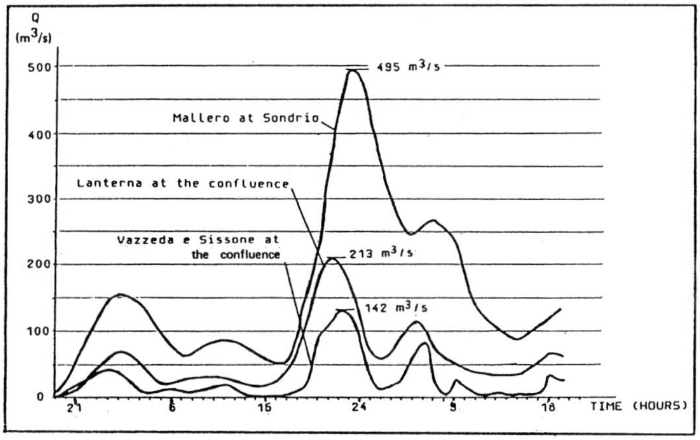

Fig.11 - Examples of waterflow hydrographs during the event of July 1987.

The input of sediment at each tributary develops in two stages:
During the first stage, the sediment transported by the water flow before the landslide events is computed by equation (2) with the morphological characteristics (slope and width) and the granulometric distribution (β io) of the final reach of the tributaries.
During the second stage, the extraordinary input of sediment due to the landslide events should be computed by one of the sediment lateral input cases (Type A or Type B, explained before in paragraph 2.4) depending on the physical phenomena to represent.
The Fig.8 shows the places in the network where the input of landslide material occurred; the one at the upper boundary (Sissone) is better represented by the Type A and all the other ones by the Type B. The chronological time in which these landslides occurred and their volumes are known with a certain approximation (see following table).
In most of the cases, granulometric samples or detailed description of the landslide material composition were also available.

Section grid point number	β* 1	2	3	4	Volume	sliding time
	%				m^3	s
0	37	21	26	16	1000000	57600
39	39	27	24	10	150000	82800
58	22	17	37	24	600000	90000
59	26	24	41	09	200000	82800
77	16	21	27	36	160000	82800
78	18	23	31	28	50000	82800

Table 1 - Characteristics of the input landslide material to the torrent Mallero, in the model simulation. The initial time of the simulation is July 17th, 1987, at 6 p.m.

The value of the parameters used in the model are: Δx=250m, θ=0.55; coefficients m=1.8, n=2.1, p=0.8 and q=1.2.

The full event is simulated, from Friday 17th (at 6pm) till Sunday 19th (at 8pm) of July, 1987. The computed bottom level of the torrent Mallero after the event is showed in Fig.12 .

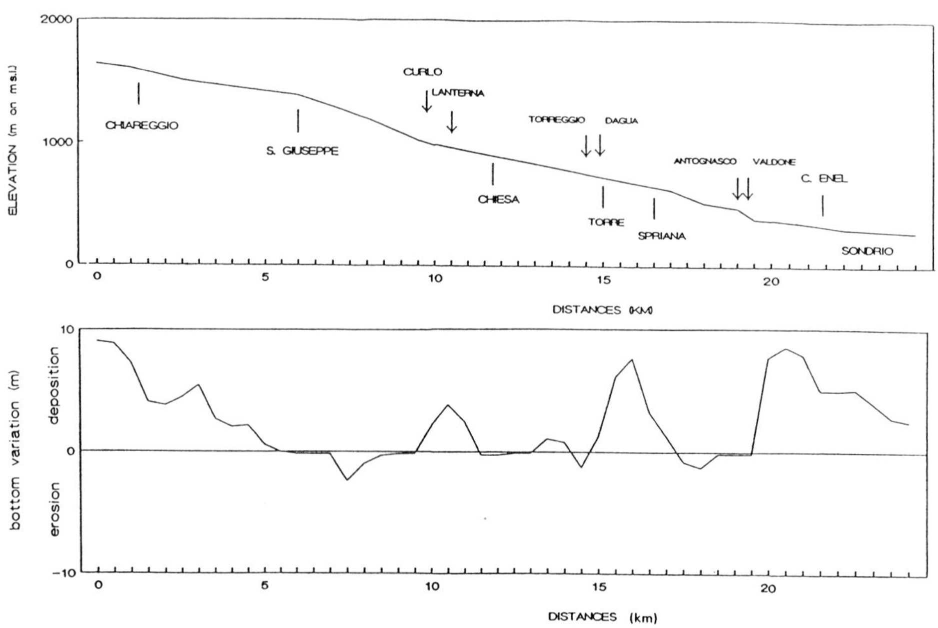

Fig.12 - Short-time scale model. Final configuration of the bottom level after the event of July 1987.

The following three available "in situ" data (from the event of July 1987) are used to calibrate the model:

a) Position and volume of the deposited material along the stream and bottom levels at some specific points.
b) Granulometric composition of the deposited material at some spcific points of the stream.
c) Cronological development of the water levels at Sondrio city.

The calibration of the model leads to select the correct values of two of the coefficients in the sediment motion equation.
They are the coefficent α and the exponent s of the exposure-correction coefficient.
The range of variation for these values, in the case of mountain rivers, is known and described in the literature (Ref.4,5,6 and 7).
An other parameter for calibration is the mixing layer thickness (δ), which should be equal to the bottom irregularities.
The table below shows the comparison between the volume of deposited material coming out from the computation and from the "in situ" observations.
There were several different overaggradation phenomena during the event of July 1987 and some of them are quite independent from the others: for example, the aggradation in the high valley (Sissone) and the aggradation in the low valley (Torreggio,Dagua,Antognasco,Valdone). As the same calibration coefficients have been applied on both regions, the calibration in one case was also a verification in the other case.

REACH (km)	NOMINATION	IN SITU m^3	COMPUTED m^3
0 to 1250	Lupo (half)	-	140000
3000 to 3900	Alpe Senevedo	47000	60000
4250 to 5000	Sabbionaccio	180000	160000
10000 to 10700	Cosi Battani	-	110000
14500 to 16800	Torre/Spriana	550000	490000
19500 to 20250	Arquino	-	83000
21500 to 24000	Ponchiera/Sondrio	320000	330000

Table 2 - Comparison between the measured and computed volumes of the deposited material.

Fig.13 shows the final configuration of the torrent bottom level at Sondrio, while Fig.14 shows in particular the cronological development of the bottom and water levels at the Garibaldi bridge.

The values used in the simulation are: s=0.8 , α =0.021 and δ=2d90. They are assumed constant in space and time, for all the phenomena taken into consideration.

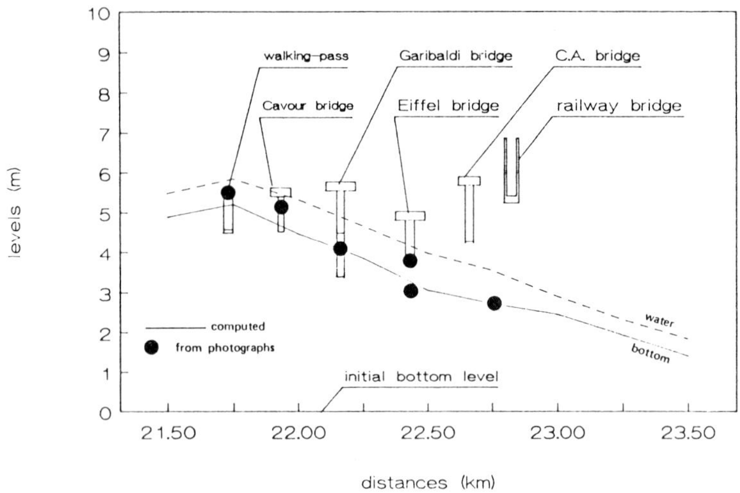

Fig.13 - Final configuration of the bottom and water levels at Sondrio (Sunday 19th of July, 8pm)

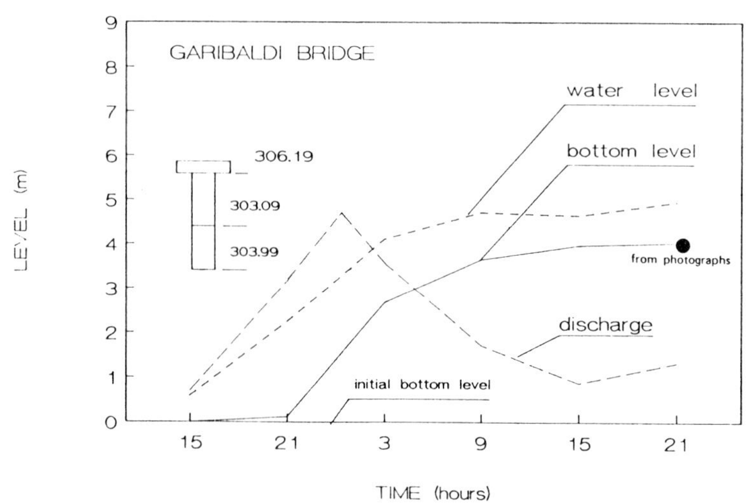

Fig.14 - Time history of bottom and water levels at Garibaldi Bridge in Sondrio during the event of July 1987

List of symbols

B	(m)	local channel width
c	(m/s)	characteristic celerity
di	(m)	representative diameter of the i-th fraction
d90	(m)	diameter overpassed by 90% of the sample material
Di	(m/s)	vertical flux of the i-th fraction of the sediment
Dphi	(-)	theoretical physical damping
I	(-)	local longitudinal slope of the channel
m,n,p,q		exponents in the sediment transport equation
Q	(m^3/s)	local discharge
Qo	(m^3/s)	annual ordinary maximum discharge
ri	(-)	exposure correction coefficient
R	(m)	local hydraulic radius
s		exponent on the exposure-correction coefficient
t	(s)	time
Ti	(m^2/s)	transport of the i-th fraction of the sediment
U	(m/s)	local mean flow velocity
Vo	(m^3/year)	annual ordinary runoff volume
x	(m)	longitudinal distance
z	(m)	level of the lower boundary of the mixing layer
Z	(m)	level of the bottom
α		coefficient on the sediment transport equation
βi	(-)	percentage of the i-th fraction on the mixing layer
βio	(-)	percentage of the i-th fraction on the mixing layer for the ordinary bottom (equilibrium)
βi*	(-)	percentage of the i-th fraction in the landslides material (perturbation)
βei	(-)	percentage of the i-th fraction in the layer just below the mixing layer (undisturbed sediment)
δ	(m)	thickness of the mixing layer
Δ t	(s)	numerical time step
Δ x	(m)	numerical space step
θ	(-)	weighting factor in the numerical scheme

References

1. Abbott M.B.,(1979) - Computational hydraulics, Elements of the theory of free surface flows, Pitman,London

2. Armanini A., Di Silvio G.,(1988) - A one-dimensional model for the transport of a sediment mixture in non-equilibrium conditions Journ. of Hydr. Res.,Vol.26,NO.3

3. Armanini A., Di Silvio G.,(1989) - On the coexistence of bedload and suspended transport for a uniform grainsize material, Symposium on Sediment Transport Modelling, New Orleans, USA

4. Brunelli S.,(1987) - Trasporto solido annuo dei corsi d'acqua in funzione delle loro caratteristiche idrologiche e morfologiche - Tesi di Laurea, Univ. degli studi di Padova

5. Di Silvio G. et al,(1982) - Relazione idraulica e bilancio del trasporto solido,TECHNITAL internal report, Verona.

6. Di Silvio G.,(1983) - Modelli matematici per lo studio di variazioni morfologiche dei corsi d'acqua a lunga e breve scala temporale.
   Studi e Richerche N. 356A, Istituto di Idraulica dell' Universita di Padova

7. Di Silvio G.,(1983) - Il trasporto solido nei corsi d'acqua alpini e problemi di difesa del territorio, Studi e Richerche N. 356A, Istituto di Idraulica dell' Universita di Padova

8. Di Silvio G., Peviani M.A., and Brunelli S.,(1989) - Report C.N.R. (to be published)

9. Hey R.D., Bathurst J.C., and Thorne C.R.,(1985)- Sediment Transport in Gravel Bed Rivers, John Wiley and Sons, London.

10. Jansen P.Ph. et al,(1979) - Principles of river engineering, Pitman,London

11. Maione U.,(1977) - Le piene fluviali - Appunti di idrologia, La Goliardica Pavese, Pavia

12. Olesen K.W.,(1981) - A numerical model for morphological computations in rivers with non-uniform sediment,Report R/1981/6H,Dept.ofCiv.Eng., Delft Univ.ofTechn.

13. Peviani M.A.,(1988) - Evaluation of numerical schemes for the computation of river bed elevation, Master Thesis, Delft Hydraulics Laboratory

14. Ribberink J.S.,(1980) - Morphological modelling for rivers with non-uniform sediment. Int.Rep.No.1-80,Delft Univ. of Techn., Dept. of Civ. Eng., Fluid Mech. Group.

15. Ribberink J.S.,(1983) - Experiments with non-uniform sediment in case of bed-load transport,Report No.83-2,Comm.on Hydr.,Dept. of Civ.Eng.,Delft Univ.of Tech.

16. Ribberink J.S.,(1987) - Mathematical modelling of one-dimensional morphological changes in rivers with non-uniform sediment, Communications on hydraulic and geeotechnical engineering - Delft University of Technology

17. Vreugdenhil C.B.,(1982) - Numerical effects in models for river morphology, Engineering applications of computational hydraulics Vol I (Abbott M.B. and Cunge J.A.),1982

18. Vries M. de,(1973) - River bed variation - aggradation and degradation, Intern. Sem. on Hydr. of Fluv. Streams,New Delhi

19. Vries M. de,(1985)- Asensitivity analysis applied to morphological computations, Delft T.H - Comunications on Hydraulics 85-2

REVIEW OF MOUNTAIN RIVER TRAINING PROCEDURES IN SWITZERLAND

Hans Peter Willi
Swiss Federal Office for Water Economy, Bern, Switzerland

ABSTRACT

The violent storms which occurred in Switzerland in 1987 caused damage to the tune of 1.2 to 1.3 billion Swiss francs. Numerous constructional measures had to be taken to protect against floods. The 1987 catastrophe asked for a fundamental review of flood protection measures from a number of different viewpoints.

After a short review and general explanations about topographical attributes, precipitations, geology, political conditions and the legal situation in Switzerland, the actual guidelines for river training procedures were presented.

The review is completed with different typical examples of protection works.

1. INTRODUCTION

The violent storms which occurred in Switzerland in 1987 caused damage to the tune of 1.2 to 1.3 billion Swiss francs. Numerous constructional measures had to be taken to protect against floods. These included restoration projects on the one hand, and additional projects on the other hand, in connection with which the concepts for protection against floods were subjected to intensive study. The current projects and construction work will keep us occupied for many years to come.

A three-year causal analysis of the events that took place in 1987 is currently in progress, and is expected to be completed by 1990. The objective of this investigation is to estimate the likelihood of such catastrophes on the one hand, and on the other hand to find out whether, and to what extent, amendments in the planning criteria regarding flood protection constructions are necessary, based on the experience gathered from the 1987 catastrophe.

It soon became clear that current knowledge regarding certain natural phenomena (landslide dynamics, erosion processes, river morphology) is insufficient. And shortcomings were also ascertained with regard to protection concepts. But it also became apparent that weak points already acknowledged had not been paid sufficient attention.

Nowadays, we have difficulty in thinking over long periods. Flooding catastrophes are already forgotten after 7 years or so. Investments in constructions aimed at providing appropriate security over the long term need to be planned with the necessary foresight. A survey of flood damage occurring during the past few years strongly emphasizes the importance of flood protection measures.

Flood damage 1975-1988 (from EAFV Report (1))

	1975	1976	1977	1978	1979	1980	1981
Damages (mio. Sfr)	170	20	280	>500	50	24	50
Deaths	2		3	9			9

	1982	1983	1984	1985	1986	1987	1988
Damage (mio. Sfr)	40	60	120	50	120	1300	130
Deaths				2	8		

The table shows the accidentally distribution of the events. Smaller damage can always occur. The objective of protection measures is to prevent catastrophic events like 1978 and 1987. On average, the annual Federal subsidies for protection works amount to approx. 40 -50 million Swiss francs. This is equivalent to a building volume of 100 - 150 million Swiss francs. These figures have doubled as a result of the 1987 storm damage. Additional significant sums are also granted for restoration and safeguarding of transport networks.

2. SURVEY OF CONDITIONS IN SWITZERLAND

2.1 Topography

Our small country, with its total surface area of approx. 41,000 square kilometres, is characterized by a high degree of variety. The central Alpine region, with its highest peaks reaching up to 4,500 metres, form the watershed of Europe. Together with the Alpine foothills, the central plains and the Jura range, they divide Switzerland up very distinctly. Its running water forms a closely-knitted network. The total length of flowing water is around $40,000 km$, i.e. a running length of 1 km per km^2.

2.2 The major rivers

There are 7 principal rivers in Switzerland

Name of river	*main river*	*sea*	*drained area*
1. Rhein 2. Aare 3. Reuss 4. Limmat	Rhein	North Sea	68%
5. Rhone	Rhone	Mediterranean	28%
6. Ticino	Po		
7. Inn	Donau	Black Sea	4%

The total length of these 7 rivers is 1400 kilometres.

2.3 Precipitations

Switzerland is a country with a high level of precipitation. Although the country is a small one, regional variations are very considerable. The mean annual precipitation is 1476 mm, of which on average 978 mm, resp. $1280m^3/s$, flow off from the surface. A statistical evaluation of the daily maximum amounts indicates how great the degree of variation is; for example, two to three times more precipitation is measured in the Alps and Alpine foothill regions than in the central lowlands and the Jura range. The most extreme levels of precipitation occur in Ticino, on the south side of the Alps. Daily maximum precipitation of up to 400 mm has been recorded here.

2.4 Geology

In addition to the topography and amounts of precipitation, the geological conditions also influence the flow characteristics of streams and rivers to a considerable extent. From a geological point of view, Switzerland has a wide variety of zones and geological characteristics. In addition to crystalline stone, there are also types of rock from both the seabed and the mainland depositions, and dating from the most recent to the oldest geological periods. The structures are of an unusual diversity: fault shifts, crevices and overthrusts over dozens of kilometres have caused the original structures to change quite fundamentally. Weathering, rivers and glaciers have thus given the country's surface its individual characteristics.

2.5 Political conditions

Switzerland is a direct democracy. The nation comprises three political levels:

Federal State
26 cantons as sovereign states
3,022 communes " republics" with limited autonomy

Cantons and communes have a high degree of autonomy, including within the water economy sector.

2.6 Distribution of tasks regarding protection against floods

The distribution of tasks in this sector is as follows:

- the cantonal level is responsible for their own water economy, and they regulate the use of water (drinking water, hydroelectric power, etc.) and flood protection works.

- the federal level supervises the following aspects:
  whether the cantons are fulfilling their obligations regarding protection against risks to water; whether the cantons sufficiently maintain the protection constructions.

The Federal State provides financial support to the cantons for the realization of protection constructions, and reviews projects from a technical point of view.

The Federal State is responsible for the provision of hydrological rudiments.

2.7 Legal fundamentals and consequences

Differing public interests have given rise to the introduction of appropriate legal stipulations. The following laws are currently in force with respect to water economy:

Federal Constitution 1848/1874
Federal Law on protection measures against floods, dated 22nd June 1877

Federal Law on Nature Conservation, dated 1st July 1966

Federal Law on the Prevention of Water Pollution, dated 8th October 1971 (Water Protection Law)

Federal Fishing Law, dated 14th December 1973

Federal Law on Land Development, dated 22nd June 1979

Federal Law on the Protection of the Environment, dated 7th October 1983

This law contains stipulations governing the testing of environmental tolerance. The according ordinance has been in force since 1st January 1989. Protection works against floods with costs of over 10 million Swiss francs are subject to this tolerance test.

Consequences of legislation with respect to the subsidization of protection measures against floods

From this summary it becomes apparent that it is not just the protection criteria which are decisive for a given project, but that a whole range of legal stipulations also have to be complied with. The Swiss Federal Office for Water Management - in its capacity as principal responsible department, and in collaboration with other federal offices connected with environment questions - determines in a joint report procedure whether a project complies with the laws or not. If shortcomings are ascertained in a project, appropriate directions can be issued, and subsidies can be cut or refused.

3. CURRENT FLOOD PROTECTION IN SWITZERLAND

3.1 Current flood protection measures

Changes over the past few decades with respect to lifestyle, space demands, and infrastructural improvements (traffic), have given rise to the fact that floods which occur today are generally far more costly than before. Nowadays, expensive equipment is kept in basements, for example deep freezers, heating units, or even very valuable computer systems. It is therefore not surprising that high demands are now made on the safety of our waters with regard to sufficiency of supply and bed stability.

During the past few decades, the situation regarding water policies has changed. Progress has been made in science and technology. Ecological connections are being recognized and acknowledged to an ever greater extent. Awareness of the environment has increased. And new objectives and appropriate new laws governing river training policies are now resulting from this newly won knowledge.

Protection against floods has the following objectives:

- to protect residential areas, agricultural regions (cultivated land), transport routes and other buildings in mountains and valleys from damage by flooding, erosion and shifting of solid matter, by means of permanent protection measures.

- to ensure protection of individual objects near bodies of water (such as bridges, dams,power plants, etc.) against high water levels.

- to create drainage for water from canalisation, roads, land improvement areas, etc.

How can sufficient protection against floods be realized?

In its publication "Protection against floods", issued in 1982, the Federal Office for Water Management pointed out that projects for protection against floods not only had to be suited to flooding protection requirements, but also had to fulfil a whole range of other public interests.

-All these interests need to be brought together under one roof, which can only occur if sufficient mutual understanding and a willingness to compromise are shown. What is necessary is an objective, careful consideration of all interests concerned.

3.2 Definition of protection objectives

The 1987 catastrophe resulted in a fundamental review of flood protection measures from a number of different viewpoints. More specific definitions of protection objectives are currently being drawn up:

- *The protection measures must be adapted to the risks involved and to the value of the objects to be protected.*
- *With regard to risk to residential areas, the degree of protection may not be less than the HQ_{100}, or the highest recorded flood level.* (Standardized minimum parameters for residential areas according to previous practice).
- *The protection measures have to be justified from the point of view of environmental tolerance and of costs (consideration of interests).*
- *A transfer of the risk to another zone outside of the project area must not occur.* (The borders of the project area must be appropriately defined)
- *Attention must be paid to a balanced sediment transport, in order that permanent protection can be assured.*
- *Existing natural overflow areas should be retained wherever possible.*

3.3 Process distinction

Dangerous processes:

Flooding: dangerous when

$$V\ H > 0.5 - 0.75 m^2/s$$

V = current velocity

H = flow depth

Strong currents or great depths of water represent danger to human life.

Fig.1 Reuss, canton of Uri, flooding 1987

Excess drift:
Build-up of coarse material.
Typical occurrence after mud flows, landslides and breaches in dams. Almost always give rise to dangerous flooding.

Lateral erosion:

Occurs mainly close to shores and banks. Greatest risk to buildings (collapse). Long term disruption of roads, railway lines and mains lines.

Since there is always a **remaining residual risk**, it is advisable to study a larger-scale event than the potential one concerned. Such a measure can often help avoid large-scale damage in the event of such a catastrophe.

Fig. 2 Reuss, lateral erosion, 1987

3.4 Priorities with regard to protection measures.

How the priorities are to be determined for a given solution has been formulated in the draft for a new law on protection measures against floods:

1. It should first be ascertained whether protection against floods could not be provided by means of improved maintenance.

Maintenance is unfortunately often carried out insufficiently. Neglect can, however, be extremely costly. Water maintenance can be compared with dental care. Repairing the damage which has usually been detected too late can often cost many times more than the costs of the neglected maintenance measures.

2. It should be ascertained in second step, whether protection could not be provided by passive protection measures. With passive protection measures we mean e.g. land use planning, direct intervention on buildings, instead of protection measures on torrents and rivers, prohibitions.

3. The third and final possibility is the instigation of active protection measures .These measures should be kept to a minimum by choosing the right correction concept, construction method and building materials. Increased knowledge regarding natural phenomena, gained

through research work, can help reduce the extent of measures taken, as well as costs. Active measures involve the following solutions:

- *widening of watercourses, stabilization measures*
- *retention of flood water, resp. bed-load*
- *diversion of flood water discharges*

Fig. 3 example for neglected maintenance.

4. EXAMPLES OF PROTECTION MEASURES

Finally I would like to give you an idea of the current status of protection measures in Swiss mountain regions by citing a few examples.

4.1 Maintenance of existing measures on mountain torrents.

Lammbach Brienz

$E = 3.2km^2$, gradient $J = 33\%$

Stabilization of slopes, stabilization of stream bottom, protection of flume cone against excess drift.

Over-use and shortage of wood since the 13th century gave rise to almost total destruction of the forest, as well as to disruption of natural vegetation. The result was severe erosion. The first major catastrophe destroyed the village of Kienholz in 1499, and filled back the upper end of Lake Brienz. A further 28 slides occurred. The last major catastrophe occurred on 27th May 1896.

Set off by a sliding mass comprising 300,000 cubic metres, which temporarily blocked the Lammbach, a gigantic mud flow caused extensive damage. The bed of the stream sunk during one night by 26.5 metres. As a result, certain protection measures were instigated. In order to stabilize the side slopes, the bed of the upper course was raised by means of 7 barrages of from 9 to 21 metres in height. After backfilling had been completed, smallish transverses were installed for the purpose of directing the flow of the stream. These constructions were mainly fixed in cemented dry masonry. The middle course was secured by means of a further 6 concrete barrages of 5 - 6 metres in height. In the lower course, the neighbouring agricultural land and the transport routes further below were protected against floods and landslides by means of dams. Further stabilization measures within the forest were also carried out at enormous expense:

Walled terraces, gabions	130,000 m^3
Reforestation	8,800,000 trees

These erosion processes have still not been successfully halted. Maintaining the existing constructional measures is vital, however, in order to avoid increasing the risks. Renovation work being carried out today is costing as much as 70shows very clearly just how costly these renovation measures can be. For the renovation of a barrage in need of repair, for instance, Sfr. 900,000.- had to be spent. Anchored pre-concreting was selected as the most economic solution.

Fig. 4 Renovation of an old barrage.

4.2 More recent barrage constructions.

Example of a simple wooden barrage.

Stabilization of beds and slopes

Construction of wooden barrages is being given increased consideration nowadays. The reasons for this are, on the one hand, ecological considerations, and on the other hand, it can represent a sensible use of material available on site. It is vital here that the climatic conditions which have such a major influence on the life time of such constructions, are given due consideration.

Concrete box barrages; Gasenzenbach (canton of St. Gall)

$E = 2.8\ km^2 \quad HQ_{DIM} = 35\ m^3/s$

Barrages in an active landslide area, stabilization of slopes.
Protection of valley floor against silt deposition.

Fig. 5 wooden model of the box barrages.

These box barrages can be described as the most solid barrage constructions, since in addition to transverses they also have continuous lengthwise training walls. Lengthwise and transverse constructions form a horizontally positioned stable framework. A graduated barrage is constructed rather like a chain. The next higher level supports itself on the level immediately below it. The complete system is constructed on a transposed layerbasis. As a result of the relocability of the frame, passive pressure is mobilized on the opposite side of that from which tension originates, and thus a new balanced position is found. Due to the high degrees of tension, the dimensions of this construction have to be limited. The maximum length of such a box construction is approx. 10 to 12 metres. This type of barrage has already proved itself to be effective and is no more costly than a conventional sideways mounted barrage.

Diversion section; Etzlibach project (canton of Uri).

$$E = 29\ km^2 \quad HQ_{DIM} = 125 m^3/s$$

Sediment, landslide and avalanche problems.
Protection of residential area, protection against silt deposition, protection of a water capture.

Since a concentrated deposit of sediment in the stream below could lead to major problems, a mud flow is intended to be diverted under controlled conditions into a alluvial zone. This zone is then intended to clear itself of the material on its own. The appropriate barrage constructions were developed in a model experiment. The natural flow direction, which had established itself following the storms of 1957, was maintained. In contrast to conventional torrent barrages, diversion sections indicate less side drift areas, which is beneficial to the sedimentation process.

4.3 Measures for retaining sediment.

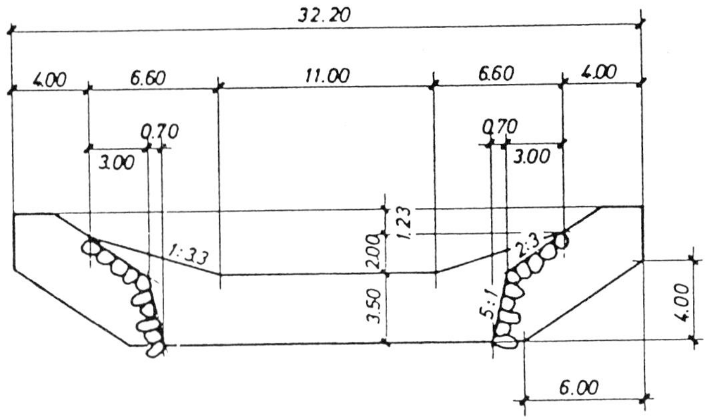

Fig. 6 type of barrage (developed in a model).

<u>Sediment-retaining barrage, Schächenbach (canton of Uri)</u>
Concrete aperture wall.

$E = 100\ km^2$, Sedimentation volume $=100 \div 200,000\ m^3$

Height of wall $= 20\ m$, coping length $80\ m$, coping width $= 5.5\ m$ at overflow section, width of aperture = 4.5 m, opening 2.5 m.

Retention of large sedimentation masses

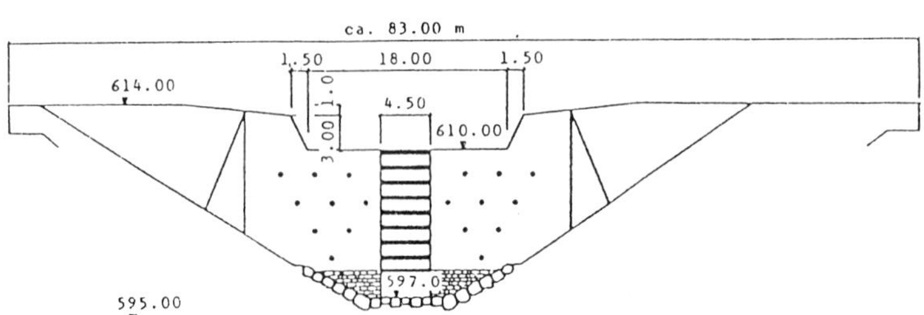

Fig. 7 Sediment-retaining barrage, Schächen (canton of Uri).

The "Schächen" caused a great deal of damage in 1977 in the "Altdorf" region, since large quantities of sediment (150 - 200,000 cubic metres) and wood filled up the river bed where it flows into the Reuss. The aim of barrage was contain such build-ups of sediments. Interferences with sediment composition can often have an effect on the stability of lower-lying sections of a river. This fact was taken into account, in that a large opening (aperture with grid) permitted the uninterrupted flow of $40 \div 50 m^3/s$. Lesser and medium floods can thus continue without interruption, and nature can run its own course to a great extent. At a flow of $110 m^3/s$, the backwater will reach the overflow section (full capacity). The project was tested and optimized in a model experiment, and it was ascertained that it was not possible to realize complete unaided clearance. Accumulations of large blocks and drifted wood prevented the clearance of more than 30% of the volume of sediment, depending on the amount of driftwood concerned. Further clearance can, however, be brought about by means of dredging (removal of the bottom paving).

Fig. 8 Sediment-retaining barrage, "Schächen" (canton of Uri).

4.4 Sediment transport channels

Lower course of Lielibach Beckenried (canton of Nidwalden).

$HQ_{DIM} = 116\ m^3/s$, bed incline $J = 10.5\%$

Protection of flume cone against drifting, securing of bed.

The 1987 floods partially destroyed the bed surface, the training walls and the bed sills. With the aim of maintaining and strengthening the existing arrangement, adaptations to the channels were carried out. In place of the existing paving, which leads to high flow speeds and thus high burden, it was decided to adopt a new concept. By means of a more coarsely formed bed, it was intended to reduce the flow speeds and achieve a continuous destruction of energy. This was achieved by means of a coarsely laid, flexible, unconcreted block "carpet" (block-weight $> 1.5t$). The dimensions are determined by the maximum water depth. With this type of construction there is a certain risk that individual blocks could be pulled out of the bed and thus cause a collapse in the channel. To prevent this, concrete bed sills were built every 15 metres.

This construction has already been able to prove itself; approximately 2 months ago it held firm during a major flood water period.

Based on the same principle, a number of different flow channel have been realized.

Diverting tunnels

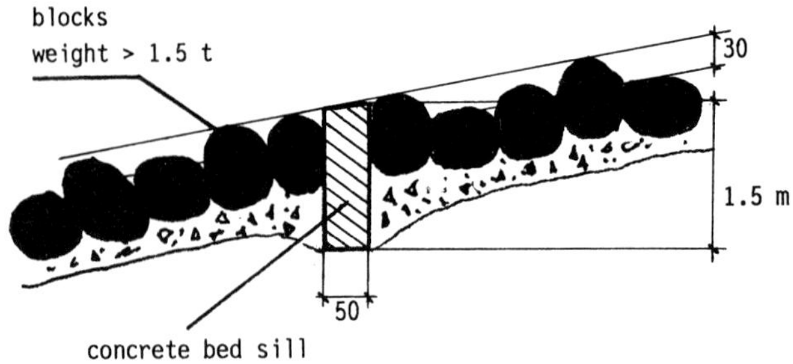

Fig. 9 longitudinal section.

Rovana diverting tunnel project (canton of Ticino).

$E = 69\ km^2$ $HQ_{DIM} = 300\ m^3/s$ diverting tunnel $L = 1.5km$. Tunnel diameter $D = 7.2m$, tunnel incline =2%.

Large-scale slides

Measures aimed at preventing a major catastrophe due to a possible landslide.

Prevention of further bed and slope erosion in the most active landslide area.

The Rovana, a tributary of the Maggia in southern Switzerland, has been causing problems for over a century. As a consequence of deforestation, the Rovana sunk by 30 to 40 metres within very short period. This led to the activation of an old sliding land mass several square kilometres in size. Several villages are affected (Campo, Cimalmotto, Cerentino). Today, reference is made to lowering of bed by 60 to 70 metres. Measurements made from the church tower of Campo during the past 80 years have indicated a displacement of 26.5 metres in a horizontal direction, and 10 metres in a vertical direction. It is now feared that in the event of a catastrophe landslide could cause a closure of the Rovana's river bed. In order to prevent such an occurrence, it is intended to construct a diverting tunnel which has been tested in model experiments.

The costs of the tunnel are currently estimated at 22 million Swiss francs. Measures costing a total of around 72 million Swiss francs are planned (hydraulic engineering, foresting, melioration, road construction).

5. CONCLUDING COMMENTS

Modern protection measures against floods sets out to work with, and not against, nature . Newly-won knowledge regarding river morphology need to be considered when determining the constructional measures to be taken. Since there is not only one type of torrent, river or small river, it is necessary to adopt a differentiated manner of consideration. A made-to-measure suit is a better fit than one off the peg.

The greatest possible protection should be aimed for with the least possible disruption. These increased demands mean that ever more specialized technical knowledge is necessary. Research operations, such as those presented at this workshop, are of major importance for later practice. They represent a valuable contribution towards differentiated assessment of the problems of rivers and mountain streams.

6. REFERENCES

Zeller,J. und Röthlisberger,G., EAFV Berichte "Unwetterschäden in der Schweiz" 1975-1988 (1), Eidgenössische Anstalt für das forstliche Versuchswesen Birmensdorf (EAFV)

Schädler,B., Wasserhaushalt der Schweiz , Mitteilung Nr. 6, Bundesamt für Umweltschutz, Landeshydrologie.

Schweizerische Geologische Gesellschaft, Geologischer Führer der Schweiz.

Zollinger, F., Die Modellversuche zum Geschieberückhaltebecken Schächen, Schweizer Ingenieur und Architekt 21/82.

Whittaker, J. und Jäggi,M., Blockschwellen, Mitteilung Nr. 91 der Versuchsanstalt für Wasserbau, Hydrologie und Glaziologie.

Eidgenössisches Amt für Strassen und Flussbau, 1877 - 1977 Hochwasserschutz in der Schweiz: 100 Jahre Bundesgesetz über die Wasserbaupolizei.

Bundesamt für Wasserwirtschaft, Hochwasserschutz an Fliessgewässern, Wegleitung 1982.

Lichtenhahn, C., Grenzen und Möglichkeiten der Vorbeugung von Unwetterkatastrophen im alpinen Raum, Sonderdruck INTERPRAEVENT 1971.

Lichtenhahn, C., Massnahmen zur Verhütung von Wildbachschäden in der Schweiz, Schweizerische Zeitschrift für Vermessung und Kulturtechnik, Heft 4/1961.

Zeller, Geiger Röthlisberger, Starkniederschläge des schweizerischen Alpen- und Alpenrandgebietes, Eidgenössische Anstalt für das forstliche Versuchswesen, Birmensdorf (EAFV).

FROM THE CHECK DAM TO THE DEVELOPMENT OF FUNCTIONAL CHECK DAMS

Aronne Armanini[1], Francesco Dellagiacoma[2] and Luigi Ferrari [2]

[1] Department of Civil and Environmental Engineering
University of Trent, Trent, Italy
[2] Provincia Autonoma di Trento, Trent, Italy

ABSTRACT

A review of the strategy for torrent control dams is presented.

The paper offers a description of the historical development of check dams in the north-east Italian Alps and southern Austrian Alps. Special attention is paid to the problem of functional check dam in order to obtain an effective control of sediment discharge during exceptional atmospheric events and of modern strategy for torrent control projects.

INTRODUCTION

The Province of Trent, (6100 km^2 wide) is a densely populated and entirely mountainous area: most of its towns, villages, communication lines and agricultural land are subject to torrent activity and a great number of flooding events have struck the area, some with dramatic social and economic effects. Therefore it is not surprising that torrent control practices have a long standing tradition and have always been given high priority by local administrations.

The town of Trent itself is located on the alluvial fan of the Fersina torrent. In 1537, to protect the town against the floods of this torrent, the prince-bishop Bernardus Clesius had a wood-and- masonry retention dam built at the end of the gorge over the town, one of the first reported torrent control structure all over the world.

Three and half centuries later, in 1884, after a major flood had claimed the lives of many people in the South-East Alps, the Austrian-Hungarian Empire (to which Trentino belonged until World War I) set up a state service for torrent control. This organisation has been maintained by the Italian State after World War I, by the Autonomous Region Trentino - South Tyrol after 1948 and later in 1973 by the Trentino Autonomous Province, during all the institutional changes of the region.

Nowadays on the more than 4000 km torrents stand over 10000 check dams, ranging from 19th century dry-masonry and wood-and-stones to modern steel-concrete dams, built in an increased effort to reestablish a certain level of security against torrent disasters after the most devastating flood of this century in 1966. Since this about 350 workers and an annual budget of about 20 billion Italian lira (approximately 15 million dollars) are involved in the torrent control in the Province.

This report deals with a short review of the evolution in the torrent control strategies during the last decades, based on the experience of A.S.S.M. (local Agency for torrent control), the one of South Tyrol and of the Austrian Forest-technical State Service for Torrent Control.

TORRENT CONTROL AND SEDIMENT TRANSPORT

The torrent control activity is essentially a problem of controlling a great quantity of sediment released during short intense floods. In a mountain stream the sediments can be transported as a dynamic sediment transport, mainly due to dynamic forces of the water, or as a sediment gravity flow (debris-flow or mud flow) depending on the morphology (and particularly the slope gradient); on the quantity (concentration) and the quality of sediments (grain-size distribution); on the modes of supply , on the chances of temporary clogging in the torrent.

According to established policies of torrent control, two different strategies were proposed in order to reduce (or often eliminate) the sediment production and movement along the torrent.

prevention of torrent erosion through the so called *consolidation check dams* as non-erodible sills in the torrent (or gully) bed subject to erosion;

prevention of sediment transport in order to avoid the over-deposition of sediments downstream control structures. This is particularly important where risk assessment is needed for developed areas sited on alluvial fans. The usual technique consisted of construction of one or more bed-load (and debris flow) retention dams.

These concepts were suggested by the idea of the existence of a equilibrium state of the streams.

THE EVOLUTION IN THE LAST DECADES

In the last 30 years torrent control activity, following the improvement in building techniques, underwent major changes, that can be viewed as the actual premises for further development. They are:

-the improvement in the building technique and the availability of new (or improved) materials has allowed the construction of highly resistant and truly monolithic structures;

-powerful and effective earth-movement machineries have allowed substantial changes in the local topography so as to reshape unsuitable conditions;

-efficient transportation facilities have made it possible to bring selected building materials even to difficult access sites;

-the analysis of flood events, as well as laboratory experiments in the end, led to a better understanding of the role of sediment transport phenomena for the purpose of an effective torrent control and to a better knowledge of its mechanism.

STABILIZATION AND CONSOLIDATION

Little has changed in torrent erosion prevention strategies: instead of large blocks, wood and masonry or squared blocks and mortar, check dams nowadays are usually built in concrete or reinforced concrete as **T**- or **L**-shaped structures. The monolithic characteristics of good concrete make it possible to take into account the anchorage of the dam at the bottom and at the side banks leading to safer and thinner structures.

The improved technique yields also the building of higher dams, thereby allowing a better consolidation on the previously eroded side slopes by elevating the torrent bed (the mere fixation of the torrent profile is defined stabilisation by Fiebiger, 1986). The consolidation check dams still continue to be the most important structures and represent by far the most common tool for torrent control.

In particular conditions wood-and-stone (or wood-earth-and-plants) structures are still built. Such structures may be an efficient design solution, because they fit very well in the landscape. The (expensive) practice of covering concrete structures with stones- and-mortar to reduce the visual impact, as in the old building technique, is often necessary in areas where tourism and a generalized demand for a *"natural environment"* are prevailing.

SEDIMENT RETENTION

The strategy of preventing the further transport of sediment along the torrent was based, in the past, on the building of high (to trap more sediments) and large (to resist to high pressure) dams. When possible, the structures were located at the end of a natural deposition basin, that is a flattened and enlarged torrent portion.
It was soon observed that:

i) the deposition space is usually filled in a very short time and the dam looses its main purpose (a secondary, but limited, effect consists in the retention due to slow adjustment of bed profile, up to be parallel to the original bed. It has often happened that the retention space had been filled by fine, unproblematic sediments carried by medium and small sized floods, so that when a major event occurred the dam had already exhausted its function and had no effect on the large bed-load or debris flow during the major flood;

ii) with a wall-type dam all the sediment is stopped, with a strong effect on sediment balance in the downstream reach of the torrent (the reappearance of erosion in sections previously subject to deposition has been observed many times).

An early and common procedure to extend the effect of retention was to enlarge the drainage slits of the dam and in addition to provide a bed-load basket, made of iron or wood poles, to prevent an all clogging exceedingly fast (Kronfellner-Kraus, 1972).

The next step was the development of open dams with different purposes and different forms. This development was seen allowed by the fast changes in the building techniques mentioned above.

On the other hand it was gradually established that the object of the torrent control is not to stop all the sediment transport but only its part that actually is the cause of damages: that is, debris flow and heavy bed load during major floods, and even during these events only for the largest grain-size fraction. Minor floods should flow undisturbed through the structure (in supercritical condition), carrying the quantity of sediments corresponding to the transport capacity of the torrent (Hampel,1969; Stauder, 1972).

At this point it should be observed that the two strategies of consolidation and sediment transport control (which is the further evolution of the retention strategy) do not represent devices. In some cases one can choose between them, but more often they appear to be complementary: every torrent regulation can stop only a part of the erosion process and neither it will be possible (nor desirable) to completely stop erosion in the basin. The continuing natural evolution and human impact will further cause formation of new erosion sources.

A great variety of forms and different structures have been designed and experimented to pursue and optimize this objective. This is due partly to different start-concepts, and partly to the characteristics of the basin, torrent and environment. Basically two main features (that sometime are combined in the same design) can be recognized:

-the **beam-dam**, with wide horizontal openings, possibly without narrowing of the torrent width, whose purpose is mostly filtering sediments and logs : the width of the slit is proportional to size of biggest boulders (Hampel, 1969);

-the **slit-dam** with one or more large, vertical and usually deep to the original bed openings, whose effect is mostly dosing the sediment transport rate. This is obtained by means of a backwater effect and hydraulic jump during floods, that reduces the stream velocity and allows most of the particles to deposit. This fraction should be subsequently eroded during minor floods, which can flow undisturbed (usually as a supercritical flow) through the slit without produce any backwater effect and thus eroding the previously deposited sediment (Üblagger, 1972).

Another possibility to temporarily stop bed-load and debris flow is the modification of the bed cross-section (widening) or bed slope (or both); recreating, artificially, what happens in nature in torrent expansion areas. The effect is to interrupt abruptly the profile, inducing the formation of secondary cones. A similar concept, based on very wide dams (up to 140 m flowing section), has been successfully applied also in Japan (Kimura and Higashi, 1984).

In comparison to wall-type retention dams the open structures have some undisputed advantages:

i) the filling space is maintained over many years: the flowing through of minor floods is granted; as it is the possibility of recreate the deposition space after a filling, if an access to excavator and trucks is given;

ii) the deposited sediment is drained, at least at some extent, thus permitting quick access to excavators;

iii) the impact on downstream morphological equilibrium and ecosystems is highly reduced, because a nearly continuous sediment discharge is allowed.

- sorting : the beam dam

One of the first large scale integrated projects of bed load management in the torrent control was carried out in Zillertal (Tyrol) in the mid-sixties: two large artificial lakes were planned as part of an hydroelectric power-plant, whose construction meant essentially the cut off of the spring snow- melt high-discharges. As a consequence of the building of the lakes the Ziller, the main river of the valley, was to be regulated: it was calculated that it could not carry cobbles larger than 28 cm. Therefore the Agency for Torrent Control was charged to develop a torrent control system with the purpose to prevent all major sediment to get in the Ziller river (Stauder, 1972), (Riccabona, 1988). In the years 1965 to 1972 on all major effluents, subject to debris flow or heavy ordinary bed-load, beam-dams were built to control the sediment discharge in to the Ziller and protect human settlements on the alluvial cones. Fig. 2 shows the beam dam built in 1968 on the Märzenbach torrent, a tributary of the Ziller.

The beams are exactly the same all over the dam. They are sized according to maximum pressure. That permits an interchanging of the beams and account for possible dynamic impacts on the upper beams. Initially it was required that the beams could be removed from downward side, to ease clearing after clogging. Nowadays it does not seem to be very important any more, because it turned out to be quite easy to send an excavator to remove the material immediately at the upstream face of the dam, thus inducing a local erosion of the deposit and allowing it through the openings.

An important question for the beam dams concerns the stability of the anchorages of the beams, that during floods can vibrate quite strongly: usually either iron plates are used, with split pinned

Fig. 1; The beam dam on the Märzenbach torrent (Tyrol, Austria). The torrent, which has a watershed is 42 Km^2, 1908 had caused huge damages during a heavy flood with more than 200000 m^3 total bed-load. The filtering part of the dam has a height of 7 m and a width of 20 m, spanning with the 4 fields, the whole bed, with a resulting filling space for sediment of 28 000 m^3; the distance between the beams is 0.5 m.

Fig. 2 ; The large beam dam on the Enterbach torrent (Tyrol, Austria) with the artificially created deposition basin. This is a well known torrent in the literature: July 1969 a debris flow brought to debris con 450 000 m^3 as reported by Aulitzky (1984/85). The strategy to control the torrent was based on both the stabilization of upper- intermediate stream, where major erosion occurred, and on the building of the big retention dam, to prevent the not consolidated sediment to go further on the cone.

bolts, or U-shaped iron joints, where the beams are held in place by distance holders. Both permit to easily change the distance between the beams thereby greatly changing the effect of the dam on the bed load.

Although there are no generally accepted rules on the distance between the beams, as a guiding framework the beam spacing varies between 1.2-1.5 (Zollinger, 1984 ; Mizuyama, 1984) and 3.0 (Üblagger, 1972) times the maximum grain size, allowed to pass downstream. Laboratory experiences of Zollinger (1984) showed that horizontal beams allow through-flow better than vertical structures. In nature, however, things are complicated by the diversity of grain sizes and, above all, by the presence of plant material: it is generally accepted that beam dams tend to clog by logs and other vegetation carried by the current, thus bringing to a complete stop even of the finest sediment (Kettl, 1984). For that reason sometimes the distance between the beams in the lower part is bigger, to allow smaller floods to flow undisturbed, then in the upper part, where the filtering effect has to be enhanced. And this is also the reason why beam filter dams are usually used on torrents where debris flow, whose debris cannot be sized anyway, are expected. Once a beam-dam is obstructed, it is necessary to remove artificially the logs and large boulders behind the screen. In this way a steep slope on the deposit is produced immediately upstream the dam opening, that triggers the erosion during minor floods.

If there is no natural enlargement or flattening of the bed torrent to create a suitable deposition basin, it can be created artificially. In Fig. 2 is shown an example of artificial deposition basins , in the Enterbach (Tyrol- Austria), created at the upper end of the debris-cone and closed downstream by a large beam dam. The combination of earth-dams with a central steel-and-concrete part bring also to a better inserting of the structure in the landscape and environment, which is at the moment a very important issue all over the Alps because of the primary economic position of tourism.

-dosing : the slit dam (hydrodynamic sorting)

This second concept for sediment-load management is based on dosing the sediment transport rate. This is obtained by means of a backwater effect in the proximity of the dam: the consequent reduction of velocity in this reach causes sedimentation of coarser grain- fraction. During subsequent minor floods, flowing undisturbed through the openings, the velocity should be sufficient to remove the most part of deposited debris, thus recreating the room for the next deposition. Therefore one (or more) large and high openings are to be found in the dam, reaching (slit) or not (window) the upper edge.

The same consideration yielding the beam spacing have been made for design the slit: for the actual objective of dosing, it is essential that the slit allows the undisturbed flow of minor floods, causing velocity sufficient to allow the erosion of the deposit.

Slit-dams cannot work properly if the slit is obstructed by logs, large boulders or when the sediment includes exceedingly cohesive fractions (clay and silt), thus resulting hardly erodible. Also the case of wide grain-size distribution is uncertain for slit-dam (e.g. in case of typical debris flow from a moraine region, carrying logs and boulders on their head with a considerable part of fine materials); in this case in fact, the obstruction of the slit by the boulders compromise very soon the dosing effect. If such a debris flow happens, it is necessary to take away the larger logs and boulders obstructing the slit with some machineries , thus inducing the erosion of the remaining (finer) sediment. A further condition for an effective operating of slit dams is the actual presence of minor floods, or clear discharges, able to eroding the deposited sediments.

A good example of a slit dam, that has successfully worked many times since its construction, is the one on the Litzbach torrent in the Austrian state Vorarlberg (Fig. 3). The deposition basin of the slit-dam has been filled and emptied more than once in the 18 years of operation.

In the effort to combine the undisturbed flowing of minor and intermediate floods (what requires wide slits) and a good efficiency in catastrophic cases (narrow slits), and to reduce the possibility of obstruction, new forms have been developed. Good results are expected by combinations of horizontal and vertical openings.

Fig. 3 ; The slit dam on the Litzbach torrent (Vorarlberg, Austria). 1970 a powerful debris flow with an estimated volume of 230 000 m^3 *took place on this torrent, caused by a landslide in the upper part of the basin. This is large, with a considerable part above the timberline: the slit dam has been located in an almost ideal place for a retention (dosing) dam: at the end of a big enlargement and flattening of the torrent (large filling space); in the upper part of the basin, close to the timberline (no logs are expected in the sediment; no possibility of clogging in narrow places, where the build up of powerful wave is possible); relatively homogeneous grain size of the sediment, without large boulders and too much fine part (easy erodibility of deposited sediments); high yearly snow-melt discharges (minor floods capable of eroding the sediments). Giving these conditions the deposition basin of the dam has been filled and emptied more than once in the 18 years the dam has been standing.*

Fig. 4; A debris flow breaker as a proper and individual structure on the Luggauerbach (Salzburg, Austria), consisting of powerful, triangular profiled, steel- concrete buttresses linked together at the top by a beam and protected by thick iron plates (from Kettl, 1984).

-wood grading

It has always been observed that logs and plant material play a key role in torrent disasters:

i) temporarily clogging the torrent and leading to abrupt breaching that cause the formation of destructive waves with extremely high impact, often described in old chronicles as "*avalanche of water and sediments* ";

ii) obstructing the openings of sorting or dosing structures, preventing them to function as designed and causing the deposition of any sediment, even the finest, thus rapidly filling the space available.

A preventive strategy is very difficult: it is surely useful to guide the evolution of vegetation on torrent banks toward a specialized plant community and maintain it in a juvenile stadium, to maximize flexibility and resistance to the water flow and avoid giant trees, although in forested areas it will be impossible, during heavy precipitation events, to prevent trees to be brought into the torrent (landslides). Therefore the purpose of trapping and filtering out logs and other plant material (roots, crowns, stumps) is very important and can be seen as preliminary in every strategy of torrent control for torrents where the presence of vegetable material is significant.

Special dams, consisting of anchored steel-wire ropes, have been developed in the past to trap logs. Fig. 4 shows an example on the torrent Flanginech in the Rendena Valley (Trentino), with anchoring in the rocks.

To protect the openings of dams from obstruction a declining, steel screens are often placed upstream of the dam, leaning on to a concrete structure: the plant material floats and is pushed to the top letting the lowest part of the screen -and the sluice way-free and water flowing through.

This effect is magnified if flow velocity is small (subcritical flow condition) and the smaller floating vegetation (crown and branch parts) are not pushed against the screen. The backwater effect can be produced by an artificial narrowing of the flowing section (implying the danger that the large logs are brought over the dam) or by slits in the central section of the dam, in front of the inclined screen. Both laboratory experiments (Zollinger, 1984) and field experience (Kettl, 1984 ; Riccabona, 1988) have shown that such declining screens can work very well and maintain the openings free from clogging, permitting bed-load management. Robust, inclined beams, parallel to water direction work better than screens, that sometimes can be clogged up, particularly if there is no water rebound. Most recently such structures have been designed with varying inclinations to improve the protection against obstruction: the results have been generally good .

-debris flow breakers

The dynamic impact of a debris flow on torrent control works is several time bigger than the water hydrostatic pressure. Lichtenahn (1971) analysing dams destroyed by debris flow has calculated pressures up to 13 times the water pressure.It is therefore necessary to design strong structures, capable of resisting the impact and reducing the energy level of the debris flow.

Debris flow breakers can either be:

i) a proper structure in form of a particularly strong and totally opened dam, consisting of single buttresses or pillars, linked together by horizontal beams;

ii) a part -usually a strong buttress- added to a dam, with the purpose to divide and split the debris flow, to protect the structure against a direct impact;

iii) an inclined beam structure, cast with beams parallel to the current direction and leaned on 2 buttresses in the central part of the dam, combined with an earth filling on the back of the dam-wings, (that can better absorb dynamic impacts than a vertical rigid wall) work as a debris flow breaker.

Debris flow breakers are subject to extreme dynamic impacts and its directly exposed parts must be accordingly protected, usually with thick iron plates.

Fig. 5; Dam with vertical openings protected from the obstruction by a strong declining iron grating on the Mörsbach torrent (Styria, Austria).

Another effective measure to resist the over-pressures is a declining upstream profile of the buttress (common to all the structures described above), with a small angle respect to the torrent bottom, to reduce the impact energy (see Fig. 5 to 8).

Purpose of debris flow breakers is not the stopping and the deposition of the debris flow, which would require large filling space that in this locations is usually not given, but rather to divide the main flow in parts to reduce its velocity and dissipate its energy, such that the subsequent deposition in a downstream suitable location is eased. Downstream the check dam often, the debris are channalized in a straight steep channel, capable to transport the materials trough the settlements without deposition or flooding (Fig. 9).

Fig. 6 shows a debris flow breaker as proper and individual structures on the Luggauerbach in Salzburg (Austria). Fig. 8 shows a debris flow breaker on Dona Torrent in Trentino integrated in dosing dams built far downstream on the same torrents, to provide a shelter for the structure against debris flow gone over the first control works or originated downstream. Downstream the check dam a steep channel was built, design to allow the transport of all the sediments passing trough the dam (Fig. 8.b).

THE STRATEGY OF BED-LOAD MANAGEMENT

In the last 20 years many different forms of both filtering (sizing) and slit (dosing) dams have been built, on the one hand experimenting new design and new combination, on the other trying to adapt the torrent control works to the variety of erosion, transport and deposition conditions of the single torrents.

Fig. 6; The dosing dam on the Luggauerbach (Salzburg, Austria) with a combination of vertical and horizontal openings to make an obstruction less probable thus improving the dosing effect.

Fig. 7; Debris flow breaker on the Rastetzenbach (Salzburg, Austria): the structure consists of very powerful buttresses (2 m thick and a between distance of 2m, covered by resistant iron plates), with a triangular profile and linked together by a horizontal beam at about half height of the buttresses.

Fig. 8 a; Double slit sorting check dam, provided with a debris flow breaker (pillar 45° upstream inclined), built at the upper end of alluvial fun. The purpose of the dam is the reduction of the debris flow velocity where there is not suitable place to deposit the sediments. Dona torrent (Fassa Valley-Trentino, Italy)

Fig.8 b ;

Steep straight channel built downstream the check dam, designed to allow the transport of all the sediments passing trough the check dam. The channel, 284 m long, has been constructed using single large blocks and is consolidated by 8 concrete groundsills.

At the same time many authors have proposed general models of sediment transport, following a procedure widely adopted in river engineering (e.g. erosion transport deposition of sediments are different stage of the same process) (Üblagger,1972 ; Kettl,1984 ; Fiebiger,1986). Fiebiger has introduced the definition of bed load management, to better stress the object of controlling (instead of just stopping) the process of erosion and sedimentation, and proposed the following scheme, which summarizes the different functions that the dam can accomplish in torrent control.

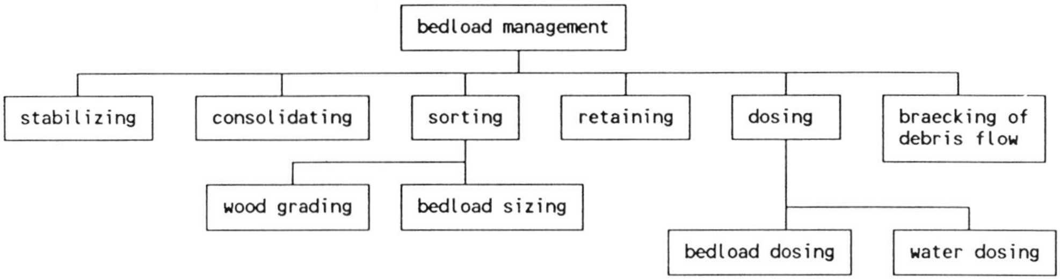

A combination of functions is possible and actually often used to optimize the effects of a dam: wooden gradings are often a premise for sizing and dosing structures; dams subject to the direct impact of debris flow need to be protected against the dynamic collisions of the largest boulders; hydrodynamic sorting (dosing) needs a previous control of bed-load, to avoid the obstruction of the openings. These objectives can be pursued using different structures or developing integrated-function-structures.

In the engineering practice a very important step toward bed-load management, and a hint to a better understanding of torrent-related processes in a watershed, has been given by the definition and the mapping of the endangered zones, experimented during the 70's, and generally introduced by the renewed Austrian Forest Law 1975. The law put the Torrent Control Agency in charge of analysing any torrent watershed under the aspects related to erosion; defining its sediments potential related to a 150 years event and expressing it in a numeric form; determining the part of it likely to reach human settlings, usually on the alluvial cone and mapping of the areas endangered by erosion, excess water or bed-load.

As a consequence, torrent control projects had to provide a better security to the human settlings, which implies the control of the sediment process (bed-load management): wood grating, sizing and/or dosing and the downstream transport of that part of the bed-load that cause no damages (or ,better, minimize it). Also the designer is required to apply the concepts of bed-load management: filtering out the largest components to prevent obstructions, reducing the energy of the current and finding places to (temporary) trap at least a part of excess sediment.

MATHEMATICAL AND PHYSICAL MODELS OF TORRENT CONTROL WORKS

Since few years, mathematical and physical modelling have been applied to torrent control engineering. This practice was not so diffused in the past probably because the cost related to such constructions was not high enough to justify any expensive analysis.

Following the economical development of mountain regions, more attention has been paid to a proper calibration and efficiency of torrent control works.

Very recently many mathematical models have been proposed, in particular devoted to the problem of gravel bed rivers and supercritical flows. All these models are based on momentum and mass balance for water and sediment transported. In the most recent models the gradation curve of the bed material is divided into a discrete number of classes. For each class mass continuity equation is enforced (Ashida and Michiue,1971 ; Andrews and Parker 1987 ; Armanini and Di Silvio,1988).

Besides the problem of the suitable calibration of sediment transport formulas, and that of the interaction among different grain size particles, the major uncertainties lies in the evaluation of sediment production in a mountain basin and the role of the wooden material.

As far as physical models are concerned the cost is much higher compared to the cost of the civil work. For this reason physical model tests are rare, and usually adopted for special projects.

The scaling law for water are generally deduced on Froude similarity, while sediment transport scales are deduced via transport formulas.

More difficulties arise in modelling debris or mud flows. The similarity lows for these materials are much more complicated and uncertain. For coarser material (Bagnold inertia grain region) a gravity Reynolds similarity can be adopted (Armanini et alii,1988). While, if the presence of finer material are important the scale is determined by a Reynolds number; in this case a correct scale reduction is practically impossible.

CONCLUSIONS

The functional check dams, discussed herein, have to resist very different ranges of pressure, ranging from drained earth pressure for consolidation check dams, to water and earth pressure for sizing and dosing dams in torrent with bed-load transport, up to high dynamic impact for the debris flow breaker.

The design of dams for bed-load management takes into account pressures higher than that of clear water. Usually, a coefficient is adopted multiplying the water pressure to account for dynamic impacts, whose value ranges from 1.3 for bed-load transport torrents to 6-7 for debris flow breakers.

There is no fixed rule for the determination of this coefficient for a single torrent. It hardly might be worth considering the variety of natural torrent conditions: only a detailed analysis of (1) the erosion processes (quantity and quality of the expected sediment) in the watershed and (2) of the flowing conditions in the torrent (flowing section, slope, narrow points where obstructions are likely) can help in this choice, based on the characteristics of the sediment transport. The two-fold torrent classification proposed by Aulitzky (1980) provides an excellent guideline in this essential study of the individual basin and torrent. Four categories of torrents are defined:

- **flood creeks**: only water discharge is expected; damage by flood and possibly erosion.

- **bed load torrents**: floods carry a considerable amount of sediment in form of bed-load (individual transport); damages by floods and deposition.

- **debris flood torrents**: sediment transport in form of debris flow, without pulsating and blocking debris waves; proper uniform flow formula must be introduced; no wave and pulsation in the flow; damages through flood, deposition and dynamic impact.

- **pulsating debris flow torrents:** debris flow in a pulsating sequence (often caused by obstructions); extremely high velocity and dynamic impacts (which is determined in a very empirical way), that cause major destructions.

REFERENCES

Andrews, E.D., Parker, G. (1987). Formation of Coarse Surface Layer as the Response to Gravel Mobility, *Sediment Transport in Gravel-bed Rivers.*, John Wiley & Ltd., pp.269-300.

Armanini,A.,and Di Silvio,G. (1988), A one-dimensional model for the transport of a sediment mixture in non-equilibrium condition, Jour. of Hydraulic Research, IAHR, vol.26,no. 3, 275-291.

Armanini A., Fraccarollo L., Trivellato F.(1988). On physical modelling of debris flow and mud flow. General Report to C.N.R.- G.N.D.C.I .

Ashida,K., Michiue,M. (1971), An Investigation of River Bed degradation Downstream of a Dam, Proc. IAHR 14th Congress, Vol.3, pp.247-255.

Aulitzky,H.(1984). Vorläufige zweigeteilte Wildbachklassifikation. Wildbach- und Lawinenverbau Sondernummer Juni 1984, pp.7-60.

Fiebiger,G.(1986). Bed-load Management on Torrent Control by Functional Check Dams. Beiträge zur Wildbacherosions- und Lawinenforschung. Mitt. der FBVA Wien vol.159 pp 325-332.

Hampel,R.(1968). Geschiebeablagerungen in Wildbächen, dargestellt am Modellversuchen. Teil 1 und 2 in Wildbach- und Lawinenverbau Nr. 1 and 2 32nd Year.

Kettl,W.(1984). Vom Verbauungsziel zur Bautypenentwicklung. Wildbach- und Lawinenverbau Sondernummer Juni 1984 pp 61-98.

Kimura,M.,Higashi,S. (1984): Über bauliche Vorkehrungen in Japan gegen Murschäden auf Schwemmkegeln. INTERPREAVENT 1984 Vol. 2 pp 259-271.

Kronfellner-Kraus,G.(1970): Über offene Wildbachsperren, Mitt. der FBVA Wien vol.88.

Leys,E.(1973): Vorschläge für die Systemeinteilung der Quer- und der Längsbauten in der Wildbachverbauung. Wildbach und Lawinenverbau Year 37 Nr 2 pp 34-55.

Mizuyama,T.(1984): Mechanism of the movement of grains and logs in Debris Flow. INTERPREAVENT 1984 Vol. 3 pp 189-196.

Puglisi,S.(1972). Nuove opere per la correzione dei torrenti.Atti delle giornate di studio di idraulica: nuovi criteri di progettazione e nuove opere di sistemazione dei torrenti.

Riccabona,B.(1988). Bisherige Erfahrungen mit Entleerungssperren und ihre Wirkungsweise bei Murstös sen und Hochwässern in den letzten 20 Jahren. INTERPRAEVENT 1988 vol.3 pp 119-129.

Senoo,K.,Mizuyama,T.(1984). Function of Structures against Debris Flow. INTERPREAVENT 1984 Vol. 3 pp 113-120.

Stauders,S.(1972). Verschiedene Konstruktionen von Balkensperren. Kolloquium über Wildbachsperren. Mitt. der FBVA Wien vol.102 pp 335-372.

Üblagger,G.(1972). Retendieren, Dosieren und Sortieren. Kolloquium über Wildbachsperren. Mitt. der FBVA Wien vol.102 pp 335-372.

Zollinger,F. (1984). Die verschiedenen Funktionen von Geschieberückhaltebauwerken. INTERPRAEVENT 1984 vol.1 pp 147-160.

Non-uniformity of Sediment:

Grain Sorting

Bed Armouring

Transport of Fine-Sediment Suspensions

DOWNSTREAM VARIATION OF GRAIN SIZE IN GRAVEL RIVERS: ABRASION VERSUS SELECTIVE SORTING

Gary Parker

St. Anthony Falls Hydraulic Laboratory
University of Minnesota, Minneapolis, Minnesota USA

ABSTRACT

Most gravel rivers show a tendency for characteristic grain size to decrease in the downstream direction over scales of tens or hundreds of kilometers. This change in grain size might be due to a) *selective sorting*, according to which finer grains are preferentially transported downstream, b) *abrasion*, by which individual grains are reduced in size, or some combination of the two. Here a framework for the simultaneous treatment of both phenomena is developed. The analysis is restricted to bedload transport of gravel. Differential sorting is treated by means of a hiding function embedded in a surface–based sediment transport relation. Abrasion is postulated to be due primarily to binary collisions between moving bedload and stationary bed particles. The model is applied to river systems undergoing wave–like aggradation such that the river profile obeys an appropriate similarity profile. The results indicate that the downstream variation in grain size is mostly due to selective sorting in the case of quartz, but that abrasion and selective sorting are of roughly equal importance in the case of limestone.

INTRODUCTION

An interesting feature of many gravel rivers is the tendency for both slope and characteristic surface grain size to decrease in a regular fashion in the downstream direction. Various researchers have hypothesized that this downstream decrease in grain size is due to selective sorting, fluvial abrasion, or some combination of the two. Until recently, however, no unified model was available to evaluate the relative importance of the two effects.

The present model is based on the following assumptions.

1. The bed is divided into a surface layer and a substrate. The thickness of the surface layer is taken to be equal to D_{s90}, i.e the surface grain size such that 90 per cent of the material is finer.

2. The surface layer is taken as the source for the bedload. Selective transport is introduced in the form of a hiding function. The transport of material finer than gravel in not considered.

3. Fluvial abrasion is treated in terms of binary collisions between bedload particles and the bed. Abrasion coefficients characteristic of various rock types are developed with the aid of abrasion mill tests.

4. Abrasion resulting from in–situ chemical or mechanical weathering is not considered.

5. For simplicity, the river is assumed to have constant width and river discharge, and to have no tributaries.

An appropriate model of gravel transport must account for the fact that the surface, or "exchange" layer of gravel, is typically somewhat coarser than the substrate. That is, the model must allow for static or mobile armor. The model must include "hiding" in some form, and account for sediment conservation on a grain–size specific basis, in order to predict selective transport.

GRAVEL TRANSPORT MODEL

The surface–based Oak Creek gravel bedload transport model of Parker (1989a, b) is employed here. The model is based on the field data of Milhous (1973) for Oak Creek, Oregon, U.S.A., and other streams. Certain minor variations from the notation of Parker (1989b) are introduced here for the sake of clarity.

In presenting the model, it is useful to introduce the ϕ scale for sediment size, and the concept of sediment content density. Where D denotes grain size in mm, the corresponding value of ϕ is given by

$$D = 2^{-\phi} \tag{1}$$

The bed sediment is divided into a surface, or "exchange" layer, and a substrate. The thickness of the surface layer, L_a, is taken to be equal to the size D_{s90} of the surface layer. Let $F(\phi)$, $f(\phi)$, and $p(\phi)$ denote the volume content density of size ϕ in the surface layer, substrate, and bedload respectively. Thus, for example, $p(\phi)d\phi$ denotes the volumetric fractional content of sediment in the size range $[\phi, \phi + d\phi]$ in the bedload.

The surface–based Oak Creek bed load relation takes the form

$$W_s^\star = 0.00218 \; G[\omega \; \xi_{sgo} \; g_o(\delta)] \tag{2}$$

where the function G is given by

$$G(\xi) = \begin{cases} 5474\left(1 - \dfrac{0.853}{\xi}\right)^{4.5} & \xi > 1.59 \\ \exp[14.2(\xi - 1) - 9.28(\xi - 1)^2] & 1 \leq \xi \leq 1.59 \\ \xi^{M_o} & \xi < 1 \end{cases} \tag{3}$$

and ξ is a dummy variable. Here

$$M_o = 14.2 \tag{4a}$$

and

$$W_s^\star = \frac{R g q_T p}{(\tau/\rho)^{3/2} F} \tag{4b}$$

is a dimensionless surface–based gravel bedload transport density per unit bed surface content of size ϕ. In (4b), q_T denotes the volume gravel bedload transport rate per unit time and channel width, g denotes the acceleration of gravity, R denotes the submerged specific gravity of the gravel, ρ denotes water density, and τ denotes the fluid shear stress at the bed.

The parameter g_o in (2) denotes a surface–based reduced hiding function, given by

$$g_o(\delta) = \delta^{-\beta} \; ; \qquad \delta = \frac{D}{D_{sg}} \tag{5a,b}$$

where the parameters β and D_{sg} are given as

$$\beta = 0.0951 \; ; \qquad D_{sg} = 2^{-\bar{\phi}} \; ; \tag{5c,d}$$

$$\bar{\phi} = \int_{-\infty}^{-1} \phi \; F \; d\phi \tag{5e}$$

That is, D_{sg} denotes the geometric mean size of the material in the surface layer. The form of the hiding function is such that it enhances the mobility of finer grains relative to coarser grains in a mixture.

Furthermore, the normalized dimensionless bed shear stress ξ_{sgo} in (2) is defined as

$$\xi_{sgo} = \frac{\tau_{sg}^\star}{0.0386} \; ; \qquad \tau_{sg}^\star \quad \frac{\tau}{\rho R g D_{sg}} \tag{6a,b}$$

The parameter ξ_{sgo} corresponds to the parameter ϕ_{sgo} of Parker (1989b); the change in notation has been made to avoid confusion with grain size ϕ of (1).

The straining parameter ω in (2) is given as

$$\omega = 1 + \frac{\sigma_\phi}{\sigma_{\phi o}} (\omega_o - 1) \tag{7}$$

Here σ_ϕ denotes the arithmetic standard deviation of the surface material on the phi scale; which can be expressed in the following form

$$\sigma_\phi^2 = \int_{-\infty}^{-1} (\phi - \bar{\phi})^2 \; F \; d\phi \tag{8a}$$

The parameter σ_ϕ is related to the more familiar geometric standard deviation σ_{sg} of the surface layer by the relation

$$\sigma_{sg} = 2^{\sigma_\phi} \tag{8b}$$

Finally, the functional relations

$$\omega_o = \omega_o(\xi_{sgo}) \; ; \qquad \sigma_{\phi o} = \sigma_{\phi o}(\xi_{sgo}) \tag{9a,b}$$

are plotted in figure 5 of Parker (1989b), and are available in tabular form in Parker (1989a). A Pascal computer program for using the Oak Creek surface bedload relation, "acronym1", is available in Parker (1989a).

In the above relations, selective sorting is driven by the effect of hiding. A perusal of (2) and (5) indicates that hiding renders coarser particles somewhat less mobile than their finer comrades.

SEDIMENT CONTINUITY

The differential form for sediment continuity of bed material load is as follows (Parker, 1989c):

$$(1-\lambda)[f_I \frac{\partial \eta}{\partial t} + \frac{\partial}{\partial t}(FL_a)] = - \frac{1}{B} \frac{\partial}{\partial x} (q_T pB) - \tilde{A} \tag{10}$$

Here η = bed elevation, B denotes the effective channel width for sediment transport, L_a = thickness of the active layer, λ = bed porosity, taken to be equal to 0.3 herein, t = time, and x = downstream distance.

Furthermore, $\tilde{A}$ denotes an abrasion rate, such that $\tilde{A}(\phi)d\phi$ denotes the rate per unit time per unit bed area at which particle volume in the range $[\phi, \phi + d\phi]$ is lost to abrasion. This term is formulated in more detail below. Finally, f_I denotes an "exchange" density at the interface between the substrate and the surface layer, such that

$$f_I = f \text{ if } \frac{\partial \eta}{\partial t} < 0 \quad \text{(degrading bed)} \tag{11a}$$

$$f_I = F \text{ if } \frac{\partial \eta}{\partial t} > 0 \quad \text{(aggrading bed)} \tag{11b}$$

Equation (10) can be generalized to the case of gravel mixtures composed of several rock types, each with a different abrasion rate. For example, the gravel might contain quartz, limestone, granite and basalt clasts.

A mixture of M rock types, k = 1..M is considered. The fraction of the surface volume content density at size ϕ composed of rock type k is denoted as $F_k(\phi)$; it then follows that

$$F = \sum_k F_k \tag{12a}$$

In a similar fashion, the density fractions or partitions p_k, $f_{I,k}$, and $\tilde{A}_k$ can be defined, each of which satisfies a relation analogous to (12a); e.g.

$$p = \sum_k p_k \tag{12b}$$

Thus $q_T p_k(\phi)$ denotes the bedload density at size ϕ of rock type k per unit width per unit time.

For mixtures of rock types, (10) takes the form

$$(1-\lambda)[f_{I,k} \frac{\partial \eta}{\partial t} + \frac{\partial}{\partial t}(F_k L_a)] = -\frac{1}{B}\frac{\partial}{\partial x}(B q_T p_k) - \tilde{A}_k \tag{13}$$

FORMULATION OF ABRASION

The formulation of abrasion employed here can be found in Parker (1989c). This formulation accounts for fluvial abrasion due to binary collisions between moving bedload particles and static particles exposed at the surface layer of the bed. The by-product of collisional abrasion is taken to be silt, which then enters the wash load and is lost to the bed. For the case of a single rock type, the following form is found for the term $\tilde{A}$ in (10);

$$\tilde{A} = q_T \beta^{\star}[(p + F_{ae}) + \frac{1}{3 \cdot ln2}\frac{\partial}{\partial \phi}(p + F_{ae})] \tag{14}$$

Here $F_{ae}(\phi)$ denotes the areal size density of the surface particles, given by

$$F_{ae}(\phi) = \frac{2^{0.5\phi} \cdot F}{\int 2^{0.5\phi} \cdot F d\phi} \tag{15}$$

and $\beta^{\star}$ denotes a coefficient of abrasion expressed in, for example, m^{-1}. In the present analysis, it is assumed that the coefficient of abrasion is independent of grain size ϕ.

For the case of mixtures of rock type. (14) generalizes to

$$\tilde{A}_k = q_T \beta_k^{\star}\left[(p_k + F_{ae,k}) + \frac{1}{3 \cdot ln(2)}\frac{\partial}{\partial \phi}(p_k + F_{ae,k})\right] \tag{16}$$

Here $\beta_k^{\star}$ denotes the coefficient of abrasion of the kth rock type, and $F_{ae,k}$

denotes the fraction of the quantity F_{ae} occupied by the kth rock type.

WAVE–LIKE AGGRADATIONAL PROFILES OF PERMANENT FORM

The streamwise profile of bed elevation of most rivers displays a characteristic concavity, such that bed slope decreases in the downstream direction over scales of tens to hundreds of kilometers. This concavity is illustrated in the case of the Red Deer River, Canada, in figure 1. Declining slope suggests a gradual decline in the competence to move a given size of gravel. As such, the concavity has long been thought to be associated with streamwise fining of bed sediment.

The tendency for bed slope to decline in the streamwise direction is likely due to a variety of causes. For example, watershed basin area increases downstream, typically resulting in increased water discharge. It is known, however, that long–term aggradational tendencies commonly result in the formation of a concave profile (e.g. Morisawa, 1968).

That this is indeed the case can best be illustrated by considering uniform material as a first step. The further constraints of constant discharge and channel width are added for simplicity. In this case, (10) reduces to

$$(1-\lambda)\,\frac{\partial \eta}{\partial t} = -\,\frac{\partial q_T}{\partial x} \tag{17}$$

Aggradation of the type illustrated in figure 2 is considered. That is, the streamwise profile of bed elevation is taken to be of self–preserving form, propagating downstream at a very slow wave horizontal speed c. Thus when (17) is transformed into an $\hat{x}$–$\hat{t}$ coordinate such that

$$\hat{x} = x - ct \;; \qquad \hat{t} = t \tag{18a,b}$$

the resulting relation is taken to be independent of $\hat{t}$. Carrying out the indicated transformation, and dropping the caret superscript on x for simplicity of notation, the following result is obtained:

$$\frac{d}{dx}\left[\frac{q_T}{(1-\lambda)}\right] = -\,c\,S \;; \qquad S = -\,\frac{d\eta}{dx} \tag{19a,b}$$

Here S denotes the bed slope.

A simple heuristic form for the sediment transport relation is

$$q_T = \alpha\, S \tag{20}$$

where α is a coefficient. Substitution of (20) into (19a,b) and solution yields the result

$$S = S_o e^{-[(1-\lambda)c/\alpha]x} \tag{21}$$

where S_o denotes the bed slope at x = 0, i.e the upstream end of the reach under consideration. The concavity of the profile is readily apparent. It is easily verified that the concavity can be similarly modelled using more realistic sediment transport models.

The above analysis is now generalized to mixtures, retaining the constraints of constant water discharge and channel width. The equilibrium case of steady, longitudinally uniform transport of gravel is realized only at a constant bed slope, for which neither profile concavity nor convexity is present. This case corresponds precisely to the absence of downstream variation in grain size. Under the stated constraints, then, it follows that streamwise variation in grain size must be associated with profile concavity or convexity. It is found that concavity is associated with downstream fining, and convexity with the rarer case of downstream coarsening (observed readily enough for geologically young tributaries of incised rivers).

Applying the transformations embodied in (18) to (13) and reducing in a fashion analogous to (19), the following relation governing streamwise variation is obtained subject to the stated constraints:

$$\frac{d}{dx}\left[\frac{q_T p_k}{(1-\lambda)} - cL_a F_k\right] = - cf_{I,k} S - \tilde{A}_k \qquad (22)$$

MODEL APPLICATION: RED DEER RIVER, CANADA

Shaw and Kellerhals (1982) provide a comprehensive compilation of data pertaining to the downstream variation of grain size in rivers in Alberta, Canada. The data were analyzed in order to extract an appropriate reach satisfying the following constraints:

1. The reach has a gravel bed, and displays a concave streamwise profile.

2. Characteristic gravel size tends to become finer in the downstream direction.

3. Channel width and bankfull water discharge do not vary strongly in the downstream direction.

4. There are no major tributaries along the reach.

No reach precisely satisfying the above constraints, and in particular the third constraint, could be found. A 111–km reach of the Red Deer River starting at a point 142 km downstream of the source (roughly between Sundre and Red Deer) was nevertheless found to be in approximate accord with the above criteria, and was selected for further analysis. The channel profile is shown in figure 1; the downstream variation in the surface median size D_{s50} and D_{s90} are shown in figure 2.

Information concerning hydraulic and geomorphic characteristics of the study reach was extracted from Kellerhans et al. 1972). Channel bed slope S on the reach varies from 0.0049 at the upstream end to 0.0012 at the downstream end; surface geometric mean size D_{sg} declines from 57 mm to 28 mm.. Average values of channel width and water discharge at the two–year flood are 91.8 m and 672 cumecs, respectively. The dimensionless Shields stress $\tau^{\star}_{sg}$ based on surface geometric mean size D_{sg} was evaluated by estimating the bed shear stress τ from the relation for (quasi–)normal flow in a wide channel,

$$\tau = \rho g H S \tag{23}$$

where H is cross–sectionally averaged depth. It was found that $\tau^{\star}_{sg}$ evaluated at the two–year flood declines from 0.0627 at the upstream end of the reach to 0.0471 at the downstream end.

Although several rock types are represented in the gravel of this reach, the two dominant components are quartz and limestone. The fraction of limestone is found to decline sharply in the streamwise direction. If the fractions are renormalized so as to approximate the bed as composed solely of quartz and limestone, it is found that some 45 per cent of the bed material is limestone at the upstream end of the reach; this value declines to 18 per cent at the downstream end. This variation in rock type fraction provides a useful means of evaluating the relative importance of abrasion and selective sorting.

Equation (13) was solved in conjunction with (2) ~ (9), (11b), and (23) in order to obtain a description of downstream fining in the reach. In order to implement the model, it is necessary to specify a bed resistance relation. A form of the Manning–Strickler relation was selected for this purpose:

$$\frac{U}{\sqrt{\tau/\rho}} = 8.1 \left(\frac{H}{k_s}\right)^{1/6} \tag{24}$$

where U denotes the mean flow velocity, and the roughness height k_s was equated to $2 \cdot D_{s90}$.

The discharge selected for the calculations was the two–year flood. Input for the model consists of the values of $\tau^{\star}_{sg}$, S, and F_k for k = 1 to 2 (limestone and quartz) at the upstream end of the reach, as well as the abrasion coefficients $\beta^{\star}_k$, the wave speed c, and an appropriate step length dx. The solution of (13) is performed in a stepwise fashion in the downstream direction. Details of the calculation procedure can be found in Parker (1989d). Output consists of the downstream variation of S, q_T, F_k, and p_k, among other parameters.

The following values were selected for the abrasion coefficients: for limestone (k = 1), $\beta^{\star}_1 = 1.25 \times 10^{-5}\ m^{-1}$; and for quartz (k = 2), $\beta^{\star}_2 = 3.0 \times 10^{-7}\ m^{-1}$. The former value was back–calculated from the abrasion studies of Maunsell (1982); it is in general agreement with the value of $1.0 \times 10^{-5}\ m^{-1}$ quoted by Krumbein (1941). The value for quartz is one of several due to Kuenen (1956). Whether or not these precise values are applicable to the limestone and quartz in the Red Deer River is open to some question; they are used in lieu of better information. The relative ease of abrasion of of limestone compared to quartz, however, is readily apparent from the numbers.

The calculations presented here were performed with a preliminary version of (7), in which $\omega = \omega_0$. The difference between the two formulations is modest.

The profile concavity for the study reach seen in figure 1 may or may not be a long–term aggradational feature. In the present analysis, it is assumed that whatever the origin of the concavity may be, if the observed longitudinal profile can be accurately reproduced by means of the selection of an appropriate value of aggradational wave speed c, then the associated downstream variation in grain size should also be accurately predicted.

The model was thus first calibrated in c in order to reproduce the streamwise bed profile. A value of c of 6.75 m/year was found to provide excellent results, as shown in figure 5. The near–linear curve on the semilogarithmic plot indicates that the concavity is approximately exponential.

Lest the value of horizontal aggradational wave speed seem too high, it must be realized that the associated vertical aggradational rate is given by

$$\frac{\partial \eta}{\partial t} = - c \frac{\partial \eta}{\partial x} = cS \tag{25}$$

The highest vertical aggradation rate is realized at the upstream end of the reach where S = 0.0049, where it takes a value of 33 mm/year. Even this value fails to give a clear indication of the true vertical aggradation rate, because the calculation is performed assuming a constant water discharge equal to the two–year flood. It is likely that such a flow continued for only several days would transport the gravel yield actually realized in an average water year. If it is assumed that the two–year flood continued for two days would transport the mean annual gravel yield, it follows that the actual maximum vertical aggradation rate would be about 18 mm per century, a value small enough to suggest a longitudinal profile that is quasi–stable in geologic time.

Predictions pertaining to downstream variation in grain size and rock type fraction were obtained with no further calibration beyond that required to reproduce the concavity. The predicted downstream variation in surface mean grain size D_{sg} and fraction limestone in the surface layer are compared against observed values in figures 5 and 6 respectively. Considering the scatter in the observed values, the agreement is rather good.

It is found that in the case of quartz, the predicted tendency for abrasion is negligible. When the calculation was performed solely with quartz, nearly all of the downstream fining was found to be due to selective sorting. On the other hand, in the case of limestone, the processes of abrasion and selective sorting were found to be of roughly equal importance in determining downstream grain size variation.

In the case presented here, abrasion is important in determining downstream fining to the extent that a significant amount of limestone remains in the bed material. As the limestone abrades to silt, however, its fraction in the bed relative to quartz declines. As a result, selective sorting of quartz can be expected to dominate beyond some point sufficiently far downstream.

CONCLUSION

A combined treatment of the mechanics of selective sorting and abrasion of river gravel by water flow allows for a comprehensive treatment of downstream variation of grain size. Selective sorting is treated in terms of a hiding function that slightly favors the mobility of fine grains at the expense of coarser ones. Abrasion is analyzed in terms of binary collisions of bedload particles with stationary bed particles, the by–product of which is taken to be silt. The model is formulated for mixtures of rock types of differing ease of abrasion.

The characteristic concavity of downstream profiles of river bed elevation can be modelled in terms of a wave of aggradation of permanent form propagating very slowly in the streamwise direction. The degree of concavity increases with increasing wave speed.

An application of the model to the Red Deer River, Alberta, Canada, indicates that for the case of quartz grains, downstream fining is dominated by selective sorting: abrasion is essentially negligible. In the case of limestone, which is typically much more readily abraded than quartz, selective sorting and abrasion contribute roughly equally to downstream fining.

In the case of a mixture of quartz and limestone, abrasion tends to reduce the fractional content of limestone in the downstream direction. Abrasion thus remains important in determining downstream fining only to the extent that a significant amount of limestone remains in the bed material.

ACKNOWLEDGEMENTS

Many fruitful discussions with C. Paola are gratefully acknowledged. This research was partially funded by the U. S. National Science Foundation (grant no. CTS 8901598).

REFERENCES

Kellerhals, R., Neill, C. R., and Bray, D. I. (1972). Hydraulic and geomorphic characteristics of rivers in Alberta. *River Engineering and Surface Hydrology Report* 72–1, Research Council of Alberta, Edmonton, Alberta, Canada.

Krumbein, W. C. (1941). The effects of abrasion on the size, shape, and roundness of rock fragments. *J. Geology*, 49, 482–520.

Kuenen, Ph. H. (1956). Experimental abrasion of pebbles. 2. Rolling by current. *J. Geology*, 64, 336–368.

Maunsell and Partners Pty. Ltd. (1982). Waste rock and sedimentation methodology. *Ok Tedi Environmental Study*, Vol. 3, 272 p.

Milhous, R. T. (1973). *Sediment transport in a gravel–bottomed stream.* Thesis presented to Oregon State University at Corvallis, Oregon, in partial fulfillment of the requirements for the degree of Doctor of Philosophy, 232 p.

Morisawa, M. (1968). *Streams: their Dynamics and Morphology.* McGraw–Hill, 175 p.

Parker, G. (1989a). The "acronym" series of Pascal programs for computing bedload transport in gravel rivers. *External Memorandum*, St. Anthony Falls Hydraulic Laboratory, University of Minnesota, Minneapolis.

Parker, G. (1989b). Surface–based bedload transport relation for gravel rivers. *Submitted to J. Hydraul. Research*, IAHR.

Parker, G. (1989c). Selective sorting and abrasion of river gravel: theory. *Submitted to J. Hydraul. Engrg.*, ASCE.

Parker, G. (1989d). Selective sorting and abrasion of river gravel: applications. *Submitted to J. Hydraul. Engrg.*, ASCE.

Shaw, J., and Kellerhals, R. (1982). The Composition of recent alluvial gravels in Alberta river beds. *Bulletin* 41, Alberta Research Council, Edmonton, Alberta, 151 p.

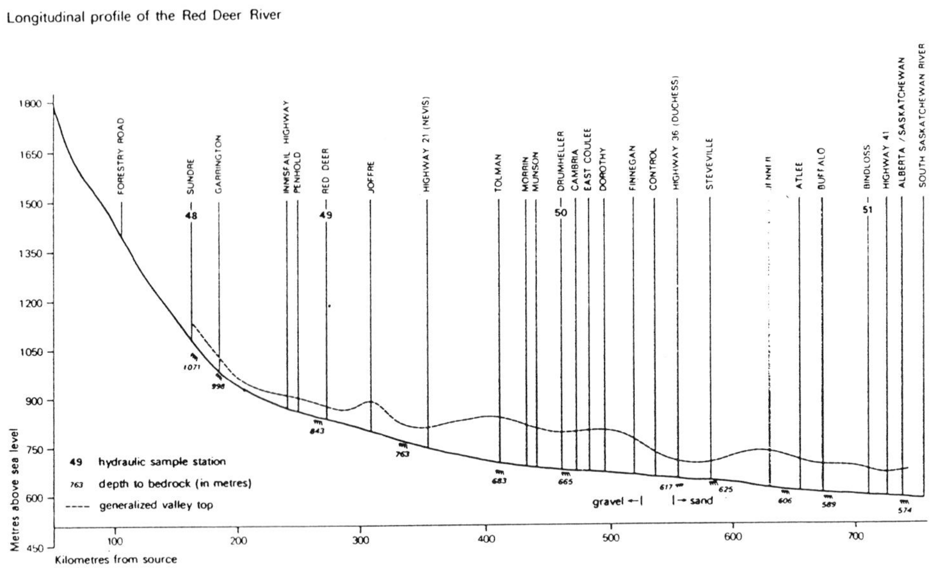

Fig. 1 Downstream profile of bed elevation in the Red Deer River; from Shaw and Kellerhals (1982).

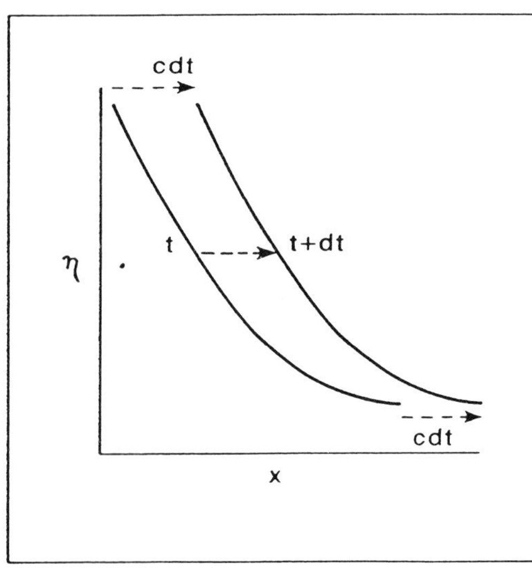

Fig. 2 Schematization of wave–like downstream aggradation of self–preserving form.

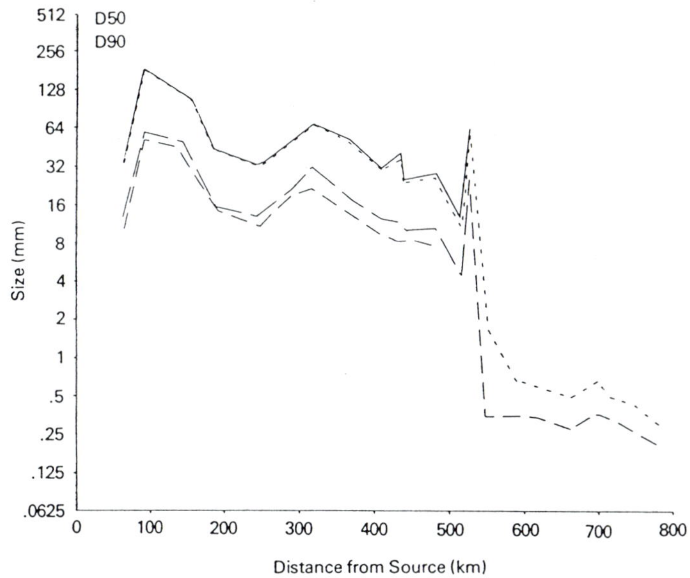

Fig. 3 Downstream variation of median and 90–percentile surface grain sizes in the Red Deer River; from Shaw and Kellerhals (1982).

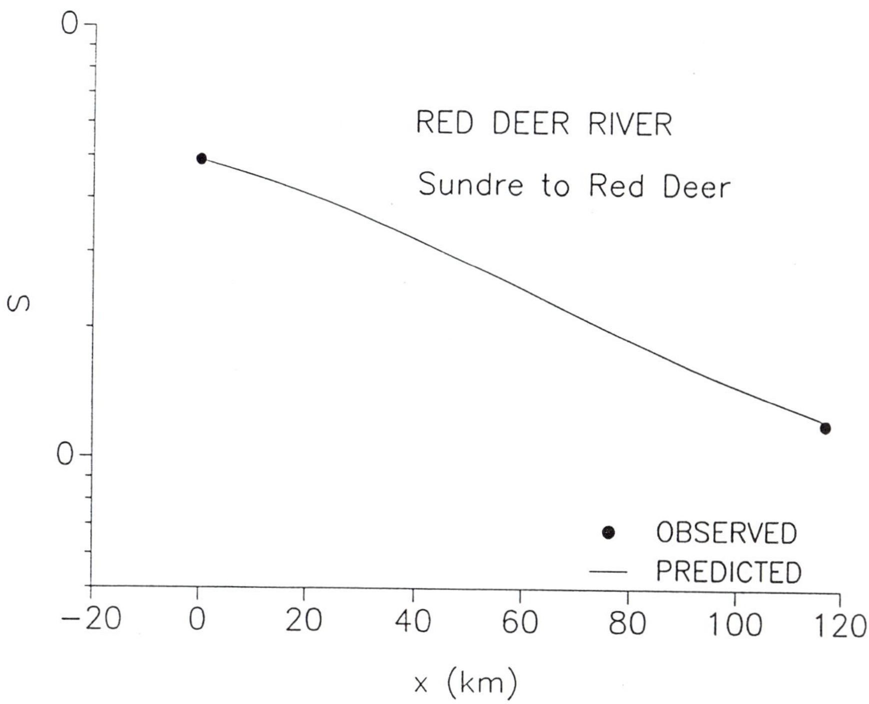

Fig. 4 Calibrated and observed downstream profiles of bed slope over the study reach.

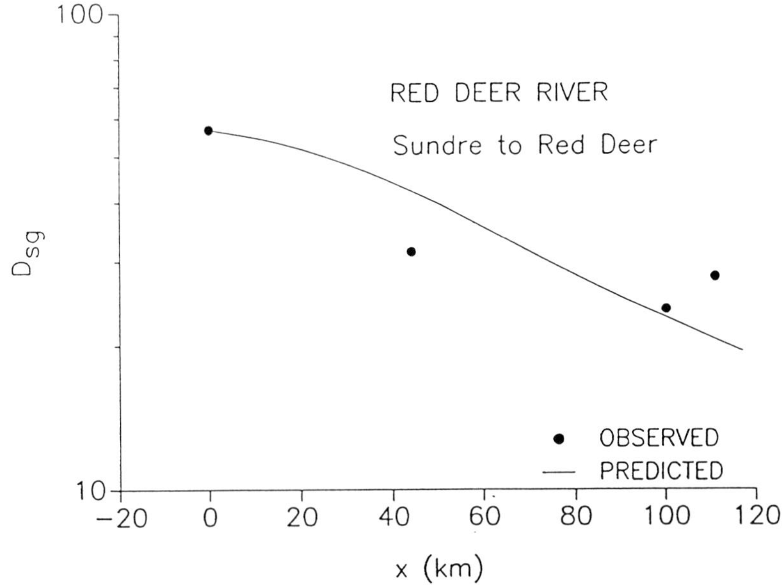

Fig. 5 Predicted and observed downstream variation of surface geometric mean size D_{sg} over the study reach.

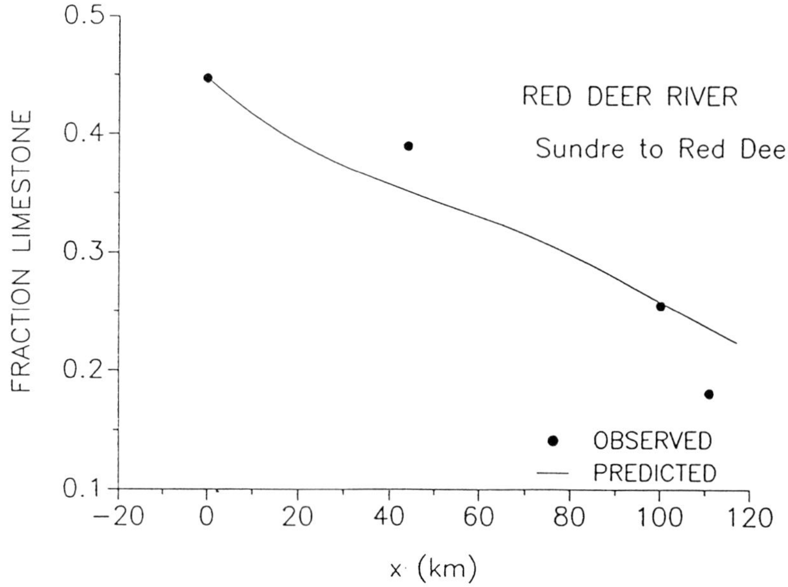

Fig. 6 Predicted and observed downstream variation of limestone fraction in the surface material over the study reach.

MORPHOLOGICAL CHANGES AND GRAIN SORTING IN MOUNTAIN GRAVEL-BED STREAMS

Domenico Pianese [(1)] and Fabio Rossi [(2)]

[(1)] Department of Hydraulics, Water Resources Management and Environmental Engineering, University of Naples, Napoli, Italy
[(2)] Institute of Civil Engineering
University of Salerno, Penta di Fasciano, Italy

ABSTRACT

Abstract : The paper shows a mathematical model for the evaluation of the morphological changes in the transitional–aggraded reaches of gravel–bed streams, characterized by a high variability of liquid discharges and a braided configuration.

The model uses the continuity and transport equations per fraction. These equations are obtained by using the hypothesis that the actual transport rate is equal to the transport capacity under geomorphoclimatic equilibrium conditions, by averaging over the year the variables involved in the phenomenon. The intra–annual distribution of liquid discharges is represented by using the flow–duration curve of the stream. The transport capacity is evaluated per each fraction of sediment, hypothesizing that, for each duration, locally steady and quasi–uniform flow conditions occur. In addition, the model takes into account the presence of the "pavement" controlling the mobility of the mixed–size fractions of the sediment.

1 — INTRODUCTION

Natural streams show a morphology that remarkably change with distance from the sediment supply areas, the primary morphology of the valley crossed, grain–size distribution and amounts of the sediment input, the time/length distribution of liquid discharges supplied by the upstream basin.

In gravel–bed streams typical of southern Italy, in the downstream direction one can schematize reaches that share common behaviours towards evolutionary processes.

In initial reaches, these streams appear carved in the bedrock. At long time scale they show a higher transport capacity than the upstream supply because of the very steep channel slopes, more than 2–3% .

In middle reaches, which are some tens of kilometers long, they show a gravel bed and slopes ranging between some units per cent and some units per mill. This configuration was attained at the end of an aggrading phase.

In final reaches instead, they resemble plain rivers, appearing meandering, degraded in fine materials, with even smaller slopes, in the order of some units per mill. In some cases, due to the close proximity to areas of coarse sediment supply, the aggraded gravel bed attains the sea, involving the final reach of the stream.

Aggraded gravel–bed streams can be divided, in turn, into two reaches with different geomorphological features.

In the upper reach, the longitudinal slopes attained almost equilibrium values, which appear to decrease downstream essentially due to the decrease in the ratio of the sediment transport rate to the liquid discharge. The grain–size distribution is practically constant along the whole length.

In the lower reach, which can be referred to as transitional, where the valley bottom widens and smoothes, both slopes and the percentages of the coarser fractions decrease drastically. Slopes and the sizes of the sediment forming the alluvial bed are, on the average, lower than the equilibrium ones typical of the upper reach. As a consequence, in the absence of training works, both the slopes and the grain–size distribution of the alluvial bed change with time, tending to reach the equilibrium condition at long time scale.

The present paper provides a model for the explanation of the changes in longitudinal bottom slopes and grain–size distribution. Such a model can be employed also for the prediction of the effects of training works to be carried out along gravel–bed streams.

For this purpose, some models available in literature will be examined, and the possibility of their application to real cases will be discussed. Afterwards, the problems peculiar to mountain gravel–bed streams will be considered.

2 — MATHEMATICAL MODELS FOR MORPHOLOGICAL CHANGES AND LONGITUDINAL GRAIN SORTING

In order to describe variation in longitudinal slopes and grain–size distribution in gravel–bed streams, it is possible to use mathematical models, based on the continuity and sediment transport equations per fraction [1].

Most of the models proposed aim at describing the bottom changes caused by flood events [2], [3], [4], [5], [6], as well as erosion phenomena downstream of a dam [7], [8]. These models take into account, in a more or less simplified way, the formation of a surface layer, called "armour layer", constituted by coarser particles than those of the underlying layer. In addition, some of these models [9], [10], [11], [12] consider also the difference between the actual sediment transport rate under non–equilibrium conditions and the theoretical transport capacity evaluated by equilibrium–transport formulas.

As a consequence, these models are generally adopted to describe the short–term evolutionary tendencies of rivers.

However, in order to evaluate the long–time scale morphological changes of a stream, one can fail to consider the actual stochastic process of liquid discharges, referring to a simplified model where the governing equations are obtained by averaging the variables over the year and by neglecting the residual terms [13].

This approach was applied by Deigaard and Fredsøe [14] and Deigaard [15] in order to describe the changes of the bed profile and grain sorting in plain rivers with non–uniform but fine sediments and dune–bed configuration. Shen [16] proposed a similar model in order to investigate the changes of the grain–size distribution in aggrading and degrading channels.

Both models refer to a very large bed, where the flow, moving with a constant liquid discharge under conditions of quasi–uniform motion, fills up the bed width.

The alluvial bed is split into an "active" surface layer, where transport and mixing are believed to occur, and an underlying layer, which is not involved in these processes.

The two models, though being similar in the general lines, differ in the way the sediment transport rate per fraction is evaluated, in the value assigned to the thickness of the mixing and transport layer, and in the upstream sediment supply boundary condition.

For each fraction, the model proposed by Deigaard and Fredsøe evaluates the transport rate by using a transport formula valid for uniform sediment. The transport rate, which is evaluated as if only the particles of a given fraction are present in the bed, is then multiplied by the percentage of that fraction contained in the mixture. This implies that the mobility would be the same whether the particles are contained in a non–uniform mixture or are the only ones to form the bed. As a consequence, the finer of two fractions present in the mixture in the same percentage, would be transported more easily than it actually happens.

Shen's model instead takes into account that fine particles, which are intrinsically more mobile, are less exposed to the flow due to the sheltering action of their coarser neighbours.

For this purpose, the author added the corrective "hiding" factor proposed by Einstein [17], to be applied to the value of the dimensionless shear stress. This coefficient varies so that higher values of the dimensionless shear stress correspond to coarser particles and lower values to finer ones.

Deigaard and Fredsøe made reference to sand–bed rivers, hence the active layer thickness is related to dune height. Less correctly, Shen assumed the active layer thickness to be constant.

Finally, the upstream sediment supply boundary condition was chosen by Deigaard and Fredsøe so that a dynamic equilibrium condition was maintained in the upstream reach. Hence, on the average, the sediment input can be considered constant (both in amount and composition). This implies that in the initial reach, both the longitudinal slope and the bed material have to be kept constant in time.

Vice versa, Shen considered only the downstream bottom elevation boundary condition. Consequently, in the upper reach, slope, bottom elevation, grain–size distribution and transport rate vary with time. Therefore, when applying this model to aggrading channels, due to the particular (exponentially decreasing) shape of the longitudinal profile chosen, the slopes of the initial reach decrease markedly, leading, in turn, to a progressive decrease in the sediment input.

The different boundary conditions chosen lead to very different results as to the time and space trend of grain–size distribution and bottom slopes.

In the examples developed by Shen for aggrading streams, at first the slopes and the median diameter of the bottom material increased progressively; subsequently, the slopes remained practically constant, whereas the median diameter continued increasing; finally, the slope always remaining constant, the grain–size distribution of the bed material changed progressively, tending to become equal to the composition of the upstream sediment supply; the median diameter tended thus to decrease.

On the contrary, in the examples shown by Deigaard and Fredsøe, the median diameter at first decreased and afterwards increased again, tending to the value of the upstream sediment supply. In addition, bottom grain–size distribution varied more quickly than did bottom slopes; therefore, during this phase, grain sorting can be considered as a quasi–steady process with respect to changes in bottom slopes [18].

More recently, Ribberink [19] proposed a mathematical model for morphological computations in rivers with non–uniform sediment, which takes into account the variability of liquid discharges. In this model, the flow is assumed to be quasi–steady. The bottom is split into a transport layer, an exchange layer and a non–

moving bed, as proposed by Armanini and Di Silvio [13]. The thickness of the transport layer, depending on the dune–bed configuration, is taken to be equal to half the dune amplitude, and, from laboratory observations, the thickness of the exchange layer equal to 25% of the dune height.

When applying it to field conditions, the annual sediment transport rate is obtained by a weighting procedure applied to the intra–annual distribution of the liquid discharge: for each month, an averaged discharge is determined which yields the observed transport rate.

Thus, Ribberink's model introduces some important improvements with respect to Shen's and Deigaard's models. However, like the latter, the former model can be applied only to very wide sand–bed rivers, where the flow fills up the width, and sediment mixing takes place in a layer with a thickness depending on the dune height.

The present work focuses on the problems arising from the predicting models for morphological changes in mountain gravel–bed streams that show the following features:

- high discharge variability (ephemeral nature);
- cross sections that widen during the rising phases and narrow during the damping phases of floods;
- gravel bed characterized by a "pavement" controlling the mobility of the mixed–size fractions of the sediment.

3 — PROBLEMS PECULIAR TO MOUNTAIN GRAVEL–BED STREAMS

Unlike what happens in sand–bed rivers, in gravel–bed streams the bed surface is formed by a layer of coarser material than that found in the underlying layer. The former is usually referred to as "pavement" [20] [21].

The pavement occurs under sediment supply conditions completely different from those observed during bed armouring.

Armouring takes place when the upstream sediment supply to a reach is practically absent, and the non–uniform bed material is subject to the action of flows characterized by shear stress values sufficient enough to entrain almost all the available particles.

The practically clear water flow starts eroding the bottom and the banks, and carry first the more exposed and finer particles, and then the less exposed and finer ones.

As erosion goes on, the percentages of the coarser fractions in the surface layer tend to increase. At the same time, due to the progressive reduction in the percentages of the finer fractions and in bottom slopes, the transport capacity decreases, leading to a decrease in the erosion capacity of the bottom material. As a consequence, the formation of a coarser surface layer ("armour") is completed very slowly.

Vice versa, the pavement is a phenomenon peculiar to an active bed, where the sediment transport is also high but is maintained by a sufficient upstream sediment supply. It occurs during transport events characterized by boundary shear stresses well above the critical value for sporadic mobilization of surface particles. When the amount of incoming material is very large, the pavement tends to disappear. By a vertical winnowing process the presence of the pavement allows the larger–sized particles to be present in a greater amount on the surface, thus counteracting the lower intrinsic mobility of the larger fractions, at least for limited values of the ratio between the diameter of the i–th fraction and the median diameter [22]. Consequently, the pavement provides a "macroscopic hiding" effect [21] different from the "microscopic hiding" effects [17], [23] related to phenomena of fine particles sheltering from the coarse ones.

Andrews et al. [22] simulated pavement formation by adopting a mathematical model where the thickness of the mixing layer is taken to be equal to the D_{90} of the surface grain–size distribution. As regards the transport layer, they observed that it cannot be defined at all, since also in the presence of high flows, the motion of bed particles is sporadic: at a certain time only a small portion of surface particles is actually in motion.

On the basis of Andrews' model, the non–equilibrium transitional phase can be described, where the fraction percentages found in the sediment carried by the flow are different from those found in the pavement and the subpavement; under equilibrium conditions they tend towards those of subpavement.

Obviously, at long time scale one has to refer to conditions of complete development of the pavement.

Interestingly, in the presence of overloading phenomena during floods that cause short–time stream overaggradation, one should refer to conditions of non–equilibrium and hence of lack of pavement. This leads to a higher transport layer characterized by a finer grain–size distribution compared to that observed under equilibrium conditions.

Another distinctive feature of gravel–bed streams is bottom configuration, which usually has braided or alternate–bar patterns.

Two different situations can be schematized, depending on whether the flow fills up the stream width or not.

In the former case, the stream forms tridimensional bars that shift slowly. In the latter case, the flow forms one or more ephemeral channels which widen during the erosional phases and narrow during the aggradation ones [24], [25]. However, a "macroscopic" mixing due to erosional and aggrading phenomena that involve bars and ephemeral channels during flood events, is superimposed on the "microscopic" mixing occurring in the active layer. As a consequence, the complete development of pavement and bed configuration requires different times of adaptation mainly depending on the liquid discharge.

Finally, it is to be noted that mountain streams display a higher discharge variability than do plain rivers; therefore, the transport rate is very discontinuous.

4 — DESCRIPTION OF THE MODEL

The long–term development of the longitudinal profile is assumed to be found by averaging the continuity equations for the sediments over the year [13]:

$$\frac{\partial(Bz)}{\partial t} = - \frac{1}{(1-n)} \frac{\partial T}{\partial x} \qquad (1)$$

where B is the width of the cross section at the bottom elevation z; T is the total transport rate obtained by adding the suspended load to the bed load; n is the porosity of the bottom material. The values (Bz) and T are averaged over the year.

The transport rate T is taken equal to the transport capacity T_c (expressed as a function of the characteristic parameters of the flow and bottom materials) assuming that the sediment transport is influenced only by local factors ("local equilibrium" hypothesis). This hypothesis is acceptable when sediments are carried mainly as bed– and wash–load, as occurs in gravel–bed streams.

Variations in the composition of the bottom material are computed by employing the continuity equations per fraction, after dividing the cross section into two areas: an area which is not involved in mixing and transport phenomena ("non–moving area"); and another area which is taken equal to that involved in erosion and deposit phenomena that occur on the average during the year ("mixing zone").

It is also hypothesized that the incoming sediment is instantaneously mixed with the material present in the mixing zone with area A. Consequently, the continuity equations per fraction is employed in the form

$$A \frac{\partial f_i}{\partial t} = - \frac{1}{(1-n)} \left[\frac{\partial T_i}{\partial x} - f_i \frac{\partial T}{\partial x} \right] \qquad (2)$$

where f_i is the volumetric percentage of the transported material belonging to the i–th class; $T_i = f_i\ T_{c_i}$ is the actual total transport rate (T_{c_i} is the transport capacity of the i–th fraction). The values f_i and T_i are averaged over the year.

The annual transport rate at each section is not evaluated with reference to a dominant discharge but by integrating the transport equations over the flow duration curve of the stream, which is the relationship between any given discharge Q and the time d that the discharge Q is exceeded.

In order to take into account the great variability of the liquid discharge in mountain reaches, the flow duration curve is expressed so that the discharge Q is constituted by three components representing surface, hypodermic and groundwater runoff

$$Q = Q_1 \exp\left[-d/K_1\right] + Q_2 \exp\left[-d/K_2\right] + Q_3 \exp\left[-d/K_3\right] \quad (3)$$

where the K_j and Q_j values ($j = 1, 2, 3$) can be evaluated on a regional scale on the basis of the available data.

For each discharge, the steady and quasi–uniform flow hypotheses are assumed.

For the evaluation of the transport rate at each section and at each discharge value, it is assumed that the flow does not fill up the valley width but tends to form one or more equal braids inside the alluvial bed (Fig. 1a). The shape of the cross sections is given to be consistent with the equilibrium bank profile found by Ikeda [26] for the cross sections of the streams with cohensionless bed sediments, as suggested by Rossi et al. [27]. The sizes of the cross sections are instead obtained in a simplified way, assuming that, regardless of the duration, the discharge forms the bottom and the banks instantaneously and completely, and employing a variational approach based on the minimum dissipation rate principle in the form proposed by White et al. [28] ($T_c = T_c$ max) . This principle has already been used by Chang [29] and Song et al. [30] in their models describing morphological changes at a short–time scale.

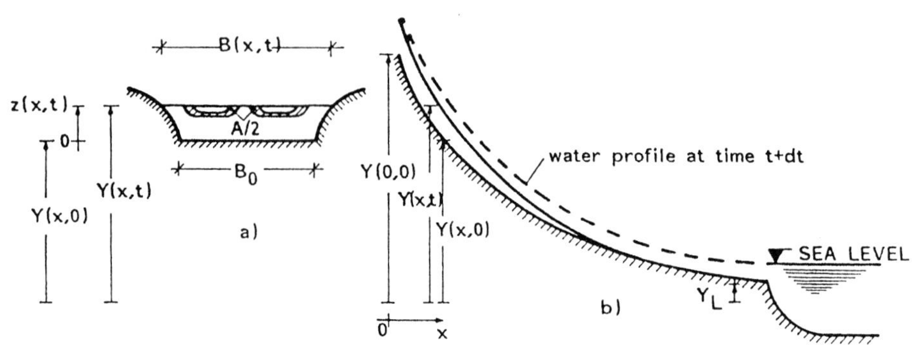

Fig.1 - Definition sketch: a) cross sectional area; b) longitudinal bed and water profile.

The mixing zone area of (2) is considered as being closely dependent on bed configuration. In fact, in ephemeral gravel–bed streams the flow forms m ($m \geq 1$) ephemeral channels, which widen during the rising phases and narrow during flood damping. The particles subject to transportation and mixing are mainly those entrained from the banks.

Consequently, it can be assumed that the mixing–zone area is computed by weighting the differences between the cross–sectional areas that form as a result of the modeling action of a discharge of given duration, Q, and the low–flow channels that are found in non–transport periods between two flood events (Fig. 1a). The weight used is the discharge duration.

The transport formula used to compute the transport capacity for each fraction is the one proposed by Parker et al. [20] obtained from measurements made on some gravel–bed streams.

The suspended load is also evaluated for each fraction by using an integration procedure similar to that proposed by Einstein [17] and taking into account, in a simplified way, the vertical distribution of the longitudinal components of the time averaged local velocities, as well as that of the concentrations of the suspended sediment [15].

In particular, the reference concentration value is measured at a distance from the bed equal to $2\,\varepsilon$, where ε is the roughness parameter, using the method proposed by Engelund and Fredsøe [31].

In order to take into account the mobility of mixed–size particles available in the mixture and the pavement, the model refers to the hiding coefficient relation proposed by Andrews [32]

$$h_i = \frac{\tau_{*c_i}}{\tau_{*c_{50}}} = \left(\frac{D_i}{D_{50}} \right)^{-r} \tag{4}$$

where h_i is the hiding factor for the i–th fraction; D_{50} is the median diameter of the subpavement; $\tau_{*c_{50}}$ the dimensionless critical shear stress appropriate for D_{50}; D_i is the mean grain–size of the i–th fraction; τ_{*c_i} is the critical dimensionless shear stress for the i–th size range; r is a positive coefficient equal to 0.872 according to Andrews and to 0.982 according to Parker et al. [20].

Note that the condition $r = 0$ implies that the hiding effects are not taken into account; whereas the condition $r = 1$ implies equal mobility of the different fractions, and hence lack of any grain sorting when the liquid discharge and the geometrical parameters of the bed (slope, width, etc.) are constant along the reach.

Finally, the roughness formula it the one by Keulegan

$$\frac{V}{u_*} = 2.5 \ln \left(11 \frac{h}{\varepsilon} \right) \tag{5}$$

where V is the averaged cross sectional flow velocity; u_* is the shear velocity; h is the depth averaged over the width; ε is the bed roughness parameter taken equal to $2\,D_{50_p}$, where D_{50_p} is the median pavement grain–size, closely related to the median diameter D_{50} of the subpavement. From observations carried out in several southern Italy gravel–bed streams, D_{50_p} is assumed to be equal to 3.5 D_{50}; hence $\varepsilon = 7\,D_{50}$.

5 — APPLICATION OF THE MODEL

5.1 — Selection of boundary and initial conditions

The model described in the previous section has been applied to the transitional reach of a hypothetical gravel–bed stream.

The longitudinal profile at $t = 0$ has been assumed to decrease exponentially with the distance from the upstream section (Fig. 1b). The cross–sectional shape of the valley where the stream has been hypothesized to flow (Fig. 1a) is described by the relation

$$B(x,t) = B_0 + 2a \left[z(x,t) \right]^b \tag{6}$$

where $B(x,t)$ is the bottom width at time t and at x ; B_0 is the bottom width at $t = 0$ (taken to be the same at all the sections); $z(x,t) = [\,Y(x,t) - Y(x,0)\,]$ is the valley bottom elevation measured at each cross section starting from an initial value; a and b are two positive coefficients. In the computations, the reach located upstream of the cross section at $x = 0$ has been assumed to be in dynamic equilibrium, i.e. the bottom elevations are variable, whereas the longitudinal slopes and the sediment supply of the various fractions do not vary with time, and are therefore equal to those at $t = 0$.

To summarize, the initial and boundary conditions chosen are the following, respectively:

- *Boundary conditions*:

$$\left[\frac{\partial Y}{\partial x} \right]_{x=0} = S_o = \text{constant} \tag{7a}$$

$$f_i\,(0,t) = P_i = \text{constant} \qquad (\,i = 1,\ 2,\ \ldots,\ N) \tag{7b}$$

- *Initial conditions*:

$$Y(x,0) = Y(0,0)\ \exp\left[\,-\,\alpha\,x\,\right]\ -\ Y_L \tag{8a}$$

$$f_i\,(x,0) = F_i = \text{constant} \qquad (\,i = 1,\ 2,\ \ldots,\ N) \tag{8b}$$

where $Y(x,t)$ is the bottom elevation at x and at t , evaluated from an initial value; $Y(0,0)$ is the bottom elevation at $x = 0$ and at $t = 0$; Y_L is the bottom elevation at $x = L$; α is a positive coefficient and N the number of the fractions contained in the mixture.

5.2 — Numerical solution

The governing equations (1) and (2), coupled with initial (8a,b) and boundary (7a,b) conditions are solved numerically using a finite difference approximation. The continuity and flow equations for water are decoupled from the continuity and transport equations for sediment, as proposed by de Vries [33].

The numerical solution of the equations (1) and (2) is obtained at each time, using a predictor–corrector scheme, as suggested by Olesen [34] (see Ribberink [19]).

The used iterative procedure based on an implicit scheme (Crank–Nicholson, with time weighting factor equal to 0.7) allows numerical stability to be enhanced, which is of particular relevance when more than one celerity has to be considered, as in the case under investigation.

In particular, the ratio of the time step to the space step has not to satisfy Courant's condition, which imposes

$$0 \leq \frac{\Delta t}{\Delta x} \leq \min_i \left(|1/c_Y| \, , \, 1/c_i \right) \tag{9}$$

where c_Y and c_i are the celerities of the disturbances of the bed and the i-th fraction, respectively.

5.3 — Numerical values of the parameters

An application of the model has been carried out referring to an aggrading reach having a length $L = 50$ km. It has been assumed that the initial upstream bottom elevation is $Y(0,0) = 250$ m, and the downstream elevation, $Y(L,t)$, is constant in time and equal to $Y_L = 55.8$ m. The latter boundary condition, which is not essential to the solution of the set of differential equations (1) and (2). has been introduced in order to simulate the flowing into the sea. The value of the constant α has been taken equal to $3 \cdot 10^{-5}$, therefore the maximum slope value in the reach is $S_0 = 0.0075$, and the slope in the final reach at $t = 0$ is 0.00167 .

The values of the constants in (6) have been assumed equal to $B_0 = 100$ m; $a = 4.5$; $b = 2$, respectively. The parameters in (3) are the following: $K_1 = 1$ day; $K_2 = 10$ days; $K_3 = 100$ days; $Q_1 = 250$ m^3 s^{-1} ; $Q_2 = 150$ m^3 s^{-1} ; $Q_3 = 60$ m^3 s^{-1} .

Finally, bed grain–size distribution at time $t = 0$ has been assumed to be log–normal, with maximum grain–size equal to 0.15 m and minimum equal to 0.0001 m.

The computations have been made by dividing the grain–size distribution of the sediment supplied to the upstream section and of the bed material into five classes, taking $\Delta x = 2.5$ km and $\Delta t = 1$ month.

6 — ANALYSIS OF THE RESULTS OBTAINED FROM NUMERICAL SIMULATIONS

A comparison has been made between the results obtained from the model, employing two different initial conditions (8b) and two different relations for the hiding factor. In particular, besides the relation (4), the hiding relation proposed by Egiazaroff [23] has been used

$$h_i = \left[\frac{\log 19}{\log \left(19 \dfrac{D_i}{D_m} \right)} \right]^2 \tag{10}$$

where D_m is the mean size of the particles available in the mixture.

6.1 — Behaviour of the model

Typical results obtained by using the mathematical model previously described are shown in Figs 2a,b, where the changes in space and in time of bed elevations and mean grain size diameter are plotted.

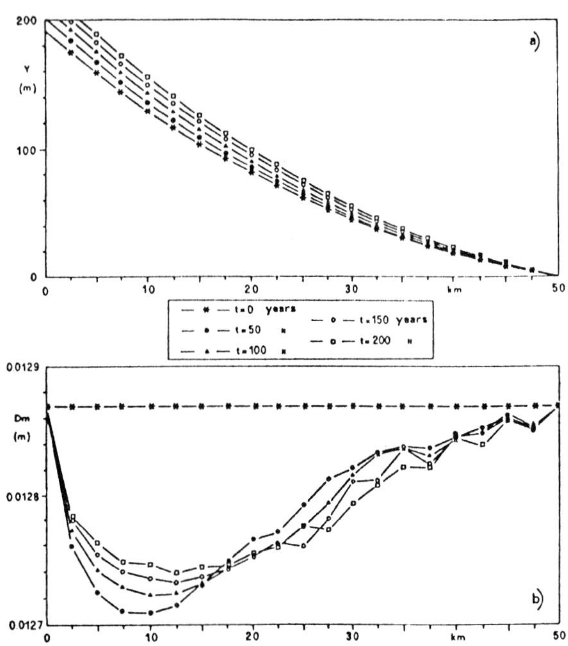

Fig.2 - Changes in space and in time of: a) bottom elevations;

b) mean grain size diameter (Andrews: F_i=P_i)

As one can observe (fig. 2a), bottom elevatio
whereas the mean grain–size diameter varies in a differ
refers to a cross section belonging to the upper or lowe

The above observations are confirmed by the diag
at $x = 10$ km, report the changes in the time of the bott
ratios D_m / D_{m_0} and σ/σ_0 for different hypotheses of the
are the mean diameter and the standard deviation, in log sc
D_{m_0} and σ_0 are the values at time $t = 0$.

The diagrams of Figs 3a,b,c show that, in agreement with
strated by Deigaard and Fredsøe [14], [15], at first a variation in
bution occurs much earlier than that in bottom elevations and longi
regardless of the boundary conditions (8b).

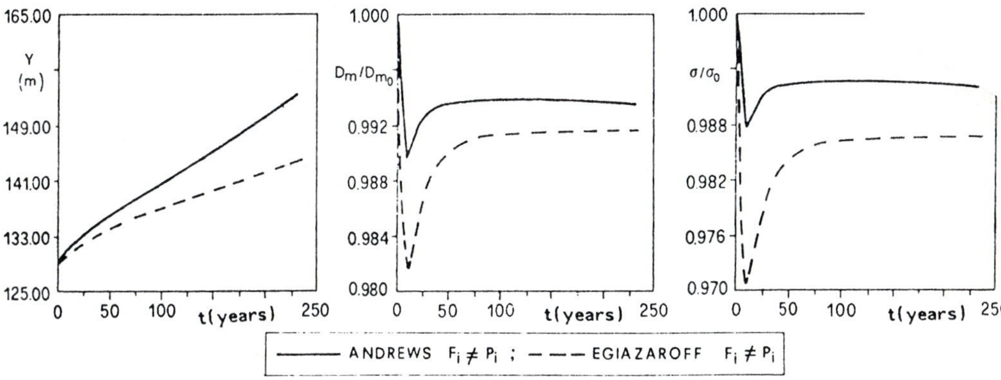

Fig.3 - Hiding factor influence

In this first phase, longitudinal grain sorting is practically independent of the aggradation process, as bottom slopes change very slowly.

In a subsequent phase, longitudinal slopes and bottom elevations vary much more markedly than the mean diameter of the transported material does. As a consequence, aggradation can be considered practically independent of longitudinal grain sorting.

In the cases considered, the slightness of the changes observed in the parameters of grain–size distribution is essentially due to the very high value of the mixing–zone area. As evidenced by Ribberink for very wide rivers, this area remarkably influences also the time scales of aggradation and grain sorting processes.

6.2 — Pavement influence

Macroscopic hiding due to the pavement plays a fundamental role in sediment

ı fact, not only does it directly affect the ratios among sediment fractions transported by the flow, but also

" by the hiding factor expressions represented in Fig. 4 .ues per fraction, $Q_{s,i}$, reported in Tab. 1, where different ompared: i) No Hiding (NH); ii) microscopic hiding, by elation (10) (EG); iii) macroscopic hiding, using the relations s (AN), Parker (PA) and equal mobility (PM) hypothesis.

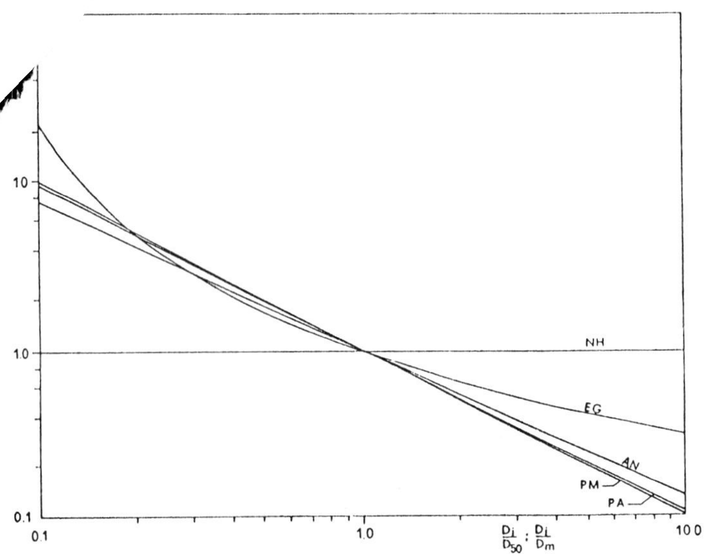

Fig.4 - Comparison of different expressions of the hiding factor

The pavement mechanism increases essentially the mobility of the bed material fractions that have the diameter D_i larger than the median diameter D_{50} (the mean diameter D_m), since the dimensionless critical shear stress values are drastically reduced because of the the very low hiding factor values.

As a consequence, very low values of the transport rate of the larger fractions occur (see Tab. 1).

Table 1 shows the mentioned influence of the hiding factor on the total amounts of the transport rate, which decrease passing from the no hiding (NH) hypothesis to the macroscopic hiding (AN or PA) and then to the microscopic hiding (EG).

This provokes a greater aggradation in the AN hypothesis than in the EG one (see Fig. 3a).

Tab. 1 — **Transport rates at $x = 0$ and $t = 0$ for different hypothesis of hiding factor h_i . ($Q_{s,i} \cdot [10^{-3}\ m^3\ s^{-1}]$)**

MEDIAN DIAMETER \ HYPOTHESIS			NH		AN		PA		PM		EG	
D_i	D_i/D_{50}	D_i/D_m	h_i	$Q_{s,i}$	h_i	$Q_{s,i}$	h_i	$Q_{s,i}$	h_i	$Q_{s,i}$	h_i	$Q_{s,i}$
0.158	0.158	0.123	1	8.37	4.98	3.08	6.10	2.22	6.31	2.09	11.99	0.23
				{ 20.11		{ 8.23		{ 6.53		{ 6.26		{ 3.65
0.398	0.398	0.309	1	11.74	2.23	5.15	2.47	4.30	2.51	4.17	2.76	3.42
1.000	1	0.777	1	8.35	1	8.35	1	8.35	1	8.35	1.20	5.62
2.512	2.512	1.952	1	0.04	0.448	3.26	0.410	4.04	0.40	4.17	0.66	0.91
				{ 0.04		{ 4.47		{ 5.99		{ 6.26		{ 0.92
6.309	6.309	4.903	1	0	0.201	1.21	0.164	1.95	0.16	2.09	0.42	0.01
TOTAL SEDIMENT TRANSPORT RATE $[m^3\ s^{-1}]$			$2.85 \cdot 10^{-3}$		$2.10 \cdot 10^{-3}$		$2.087 \cdot 10^{-3}$		$2.086 \cdot 10^{-3}$		$1.020 \cdot 10^{-3}$	

NH: No Hiding
AN: Andrews; relation (4) with $r = 0.872$
PA: Parker; relation (4) with $r = 0.982$
PM: equal mobility; relation (4) with $r = 1$
EG: Egiazaroff; relation (10)

The mean diameter of the sediment transported in the AN hypothesis is, asymptotically, always larger than that of the sediment transported in the EG hypothesis, since a greater transport capacity of larger–sized particles prevails in the former (Fig. 3b). In addition, owing to the restrictions imposed by the Egiazaroff relation on the transport rate of both smaller– and larger–sized fractions than the mean diameter, the standard deviation are always lower than in the AN hypothesis (Fig. 3c).

6.3 — Influence of the sediment bed initial condition

The original bed grain–size distribution not only influence the bed material composition consequent on longitudinal sorting, but also the rates of the sediment transport and aggradation, since the transport rate is low and the sediment available for mixing with the incoming material is abundant.

This is confirmed by the diagrams of Figs 5 and 6, where two cases are compared: i) original composition of the transport layer material finer than (case A) or ii) equal to the composition of the incoming material (case B).

The curves of fig. 5 depict the time variations in bottom elevations (Fig. 5a), mean diameter (Fig. 5b) and the standard deviation in log scale (Fig. 5c) at $x = 10$ km (AN hypothesis).

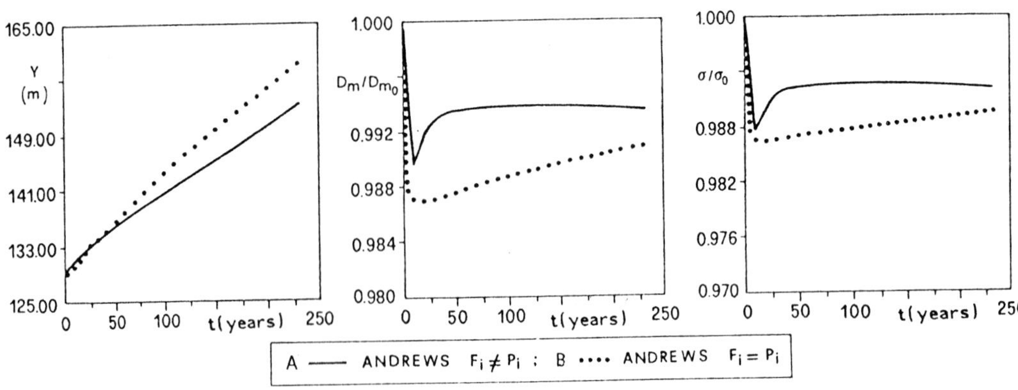

Fig.5 - Influence of the sediment bed initial condition

In this section, which is quite close to the sediment supply area, longitudinal grain sorting leads quickly enough to a grain–size distribution depending mainly on that of the supplied material. In fact, in case A, where D_{m_0} is smaller than in case B, a practically asymptotical value is more rapidly attained (Fig. 5). By contrast, in farther regions, the mean diameter depends on the initial bed composition for a longer time (see also Fig. 2b).

Differences in the aggradation process are more complex when the original composition of the transport layer material is finer (case A).

The transport rate in the upstream reach is higher, so that there is a greater tendency for sediment accumulation in downstream reaches. It ensues that longitudinal slopes increase less fast downstream, and at first even decrease upstream (Fig. 6). Therefore, the aggradation rate depends on the different mobility of the fractions (see Fig. 6a and 6b), as well as on the different role played by the sediment storage and the variation in slopes.

In this regard, case B, which is reported in Fig. 2b and is generally considered in the literature [14] [15] [19], is the less probable for mountain gravel–bed streams.

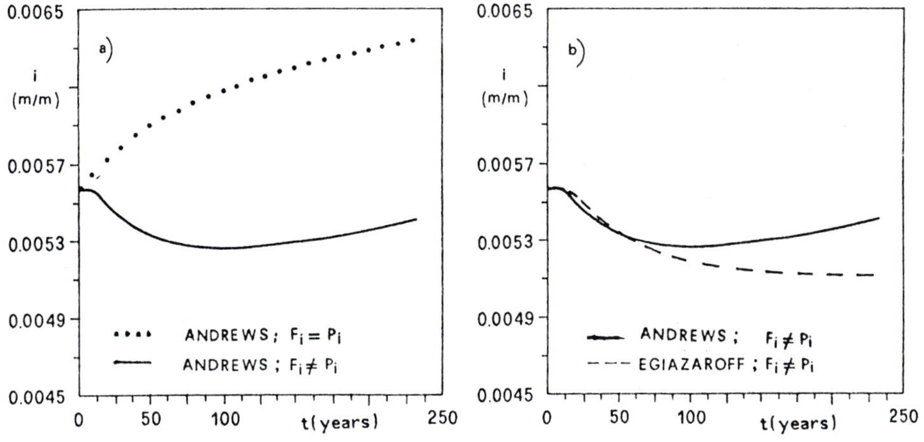

Fig.6 - Changes in time of longitudinal bottom slopes at x=10 Km

7 — CONCLUSIONS AND FURTHER DEVELOPMENTS OF THE INVESTIGATIONS

In the previous sections a mathematical model has been provided for the study of long–time scale changes in bed morphology and grain–size distribution along the transitional reaches of mountain gravel–bed streams.

The model proposed considers the main features of these streams, i.e.: high variability of the liquid discharge; time and space variability in the flow cross section area due to variations in liquid discharge, grain–size distribution of the bed material, and longitudinal slopes; presence, on the bed surface, of a layer, referred to as "pavement", constituted by coarser particles than the underlying layer.

The governing equations have been obtained by averaging the variables involved (longitudinal slopes, bottom widths, grain–size distributions, etc.) over an intra–annual scale, and by approximating the actual transport rate by the transport capacity under equilibrium conditions.

The discharge–duration relationship replaces the actual stochastic process of the liquid discharge.

The shape and size of the flow cross sections have been determined by using a variational approach and the hypothesis that flows with different discharges exert their modeling action on the bottom and the banks instantaneously and completely. For this purpose, water flow has been hypothesized to be locally steady and uniform.

The pavement has the effect of substantially decreasing the differences in particle mobility. This is taken into account by an empirical relation resulting from field observations in gravel–bed streams.

Some applications of the model show that it is able to explain the changes in longitudinal bed slopes and in the grain–size distribution of the bed material in mountain gravel–bed streams.

In particular, it has been shown that two phases can be distinguished: at first, the grain–size distribution of the transported material changes more quickly than longitudinal bed slopes do; on the contrary, in a subsequent much longer phase the latter change more quickly.

As a consequence, the process of grain sorting develops in a much shorter time than the aggradation phenomena.

In addition, it has been shown that the practically equal mobility induced by the pavement increases the mobility of larger–sized particles and hence leads to an increase in bottom elevation.

Finally, stress has been laid on the influence of sediment bed initial condition on the bed composition due to longitudinal grain sorting, mainly in the farthest reaches from the sediment supply area.

Beside the semplifications made to take into account the intra–annual flow variability, two hypotheses have been adopted:

i) the instantaneous adaptation of the flow cross sections to the liquid discharges, which would involve a much greater amount of sediment available for mixing than it actually occurs, and, therefore, may lead to underestimate the rate of changes in grain sorting;

ii) the presence of the pavement, which is found when the already high transport rate is maintained by a sufficient upstream sediment supply; actually it tends towards an armour layer when the ratio of transport rate to liquid discharge decreases, whereas it tends to disappear in the presence of very large transport events.

In order to provide a model for short–time scale morphological changes, account should be necessarily taken of these two hipotheses.

Acknowledgment

This research was partially supported by Ministero Pubblica Istruzione (fondi M.P.I. 40%) and by G.N.D.C.I. — Gruppo Nazionale per la Difesa dalle Catastrofi Idrogeologiche — contributions.

REFERENCES

[1] HIRANO, M.: *River bed degradation with armouring* — Trans. of JSCE, Vol.3, Parte 2, 1971.

[2] THOMAS, W.A.: *Mathematical modelling solutions* — Reported in "Application of stochastic processes in sediment transport", Ed. by H.W. Shen and H. Kikkawa. Water Resources Publications, USA, Chap. 16, 1980.

[3] THOMAS, W.A.: *Mathematical Modelling of Sediment Movement* — In "Gravel-bed Rivers" , Ed. by R.D. Hey, J.C. Bathurst and C.R. Thorne, John Wiley & Sons Ltd., pp. 487 ÷ 508, 1982.

[4] HOLLY, F.M. , KARIM, M.F.: *Simulation of Missouri River Bed Degradation* — Jour. of Hydr. Eng., ASCE, Vol.112, No. 6, pp. 497 ÷ 517, 1986.

[5] KARIM, M.F. , HOLLY, F.M.: *Armoring and Sorting Simulation in Alluvial Rivers* — Jour. of the Hydr. Eng., ASCE, Vol.112, No. 8,pp. 705 ÷ 715, 1986.

[6] LI, R.M. , FULLERTON, W.T.: *Investigation of Sediment Routing by Size Fractions in a Gravel-bed River* — Reported in "Sediment Transport in Gravel-Bed Rivers" , Ed. by C.R. Thorne, J.C. Bathurst and R.D. Hey, John Wiley & Sons Ltd., pp. 421 ÷ 442, 1987.

[7] BETTESS, R. , WHITE, W.R.: *Mathematical simulation of sediment movement in streams* — Proc. Inst. Civ. Engrs., Part 2, 71, 1981.

[8] MICHIUE , M. , SUZUKI, K. , HINOKIDANI, O.: *Sediment transport and river bed variations* — Proc. of the ROC–JAPAN Joint Seminar on Multiple Hazards Mitigation, pp. 651 ÷ 667, Taipei, ROC, 1985.

[9] BORAH, D.K. , ALONSO, C.V. and PRASAD, S.N.: *Routing graded Sediments in Streams: Formulations* — ASCE, Jour. of the Hydr. Div., Vol. 108, No. HY12, pp. 1486 ÷ 1503, 1982.

[10] BORAH, D.K. , ALONSO, C.V. and PRASAD, S.N.: *Routing graded Sediments in Streams: Applications* — Jour. of the Hydr. Div., ASCE, Vol. 108, No. HY12, pp. 1504 ÷ 1517, 1982.

[11] ARMANINI, A. , DI SILVIO, G.: *Transport of Suspended Sediments along Channels with a Thrench* — XXI Congress IAHR, Melbourne, Australia, 1985.

[12] BELLEUDY, P.H. , RAHUEL, J.L. and GOULU, Y.: *CARICAR - Mobile bed modelling of graded sediments in unsteady flow* — XXII Congress IAHR, Topics in Hydr. Mod., pp. 138 ÷ 143, 1987.

[13] ARMANINI, A. and DI SILVIO, G.: *A one-dimensional model for the transport of a sediment mixture in non-equilibrium conditions* — Jour. of Hydr. Res., Vol. 26, No. 3, pp. 275 ÷ 292, 1988.

[14] DEIGAARD, R. and FREDSØE, J.: *Longitudinal Grain Sorting by current in Alluvial Streams* — Nordic Hydrology, 9, pp. 7 ÷ 16, 1978.

[15] DEIGAARD, R.: *Longitudinal and tranverse sorting of grain sizes in alluvial rivers* — Techn. Univ. of Denmark, Lyngby, 1980.

[16] SHEN, H.W.: *Sediment sorting processes in certain aggrading and degrading streams* — Proceedings of the First International Symposium on River Sedimentation, Vol. 1, pp. 273 ÷ 300, Beijing, China, 1980.

[17] EINSTEIN, H.A.: *The bed load function for sediment transportation in open channels* — Techn. Bull. 1026, United States Department of Agriculture, Soil Conservation Service, Washington, 1950.

[18] RANA, S.A. , SIMONS, D.B. and MAHMOOD, K.: *Analysis of Sediment Sorting in Alluvial Channels* — Jour. of the Hydr. Div., ASCE, Vol. 99, No. Hy11, pp. 1967 ÷ 1980, 1973.

[19] RIBBERINK, J.S.: *Mathematical modelling of one-dimensional morphological changes in rivers with non-uniform sediment* — Delft Univ. of Techn., Rep. No. 87-2, 1987.

[20] PARKER, G. , KLINGEMAN, P.C. and McLEAN, D.G.: *Bedload and Size Distribution in Paved Gravel-Bed Streams* — Jour. of the Hydr. Div., ASCE, Vol. 108, No. HY4, pp. 544 ÷ 571, 1982.

[21] PARKER, G. and KLINGEMAN, P.C.: *On Why Gravel Bed Streams Are Paved* — Water Resources Research, Vol. 18, No. 5, pp. 1409 ÷ 1423, 1982.

[22] ANDREWS, E.D. , PARKER, G.: *Formation of a Coarse Surface Layer as the Response to Gravel Mobility* — Reported in "Sediment Transport in Gravel-bed Rivers" , Ed. by C.R. Thorne, J.C. Bathurst and R.D. Hey, John Wiley & Sons Ltd., pp. 269 ÷ 300, 1987.

[23] EGIAZAROFF, I.V.: *Calculation of nonuniform sediment concentrations* — Jour. of the Hydr. Div., ASCE, Vol. 91, No. HY4, pp. 225 ÷ 247, 1965.

[24] ANDREWS, E.D.: *Bank Stability and Channel Width Adjustment, East Fork River, Wyoming* — Water Resources Research, Vol.18, No.4, pp. 1184 ÷ 1192, 1982.

[25] PIANESE, D. , ROSSI, F. and SILVAGNI, G.: *Analisi stocastica delle variazioni di fondo alveo di corsi d'acqua in ghiaia* – Atti del Seminario "Modelli dei Fenomeni Idraulico-Fluviali", pp. 62 ÷ 78, Bologna, 1986.

[26] IKEDA, S.: *Self-Formed Straight Channels in Sandy Beds* — Jour. of the Hydr. Div., ASCE, Vol. 107, No. HY4, pp. 389 ÷ 406, 1981.

[27] ROSSI, F. , SILVAGNI, G.: *Regime of ephemeral braided streams* — Excerpta, Vol. 2, pp. 135 ÷ 149, 1987.

[28] WHITE, W.T. , BETTESS, R. and PARIS, E.: *Analytical Approach to River Regime* — Jour. of the Hydr. Div., ASCE, Vol. 108, No. HY10, 1982.

[29] CHANG, H.H.: *Modelling of River Channel Changes* — Jour. of the Hydr. Eng., ASCE, Vol. 110, No. 2, pp. 157 ÷ 172, 1984.

[30] SONG, C.C.S. , ZEN, Y. and YANG, C.T.: *Some computational aspects of USBR's GSTARS model* — Proc. XXII Congress IAHR, Topics in Hydr. Mod., pp. 144 ÷ 149., 1987.

[31] ENGELUND, F. and FREDSØE, J.: *Hydraulic Theory of alluvial rivers* — Advances in Hydroscience, Vol. 13, pp. 187 ÷ 215, 1982.

[32] ANDREWS, E.D.: *Entrainment of gravel from naturally sorted riverbed material* — Geol. Soc. Amer. Bull., Vol. 94, pp. 1225 ÷ 1231, 1983.

[33] DE VRIES, M.: *River-Bed Variations - Aggradation and Degradation* — International Association for Hydraulic Research, International Seminar on Hydraulics of Alluvial Streams, New Delhi, India, 1973.

[34] OLESEN, K.W.: *A numerical model for morphological computations in rivers with non-uniform sediment* — Rep. R/1981/6H, Dept. of Civ. Eng., Delft Univ. of Tech., 1981.

DIVERSION STRUCTURE FOR THE "VALABRES" WATER PROJECT ON THE TINEE RIVER - FRANCE

REINTERPRETATION OF 1953 PHYSICAL MODEL STUDIES IN LIGHT OF CURRENT KNOWLEDGE ON GRAIN SORTING

Maurice Bouvard

Institut de Mècanique de Grenoble, Grenoble, France

ABSTRACT

Findings from a 1953 report by A. NIZERY and G. BRAUDEAU are summarized, and reexamined from the present state of knowledge concerning grain sorting. In their comments on the importance and expected effects of grain sorting, these authors appear to have been remarkably ahead of their time. Their execution and interpretation of physical model studies also demonstrates they were aware of the essential role played by the transience of streamflow. They suggested that two flow rates, the limiting discharges for the finest and coarsest material, should characterize the importance of sorting in a given river. We further discuss their findings, and compare them to results from current numerical models.

I/ THE PROBLEM. THE PHYSICAL MODEL.

This relatively modest structure was destined to feed a hydroelectric plant to be built in the Southern French Alps, of 30 MW installed capacity. The maximum diverted flow was only 15 m3/s. Depite this it was decided to build a physical model of the structure, given the benefits that this technique (relativeley new at that time), together with new knowledge in the field of bedload transport, could be expected to provide. Another justification was provided by the great number of similar projects being built at that time. The model was generously designed at a scale of 1/12.5, and was non-distorted, so as to allow the maximum amount of information to be gathered; the scale allowed the size distribution of the bed material to be well represented under proper modelling conditions despite the wide range of particle size: d90 was about 500 mm, while d50 was 135 mm and d10 was on the order of a few millimeters. The limiting discharge was only 10 m3/s, and so was apparently less than the maximum diverted flow. In fact this very low value resulted from the definition of limiting discharge as that capable of moving any particles, however fine, rather than the discharge that would move particles lying at some given point in the size distribution curve. The slope of the river was roughly 10%.

Consequently, the size of bed particles covered a sufficiently broad range that marked grain sorting was immediately noticed during model studies, its importance depending on the flow rate being simulated in a given trial. Most of the results obtained in this connection were discussed in a paper given by A. NIZERY and G. BRAUDEAU at the fifth IAHR Congress (September 1953, Minneapolis), "Variation of the size distribution of bedload material along a stretch of river." A rereading of this report holds great interest, now that 35 years' worth of data collected from model or field studies has provided (despite the absence of well-defined measurements in natural rivers) a substantial body of literature, and hence the basis for some serious debate. From this, at least one commonly held conclusion seems to have emerged; bedload formulae, and the model studies themselves, lead to significant overestimates of the quantity of transported sediment.

Another technique, that of numerical modelling, has by now established its importance and potential. Although the basis for these models can sometimes appear questionable, the limitations on their use and the difficulties encountered in their implementation are far less than is the case for physical models (some would say excessively so). In particular, the computer analysis of bedload transport already performed by a good number of workers, notably in Grenoble, seems to clearly demonstrate that the phenomenon of grain sorting plays a very important role in one's very conception of bedload transport. This conclusion arises independently of the particular algorithms used for a given model, even if they are quite rough.

The frequent remarks by the authors of the 1953 paper on the physics and the consequences of grain sorting show they anticipated the importance of the phenomenon. The uncertainties that have emerged during 35 years of discussion about bedload transport, results obtained in the field and on physical models, and the initial conclusions from numerical modelling work, both complement and demonstrate the basic soundness of their remarks, thus making this document a useful starting point from which to pursue an analysis of a subject too long ignored.

II/ INITIAL RESULTS FROM THE MODEL TESTS

1 Calibration of the model

From the start, particular phenomena were noticed. The calibration of the model had been planned according to the usual method. It was considered that the rate of sediment transport associated with a given streamflow was that which gave the same slope as the river, as well as an identical size distribution of bed material. By varying the discharge of water one would therefore find a well-defined relationship between bedload transport and streamflow, the existence of which was the fundamental notion in the theory of bedload transport at that time. However, it was quickly noticed that the response of the model was more complicated. It was impossible to obtain the natural slope if the size distribution of material introduced upstream was the same as that of the bed, and if it did not vary with the water flow.

It was therefore necessary to consider variations in the bedload size distribution, not usually accounted for in mobile bed models, in defining the bedload transport relationship. But the size distribution seemed a very unwieldy variable, and it had to be studied how to determine the grain size distribution associated with a given streamflow. It was assumed that the appropriate size distribution in the model followed from a process of iteration; the distribution associated with a given flow rate was defined by the equilibrium condition that material recovered at the outflow, when reintroduced upstream, resulted in a stable bed and water stream, for which the net exchange between the sediment charge and the bed was obviously nil.

Important observations were made during these trials, first of a qualitative nature:

- "A flow rate of 50 m3/s (five times the limiting discharge for bedload movement) transports a mixture whose mean diameter is not yet a third that of the bed material."
- "At the maximum discharge tested, the d90 for the transported material (12 mm) is still not quite the average diameter of the bed material."
- According to the report, the bedload transport is characterised by two values for limiting discharge. The lower of these, Q1, corresponds to the finest material, defined for example by d10 , which is about 3mm in this case. The higher value Q2 is appropriate to the coarsest material, defined by d90.

Not all rivers behave in the same fashion with regard to the relative difference between these two values of limiting discharge. The theory of bedload transport had been established for the case of rivers on plains, where the grain size distribution is narrow, thus leading to little difference between the two extreme limiting discharges mentioned above. Furthermore, natural discharges are usually much larger than these critical values, thus overshadowing any difference between them.

By contrast, "things will be very different in the case of a torrential river, whose bed will include an extremely wide range of particle sizes, this being due as much to the diversity and proximity of the contributing zones as to the variability of discharge. Here we will find fine and coarse sand as well as boulders several tens of centimeters in diameter. The largest material can then only be transported by exceptionally high flows corresponding for example to the annual flood, or even to the five- or ten-year flood. Most of the transported material will therefore be moved by flows lying between Q1 and Q2, and the phenomenon may thus depend on laws very different from those established in the laboratory. We will almost always find ourselves in the regime of partial transport, and not that of general bedload movement."

The authors also do not fail to evoke "armoring, whose existence and, in certain cases, preponderant influence is well known."

Quantitatively, the following formulae were proposed:

$$d = 18*10^{8}\ \tau^{4}$$

$$g = 7.3*10^{4}\ (d^{1/4} - d_{o}^{1/4})^{4}\ (\tau - \tau_{o})^{3/2}$$

where d represents the average diameter of transported material, and g is the discharge of sediment given in M.T.S. units.

These formulae show that the size of transported material varies greatly with stream discharge, as in the graph below.

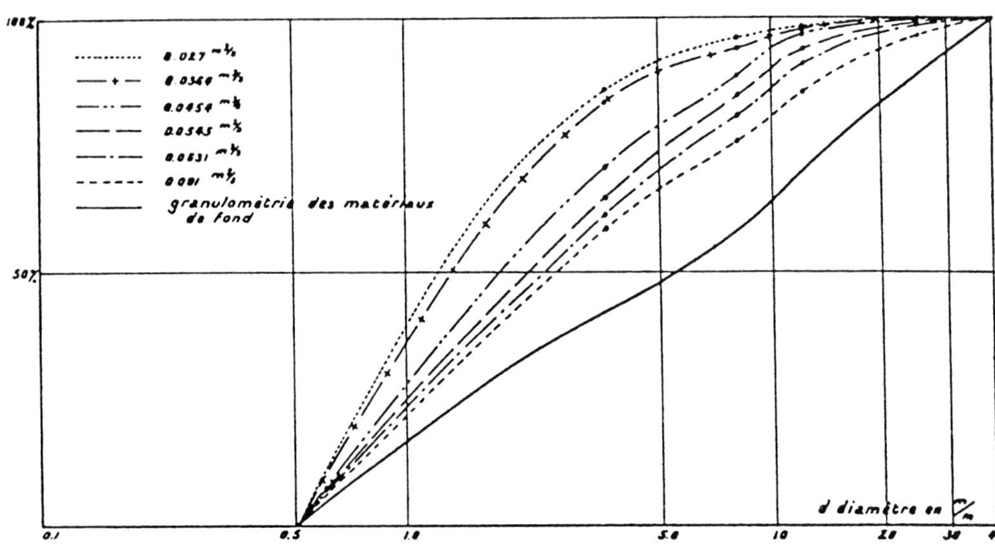

Size distribution curves for transported and bed material

The model tests also demonstrated important secondary effects:

- The interaction between the spatial distribution of shear force on the river bed and the range of transported particle sizes. The boundary shear stress can only be considered uniform for a flow in a rectangular channel, (and even then one can expect significant fluctuations with time). The size of transported material will therefore vary along the cross section of a river.

- The "very wide scatter, in the zone corresponding to low bedload transport, around the (bedload versus flow) curve proposed by Professor Meyer-Peter, which has the effect of almost completely hiding the influence of the variation of the diameter of transported material."

- The variation of boundary roughness as a function of the composition of bed material, and its effect on the quantity of bedload sediment transport.

It is important to note however that partial bedload transport was always considered as a completely steady, spatially unvarying phenomenon.

III/ CRITICAL ANALYSIS OF THE RESULTS FROM THE PHYSICAL MODEL, IN LIGHT OF RESULTS FROM CURRENT NUMERICAL MODELS.

A) Fundamental assumptions and limitations for the model studies.

In practice these model tests, although in principle faithful to the natural situation, are limited by certain technological constraints, even if these do not appear to lend themselves to extensive discussion.

First of all, they can only be used to treat steady-state situations. Usually these models immediately recycle the material leaving the downstream end of the model, and reinject it upstream. This was in fact the solution employed for these studies. The model therefore represents a section of river under steady-state conditions. To our knowledge, no model has ever been equipped with a sorting mechanism by which the size distribution of material introduced upstream could be varied as a function of flow rate, which would avoid the necessity for sediment recirculation. It seems likely that considerations of cost, and the complications potentially introduced in model operation, would explain why these questions were not pursued further, to the point of influencing the methods employed in physical model studies. Besides, no particular requirement seemed to justify such effort, in the absence of results showing it to be necessary.

This limitation to the steady case is confirmed several times in the report. Regarding in particular the assumption introduced in calibrating the model (identical composition of entering and exiting bedload, stability of the streambed), it states "(this assumption) permits one to render the state defined by flow rate, bed load, composition and configuration on the bed, independent of the history of these different factors. This method practically eliminates the influence of the history of the bed, and so allows only the study of steady bedload transport phenomena."

As we shall see, the transitory nature of bedload transport caused by sorting, and the dependence of sediment transport at a given moment on the history of flows in a river, are of fundamental importance. The authors of the report indeed cited the important results concerning these aspects obtained by the Swiss Federal Water Service on the Aar river in 1939, where widely varying values of bedload transport were observed for the same value of stream discharge.

This is rather obvious, at least for certain size fractions. If there is fine sediment at the bed surface for a flow rate slightly greater then the limiting discharge appropriate to the finest material (about 10 m3/s), then it will surely be transported, at least initially. But it is clear that the volume of material thus available will be fairly small; this sediment will be protected from entrainment even at very shallow depths below the bed surface, and the solid transport will soon cease through lack of source material.

The essential point is to know how great an influence this phenomenon will have. This will be determined by the volume of fine material that is mobile during partial transport, whence we can deduce the time during which the flow can be fed by transportable solid material. This duration must then be compared with the return period of flows capable of entraining coarser material.

But the physical model cannot reply to this question which, by referring to transitory phenomena, lies outside the model's range of application.

B) Results of current numerical models

The discussion offered by the authors of this report was certainly pushed to the extreme limit of what one could imagine in the framework of steady-state phenomena. In any case, another characteristic of the physical model comes into play; for several reasons, it is necessarily (very) short compared to the lengths of the stretches which extend both up- and downstream of the section actually represented in the model. Now, when in a transitional regime, there must be storage occurring in one zone or another. Including these several zones would require an extended length of the river to be modelled, which is not feasible for several reasons: available area, scale of the model versus sediment sizes, cost of the model and of the studies. It has therefore been necessary to wait for numerical models (3) in order to study transitory effects in bedload transport, such as the degradation of the Rhine downstream of the Alsacian plain, due to the cutoff of sediment from upstream and to its canalization (which considerably increases its transport capacity).

We now describe in a simplified way the basic idea behind the numerical models:

- The sediment transport itself is obviously modelled on the basis of results from experiments on physical models. Use is thus made of the Shields diagram, the Meyer-Peter laws, etc.

- On the other hand, all transitory aspects of transport are simulated numerically. It is therefore possible to consider the effect of storage and depletion of material in the bed, and the evolution of phenomena over practically unlimited lengths of river.

Partial bedload transport is now treated by dividing the sediment size distribution into classes, each having a representative particle size, to which a well-defined critical shear stress is assigned.

One notes results which, for simple qualitative reasons, seem very robust with respect to the basic assumptions. The outcome will clearly depend on the width of the particle size distribution under consideration. But in this connection the error cannot be considerable since we are basing ourselves on the results of physical studies (which could be those represented by the empirical formulae mentioned above) almost without interpretaion. One can certainly debate the thickness appropriate to the layer of material from which the finest material can be extracted for partial bedload transport; d90 is about 450 mm, and it is hard to see how the fine material can emerge from the bottom of a layer on the order of 1.5 meters thick (this being two or three times the coarsest particle diameter, as typically assumed in these numerical models) to be transported by the flow. It would not seem absurd to consider that the 'exchange layer' might even have a thickness less than 450 mm, given the imbrication of the different sized particles.

Numerical modelling of the phenomenon very quickly shows that one must account for a new parameter, the surface grain size distribution over the length of river being studied. This in turn depends on the history of flow rates, reflecting the essentially transitory nature of the problem. We should note though that the experimental determination of this new parameter in the field would pose considerable problems, which the classical technique did well to ignore; with the background of uncertainty against which it was undertaken, there was no need for a new parameter, even more difficult to measure, or even to define, than those usually considered.

C) Qualitative conclusions.

Concerning the size of the drainage basin and the importance of grain sorting, it is interesting to compare the conclusions of numerical studies with those presented in the 1953 report.

Armoring plays a capital role, and none of its aspects can be ignored, especially regarding the transitory nature of bed load when the grain size distribution is as wide as is the case here. The report specifies that "each combination of flow rate, bed load, grain composition and configuration (slope) of the bed (given by the model, was) independent of the history of these different factors" (cf. above), given the steady character of the process represented. The insistence with which the authors express their reservations confirms that they were well aware of the importance of the transitory nature of this effect, and of the doubt cast on the value of the results given by the physical model.

By contrast, for a large river on a plain, "the range of natural flow rates is very great compared to the difference between the limiting discharges" (for the extreme particle sizes found in the river bed, as the size distribution is narrow), so that "the scatter that might result in a very limited zone of the curve barely adds to the great uncertainty inherent in estimates of the initiation of particle movement."

This last paragraph defines a parameter of general importance in sediment sorting, notably in the extent to which it affects large rivers.

- On the amounts transported.

Many experimental results confirm that the usual bedload formulae significantly overestimate the quantities of transported sediment. This is particularly the case for mountain streams. Reference (1) mentions overestimates by a factor of one to eight in the experimental case there considered, and attributes at least part of this to armoring. Numerical models, based on perfectly plausible assumptions, easily confirm this result.

- On the existence of a biunique bed load law.

If one admits that the stream discharge and the bedload sediment discharge are related by a well defined law, one arrives at the conclusion that the diversion of a certain flow from a stream should systematically result in serious imbalances, and notably in a rise in the bed level downstream of the diversion. Hydraulic engineers were in fact rather worried in this connection around 1950, shortly after the appearance of the first theories of this kind. These concerns have since disappeared completely, since no difficulty of this kind has ever been encountered in the course of several decades' development of surface water diversion projects.

IV/ OUTLOOK FOR FUTURE DEVELOPMENTS

A) The present situation

Grain sorting therefore constitutes a very important phenomenon, to the point of significantly modifying the very conception of bedload transport. It therefore poses tough questions, if only by introducing new parameters (which are moreover difficult, if not impossible, to measure) intervening in the physical process.

- One satisfying observation; the transported quantities turn out to be lower than was anticipated, and part of the explanation for the difference between predictions of empirical formulae and experimental results has been found.

- For mountainous drainage basins, of limited area and stream discharge, the basic mechanisms of bedload transport are now better understood. But the applicability of bedload formulae in their classical form has lost all credibility. We can be consoled that their use leads to overestimates, as mentioned above. But it remains to fill this void by developing new methods within the framework of the phenomena discussed here.

- In rivers with large discharges, not having beds formed of sand nor silt material, it is still unknown if sediment sorting remains important; if, due to the narrow grain size distribution, the difference between limiting discharges for the finest and coarsest material is much less than typical natural discharges, then the influence of sorting would be diminished.

B) The necessity for further basic study.

A new look at bedload transport theories now appears unavoidable. This will necessarily require great effort in several areas of research.

- The quality of the discussion of the physics in the interpretation of results from the Valabres model shows this kind of model will remain very useful. The authors certainly came very close to conclusions well in advance of techniques in practice at the time, and one might even argue that their remarks have renewed value now that consideration of armoring is entering common engineering practice. On the other hand, it would be very useful to develop new, more refined techniques in the operation of mobile bed models, and to go further in interpreting their results.

- Research to be undertaken should particularly concentrate on the transitory aspect of these phenomena, which are doubtless beyond the scope of physical model studies.

- Also, linking results from physical and numerical models will be indispensable. In this area, studies intended to establish the fundamental algorithms of physical modelling, such as those performed by J. Gessler at Zurich from 1965 onwards, will have to be further developed.

- One must not forget experiments on natural rivers. The studies on the Aar in Switzerland represent the only available documents on this aspect, but they are very important because they were the first, and almost the only, to prove that the concept of a biunique law relating stream discharge and bed load did not have a basis in reality for the case they studied.

- In any case, a series of lengthy and difficult research projects is necessary. One can ask if a program of international cooperation such as this one could not provide essential advantages, either by allowing costs to be shared (and so reduced), or by facilitating the exchange of ideas in this very complex domain. Perhaps the F.H.C. could play a fundamental role in establishing such a structure, at least concerning the aspect of intellectual exchanges.

Within certain limits, our ambitions for prediction of bedload transport can only be limited. The accuracy conceivably obtainable is certainly rather low, at best 50%, though incidentally this would be quite acceptable for engineering purposes. But we should no longer have the pretention to present, as has occurred in the past, results from field experiments, formulae, and physical model studies, which agree to within a few percent.

BIBLIOGRAPHY

(1) H. LAUFFER & N. SOMMER. Studies on sediment transport in mountain streams of the Eastern Alps. International Commission on Large Dams, 14th Congress, Rio de Janeiro, 1982

(2) D. JEANPIERRE. L'érosion des fonds, conséquence secondaire de la protection de la plaine d'Alsace par l'endiguement longitudinal du Rhin. Xèmes Journées de l'Hydraulique, Paris, 1968

MOBILE ARMOURING OF BED SURFACE IN STEEP SLOPE RIVER WITH GRAVEL AND SAND MIXTURE.

Koichi Suzuki and Koichi Kato

Department of Civil Engineering,
Ehime University, Matsuyama, Japan

ABSTRACT

Characteristics of mobile armoring, which is the sorting of surface sand on a river bed in a state of dynamic equilibrium are discussed through experiments on a steep slope channel with a gravel and sand mixture bed. The bed surface with a gravel and sand mixture is usually flat and the bed surface sand is proved to be coarser than the transported sand, even if all the particles on the bed surface are moving in an equilibrium state of sediment discharge. Also a method for the estimation of grain size composition of the bed surface is proposed.

1. Introduction

Mountain rivers are usually steep in slope and their bed consists of a gravel and sand mixture. In these rivers, the bed surface is often covered with coarse sand and gravel. As Kellerhals & Bray[1)] pointed out, at a low to intermediate stage of water discharge, a distinct pavement of the bed with a gravel layer one grain thick is formed in the case of a gravel sand bed. In most gravel beds the grain population making up the surface layer is different from the population of the underlying material, due to paving effects.

So called armoring phenomena have been discussed extensively, starting with Gessler's work[2)] under the hydraulic conditions that the sediment discharge is nonequilibrium and the bed is almost stable, which may be called " static armoring". On the other hand, Parker et.al[3),4)]suggested the possibility of " mobile armoring" which is the vertical sorting of bed surface sand in a dynamic equilibrium condition of sediment discharges. Recently, Kuhnle & Southard[5)] and Ikeda & Iseya[6)] also discussed experimentally the mobile armoring and showed the fluctuation of sediment discharges with time on the bed with a gravel and sand mixture. Suzuki & Michiue[7)] found that the bed surface sand becomes coarser than the transported sand mixture even if the sediment discharge is in an equilibrium condition where only fine sands are moving and some coarse particles stand still, which may be called "semi-static armoring". The above mentioned armoring phenomena can be classified as in Table 1.

Table 1 Classification of armoring phenomena.

armoring	sediment discharge	movement of coarse sand	direction of sorting
static	nonequilibrium	no	horizontal
semi-static	equilibrium	no	vertical
mobile	equilibrium	yes	vertical

In the case of static armoring, all the sand particles on the bed surface stand still at the final stage after the long sorting process along the longitudinal direction of the channel. But mobile or semi-static armoring is assumed to be obtained in a relatively short time because the sorting occurs only vertically near the bed surface.

In this paper, the mobile armoring phenomena of the bed surface in a steep slope channel with a gravel and sand mixture of a widely distributed grain size is discussed both experimentally and theoretically.

2. Summary of Experiments

Experiments were carried out in order to know the composition of the bed surface sand under dynamic equilibrium conditions of sediment discharges where all the sediment particles are moving.

The flume used for the experiments was 10m long and 40cm wide, and its slope was adjustable within the range from 0 to 1/5 by raising the upstream end of the flume with a chain block as shown in Figure 1. The gravel and sand mixture used in the experiments has a mean diameter d_{mo} of 0.414cm and a standard deviation $\sigma(=\sqrt{d_{84}/d_{16}})$ of 3.00, and the grain size accumulation curve is shown by Figure 2.

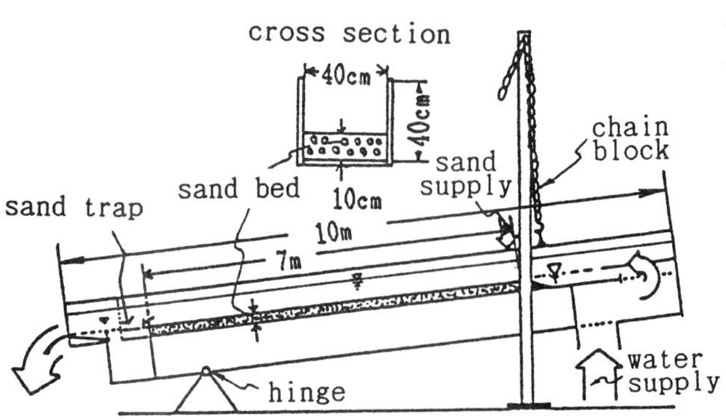

Fig.1 Experimental flume.

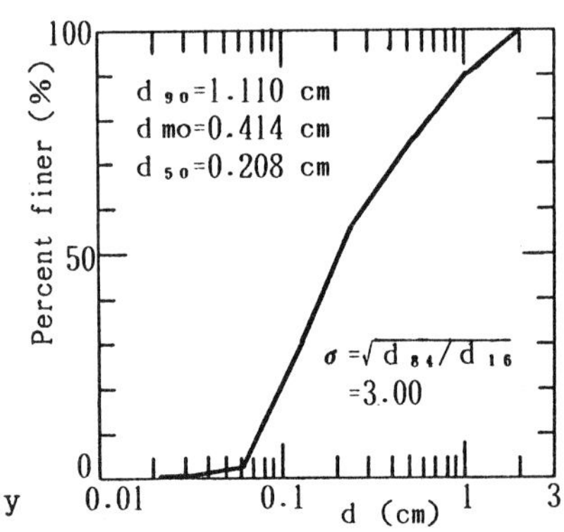

Fig.2 Grain size distribution curve.

Experimental procedures are as follows: The gravel and sand mixture which is the same as that of the channel bed was continuously supplied manually from the upstream end of the flume, adjusting the sediment supply rate so as to keep the bed slope constant, until the transported sediments measured at the end of the flume became almost equal to the supplied sediments both in volume and in size. After achieving a bed in dynamic equilibrium, sands of the bed surface layer were sampled with a thickness of the maximum grain size from the top of the most exposed gravel.

Experimental conditions are shown in Table 2. For each bed slope of 1/100, 1/50 and 1/30, flow discharges were changed by five steps under the condition that sand particles of every grain size on the bed surface can be transported. The total number of experimental runs is 15. Resulting bed forms of all the experimental runs were planes which may be called bed load sheets[5]. Resulting mean diameters and standard deviations of the grain size of the bed surface are also shown in the table.

Table 2 Experimental conditions and results.

Run No.	Bed slope	Flow discharge	Flow depth	Sand mixture	Bed forms	Bed surface sand	
						Mean diameter	Standard deviation
	I	q (cm²/s)	h(cm)			dm(cm)	$\sigma=\sqrt{d_{84}/d_{16}}$
1	1/100	400	5.56	mean diameter dmo=0.414cm standard deviation $\sqrt{d_{84}/d_{16}}$=3.00	plane (bed load sheets)	0.603	3.47
2	〃	450	5.92			0.589	3.50
3	〃	500	6.42			0.597	3.62
4	〃	600	6.92			0.409	3.16
5	〃	700	7.64			0.367	2.95
6	1/50	200	3.12			0.466	3.37
7	〃	225	3.50			0.496	3.39
8	〃	275	3.90			0.518	3.39
9	〃	300	4.08			0.440	3.28
10	〃	325	4.40			0.419	3.01
11	1/30	100	1.94			0.412	3.24
12	〃	125	2.40			0.414	3.24
13	〃	150	2.70			0.394	3.08
14	〃	175	3.12			0.348	2.80
15	〃	200	3.44			0.416	3.23

3. Experimental Results

3.1 Bed form

Dimensionless bed shear stresses $\tau_{*mo}(=u_*^2/sgd_{mo})$ were changed from 0.08 to 0.15 in the experiments and all sizes of sand particles on the bed surface were transported, where d_{mo} is the mean diameter of the supplied sand, u_* is the friction velocity, $s=\sigma/\rho-1$, σ and ρ are the densities of sand and water, respectively, and g is the acceleration due to gravity. Although all the particles on the bed surface were thus moving actively, no obvious sand waves were formed. Kuhnle & Southard[5] pointed out that sand waves are not easily formed and the bed is usually flat in channels with a gravel and sand mixture. He called this plane bed with an active movement of all the particles on the bed surface as "a bed load sheet", showing his experimental data of flat beds for the values of τ_{*mo} less than 0.22. Figure 3 shows the Garde-Raju diagram[8] for the classification of bed forms and data obtained by the experiments in which all the beds were flat, where I is the bed slope, R is the hydraulic radius and d is the sand diameter. A diagram proposed by Ashida & Michiue[9] which classifies bed forms with τ_* and R/d is essentially identical with the Garde-Raju diagram (Figure 3) which shows the relationship between I/s and R/d, because $\tau_*(=u_*^2/sgd)$ can be written by RI/sd. In the case of a gravel and steep slope bed as often seen in mountain rivers where the sand diameter is relatively large and the water depth is small, the values of R/d are usually small. As shown in Figure 3, ripples and dunes can not be formed for the value of R/d less than about 10. If the classification lines are extended to the region of smaller values of R/d and larger values of I/s, all the experimental data fall in the transition region when the mean diameter of the supplied sand mixture is used as a representative sand diameter. If fifty percent diameter d_{50} is used instead of the mean diameter as a representative diameter, data in the diagram move towards larger values of R/d and some data fall in the anti-dunes region which does not coincide with the observed data. It may be concluded that for a small value of R/d the bed form is usually flat and can be estimated by the Garde-Raju diagram of the bed form classification, using the mean diameter as a representative sand diameter even if the bed is composed of a gravel and sand mixture.

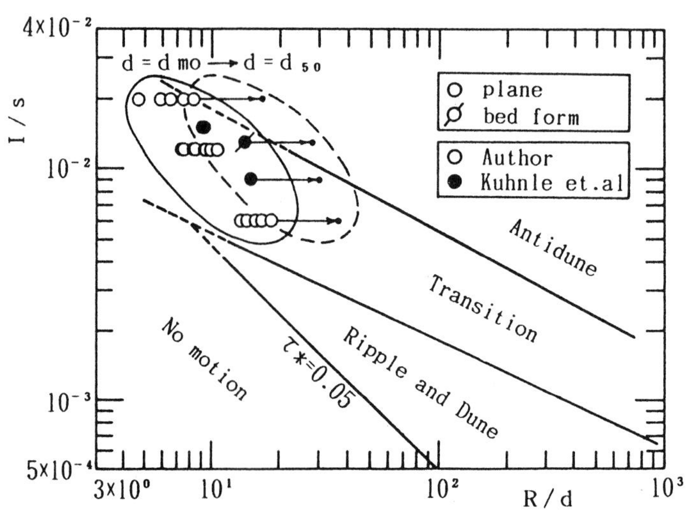

Fig.3 Garde-Raju's diagram of bed form classification.

3.2 Grain size composition of bed surface

Figure 4 shows the resulting grain size accumulation curves of the bed surface in the experiments under the dynamic equilibrium conditions of sediment discharges. As shown in Figure 4(a) for I=1/100, bed surface sands are coarser than the transported (supplied) sand mixture for relatively small bed shear stresses such as τ_{*mo}=0.081, 0.087 and 0.094, whereas they are almost the same as the supplied sand mixture for much larger bed shear stresses than the critical shear stress of the bed sand such as τ_{*mo}=0.101 and 0.112. In the case of I= 1/50 as shown in Figure 4(b), bed surface sands are slightly coarser than the supplied sand mixture for τ_{*mo}=0.091 and 0.114, but they are not different from the supplied sand for relatively large bed shear stresses such as τ_{*mo}= 0.102, 0.120 and 0.129. From Figure 4(c) for I= 1/30, it can be seen that bed surface sands do not become coarser than the supplied sand mixture for extremely large bed shear stresses. From the above mentioned experimental results, it can be concluded that the obvious mobile armoring occurs only when the dimensionless bed shear stress τ_{*mo} is less than about 0.10 and the bed surface sand becomes coarser as τ_{*mo} becomes smaller. But, according to the experiments by Kuhnle & Southard[5] as shown in Figure 5, this criteria of τ_{*mo} for the mobile armor coat formation is somewhere between 0.136 and 0.178 which is much larger than 0.10. This difference may

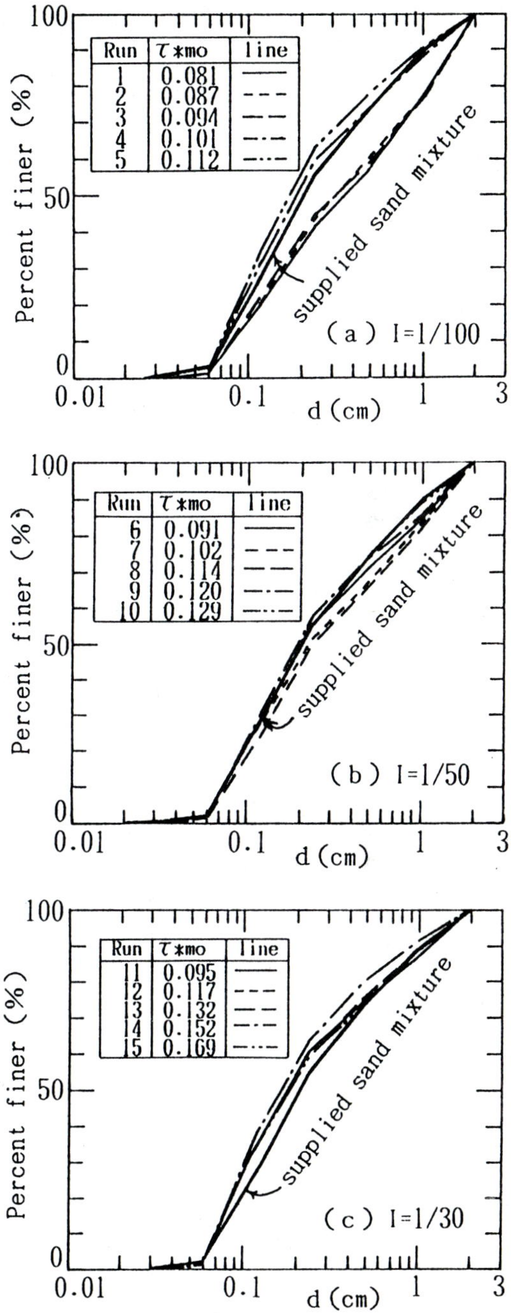

Fig.4 Grain size accumulation curves of bed surface layer.

result from the different sampling method of the bed surface sands: Kuhnle et.al sampled bed surface sands by a piston sampler 13cm in diameter, which was coated with a stiff mixture of clay and water and when pushed into the bed, picked up essentially only the surface layer of grains. Small sand particles in the space beneath exposed large grains may not be picked up by this kind of sampler, resulting in more armoring than sampling by a spatula. Mean diameters of the bed surface sands d_m obtained from the grain size accumulation curves in Figure 4 are shown versus bed shear stresses in Figure 6 in a dimensionless form. The curve in the figure is a theoretical curve discussed later. Although there is a large spread in the data, there is little data of d_m/d_{mo} slightly larger than unity with the larger value of τ_{*mo} of about 0.12. As τ_{*mo} becomes smaller than this value, d_m/d_{mo} becomes larger than unity which means that bed surface sands become coarser.

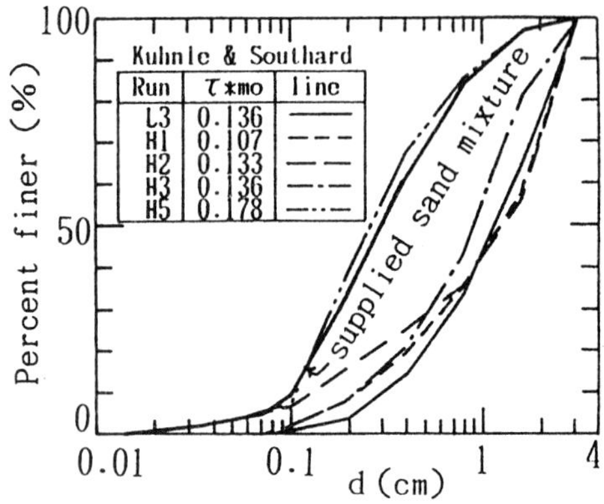

Fig.5 Grain size accumulation curves of bed surface. (Kuhnle et.al[5])

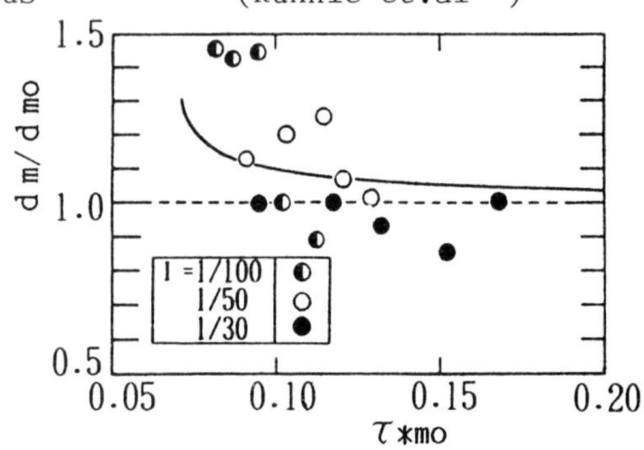

Fig.6 Variation of mean diameter of surface sand with bed shear.

4. Hypothesis on Mechanism of Armor Coat Formation

Here the semi-static or mobile armoring which is the vertical sorting of bed surface sands in an equilibrium state of sediment discharges will be considered. For the semi-static armoring, Suzuki & Michiue[7] proposed the hypothesis on the sorting mechanism that the armor coat is formed by trapping smaller particles beneath the larger ones which stand still under the hydraulic conditions of bed shear stresses smaller than the critical shear stress of the larger particles. This hypothesis is based on the concept of hiding effects of the smaller particles by the larger ones. Figure 7 shows a schematic figure of an active layer on the bed surface which may be divided into two layers: one is a transport layer and the other an exchange layer. The transport layer is exposed directly to the water flow and sand particles smaller than the critical size d_c in the layer are moving mainly toward downstream. In the exchange layer, the smaller particles are sheltered by the larger ones but sometimes move vertically and are exchanged with particles in the transport layer. In the case of this semi-static

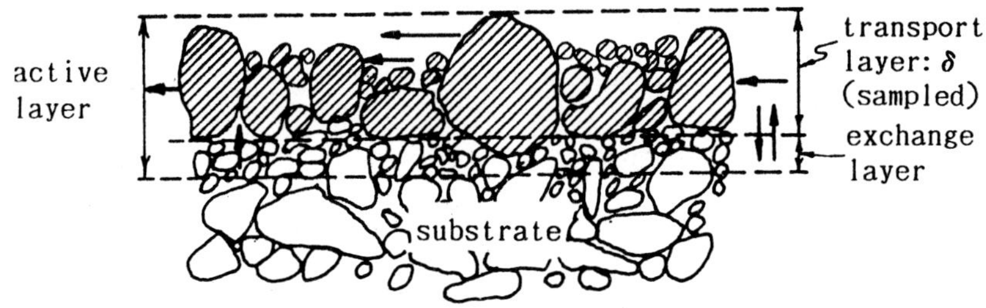

Fig.7 Schematic figure of bed surface sands.

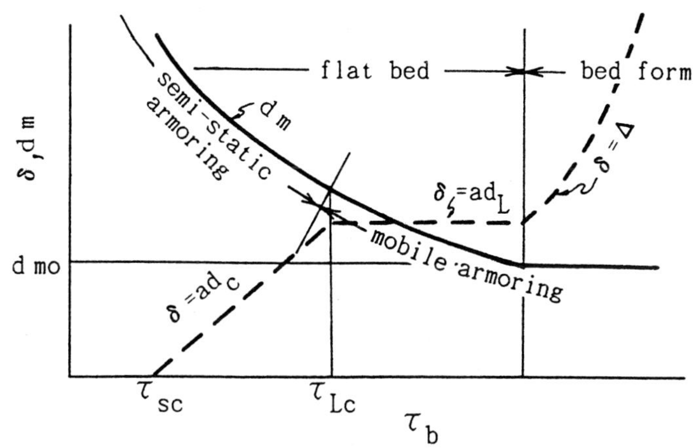

Fig.8 Schematic figure of mean diameter of bed surface sands d_m and active layer thickness δ versus bed shear stress.

armoring, the thickness of the transport layer δ where sand particles are moving mainly toward downstream may be assumed to be proportional to the largest size d_c among the moving particles($\delta=ad_c$, a:proportional constant). As d_c is proportional to the bed shear stress τ_b in the flow range of the constant Shields' parameter, δ can be assumed to be proportional to τ_b until τ_b exceeds the critical shear stress τ_{Lc} of the largest sand particle d_L or until bed forms such as sand waves appear. The transport layer thickness seems to be almost constant when τ_b is larger than τ_{Lc} as long as the bed is flat. Furthermore, it will be equal to the sand wave height Δ if sand waves are formed with larger value of τ_b. A schematic figure of this concept is shown by a broken line in Figure 8.

According to the above mentioned hypothesis for semi-static armoring, the mean diameter of the bed surface sands which are exposed in the transport layer become larger as the bed shear stress becomes smaller as shown by the solid line in Figure 8. Even in the case of mobile armoring in which all sizes of particles on the bed surface are moving, these hiding effects of the smaller particles by the larger ones can probably still remain if the bed shear stress is not so much larger than the critical shear stress of the largest sand particle, because selective transport effects among different sizes of particles are assumed to remain. This means that the hypothesis proposed above for the semi-static armoring mechanism is also true for mobile armoring and that its essential vertical sorting mechanism is the hiding of the smaller particles by the larger ones.

The mean diameter of the transport layer may be considered as that of the supplied sand mixture when sand waves are formed with relatively large bed shear stress although sand waves are not easily formed on the bed with a gravel and sand mixture as mentioned in the previous section.

5. Grain Size Estimation of Bed Surface Layer

5.1 Method for grain size estimation

In the case of a dynamic equilibrium state of sediment discharges, the difference of grain size composition between sand in the transport layer on the bed surface and the transported (supplied) sand mixture out of the flume end comes from the velocity difference of each size of sand particles. If the finer sand is transported faster than the coarser sand, the composition of the finer sand in the transported sand mixture can be greater than that in the transport layer on the bed surface. This means that if the effect of the selective transport by which the finer sand is more easily transported than the coarser one exists, the coarser sand should be in greater abundance in the transport layer on the bed surface than in the transported sand mixture, which results in the mobile armoring on the bed surface. The selective transport of each grain size in the sand mixture is obvious when the bed shear stress is smaller than the critical shear stress of the coarser sand and larger than that of the finer sand. But selective transporting will almost disappear when the bed shear stress is much larger than the critical shear stress of the coarser sand, because all sizes of sand particles interact with each other and individual grains cannot move independently when all the grains are actively moving.

Sediment discharges of each grain size in a sand mixture may be assumed to be estimated by one of the already proposed formulae for uniform sands, $q_{s*}=f(\tau_*,\tau_{*c})$, modifying the dimensionless critical shear stress τ_{*c} to the dimensionless critical shear stress of each grain size τ_{*ci} in the sand mixture as follows[10]:

$$i_o \, q_{s*i} = i_s \, f(\tau_{*i},\tau_{*ci}) \qquad (1)$$

where i_o and i_s are the frequencies of sand d_i in the transported sediment and in bed surface sands, respectively; $q_{s*i}=q_s/\sqrt{sgd_i^3}$, q_s is the total sediment discharge per unit width, $\tau_{*i}=u_*^2/sgd_i$, $\tau_{*ci}=u_{*ci}^2/sgd_i$, and u_{*ci} is the critical shear velocity of sand d_i. When the surface sands of every particle size are actively moving, q_s can be approximately estimated using the mean diameter d_m of the transport layer on the bed surface as a representative particle size as follows:

$$q_{s*m} = f(\tau_{*m}, \tau_{*cm}) \qquad (2)$$

From Equations (1) and (2)

$$q_s = (i_s/i_o)\sqrt{sgd_i{}^3}f(\tau_{*i},\tau_{*ci}) = \sqrt{sgd_m{}^3}f(\tau_{*m},\tau_{*cm})$$

is obtained. Further, from this equation, the following relationship between i_s and i_o is obtained:

$$i_s/i_o = (d_m/d_i)^{3/2}f(\tau_{*m},\tau_{*cm})/f(\tau_{*i},\tau_{*ci}) \tag{3}$$

where τ_{*cm} can be given by Shields' curve and assumed to be constant. As for the dimensionless critical shear stress τ_{*ci} of sand d_i in the sand mixture with the mean diameter d_m, Egiazaroff[11] proposed the following equation: $\tau_{*ci}/\tau_{*cm}=(\log 19/\log(19(d_i/d_m))^2$. This equation may be correct for the initiation of sand movement, but it is not always adequate when used in the form of sediment discharges as in Equation (1). Based on the discharges of each grain size in the form of Equation (1), Suzuki and Michiue[7] proposed the following experimental equation:

$$\tau_{*ci}/\tau_{*cm} = (d_i/d_m)^{-0.93} \tag{4}$$

Therefore, i_s/i_o can be expressed by a function of d_i/d_m and τ_{*m} only. If Ashida-Michiue's formula of sediment transport

$$q_s/\sqrt{sgd^3} = 17\ \tau_{*e}{}^{3/2}(1-\tau_{*c}/\tau_*)(1-\sqrt{\tau_{*c}/\tau_*}) \tag{5}$$

is used, for $\tau_{*m}>\tau_{*cm}$ Equation(3) becomes:

$$\frac{i_s}{i_o} = \frac{(1-\tau_{*cm}/\tau_{*m})(1-\sqrt{\tau_{*cm}/\tau_{*m}})}{\{1-(d_i/d_m)^{0.07}(\tau_{*cm}/\tau_{*m})\}\cdot\{1-\sqrt{(d_i/d_m)^{0.07}(\tau_{*cm}/\tau_{*m})}\}} \tag{6}$$

$$d_m = \sum_i i_s d_i \tag{7}$$

where the dimensionless effective shear stress τ_{*e} is assumed to be τ_*, because the bed is assumed to be flat. Equation (6) is the relationship between frequencies of sand d_i in the transported (supplied) sand i_o and in the transport layer on the bed i_s. When i_o and the shear velocity u_* are given, i_s can be estimated with Equations (6) and (7). As a first approximation in the estimation, i_s is calculated assuming the mean diameter d_m in the transport layer on the bed surface to be d_{mo} which is

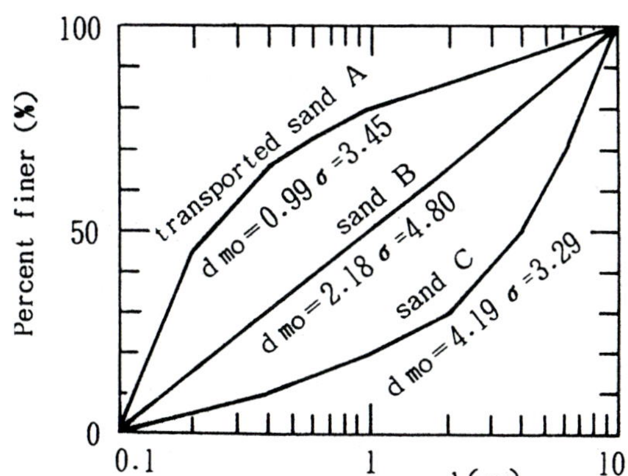

Fig.9 Grain size accumulation curves of model sand mixtures.

the mean diameter of the transported sands. Then the calculated i_s is used to estimate d_m by Equation (7) for the second approximation. This procedure is repeated until i_s does not differ from i_s of the previous approximation. For three types of sand mixtures shown in Figure 9, relationships between d_m/d_{mo} and τ_{*mo} are calculated with $\tau_{*cm} = 0.04$ in Figure 10. Figure 11 shows the effects of τ_{*cm} on the relationships between d_m/d_{mo} and τ_{*mo} for sand B in Figure 9. It can be seen from these figures that d_m/d_{mo} is strongly dependent not only on τ_{*mo} but also on the composition i_o of the supplied sand mixture and τ_{*cm}.

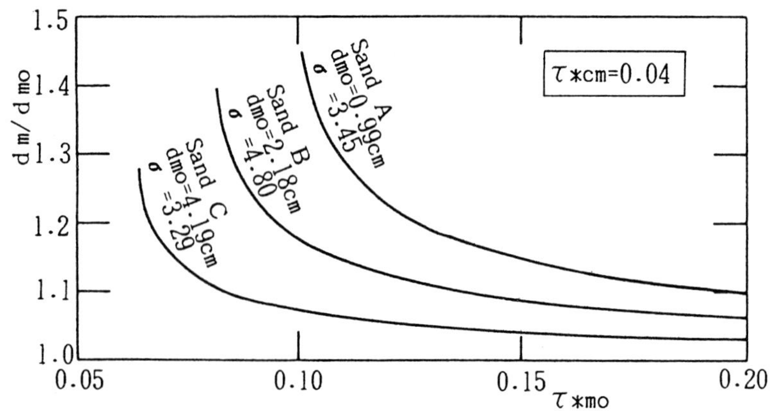

Fig.10 Relationships between d_m/d_{mo} and τ_{*mo} for the model sand mixtures with τ_{*cm}=0.04.

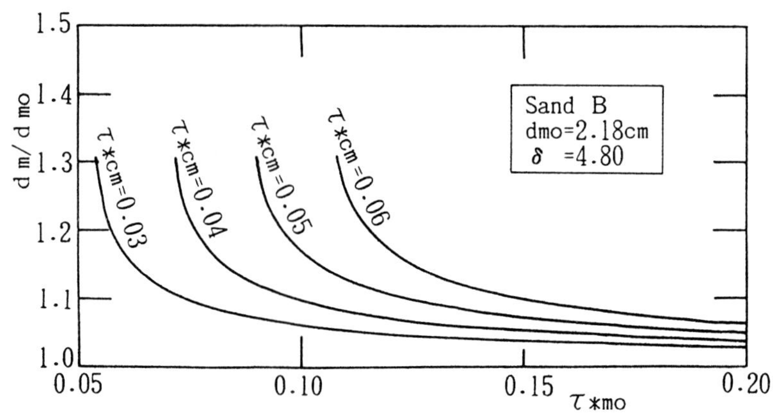

Fig.11 Effects of τ_{*cm} values on the relationship between d_m/d_{mo} and τ_{*cmo} for sand B.

5.2 Comparison with experimental data

Figure 12 shows the calculated grain size accumulation curves of the transport layer on the bed surface for the supplied sediment mixture used in the experiments. This figure corresponds to Figure 4. Although detailed comparison with these figures is difficult because of the large spread in the experimental data, the characteristics of mobile armoring can be seen for smaller values of τ_{*mo}. Armoring phenomena are not so remarkable in the case of mobile armoring ($\tau_{*mo}>\tau_{*cL}$) as in the case of semi-static armoring[7] ($\tau_{*mo}<\tau_{*cL}$). The solid lines in Figure 6 are also calculated by the above mentioned method using the value 0.04 for τ_{*cm}. The estimated d_m/d_{mo} seem to follow the observed data to some extent as shown in Figure 6.

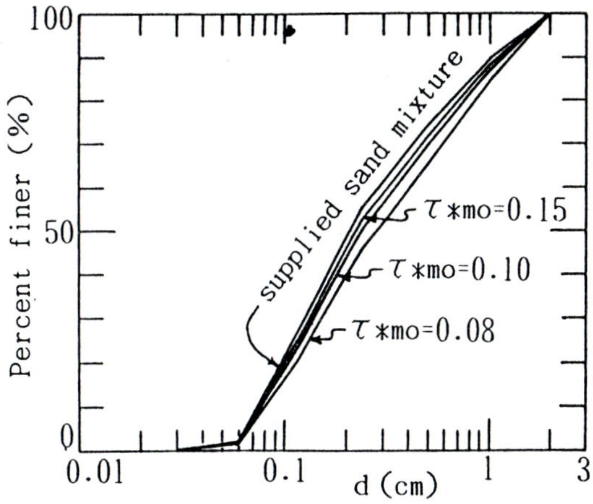

Fig.12 Calculated grain size accumulation curves for sand mixture used in the experiments.

6. Conclusions

The main conclusions about the mobile armoring of the bed surface obtained by experiment are as follows:

1) Sand waves are not easily formed in a channel with a mixture of sand and gravels. And the flat bed form can be estimated in the category of transition on the Garde-Raju diagram with the use of the mean diameter of the supplied sand mixture as a representative sand diameter.

2) When the bed shear stress is moderately large, the bed surface sand becomes coarser than the supplied sand mixture even if all the particles on the bed surface are moving.

3) A method for the estimation of sand composition of the bed surface is proposed and proved to be useful.

References

1) Kellerhals, R. and Bray, D.I.: Sampling procedure for coarse fluvial sediments, Proc.ASCE, Vol.97, HY8, pp.1165-1180, 1971.

2) Gessler, T.: Self-stabilizing tendencies of alluvial channels, Proc. ASCE, Vol.96, WW2, pp.235-249, 1970.

3) Parker, G., Klingeman, P.C and Mclean, D.G.: Bed load and size distribution in paved gravel-bed streams, Proc. ASCE, Vol.108, HY4, pp. 544-571, 1982.

4) Parker, G. and Klingeman, P.C.: On why gravel bed streams are paved, Water Resources Research, Vol.18, No.5, pp.1409-1423,1982.

5) Kuhnle, R.A. and Southard, J.B.: Bed load transport fluctuations in a gravel bed laboratory channel, Water Resources Research, Vol.24,

No.2, pp.247-260, 1988.

6) Ikeda, H. and Iseya, F.:Experimental study of heterogeneous sediment transport, Environmental Research Center Papers, The University of Tsukuba, No.12, pp.1-50, 1988.
7) Suzuki, K. and Michiue, M.: Change of river regime and graded sediment discharge during flood, Proc. of International Conference on River Regime, Hydraulics Research, Wallingford, pp.231-240, 1988.
8) Garde, R. and Raju, K.G.R.: Regime criteria for alluvial streams, Proc.ASCE, Vol.89, HY6, pp.153-164, 1963.
9) Ashida, K. and Michiue, M.: Basic study on flow resistance of alluvial bed and bed load transport, Proc.JSCE, No.206, pp.59-69, 1972. (in Japanese)
10) Ashida, K. and Michiue, M.: An investigation of river bed degradation downstream of a dam, Proc.IAHR,14th Congress,Vol.3,pp.247-255, 1971.
11) Egiazaroff, I.V.:Calculation of nonuniform sediment concentrations, Proc.ASCE, Vol.91, HY4, pp.225-249, 1965.

EXPERIMENTAL ANALYSIS OF ARMOURING PROCESS

Alberto Lamberti [(1)] and Ennio Paris [(2)]

[(1)] Hydraulics Institute, University of Bologna, Italy
[(2)] Department of Civil Engineering, University of Florence, Italy

ABSTRACT

Preliminary results from an experimental investigation on armouring processes are presented.

Particularly, the process of development and formation of the armour layer under different steady flow conditions has been analyzed in terms of grain size variations and sediment transport rate associated to each size fraction.

1. INTRODUCTION

Mountain streams often show a surface layer markedly coarser than the substrate. The characteristics of this surface layer, here indicated as armour layer, are of great importance in the dynamics of mountain streams; particularly, formation and stability of the armour layer reduces considerably both the amount of sediment material delivered downstream and the bed degradation phenomena.

Several aspects of the armouring process have been experimentally investigated:
- the effects on incipient motion and sediment transport related to non uniformity of grains (hiding): Eintein & Chien, Egiazaroff, Gessler '65, Day, Ashida & Michiue '73, Misri & al., Proffitt & Sutherland, Samaga & al., Falcon, Ribberink;
- the relation between the final grain-size distribution of the armour

layer and the characteristics of the parent bed material: Gessler '65, Ashida & Michiue '71, Little & Mayer '72, Proffitt & Sutherland, Saad;
- the erosion associated with the armour layer development: Little & Mayer, Proffitt, Saad unpublished data;
- the temporal evolution of the surface layer and/or tranport during the armouring process: Little & Mayer, Proffitt.

Mathematical models for the interpretation of the phenomenon have also been proposed: Hirano, Bayazit, Deigaard & Fredsoe, Borah & al, Shen & Lu, Karim & Holly, Falcon, Ribberink, Lee & Odgaard, Willetts & al., Palaniappan & Godbole, Armanini & Di Silvio.

Nevertheless, many aspects related to armouring process still remain to be deeply investigated or are open for discussion. Between these we mention: -sampling and analysis techniques of the armour layer, -hiding (sheltering) effect, i.e. the redistribution of the actual and critical (for incipient motion) bed shear stress over grain size, -clustering of the grains, i.e. phenomena not fully represented by the grain size distribution.

Aiming to better understanding the armour layer dynamics under steady and unsteady flow conditions, an experimental work has been initiated at the Laboratory of the "Istituto di Idraulica" of the University of Bologna; particularly, the process of formation of the armour layer under steady flow conditions, the effects of unsteady flow on bed stability and the phenomenon of distruction of the armour layer are experimentally investigated.

In this paper the methods of analysis and the first results on the development of the armour layer under steady flow conditions are presented by analysing the temporal variation of the amount and grain size distribution of the eroded sediments, the changing bed roughness and the eventual size distribution of the armour layer.

2. EXPERIMENTAL SETUP AND PROCEDURE

The experimental equipment consists of a recirculating and tilting flume, 12 m long, 0.30 m wide and 0.50 m deep; side walls and bottom are smooth steel, except the central part where the flume has glass walls.

The flume is supported on a pivot upstream and by two columns of adjustable height downstream, allowing a range of bottom slopes from 0 to 10%.

The maximum recirculating discharge is about 100 l/s.

The testing reach, about 6 m long, is provided with 4 pitot tubes, two point gauges and two piezometers in fixed locations, while a mobile micropropeller apparatus is used for measuring the local flow velocity.

An initial flat bed is placed on the flume bottom of the testing reach, using bed sediment with a grain size distribution obtained by mixing commercially available material. The sediment characteristics are summarized in table 1, where: D_{nn} are the sieve sizes for which nn% of the sampled material is finer; σ is the standard deviation in ϕ units of the size distribution; S_g is the specific gravity of sediments (γ_s/γ , with γ_s and γ , specific weight of sediment and water, respectively) and ε is the porosity. Sediment 1 is normally distributed in ϕ units, sediment 2 shows a bimodal distribution, while sediment 3 is a 1 to 1 mixture of sediment 1 and 2 with flat distribution between 0.7 and 11 mm.

Table 1. - Sediment characteristics

	D_{16} mm	D_{50} mm	D_{84} mm	σ	Sg	ε
1	1.45	2.69	5.27	.90	2.58	0.35
2	1.00	3.68	7.73	1.38	2.59	0.34
3	1.10	2.36	6.54	1.13	2.58	0.35

A uniform flow is obtained for mild slope using a control gate located at the downstream end of the flume; in the case of steep slopes no artificial control was used as the backwater effect were hardly achievable. Armour layers are generated by running sediment-free water through the flume; sediment transport leaving the testing reach is entirely collected in a plastic basket and periodically measured (weight and size distribution); the run is prolonged until a small fraction of the initial sediment trasnport is collected in the basket (generally no more than 5% of the initial transport). To this point the bed is assumed to be armoured.

Unsteady flow conditions are then imposed by using a sudden increase of discharge until the armoured layer appears completely removed.

Three series of runs have been carried out by using the bed material shown in table 1 and bed slope varying between 0.2% to 1% and discharge between 6 and 40 l/s for steady armouring conditions. Only the results of the first series are referred in this note.
Run 1 and run 2 have been carried out under the same hyadraulic conditions in order to test the repeatibility of the experiments. Practically identical results have been obtained by both runs; therefore, only one set of results is here presented and in the following is indicated as run 2.
Run 3 and 4 have also been performed under the same hydraulic conditions, with the difference that run 3 has been partially fed with bed material during the test. As differences were observed, only the results of run 4 are referred (no sediment feed).

During each run, the following measurements have been made:

- liquid discharge, by using a calibrated orifice installed in the recirculating system;
- water temperature;
- piezometric and total head, by two piezometric intakes at the upstream and downstream sections of the reach and by four Pitot-Prandtl tubes located approximately 1.5, 3.0, 1.5 m apart each other;
- velocity profiles, by a movable micropropeller apparatus;
- solid discharge, by periodically collecting and weighting the amount of sediment leaving the testing reach;
- size frequenzy distribution of the eroded bed sediment, by sieve analysis of the collected material;
- bed level variations, by measuring at five sections the bed level with respect to a fixed reference;
- bed surface grain size analysis by photograph technique.

In tables 2 - 6 is a summary of the experimental data referring to the 1st series; symbols are:

S_b , the initial bed slope,
t , run time (hours),
Q , liquid discharge (l/s),
T , water temperature (°C),
h , water depth (cm),
S_f , energy slope,
U_v , mean velocity derived from velocity profiles (cm/s),
U_p , mean velocity derived from Pitot measurements (cm/s),
v_{*vp} , bed shear velocity obtained from velocity profile (cm/s),
v_{*wc} , bed shear velocity obtained from side wall correction (cm/s),
R_b , bed hydraulic radius (cm),

K_{sb} , bed hydraulic roughness (cm),
Q_s , sediment discharge (kg/h);

the suffixes $_e$ and $_a$ for the sediment diameter D_{84}, D_{50} and D_{16} refer to the eroded material and the armoured layer, respectively.

3. ANALYSIS OF EXPERIMENTAL RESULTS

Although the experimental investigation is still in progress, preliminary analysis can be made already, on the basis of the first set of available data.

Due to the low values of the witdh to depth ratio (up to 1 in runs 1 and 2), the side wall effects have been taken into account according to the procedure described below.

Afterwards, the following aspects have been considered:
- grain sizes temporal evolution and its simulation;
- effects of non-uniformity of grains on sediment transport.

3.1 Side wall correction

In order to account for the side walls effects, the following procedure has been adopted:

1) the cross section A has been subdivided to obtain a central region where the bed effects are dominant, and two lateral regions where the effects of wall prevail.
The definition of hydraulic radius and trivial geometric considerations lead to the following equation:

$$R_w \cdot 2h + R_b \cdot B = A \qquad (1)$$

where R_w, R_b are the hydraulic radius ascribed to walls and bed, respectevely, and B is the channel width.

Assuming mean velocity and slope be equal for all the regions of flow, i.e.:

$$U = U_w = U_b \qquad \text{and} \qquad S_w = S_b = S_f$$

we have:

Table 2 - Summary of data from RUN 1. S_b = 0.002

t (h)	Q (l/s)	T (C)	h (cm)	S_f *1000	U_p (cm/s)	U_v (cm/s)	V_{*pv} (cm/s)	V_{*wc} (cm/s)
0.00	40.8	18.8	18.8	2.45	-	-	-	5.4
2.00	"	"	18.9	2.40	-	-	-	5.3
5.00	"	18.9	19.1	2.25	-	-	-	5.2
11.00	"	19.4	19.4	2.15	-	-	-	5.1
13.00	"	19.3	"	2.10	-	-	-	5.1
18.50	"	19.6	19.8	2.08	-	-	-	5.1
21.50	"	19.9	20.0	2.01	-	-	-	5.1
28.50	"	20.0	20.0	2.00	-	-	-	5.1
43.50	"	20.4	20.2	1.95	-	-	-	5.0
59.50	"	20.8	20.3	1.85	-	-	-	5.0
84.50	"	21.1	20.3	1.95	-	-	-	5.0
85.50	68.7	20.2	30.0	1.45	-	-	-	4.5
86.00	"	20.3	-	-	-	-	-	-
87.00	"	20.2	-	-	-	-	-	-

t (h)	Rb (cm)	ksb (cm)	Qs (kg/h)	d84e (mm)	d50e (mm)	d16e (mm)	d84a (mm)	d50a (mm)	d16a (mm)
1.00	12.5	.80	0.75	5.28	3.40	2.13	4.81	2.80	1.46
2.00	12.5	.75	2.38	4.75	2.94	1.82	4.18	2.66	1.33
7.00	12.6	.71	2.12	4.51	2.60	1.55	6.34	3.00	1.30
12.00	12.7	.72	1.22	4.59	2.54	1.49	6.80	3.30	1.21
19.50	13.0	.75	0.67	4.28	2.60	1.55	7.25	3.50	1.11
28.50	13.0	.75	0.40	4.20	2.55	1.57	6.98	3.63	1.07
43.50	13.0	.79	0.41	4.63	2.58	1.48	7.45	3.73	1.05
63.50	13.0	.79	0.64	4.35	2.21	1.25	7.78	4.10	1.52
85.50	14.1	1.89	0.12	3.94	2.39	1.48	7.87	4.20	1.54
86.00	-	-	2.98	-	-	-	-	-	-
87.00	-	-	1.81	5.24	2.77	1.39	-	-	-

Table 3 - Summary of data from RUN 2. S_b = 0.002

t (h)	Q (l/s)	T (C)	h (cm)	S_f *1000	U_p (cm/s)	Uv (cm/s)	V_{*pv} (cm/s)	V_{*wc} (cm/s)
0.00	40.8	15.2	19.0	2.40	-	-	-	-
1.00	"	"	"	2.38	-	-	-	4.8
3.50	"	"	19.5	2.36	68	66	5.0	5.0
4.00	"	16.1	19.4	2.37	-	-	-	4.9
20.00	"	17.4	"	2.29	-	-	-	5.2
43.00	"	"	19.8	2.22	68	-	-	5.2
47.00	"	19.2	19.8	2.10	67	-	-	5.3
54.00	"	18.5	19.5	2.00	65	67	5.0	5.3
65.50	"	20.1	21.3	1.70	65	66	5.3	5.3
68.50	68.0	20.0	31.1	1.65	-	-	-	6.0
69.00	"	20.0	31.2	1.65	72	-	-	6.0
69.50	"	19.0	31.1	1.65	-	-	-	6.1
71.50	"	19.0	31.1	1.65	-	-	-	6.1

t (h)	Rb (cm)	ksb (cm)	Qs (kg/)	d84e (mm)	d50e (mm)	d16e (mm)	d84a (mm)	d50a (mm)	d16a (mm)
1.00	12.0	.56	2.56	4.59	2.88	1.71	5.35	2.77	1.39
4.00	12.4	.75	2.28	4.59	2.52	1.51	6.11	3.01	1.32
20.00	13.8	1.46	1.41	4.92	2.57	1.51	7.70	3.73	1.16
43.00	14.0	1.60	0.04	4.00	2.41	1.46	7.70	3.77	1.16
47.00	14.0	1.67	0.34	4.36	2.46	1.43	7.60	3.81	1.16
54.00	14.2	1.76	0.49	6.64	3.20	1.52	7.50	3.61	1.16
68.50	14.4	1.82	0.17	4.23	2.43	1.43	7.14	3.73	1.16
69.00	18.8	2.74	1.16	4.85	3.01	1.48	-	-	-
69.50	19.1	2.91	1.38	4.92	2.75	1.45	-	-	-
71.50	19.1	2.91	3.48	4.53	2.57	1.47	-	-	-

Table 4 - Summary of data from RUN 3. S_b = 0.005

t (h)	Q (l/s)	T (C)	h (cm)	Sf *1000	U_p (cm/s)	U_v (cm/s)	V_{*pv} (cm/s)	V_{*wc} (cm/s)
1.00	11.9	18.5	6.2	7.00	53	54	-	6.1
2.00	"	"	6.4	6.90	53	52	-	-
3.00	"	-	6.4	6.75	-	-	-	6.1
4.00	"	-	6.4	6.50	53	-	-	6.0
5.00	"	-	-	-	-	-	-	-
7.00	"	-	-	-	-	-	-	-
8.00	"	18.7	-	-	-	-	6.5	-
9.50	"	18.8	6.6	6.30	53	-	-	6.0
11.00	"	18.9	-	-	-	-	-	-
13.00	"	19.2	6.6	6.30	53	-	-	-
15.00	"	19.3	6.6	6.30	53	-	-	-
18.00	"	19.4	6.6	6.30	53	-	6.7	6.0
18.50	15.6	19.5	8.0	6.00	55	-	-	-
19.00	"	19.0	-	5.50	-	-	-	-
19.50	19.6	19.7	9.4	5.00	64	-	-	6.3
20.00	"	-	-	-	-	-	-	-
20.30	19.6	19.7	9.6	-	-	-	-	-

t (h)	Rb (cm)	ksb (cm)	Qs (kg/h)	d84e (mm)	d50e (mm)	d16e (mm)	d84a (mm)	d50a (mm)	d16a (mm)
1.00	5.4	1.96	17.56	4.96	2.93	1.74	6.16	2.40	1.08
2.00	-	-	18.16	3.83	2.23	1.41	7.75	3.78	1.04
3.00	5.7	2.48	13.06	4.22	2.44	1.47	7.83	4.24	1.02
4.00	5.7	2.33	4.66	4.00	2.34	1.38	7.94	4.49	1.08
5.00	-	-	2.21	4.31	2.63	1.52	8.01	4.63	1.03
7.00	-	-	2.21	4.03	2.30	1.32	8.18	4.86	1.19
9.50	5.8	2.52	1.71	4.20	2.30	1.30	8.13	4.93	1.35
11.00	-	-	1.25	4.46	2.58	1.69	8.17	4.98	1.19
13.00	-	-	2.71	4.53	2.49	1.47	8.20	5.02	1.02
15.00	-	-	1.30	4.69	2.75	1.58	8.38	5.02	1.01
18.00	5.8	2.52	0.54	4.25	2.41	1.44	8.22	5.07	1.00
18.50	-	-	0.56	5.65	2.77	1.29	-	-	-
19.00	-	-	5.24	5.20	2.77	1.47	-	-	-
19.50	8.0	2.97	4.03	5.24	2.43	1.26	-	-	-
20.00	-	-	23.98	4.56	2.51	1.39	-	-	-
20.30	8.1	3.00	30.61	4.44	2.39	1.36	-	-	-

table 5 - Summary of data from RUN 4. S_b = 0.005

t (h)	Q (l/s)	T (C)	h (cm)	S_f *1000	U_p (cm/s)	U_v (cm/s)	V_{*pv} (cm/s)	V_{*wc} (cm/s)
1.00	11.9	17.6	6.0	5.30	56	57	5.8	5.2
2.00	"	17.8	6.5	5.30	56	57	-	5.4
3.00	"	-	-	-	-	-	-	-
7.00	"	18.0	6.5	5.00	57	-	-	5.3
10.00	"	18.2	6.5	-	-	-	-	-
22.50	"	18.1	6.5	-	-	-	-	-
23.50	"	17.9	6.5	5.00	-	-	-	-
24.00	19.7	18.4	8.8	4.80	-	-	-	5.8
24.50	"	18.5	-	4.60	-	-	-	-
26.00	"	-	9.5	4.50	66	65	6.7	5.8

t (h)	Rb (cm)	ksb (cm)	Qs (kg/h)	d84e (mm)	d50e (mm)	d16e (mm)	d84a (mm)	d50a (mm)	d16a (mm)
1.00	5.2	0.90	12.44	4.96	2.91	1.63	5.67	2.65	1.35
2.00	5.7	1.56	-	-	-	-	-	-	-
3.00	-	-	4.13	4.22	2.53	1.54	6.87	2.99	1.17
7.00	5.7	1.46	1.33	3.94	2.43	1.48	7.00	3.48	1.12
10.00	-	-	0.60	3.92	2.31	1.40	7.20	3.63	1.11
22.50	-	-	0.30	4.35	2.53	1.47	7.63	3.76	1.10
23.50	-	-	0.10	3.81	2.26	1.44	7.65	3.78	1.10
24.00	7.2	0.80	15.88	5.24	2.67	1.37	-	-	-
24.50	-	-	13.24	4.95	2.69	1.49	-	-	-
26.00	7.7	1.22	6.74	4.72	2.58	1.46	-	-	-

Table 6 - Summary of data from RUN 5. S_b = 0.01

t (h)	Q (l/s)	T (C)	h (cm)	S_f *1000	U_p (cm/s)	U_v (cm/s)	V_* (cm/s)
1.00	6.4	17.4	4.0	10.30	44	45	6.3
3.00	"	17.4	-	-	-	-	-
7.00	"	17.5	4.0	9.80	43	43	6.2
10.00	"	17.5	4.0	9.80	-	-	-
22.00	"	-	-	-	-	-	-
24.00	"	17.5	4.0	9.20	44	45	6.0
24.50	11.3	17.5	5.2	8.20	68	69	6.3
25.00	"	17.5	5.2	8.20	-	-	-
26.00	"	17.5	-	-	-	-	-

t (h)	Qs (kg/h)	d84e (mm)	d50e (mm)	d16e (mm)	d84a (mm)	d50a (mm)	d16a (mm)
1.00	10.82	4.86	2.93	1.71	5.94	2.61	1.33
3.00	4.77	4.35	2.60	1.57	7.04	2.91	1.11
7.00	1.57	4.16	2.48	1.45	7.19	3.42	1.05
10.00	0.65	4.08	2.31	1.41	7.32	3.55	1.05
22.00	0.50	4.22	2.34	1.39	7.84	3.84	1.07
24.00	0.39	4.08	2.38	1.43	7.86	3.86	1.07
24.50	22.26	5.31	2.96	1.61	-	-	-
25.00	21.72	4.72	2.54	1.41	-	-	-
26.00	12.17	4.82	2.54	1.40	-	-	-

$$C_w \cdot \sqrt{(gR_w S_f)} = C_b \cdot \sqrt{(gR_b S_f)} \qquad (2)$$

where C_w, C_b are the nondimensional Chezy coefficient; for fully turbulent flows:

$$C_w = 8(R_w/K_{sw})^{1/6} \quad \text{and} \quad C_b = 8(R_b/K_{sb})^{1/6} \qquad (3)$$

where K_{sw}, K_{sb} are the hydraulic roughness of the walls and the bed.

Substituting 3 in 2 and rearranging, we obtain:

$$R_w = R_b \cdot (K_{sw}/K_{sb})^{1/4} \qquad (4)$$

Substituting 4 in 1, the bed hydraulic radius is:

$$R_b = A/(2h(K_{sw}/K_{sb})^{1/4} + B)$$

Furthermore, the condition $U_b = U = Q/(B \cdot h)$ yields:

$$C_b \cdot \sqrt{(gR_b S_f)} = Q/(B \cdot h) \qquad (5)$$

Equations 1,4,5 allow the calculation of R_b, R_w and K_{sb}, given: side wall roughness, slope, flow depth and discharge.
Results are reported in tables 2 to 6, in terms of bed hydraulic radius, R_b, and bed roughness, K_{sb}.
The reliability of the side wall correction procedure has been tested by comparing the shear velocity $v_{*wc} = \sqrt{(gR_b S)}$ with the values of shear velocity obtained by analysing the velocity profiles, v_{*vp}. Results , shown in figure 1, are quite satisfactory.

3.2 Bed surface grain size analysis

Aiming to preserve the integrity of the armour layer during the experiment, the size frequency distribution has been obtained by photograph recording of the bed surface analyzed according to the grid sampling method (fig.2).
To this purpose, photographs of the surface layer at the upstream, the middle and the downstream end of the testing reach have been taken at the beginning of the run, at the end of the armoured layer development and after its destruction.

Size analysis by photographs has been already used and tested by several researcher (see Adams and the referred bibliography). Results show that no significant correction is needed if the minor apparent axis

is used as the particle sieve diameter. In order to verify the original bed surface composition, and secondarly, to provide an additional confirm of the method, a comparison between bulk composition and size analysis from photographs taken at the beginning of each run has been made. In figure 3, the significative sieve diameters D_{84}, D_{50}, D_{16} have been plotted against the corrisponding diameters obtained by photograph analysis, for the initial bed material.
As it can be seen, photograph-grid analysis applied to the initial bed show a satisfactory agreement with bulk analysis.

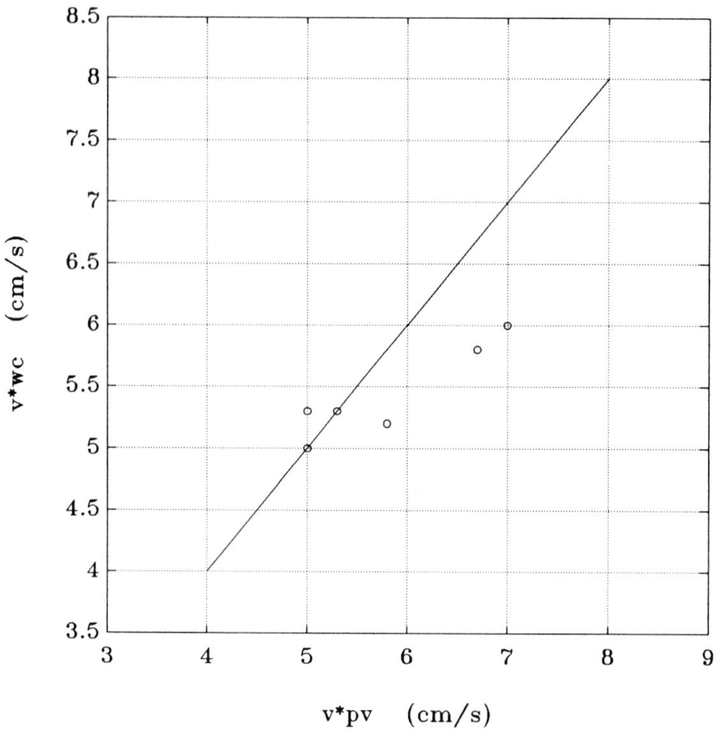

Figure 1.- Shear velocity from velocity profile, v_{*vp}, versus shear velocity from side wall correction, v_{*wc}.

3.3 Temporal variation of bed material composition

In order to estimate the changes in grain size distribution wich occur in the bed material during the armouring process, the sediment continuity equation must be applied to each size fraction, by assuming an adequate control volume and an erosion/deposition model.

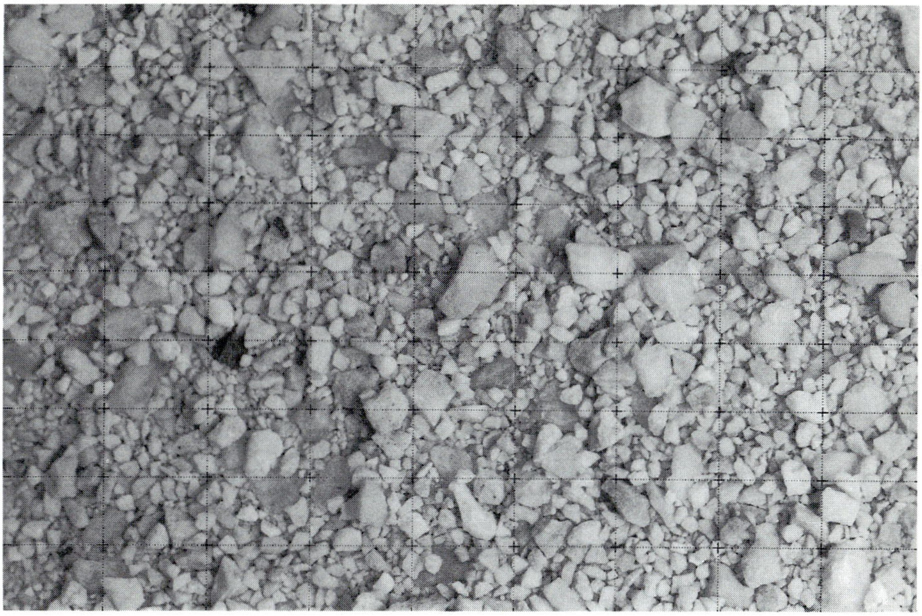

Fig. 2. - An example of analysis of the surface layer photograph by the grid sampling method.

In this study, the simple Bayazit scheme for the exchange of grains between a surface and a subsurface layer of bed sediment is assumed.

In particular, denoting with p_i and t_i the fractions of grains in class interval i respectively in the parent bed and in the eroded material, δm_i the variation of the fraction i in the mixing layer during the time interval δt, the continuity equation yields:

$$M \cdot \delta m_i = \delta M \cdot (p_i - t_i) \qquad (6)$$

where M is the mass of sediment in the mixing layer and δM is the mass of sediment which is removed from the bed during the time interval δt.

Equation 6 is derived under the following assumptions :
- the material eroded from the mixing layer (δM) is replaced by bed material of the same weight from the layer below the mixing layer (parent bed material), that is, the thickness of the mixing layer is

independent from size class and constant with time.
- the material lying beneath the mixing layer is not moving, therefore has the parent bed size distribution.

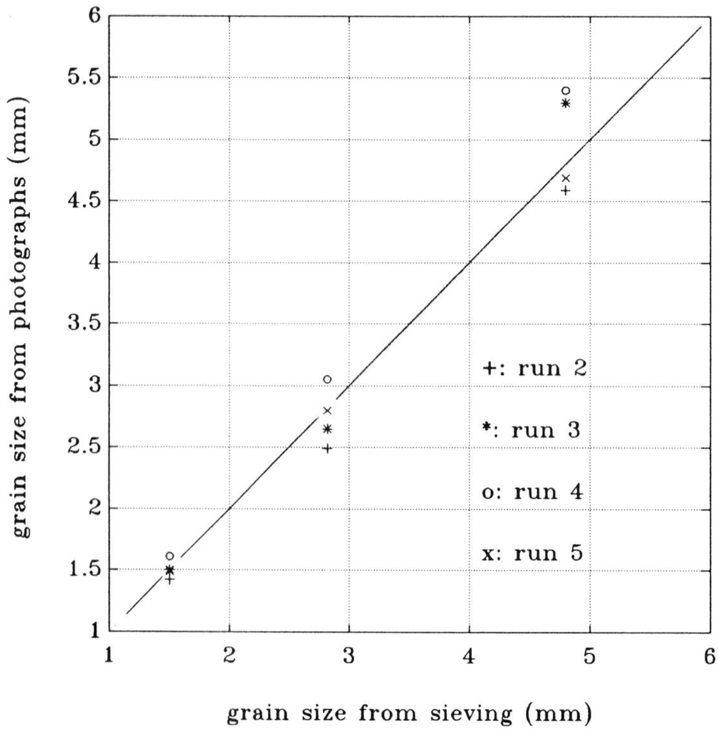

Figure 3.- Comparison between sieve diameters and sizes from photograph, for the initial bed conditions.

The appropriate value of the thickness of the mixing layer has been deduced by comparing results obtained by eq. 6 with those obtained by the analysis of photographs.

The value of the mixing layer thickness which provides the best fitting with the experimental data has been found to be equal to the D_{50} of the armoured layer. The mixing layer thickness has been estimated through eq. 6 as follows:
multiplying by the current grain-size (D_i in metric or ϕ units) and adding up over all sizes, we obtain a similar relationship between:
-mean sizes of parent, armoured and eroded material (known from sieve analysis), -the eroded mass (measured) and -the mixing layer mass (unknown).

The agreement is generally good, except for run 3 in which the supply of sediment during the run is not accounted for in equation 6. For this reason, data from run 3 have not been included in the following analysis.

Figure 4 refers to the armoured conditions, where D_{84}, D_{50} and D_{16} calculated by equation 6 are plotted against photograph estimated grain sizes.

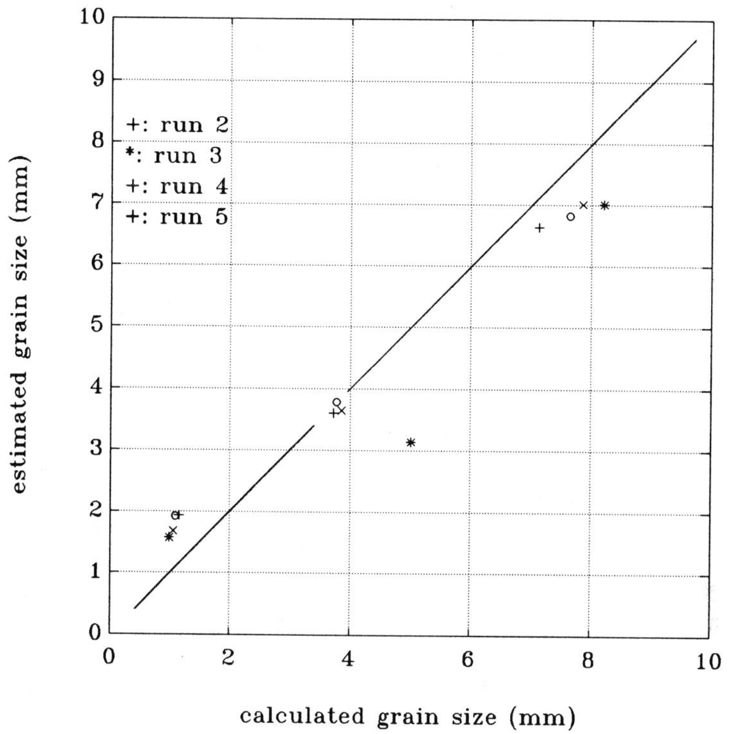

Figure 4. - Calculated (eq. 6) against observed (photograph technique) grain size for armoured conditions.

3.4 Effects of non-uniformity of bed material on sediment transport

Conditions under which an armoured layer can develop are characterized by a progressive coarsening of the top layer induced by a selective sediment removal of the size fractions from the bed surface.

The sediment transport processes related to this phenomenon are considerably affected by the non-uniformity of bed material, as pointed out by many investigators (Einstein & Chien, Egiazaroff, Gessler, Day, Ashida & Michiue), who proposed relationships to estimate the bed-load for the non-uniform bed material by calculating the transport rate per size fraction.

Such relationships are commonly derived from existing formula for uniform sediments by introducing corrections for the availability of that size fraction in the bed material, for the shear stress acting on the particle and for the resistance to motion of the particle.

The former type of correction generally consists in reducing the transport rate obtained for uniform sediment under the same hydraulic conditions by the percentage of the corrisponding size fraction actually present in the bed material.

The second correction, which is here referred as sheltering effect, accounts for the fact that bigger particles are more exposed to the flow than the finer.

The third correction, here referred as hiding effect, accounts for the effects of the surrounding particles building the hollow where the particle stays.

Practically, two approaches have been proposed so far to describe this phenomenon; one introduces a correction for the effective shear stress acting on grains, τ', by reducing its value for the finer fractions and increasing it for the coarser fractions (sheltering).

The other makes a correction for the critical bed shear stress, τ_{cr}, increasing its value for the finer fractions and reducing it for the coarser fractions (hiding).

The first approach, originally proposed by Einstein, has been followed by Day, Profitt & Sutherland, Misri et al.. The second one was first introduced by Egiazaroff and then adopted also by Ashida & Michiue. Both approaches have similar effects when sediment transport rate, s, is expressed as:

$$s = f(\tau - \tau_{cr})$$

In order to test the validity of the criteria proposed by Enistein (first approach) and by Egiazaroff (second approach) the experimental data obtained by the present investigation have been used in conjunction

with the formula of Meyer-Peter & Muller.
This formula, by introducing the sheltering factor β, yields:

$$\phi_i = 8(\beta_i \cdot \tau_{*i} - \tau_{cr})^{3/2} \tag{7}$$

for the Einstein approach; introducing the hiding factor α, the Egiazaroff approach gives:

$$\phi_i = 8(\tau_{*i} - \alpha_i \cdot \tau_{cr})^{3/2} \tag{8}$$

In 7 and 8, ϕ_i represents the dimensionless sediment transport rate for the i-th fraction, per unit width, given by:

$$\phi_i = s_i / \sqrt{[(S_g - 1) g D_i^3]} \tag{9}$$

where d_i is the representative diameter for the i-th fraction, and s_i is the solid discharge (weight per unit time and unit width) for the same fraction; τ_{*i} denotes the Shields parameter, defined as:

$$\tau_{*i} = v_*^2 / [(S_g - 1) g D_i] \tag{10}$$

while τ_{cr} is the critical value of the Shields parameter; in the formulation of Meyer-Peter & Muller τ_{cr} has a constant value of 0.047.

From the experimental sediment transport data, three size classes, represented by D_{16}, D_{50} and D_{84} have been considered from run 2, 4, 5, relatively to the armouring layer formation.

A first inspection of data can be made in terms of the mobility number, τ_{*i}, compared with the corrisponding critical values, τ_{cr}, given by the Shields criterium here expressed in terms of the dimensionless grain size, D_*:

$$D_* = D \cdot ((S_g - 1) g / \nu^2)^{1/3}$$

Figure 5 shows the finest grains to have a mobility number considerably greater than the critical one, while for the coarsest fraction τ_{*i} is always lower than τ_{cr}.

It is worth to note that these results refer to sediment transport gradually decreasing during the armoring process, under practically steady flow conditions.

Therefore, the application of a sediment transport formula without correction through the hiding factor would give unrealistic results.

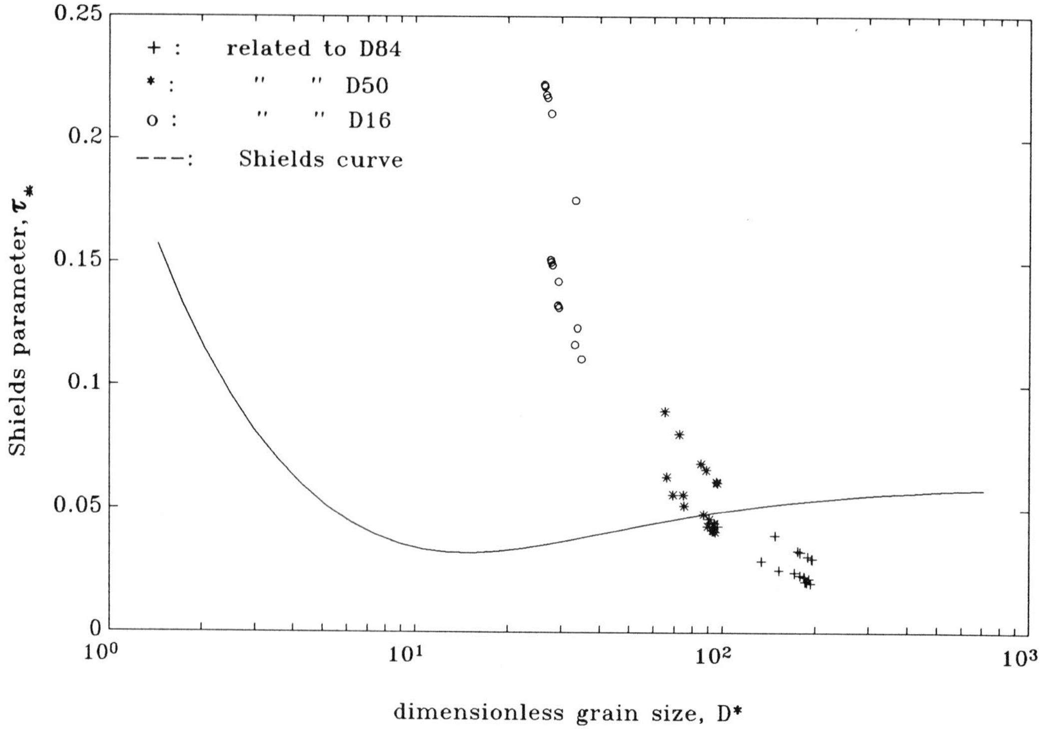

Figure 5. - Mobility parameter of different size fractions compared with critical Shields values.

A successive analysis is made by comparing the sheltering and hiding factors, β and α, obtained by using experimental data in equations 8,9, with the corrispondig values derived from Einstein and Egiazaroff criteria.

The calculated Einstein sheltering factor (the method is not reported here for brevity) is plotted against the estimated values by eq. 8 in figure 6; it can be noticed that the agreement is satisfactory only for low values of β (< 5); for the higher values of β, the method tends to overestimate considerably.

This is in accordance with the results obtained by Lee and Odgaard , who modified the orginal hiding factor curve by putting an upper limit of β = 5. Shen & Lu and Bayazit also report similar results.

The Egiazaroff criterium for α can be expressed as:

$$\alpha_i = (\ln 19/\ln(19 \cdot D_i/D_{50}))^2 \qquad (11)$$

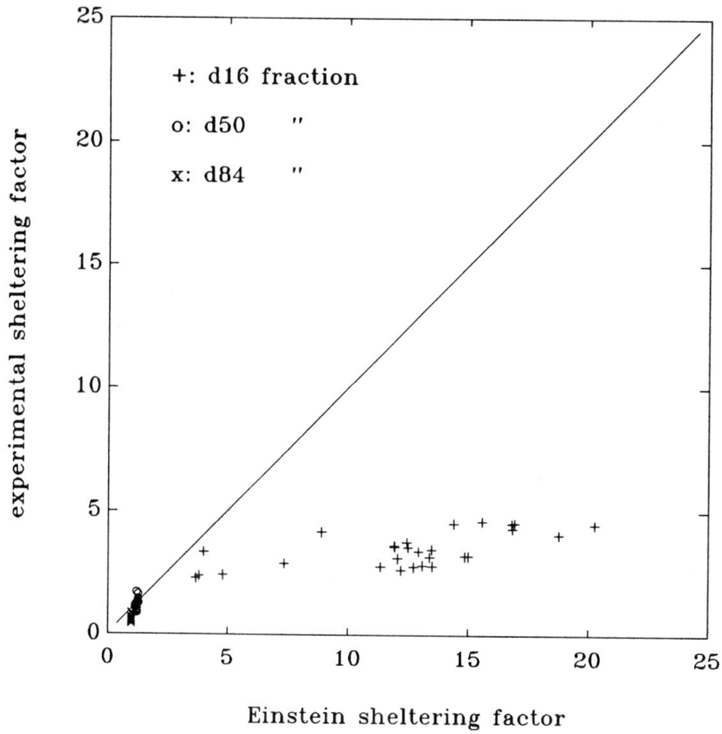

Figure 6.- Sheltering factor estimated from sediment transport against sheltering factor evaluated according to Einstein criterium.

Calculated values of α by eq. 11 have been compared with the ones estimated from eq.10; as it can be seen from fig. 7, the agreement in this case is much more satisfactory.

4. CONCLUSIONS

An experimental investigation on armouring process has been initiated at the University of Bologna.

A first set of laboratory data have been here presented and analysed by focusing the attention on the temporal evolution of the grain sizes in the armoured layer, and the effects of non-uniformity of sediments on the sediment transport of the individual fractions.

Particularly, the one-layer scheme proposed by Bayazit gives quite good results when associated to a mixing layer thickness equal to D_{50} of the armoured material.

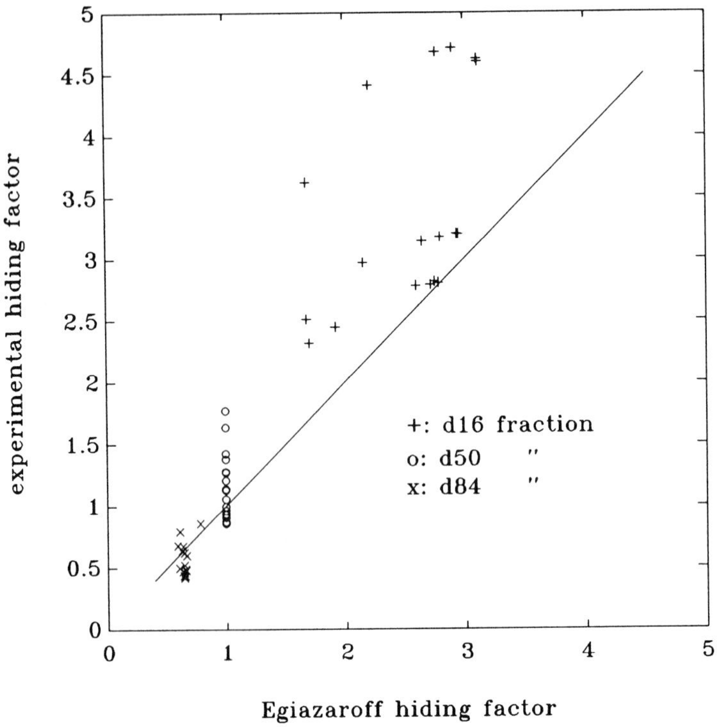

Figure 7.- Hiding factor estimated from sediment transport against hiding factor calculated according to Egiazaroff criterium.

As far as sediment tansport is concerned the introduction of an hiding factor in the sediment transport formula is necessary in order to obtain reasonable results.

Comparison of the estimated hiding factor and the calculated values using the two criteria of Einstein and Egiazaroff leads to the conclusion that the criterium proposed by Einstein overestimats considerably the hiding factor while the the criterium of Egiaroff seems to be more reliable.

Other aspects of armoring process (stability of the armored layer, effects of unsteadyness) will be investigated in a successive paper.

BIBLIOGRAPHY

Adams J., "Gravel size analysis from photographs", ASCE J. Hydraulics Div., 105 HY10, oct 1979.

Armanini A., Di Silvio G., "A one-dimensional model for the transport of a sediment mixture in non-equilibrium conditions", J. Hydraulic Res., vol.26 no.3, 1988.

Ashida K., Michiue M., "Studies on bedload transport rate in open channel flows", Proc. Int. Symposium on River Mechanics, Bangkok, Vol. 1, pp.407-418, 1973.

Ashida K., Michiue M., "An investigation of river bed degradation downstream of a dam", Proc. 14th IAHR Congress, vol.3, pp.247-256, 1971

Bayazit M., "Simulation of armor coat formation and destruction", Proc. 16th IAHR Congress , Sao Paulo, Brazil, paper B10, pp.73-80, 1975.

Borah D.K., Alonso C.V., Prasad S.N., " Routing graded sediments in streams: formulations" , " Routing graded sediments in streams: applications", ASCE J. Hydraulics Div., 108 HY12, pp. 1486-1517, dec 1982.

Day T.J., "A study of the transport of graded sediments", HRS Wallingford, rep. IT 190, apr 1980.

Deigaard R., Fredsoe J., "Longitudinal grain sorting by current in alluvial streams", Nordic Hydrology, vol.9, pp.7-16, 1978.

Egiazaroff P.I., "Calculation of non-uniform sediment concentrations", ASCE J. Hydraulic Div., 91 HY4, jul 1965.

Einstein H.A., Chien N., "Transport of sediment mixture with large range of grain sizes", MRD Sediment Series No. 2, U.S. Army Eng. Div., Missouri River, Corps of Engineers, Omaha, Neb., 1953.

Ettema R., "Sampling armor-layer sediments", ASCE J. Hydraulics Div., 110 HY7, pp. 992-996, jul 1984.

Falcon M., "Interaction between beload and armoring", Proc. 3rd Int. Symposium on River Sedimentation, Jackson, Mississippi, U.S.A., 1986.

Gessler J., "The beginning of bedload movement of mixtures investigated as natural armoring in channels", rep.no. 69 Lab. Hydraulic Research and Soil Mechanics, Swiss Fed. Inst. of Technology, Zurich, 1965.

Gessler J., "Self-stabilizing tendencies of alluvial channels", ASCE J. Waterways Harbors Div., 96 WW2, pp.235-249, may 1970.

Hirano M., "On phenomena of river bed lowering and armouring below reservoirs" 14th Proc. of Hydr. Lect. Meeting, Hatsumei, Kaikan, 1970.

Hirano M., "River bed degradation with armouring" Trans. of JSCE vol.3 part 2, 1971.

Hirano M., "Studies on variation and equilibrium state of a river bed composed of non-uniform material" Trans. of JSCE vol.4, 1972.

Holly F.M., **Karim M.F.**, "Simulation of Missouri river bed degradation", ASCE J. Hidraulic Eng., 112 HY6, pp.497-517, jun 1986.

Karim M.F., **Holly F.M.**, "Armoring and sorting simulation in alluvial rivers", ASCE J. Hydraulic Div., 112 HY8, pp.705-715, aug 1986.

Lee H.Y., **Odgaard J.A.**, "Simulation of bed armoring in alluvial channels", ASCE J. Hydraulic Div., 112 HY9, pp.794-801, sep 1986.

Little W.C., **Mayer P.G.**, "The role of sediment gradation on channel armouring", publ. no. ERC-0672, Georgia Institute of Technology, Atlanta, 1972.

Little W.C. , **Mayer** P.G., "Stability of channel beds by armouring", ASCE J. Hydraulics Div., 102 HY11, pp.1647-61, nov 1976.

Misri R.L., **Garde** R.J., **Ranga Raju** K.G., "Bedload transport of coarse non-uniform sediments", ASCE J. Hydraulics Div., 110 HY3, pp.312-28, mar 1984.

Odgaard A.J., "Grain-size distribution of river-bed armor layers", ASCE J. Hydraulics Div., 110 HY10, pp.1479-84, oct 1984.

Palaniappan A.B., **Godbole P.N.**, "Modeling of armouring process in alluvial streams", Proc. 3rd International Workshop on Alluvial River Problems, TIWARP-89, pp.63-72, Roorkee, India, 1989.

Parker G., Klingeman P.C., McLean D.G., "Bedload and size distribution in paved gravel-bed streams", ASCE J. Hydraulics Div., 108 HY4, pp.544-571, apr 1982.

Proffitt G.T., "Selective transport and armouring of non uniform alluvial sediments", Dept. Civil Eng. Univ. of Canterbury N.Z., Res. Rept. 80/22, pp.1-203, 1980.

Proffitt G.T., Sutherland A.J., "Transport of non-uniform sediment", IAHR J. Hydraulic Res., 21-1, pp.33-43, 1983.

Raudkivi A.J., Ettema R., "Stability of armour layers in rivers", ASCE J. Hydraulics Div., 108 HY9, pp.1047-57, sep 1982.

Ribberink J.S., "Mathematical modelling of one-dimensional morphological changes in rivers with non-uniform sediments", Rep. no. 87-2, Delft Univ. of Technology, 1987.

Saad M.B., "Armouring of alluvial channel beds", Proc. IARH Int. Conf. Fluvial Hydraulics, pp.52-57, Budapest, 1988.

Samaga B.R., Ranga Raju K.G., Garde R.J., "Bedload transport of sediment mixtures", ASCE J. Hydraulics Div., 112 HY11, pp.1003-18, nov 1986.

Shen W.H., Lu J.Y., "Development and prediction of bed armoring", ASCE J. Hydraulic Div., 109 HY4, apr 1983.

Willetts B.B., Maizels J.K., Florence J., " The simulation of stream bed armouring and its consequences", Proc. Instn Civ, Engrs, Part 1, 82, pp.799-814, aug 1987.

BED LOAD TRANSPORT AND HYPERCONCENTRATED FLOW AT STEEP SLOPES

Dieter Rickenmann

Laboratory of Hydraulics, Hydrology and Glaciology
E.T.H.- Zentrum, Zürich, Switzerland

ABSTRACT

As an extension of the study of Smart and Jäggi (1983), bed load transport tests were performed in a steep flume with a clay suspension, in order to simulate the fine material slurry of a hyperconcentrated flow in a torrent. Bed load transport rates were found to increase with increasing fluid density, if the flow around the grains is not laminar. The experiments in which density effects dominated over viscous effects, are analysed together with other bed load transport data, resulting in new transport equations.

1 INTRODUCTION

In a debris flow, big boulders and stones are transported in a slurry of finer material, which typically consists of a mixture of sand, silt and clay particles. At conditions intermediate between a "normal" water flood and a debris flow the same viscous slurry may be present but the flow can be more or less steady and uniform. It has been found that these slurries often show a Bingham type rheological behaviour (e.g. Costa, 1984), which is different from that of a Newtonian fluid such as clear water.

As an extension of the study performed by Smart and Jaeggi (1983) experiments were conducted at the same hydraulic laboratory in order to examine the effect of a hyperconcentrated flow on bed load transport in a steep flume. The objective of these experiments was to examine the effect of an increasing fluid density and viscosity on the flow and bed load transport behaviour.

2 EXPERIMENTS

The experimental setup used by Smart and Jaeggi (1983) was slightly modified for this study. A clay suspension of increasing concentrations was recirculated in the flume system in order to simulate the viscous slurry which is made up by the suspended fine material in a hyperconcentrated flow. The rheology of the clay suspension can be described by a Bingham model.

In a first test series, velocity and depth measurments of the clay suspension flows were made without any bed load transport. The analysis showed that the flow resistance does practically not change in the hydraulically turbulent and transitional regime. By defining a suitable viscosity μ_{e2}, conventional formulae developed for Newtonian fluids could be applied (Rickenmann, 1988). Here $\mu_{e2} = \eta_B + \tau_B h/(2 \cdot V)$, where τ_B is the Bingham yield stress, η_B is the Bingham viscosity, V is the mean fluid velocity, and h is the flow depth including the part occupied by transported solids.

In the main experiments, the equilibrium bed load transport rate was determined for a given flume slope S, fluid flow rate q and suspension properties. A rather uniform gravel mixture with a mean diameter of 1.0 cm was used as bed material; this was almost identical to material no. IV used by Smart and Jaeggi, and thus allowed a direct comparison with their corresponding tests. The experimental parameters were varied within the following ranges: S was set from 7 % up to 20 %, and q varied from 50 l/s·m to 150 l/s·m; the fluid properties of the clay suspension ranged from those of clear water up to a density of about 1.4 g/cm^3 and to a maximum effective viscosity μ_{e2} of the order of 1000 cps.

3 RESULTS AND ANALYSIS

For a given slope and flow rate, the main experiments showed a clear increase in the bed load transport rates with increasing clay concentration of the suspension, as long as the thickness of the laminar sublayer δ was less than the mean grain diameter d_m; however, if the flow around the bed load grains was laminar, transport rates were found to decrease strongly (Rickenmann, 1988). In the following paragraphs only those clay suspension tests are considered, for which δ was less than d_m.

If it is assumed that in the turbulent or transitional regime a change in the fluid viscosity does not greatly affect the bed load transport of grains of uniform size, then the change in fluid density alone must be responsible for the change in the bed load transport capacity. It can be seen from Fig. 1 that the ratio q_B/q_r increases more than linearly with the decreasing density term (s-1); q_B is the volumetric bed load transport rate per unit width, q_r denotes the flow rate per unit width corrected for side wall influence, and $s = \sigma/\rho$ is the ratio of grain to fluid density. A regression calculation resulted in the following equations:

$$q_B = 25.2\ q_r\ S^{2.3}\ (s-1)^{-2.0} \qquad (1)$$

and

$$q_B = 16.7\ q_r\ (S-S_{cr})^{2.1}\ (s-1)^{-1.6} \qquad (2)$$

where $S_{cr} = \theta_c(s-1)d_m/h_r$ is the critical slope (for initiation of motion), θ_c is the dimensionless critical bed shear stress at initiation of motion, and h_r is the mixture flow depth corrected for sidewall influence.

It may be noted that in equ. (2) the same parameters appear as in the bed load transport formula proposed by Smart and Jäggi (1983):

$$q_B = 4.0\ (d_{90}/d_{30})^{0.2}\ q_r\ S^{0.6}\ (S-S_{cr})\ (s-1)^{-1.0} \qquad (3)$$

where d_{90} and d_{30} are the grain sizes for which 90% and 30%, respectively, of bed material by weight are finer. Obviously the exponent of the density term (s-1) is different in equ. (3) but the corresponding tests did not include a variation of s. During the analysis (Rickenmann) it was found that the exponent of the slope term in equ. (3) should be larger, the reason being a difference in the data preparation for the derivation of equ. (3) and the new equations presented here.

Therefore, a combined analysis of the clay suspension tests (data set RI) and all experiments of Smart and Jäggi (data set SJ) was made (total number of observations N = 115), yielding the equation:

$$q_B = 13.0\ (d_{90}/d_{30})^{0.2}\ q_r\ (S-S_{cr})^{2.0}\ (s-1)^{-1.6} \qquad (4)$$

Schoklitsch (1950) developed an expression for the critical discharge at initiation of motion, q_{cr}, as a function of a characteristic grain size, the slope and the density term (s-1). Based on a great number of experimental data, including a slope range $0.25\% \le S \le 20\%$, Bathurst et al. (1987) slightly modified Schoklitsch' equation. If the density term is reintroduced, the following formula is obtained (Rickenmann):

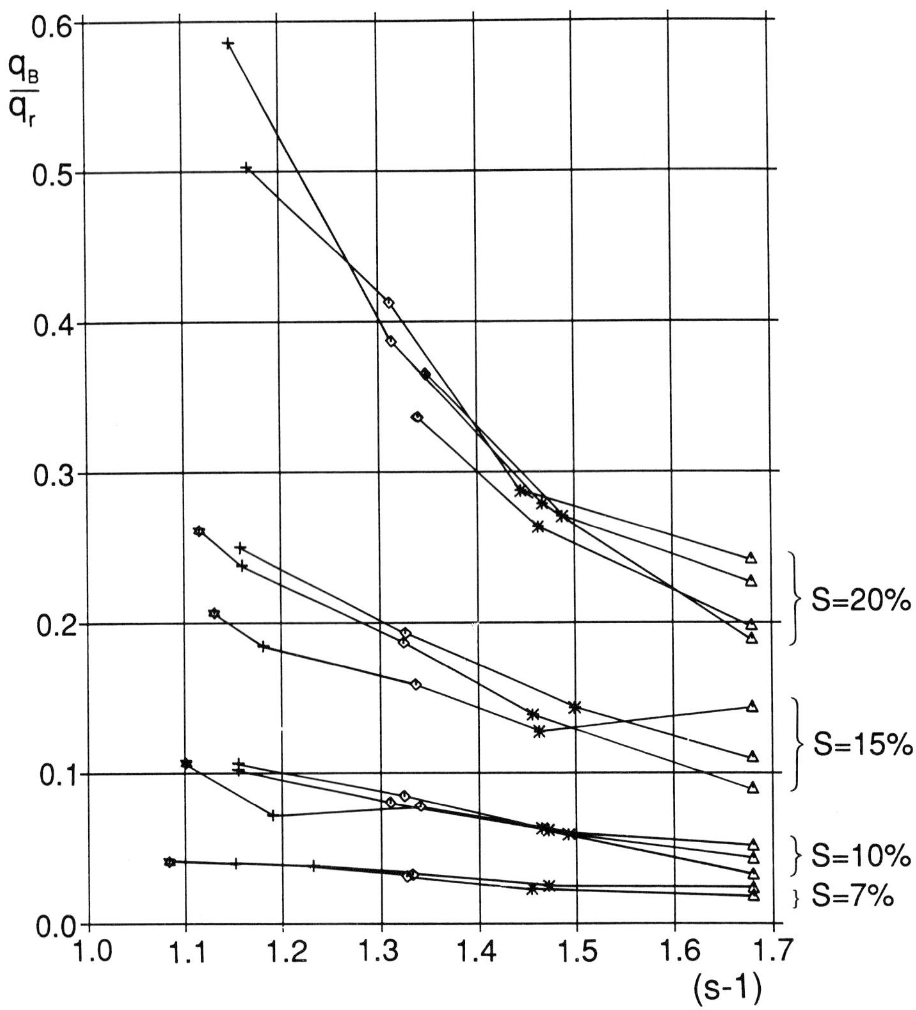

Fig. 1 : Ratio of bed load transport rate to corrected fluid discharge, q_B/q_r, as a function of the density term (s-1), showing the slope S as additional parameter; data points with equal flow rate and equal slope are connected by straight lines.

$$q_{cr} = 0.065\ (s-1)^{1.67}\ g^{0.5}\ d_{50}^{1.5}\ S^{-1.12} \tag{5}$$

where g is the gravitational acceleration, and d_{50} is the grain size for which 50% of bed material by weight are finer. Since the use of the critical slope S_{cr} requires a knowledge of the flow depth, a transport equation in terms of q_{cr} would be more convenient. A regression analysis with the steep flume data (SJ and RI) gives the alternative bed load transport formula:

$$q_B = 12.6\ (d_{90}/d_{30})^{0.2}\ (q_r - q_{cr})\ S^{2.0}\ (s-1)^{-1.6} \qquad (6)$$

Equ. (3), (4) and (6) all have the same correlation coefficient between measured and predicted transport rates ($r^2 = 0.95$), and they also give a similar standard error ($S_E \approx 21\%$). The application of all the above bed load transport equations should be limited to a slope range $3\% \leq S \leq 20\%$, and to essentially flat beds; it is further assumed that there is a full availability of bed material.

If the slurry data is analysed in terms of the dimensionless parameters Φ_B and θ, it is found that the data points follow the same general curve as defined by the clear water data obtained by Smart and Jaeggi; here $\Phi_B = q_B/[g(s-1)d_m^3]^{0.5}$ and $\theta = hS/[(s-1)d_m]$. For a given slope and flow rate, a slurry data point can be traced to move upward along the curve with increasing clay concentration. A Φ_B vs. θ representation of the steep flume data is shown in Fig. 2, together with the experiments of Meyer-Peter and Müller (1948) (N = 137). Their data set (MPM) is given in Smart and Jäggi (1983). Different symbols are used to distinguish between experiments with relatively uniform bed material and those with a mixture of grain sizes; experiments with a density ratio s other than 2.68 are also marked with special symbols. Considering the general trend defined by the data points in in Fig. 2, a combined analysis with all three data sets seemed promising.

A regression analysis was performed correlating Φ_B with the parameters θ, Fr, S and (s-1), where the Froude number is defined as $Fr = V/\sqrt{gh}$. The following equation was obtained:

$$\Phi_B = 3.1\ (d_{90}/d_{30})^{0.2}\ \theta^{0.5}\ (\theta - \theta_c)\ Fr^{1.1}\ (s-1)^{-0.5} \qquad (7)$$

where θ_c was taken as 0.047 for the (MPM) data, and θ_c was determined according to the procedure given in Smart and Jäggi (1983) for the steep flume data (SJ and RI). For equ. (7), the correlation coefficient squared is $r^2 = 0.96$ and the standard error $S_E = 35\%$, determined between measured and predicted transport rates q_B. The performance of equ. (7) can also be judged with Fig. 3. It is seen that the scatter of the data points is considerably larger in the low intensity transport region; this is associated with the assumption of a constant θ_c value, which greatly affects predicted transport rates in this region.

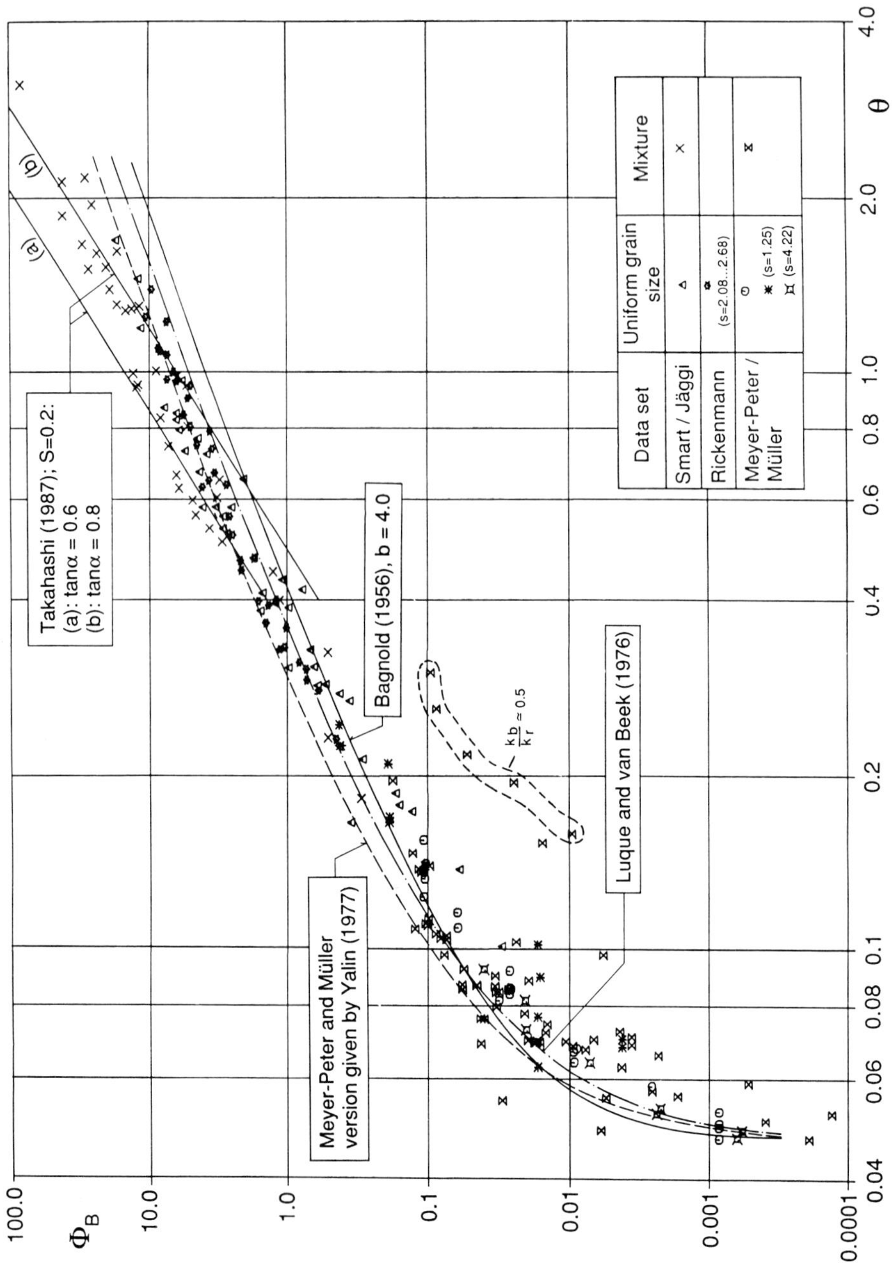

Fig. 2: Experimental data of the three sets used in this study, in terms of the dimensionless parameters Φ_B and θ. The term k_b/k_r represents the ratio of grain friction to total friction (including form drag); five tests with an exceptionally low ratio are marked in the figure. Also shown for comparison are a few other bed load transport equations.

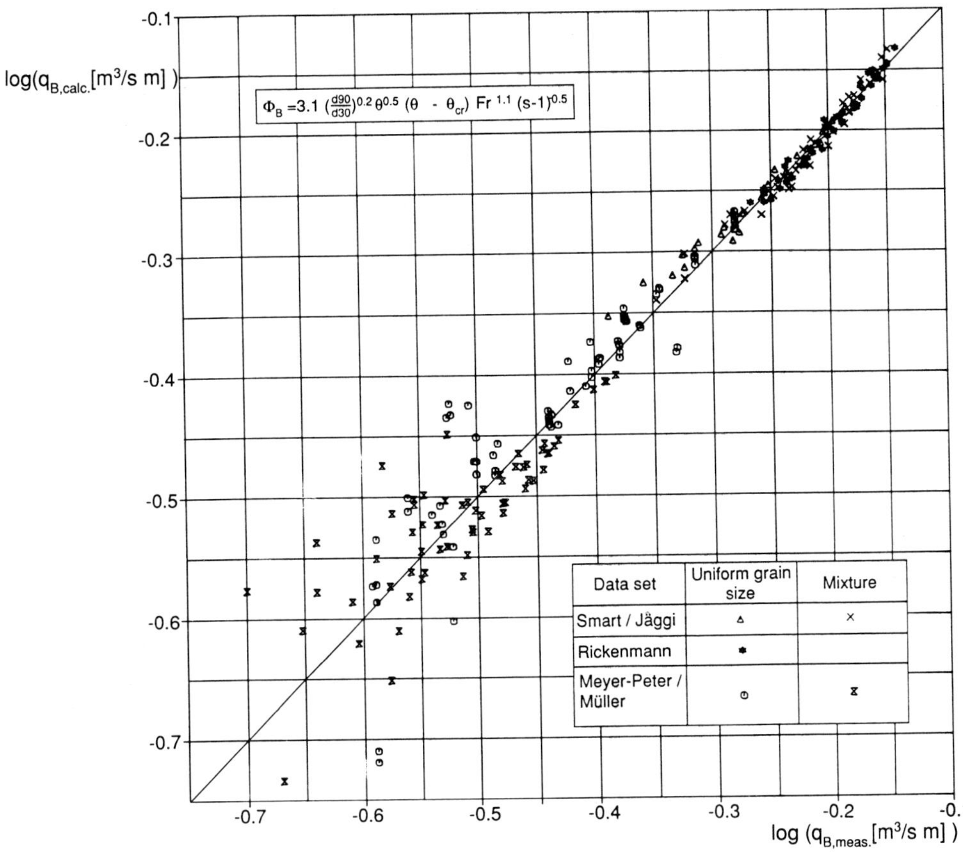

Fig. 3: Comparison between measured bed load transport rates, $q_{B,meas.}$, and those calculated with equ. (7), $q_{B,calc.}$, for the three data sets used in this study.

It should be noted that equ. (7) is valid over a slope range 0.2% ≤ S ≤ 20%, and it can also be applied to situations with moderate form drag losses as in the experiments of Meyer-Peter and Müller (1948). In contrast to the other bed load transport equations presented above (in terms of q_B, q and S), equ. (7) contains two parameters representing the flow stage, and thus it is thought to be of more general use.

4 COMPARISON WITH OTHER FORMULAE AND DISCUSSION

Mizuyama and Shimohigashi (1985) performed steep flume tests on bed load transport using a suspension with fine sand as fine material slurry. For a slope range between about 10% and 20 %, they found the following equation to be valid (Mizuyama, 1988):

$$q_B = A\, q\, S^2\, (s-1)^{-2} \qquad (8)$$

in which A = 20 for clear water and A = 25 for the experiments with the fine sediment suspension. It is interesting that there is the same dependence of q_B on the term (s-1) as in equ. (1).

Based on earlier, similar steep flume tests tests with clear water as transporting fluid, Mizuyama (1981) presented the equation:

$$q_B = 5.5\, q\, S^2 \qquad (9)$$

For comparison, a regression calculation was performed in this study (with the data set RI) fixing the term S^2, which yielded the relationship:

$$q_B = 13.4\, q_r\, S^2\, (s-1)^{-2} \qquad (10)$$

In the case of clear water (s-1 = 1.68) equ. (10) reduces to $q_B = 4.9\, q_r \cdot S^2$ thus being not much different from equ. (9).

Low (1989) performed flume tests at low slopes to study the effect of different particle densities on the bed load transport rates. He compared his results with existing transport formulae and found that most of them adequately account for a change in the density term (s-1); but he proposed that the density term in equ. (3) of Smart and Jäggi should have an exponent of -1.5 instead of -1.0. This finding is in agreement with equ. (6) and (7) where there is also a stronger dependence of the transport rate on (s-1) than in equ. (3). A few experiments of Meyer-Peter and Müller (1948) were performed with sediment particles lighter (coal) and heavier (baryt) than ordinary gravel; these tests also confirm the exponent of (s-1) in equ. (7). In many bed load transport equations in the form of Φ_B and θ, the term (s-1) does not appear explicitly; however, if the Froude number Fr is included (such as in equ. 7), it seems that the additional factor $(s-1)^{-0.5}$ is required.

From the analysis of a number of experiments with plane beds, Bridge and Dominic (1984) suggested that the dimensionless bed load transport rate Φ_B should not only depend on θ but also on the flow stage. In fact, the Froude number Fr appears in the transport equation of Smart and Jaeggi (1983), in a transformed version of equ. (3). They proposed that Fr is to some extent a measure of the additional flow resistance caused by form drag and transported sediment; Fr is thought to partly account for the form drag parameter $(k_b/k_r)^{1.5}$ in the original formula of Meyer-Peter and Mueller (1948); k_r denotes the Strickler coefficient due to grain roughness only, and k_b is the corresponding coefficient including also bed form drag.

Concerning the experiments performed at steeper slopes, form drag is not expected to play an important role since the bed forms were always plane beds or antidunes. At the higher bed load concentrations the dimensionless bed shear stress θ (determined with the mixture flow depth h) alone may no longer be a sufficient measure for the forces causing bed load transport, since the moving solids layer extends over a considerable part of the depth. Then the Froude number (or the fluid velocity) may be looked at as a measure of the forces acting on the grains within the flow.

Semi-theoretical considerations on bed load transport lead to the conclusion that at high shear stresses the equation $\Phi_B = B\ \theta^{3/2}$ becomes a good approximation (Bagnold, 1956; Yalin, 1977; Bridge and Dominic, 1984). In equ. (7), Φ_B also tends to get proportional to $\theta^{3/2}$ at high transport stages. Takahashi (1987) analysed experiments in the "immature debris flow" region, where a grain layer starts to move as a whole. Based also on observed grain velocity profiles, he developed a "bed load" transport equation which can be given as:

$$\Phi_B = \frac{2}{3} \frac{4.2-0.3C_*}{\cos^2\beta\ (\tan\alpha-\tan\beta)^2}\ \theta^{2.5}\ \left(1-\frac{\theta_c}{\theta}\right)^2 \qquad (11)$$

where C_* is the maximum possible packing concentration of the grains, β is the slope angle and α is the (dynamic) angle of internal friction. It is seen that for high shear stresses, Φ_B becomes proportional to $\theta^{2.5}$. Interestingly, Hanes and Bowen (1985) proposed a similar relationship for intense bed load transport, based on a theoretically developed granular fluid model; an approximate expression resulting from this model is given by Hanes (1986) as: $\Phi_B = 6\ \theta^{2.5}$. These findings indicate a stronger dependence of Φ_B on θ at very high transport intensities, than one may expect from extrapolation of bed load transport equations based on a limited slope range.

Fig. 2 shows the three experimental data sets used in this study in a Φ_B vs. θ representation. A few other bed load transport equations are included for comparison. Neglecting the influence of form drag, Yalin (1977) gave a simplified version of the original formula of Meyer-Peter and Müller (1948):

$$\Phi_B = 8\ (\theta - \theta_c)^{1.5} \qquad (12)$$

Bagnold (1956) theoretically developed a bed load transport formula which he verified with extensive experimental data. His equation is given by Yalin (1977) as:

$$\Phi_B = b\ \theta^{0.5}\ (\theta - \theta_c) \tag{13}$$

where b = 4.0 was taken for the comparison in Fig. 2; a value of b = 4.25 is given by Yalin (1977) for flows in the rough turbulent regime. Luque and van Beek (1976) performed bed load transport experiments with particles of different densitites. Their transport rates were relatively low ($\Phi_B < 0.1$), and they proposed the empirical equation:

$$\Phi_B = 5.7\ (\theta - \theta_c)^{1.5} \tag{14}$$

It can be seen from Fig. 2 that the equs. (12), (13) and (14) give fair predictions of Φ_B for θ values below about 0.4; above this value they seem to underpredict the measured transport rates. It is interesting to note that $\theta \approx 0.4$ is an approximate limit above which bed load concentrations in the steep flume tests (data sets SJ and RI) were no longer negligible, i.e. the transport layer occupied a considerable part (say more than about 10%) of the flow depth.

The angle of internal friction for the dynamic case of bed load transport is difficult to determine. Bridge and Dominic (1984) summarised values from the literature: A range of $\tan\alpha = 0.4 \ldots 0.7$ seems to apply to inertial grain shearing conditions, while $\tan\alpha = 0.75$ was proposed by Bagnold (1954) for macroviscous conditions (where the influence of viscosity is important). Some of the clay suspension tests represented in Fig. 2 where just at the limit to macroviscous flow (Rickenmann). $\tan\alpha$ values of 0.6 and 0.8 were used, together with $C_* = 0.65$ and S = 0.2, in Takahashi's equ. (11) for comparison with the data of this study in Fig. 2. According to Takahashi (1987), some of the experiments of Smart and Jäggi (1983) fall into the category of "immature debris flows". Equ. (11) is seen to be sensitive to the choice of $\tan\alpha$, which is difficult to determine or estimate.

It is interesting to compare the steep flume data with the stream power approach. Putting $Fr^{1.1} \approx Fr$ and considering $\theta \gg \theta_c$, equ. (7) can be simplified and rearranged to:

$$i_B = B\ \omega\ S^{0.5} \tag{15}$$

where i_B is the bed load transport rate by immersed weight of solids per unit width, B is a constant, $\omega = V\tau_o$ is the stream power, and τ_o is the bed shear stress. According to Bagnold (1973) the maximum transport efficiency at steep slopes can be expressed as:

$$i_B = (\tan\alpha - \tan\beta)^{-1}\,\omega \qquad (16)$$

where the first factor on the right hand side is the efficiency factor. The steep flume data (SJ and RI) of this study is shown in Fig. 4 in terms of the stream power parameters. Equ. (16) is included for comparison where the efficicency factor has been determined for a slope of 20% ($\beta = 11.3^0$), and for $\tan\alpha = 0.4$ (inertial grain shearing conditions, line A) and $\tan\alpha = 0.75$ (macroviscous grain shearing, line B), respectively. It is interesting that the data points of the clay suspension experiments which approached the limit of macroviscous grain shearing conditions, lie close to the 100% efficiency line (B).

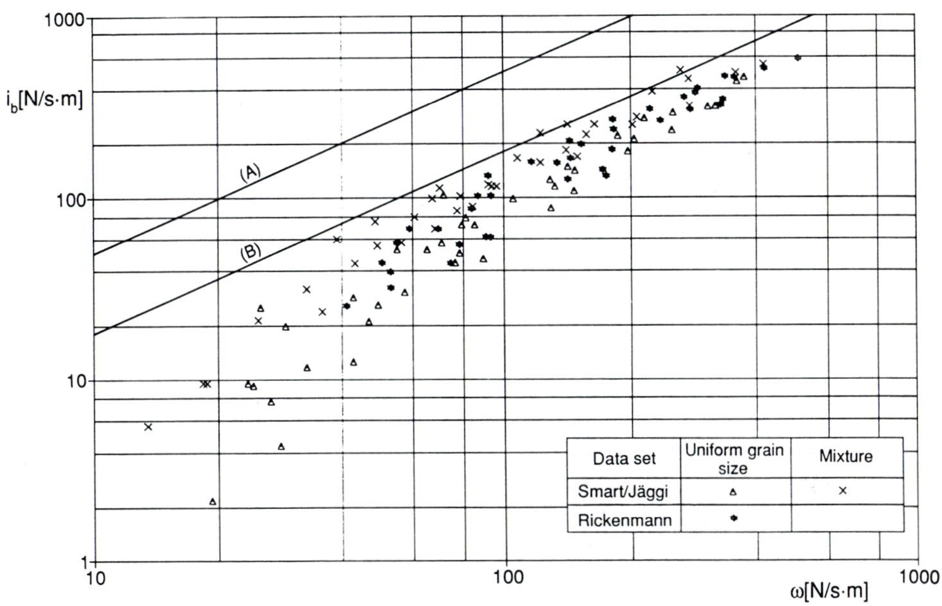

Fig. 4: Experimental data from the steep flume data sets (SJ and RI), shown in terms of the stream power parameters i_B and ω. The straight lines represent the 100% efficiency relationship, equ. (16), for a slope of 20% and for inertial grain shearing (line A) and for macroviscous grain shearing (line B), respectively.

5 SUMMARY

Bed load transport experiments were performed with a clay suspension at steep slopes up to 20%. As compared to the corresponding clear water flows, an increase in q_B was observed with increasing density of the clay suspension. The effect of an increased viscosity becomes important if the thickness of the laminar sublayer is larger than the grain diameter.

In the case where density effects are dominant, the clay suspension experiments could be analysed together with other bed load transport data (with clear water as transporting fluid). Two new transport equations were developed which represent slight modifications of the formulae proposed by Smart and Jäggi (1983). The first one is limited to the steep slope range; it requires a knowledge of the flow rate but not of the flow depth. The second one is of more general applicability (both steep and low slopes); the dimensionless bed load transport rate Φ_B is given as a function of bed shear stress θ, the Froude number Fr, the density term (s-1) and characteristic grain sizes.

REFERENCES:

Bagnold, R.A. (1956):
"The flow of cohesionless grains in fluids", Phil. Trans. Royal Soc. London, 249 A, pp 235-297.

Bagnold, R.A. (1973):
"The nature of saltation and of 'bed-load' transport in water", Proc. Royal Soc. London, Vol. 332 A, pp 473-504.

Bathurst, J.C., Graf, W.H. and Cao, H.H. (1987):
"Bed load discharge equations for steep mountain rivers", in: Sediment transport in gravel bed rivers, eds. Thorne, Bathurst, Hey, John Wiley and Sons, pp 453-477.

Bridge, J.S. and Dominic, D.F. (1984):
"Bed load grain velocities and sediment transport rates", Water Res. Res., Vol. 20, No. 4, pp 476-490.

Costa, J.E. (1984):
"Physical geomorphology of debris flows", Developments and applications of geomorphology (eds. Costa and Fleisher), Springer-Verlag, Berlin and Heidelberg.

Hanes, D.M. (1986):
"On the use of grain-flow dynamics to model intense bedload sediment transport", ASCE Spec. Conf. on advances in aerodynamics, fluid mechanics, and hydraulics, Minneapolis, pp 346-352.

Hanes, D.M. and and Bowen, A.J. (1985):
"A granular-fluid model for steady intense bed-load transport", J. Geophys. Res., Vol. 90, No. C5, pp 9149-9158.

Low, H.S. (1989):
"Effect of sediment density on bed-load transport", ASCE, J.Hydr.Eng., Vol. 115, No. 1, pp 124-137.

Luque, R.F. and van Beek, R. (1976):
"Erosion and transport of bed-load sediment", IAHR, J.Hydr.Res., Vol. 14, No. 2, pp 127-144.

Meyer-Peter, E. and Müller, R. (1948):
"Formulas for bed-load transport", 2nd meeting IAHSR, Stockholm, Sweden, pp 1-26.

Mizuyama, T. (1981):
"An intermediate phenomenon between debris flow and bed load transport", Symp. on Erosion and Sed. Transp., Pacific Rim, Christchurch, New Zealand, IAHS Publ. No. 132, pp 212-224.

Mizuyama, T. and Shimohigashi, H. (1985):
"Influence of fine sediment concentration on sediment transport rates", Jap. Civil Eng. Jour. 27-1, pp 46-49 (in Japanese).

Mizuyama, T. (1988):
written communication

Rickenmann, D. (1988):
"Experimental study on debris flow like sediment transport", Interpraevent, Graz, Austria, Bd. 3, pp 167-175.

Rickenmann, D. :
"Bedload transport capacity of slurry flows at steep slopes", Mitteilung der Versuchsanstalt für Wasserbau, Hydrologie und Glaziologie, ETH Zürich, in preparation.

Schoklitsch, A. (1950):
"Handbuch des Wasserbaus", 2. Auflage, Springer Verlag, Wien.

Smart, G.M., and Jäggi, M.N.R. (1983):
"Sediment Transport on Steep Slopes", Mitteilung der Versuchsanstalt für Wasserbau, Hydrologie und Glaziologie, ETH Zürich, Nr. 64.

Takahashi, T. (1987):
"High velocity flow in steep erodible channels", Proc. 22nd IAHR Congress, Lausanne, Switzerland, Tech. Session A, pp 42-53.

Yalin, M.S. (1977):
"Mechanics of sediment transport", 2nd ed., Pergamon Press, Oxford.

EXPERIMENTAL INVESTIGATIONS ON BED-LOAD AND SUSPENDED TRANSPORT IN MOUNTAIN STREAMS

Giampaolo Di Silvio and Stefano Brunelli

Hydraulics Institute "G.Poleni"
University of Padua, Padua, Italy

ABSTRACT

Flume experiments carried out in equilibrium conditions with a mixture of sand and gravel show that the surface layer is coarser than the layer below; the composition of both layers, however, is coarser than the transport composition. Pavement may also occur during agrradation.

1. Introduction

Solid discharge of mountain rivers during exceptional events is often extremely abundant when compared to the transport in ordinary conditions: indeed, the volume of sediments conveyed through a certain cross-section during a few hours of long and intense rainfall, can be one order of magnitude larger than the average amount of sediment transported in one year.

In these circumstances the sediment supplied to the hydrographic network by landslides and debris flow is also particularly large (overfeeding) and sudden deposition (overaggradation) takes usually place along the streams up to several tens of kilometers away from the input location . In fact, since the material fed to the stream is relatively rich of fine fractions, there is an increase of transport capacity as well as a selective deposition all over the hydrographic network. A mathematical model to describe this process at relatively large space-scale is presented in another contribution to this Workshop [2].

In the present paper, an experimental investigation is described to have a better insight of the phenomena at local scale, especially as far as the co-presence of bedload and suspended transport is concerned. Particular attention has been given to the vertical composition of the bed (pavement and substratum) and to its relationship with the composition of the sediments transported by the stream.

2. Experimental set-up.

The driving idea of the experiments is to investigate in a laboratory flume the sediment transport of a typical alpine torrent, both in ordinary and exceptional conditions.

A quasi-uniform reach (few hundred meters long) of a middle-size stream belonging to a mountain watershed of a few hundred square kilometers, has been considered. In such a situation, the typical width of the stream is 30-50 m, the bed slope 1-3% and the annual flood discharge of the order of 100 m^3/s. In flood conditions, the depth-to-width ratio is usually large enough to have no braiding nor large alternate bars in the torrent.

The granulometric composition of the bed is generally quite coarse as it includes about 90% of boulders, pebbles and gravel and only 10% of sand (grainsize less than 2mm). By contrast, the finer fractions are much more abundant in the transported material, as shown by the composition of landslides feeding the stream as well as by the size of particles trapped in the reservoirs which may have built across the torrent.

As for the transport concentration during the exceptional events, it has been estimated that, for the same water discharge, it may become one order of magnitude larger than during ordinary floods.

The characteristics of the flume (10 m long, 0.5 m wide) correspond to the above-mentioned hypothetical field-conditions (if half-width of the bed is considered) with a lenght scale between 1:30 to 1:50.

The sediment fed into the flume is a mixture of two classes: sand (practically uniform with a d_{50} = 0.120 mm) and gravel (from 6 to 8 mm). Tests have been conducted with different ratios of sand and gravel in the input mixture: the mixture is supplied to the flume by a distributor made by a roller scraping under the feedbox.

A movable sieve, made by a trolley provided with a net, is used to collect sediment transport (bedload and suspension) at the downstream-end of the flume. Several measurements are necessary not only to describe the evolution of sediment transport in unsteady conditions (aggradation process), but also to have a statistically significant description in steady conditions (equilibrated stream), due to the strong pulsations of the solid discharge.

3. Experimental simulation of ordinary and exceptional conditions.

To simulate the ordinary conditions of the river, an arbitrary bed-composition (close to the expected substratum of the "equilibrated stream") is initially set in the flume with a slope conservatively milder than the final one. By feeding a constant water- and sediment discharge, the flume builds-up progressively its bottom until the equilibrium configuration is attained, through a transition period that may take a few hours. In equilibrium conditions the time-averaged bed slope remain constant, while the time-averaged output is equal to the input.

Sediment output from the flume, during the adaptation period, is recorded together with water and bottom profile. Some "stops" of water- and sediment input at certain instants permit to take samples of bottom granulometry and evaluate its space- and time distribution.

Tests are subsequently performed to simulate the deposition process produced by a sudden augmentation of the sediment input to the flume, typical of catastrophic events.

With respect to a given equilibrium configuration (already singled out in the previous experiments), the sediment concentration of the water flow entering the flume is sharply augmented and kept constant for a certain period of time. Also in this case, sediment output and bed composition are recorded at short time intervals, toghether with the rapid aggradation of the bottom.

Although the boundary conditions of these experiments are not quite the same as in nature (since the flume is short and has a fixed base level, there is an exaggerated increase of the bottom slope due to the deposition), the laboratory results have shed light on some interesting features of the sediment transport of a mountain river during normal floods and during catastrophic events.

4. Results of the experiments.

Four experiments, as described above, have been carried out in "ordinary conditions" (1, 2, 3 and 4) and two experiments in "exceptional conditions" (5 and 6). The relevant quantities of the (final) equilibrium configurations for experiments 1, 2, 3 and 4, and the input characteristics for experiments 5 and 6, are resumed in the following tables. Experiments 5 and 6 have been initiated, respectively, from the (final) equilibrium configurations of experiments 3 and 4; during the experiments in exceptional conditions the flume has been fed with the same water discharge and the same sediment composition as in ordinary conditions, yet with a sediment input 10 and 7 times larger. These experiments, however, have been interrupted before reaching the equilibrium configuration that corresponds to the increased sediment input.

Experiments in ordinary conditions
(equilibrium configurations)

	(1)	(2)	(3)	(4)
Water discharge (l/s)	11.8	15.0	20.3	20.3
Total sediment transport (cm^3/s)	45	33	40	75
Sand percentage in the transport (%)	75	82	82	89
Sand percentage in the pavement (%)	6.0	8.0	7.5	9.0
Sand percentage in the substratum (%)	23	25	24	27
Bottom slope (%)	2.40	1.75	1.50	1.30
Water depth (cm)	4.49	4.33	5.30	5.26
Chézy coefficient ($m^{1/2}s^{-1}$)	17.8	26.2	27.7	29.0
Froude number (-)	0.80	1.07	1.06	1.08

Experiments in exceptional condition
(input characteristics)

			(5)	(6)
Water discharge (l/s)	-	-	20.3	20.3
Total sediment input (cm^3/s)	-	-	420	525
Sand percentage in the input (%)	-	-	82	89

The measurements made during the six experiments are reported in the attached figures. For each experiment they provide, as a function of time: the sediment input- and output rates of both sand and gravel; the slope of the bottom and of the waterstream energy (practically coincident); the composition of input and output, as well as the composition of pavement and substratum in the bottom.

Although the sediment input rate was meant to be constant during each experiment, a certain variation in time was recorded due to slightly different consolidations of the mixture in the "feeder". The sediment output rate, by contrast, appeared to be highly pulsating mainly because of the bedforms (however small) propagating along the flume.

The "pavement" is defined as the upper layer of the bottom having the thickness of about 7 millimeters (gravel diameter); the "substratum" is defined as the layer immediately below.

As the initial conditions of the flume are not in equilibrium with the water-and sediment input, an adaptation process takes place, until the steady uniform conditions are eventually reached.

Experiment n.1 (Q=12 l/s)

Input and Output Rates

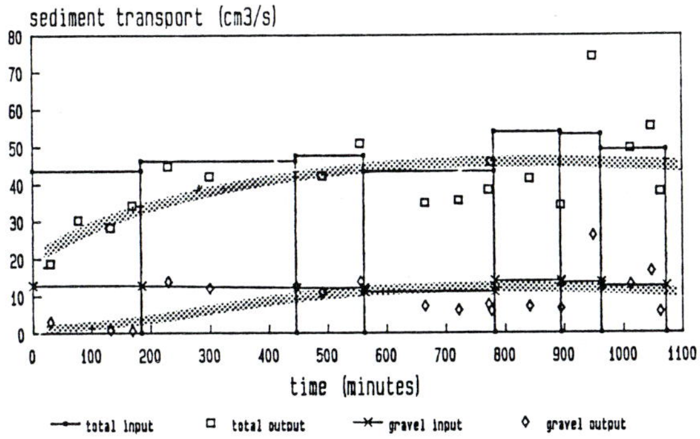

Bed and Energy Slope

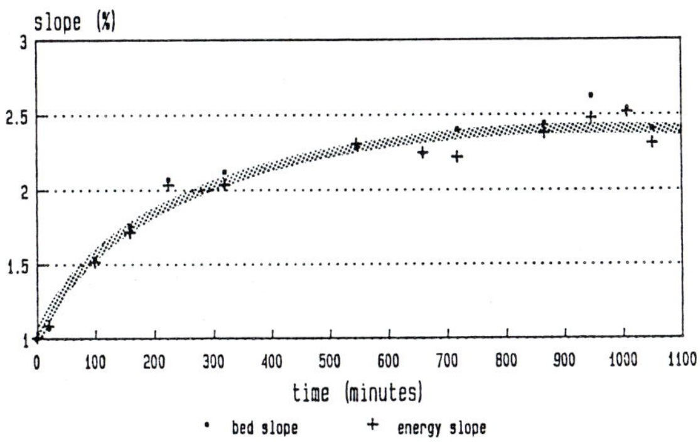

Composition of Transport and Bottom

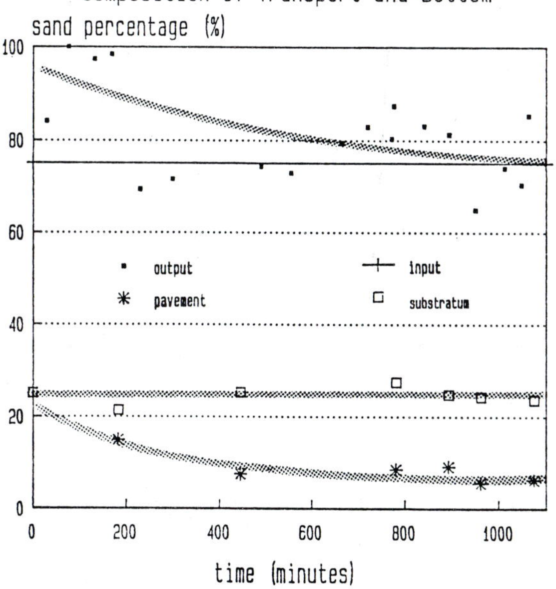

Experiment n.2 (Q=15 l/s)

Input and Output Rates

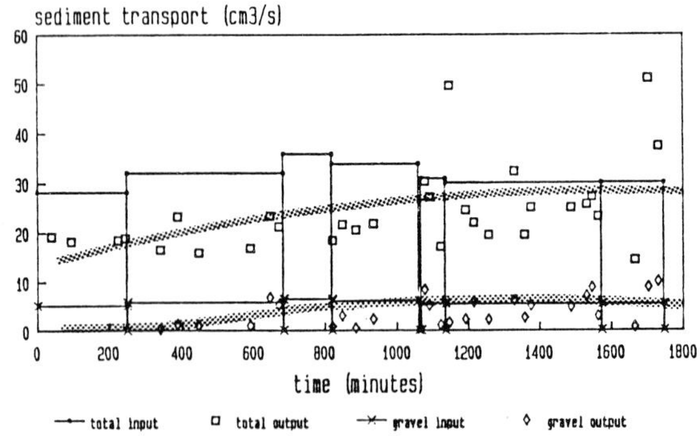

Bed and Energy Slope

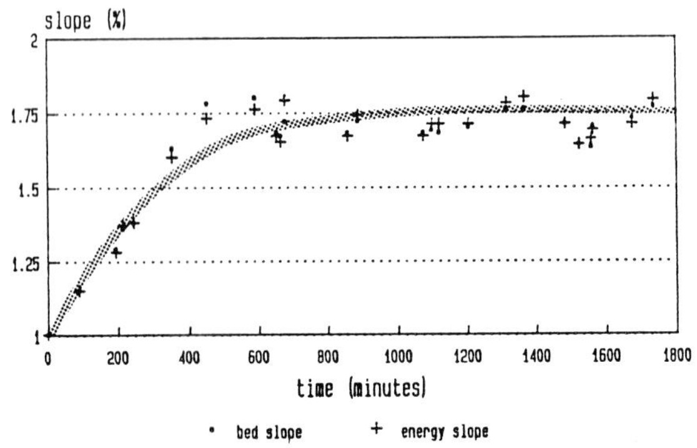

Composition of Transport and Bottom

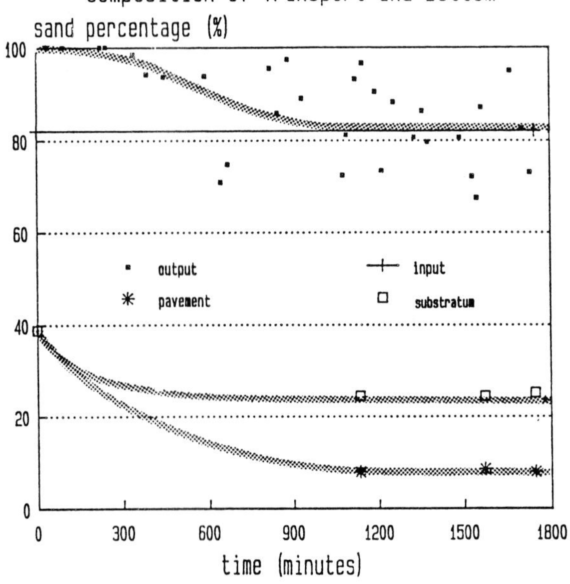

Experiment n.3 (Q=20 l/s)

Input and Output Rates

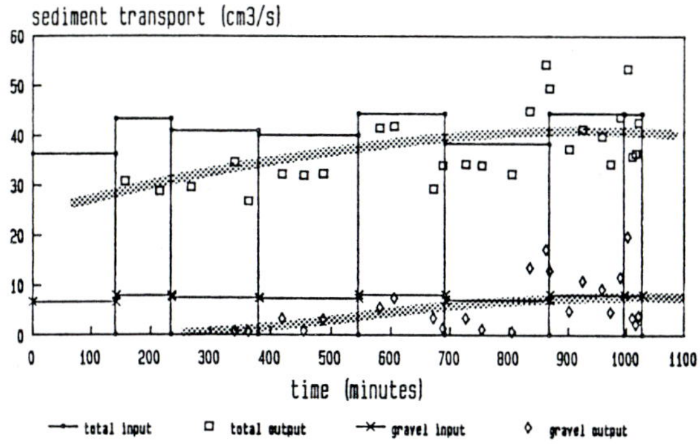

Bed and Energy Slope

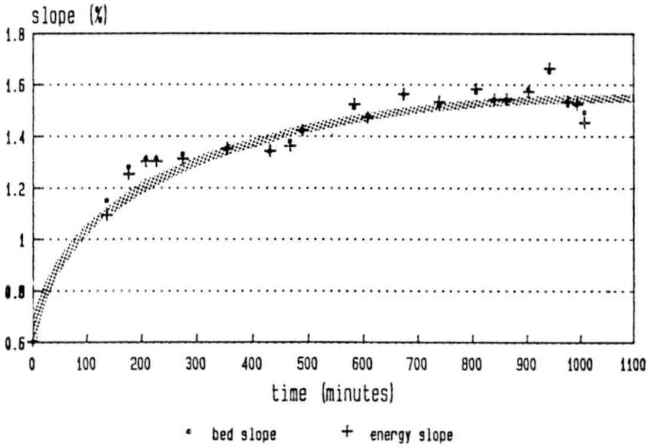

Composition of Transport and Bottom

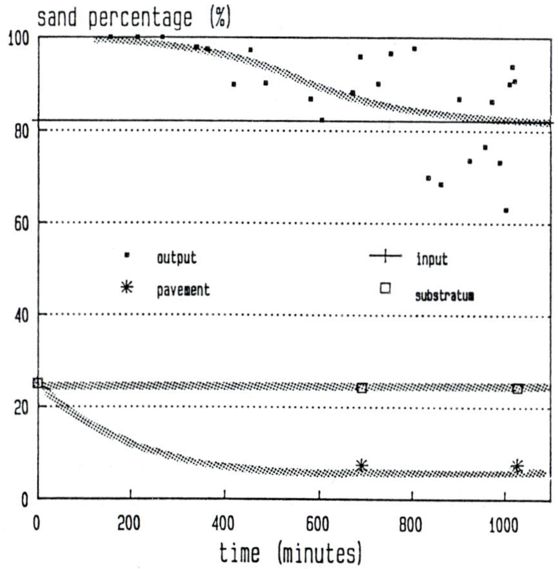

Experiment n.4 (Q=20 l/s)

Input and Output Rates

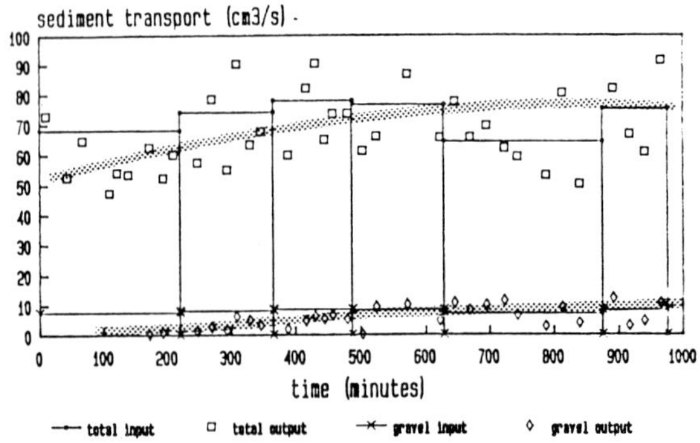

Bed and Energy Slope

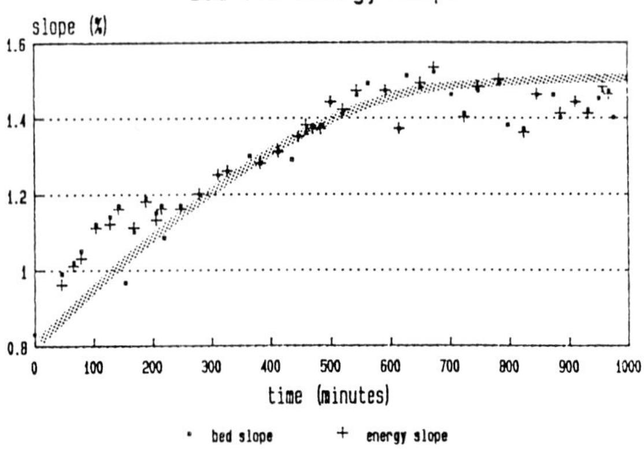

Composition of Transport and Bottom

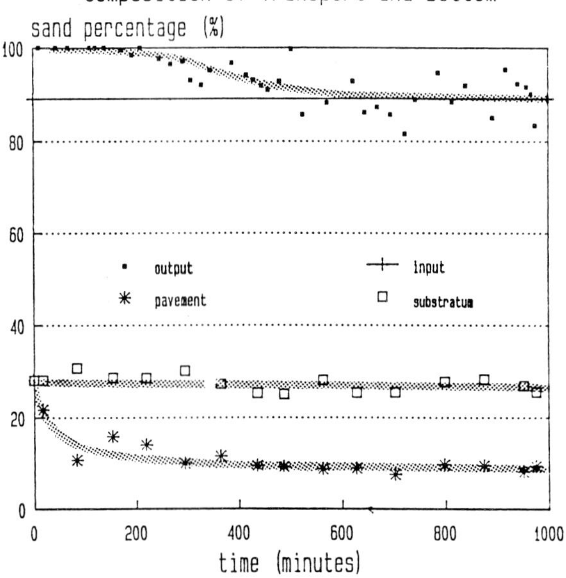

Experiment n.5 (Q=20 l/s)

Input and Output Rates

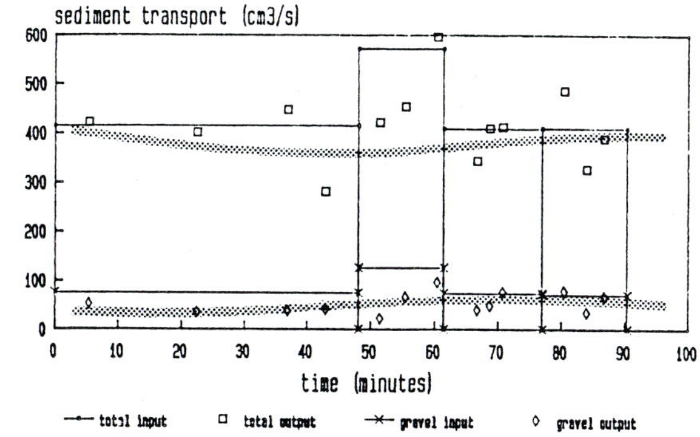

Bed and Energy Slope

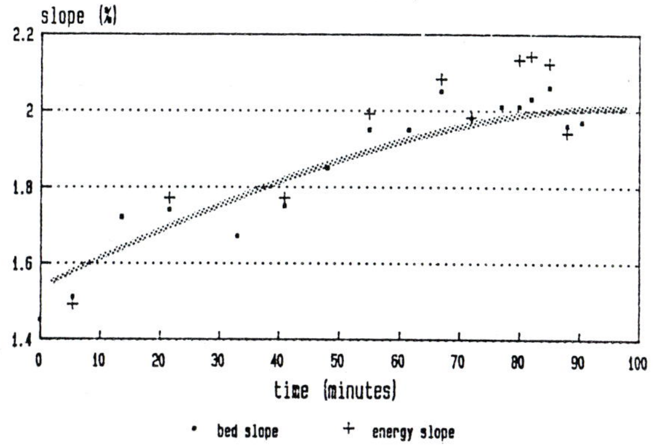

Composition of Transport and Bottom

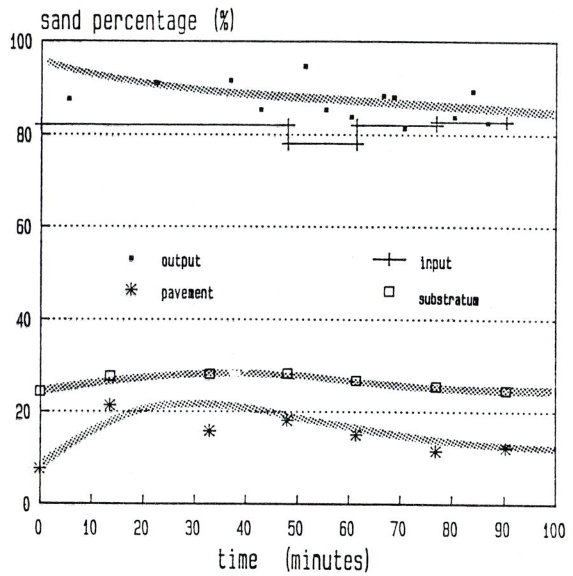

Experiment n.6 (Q=20 l/s)

Input and Output Rates

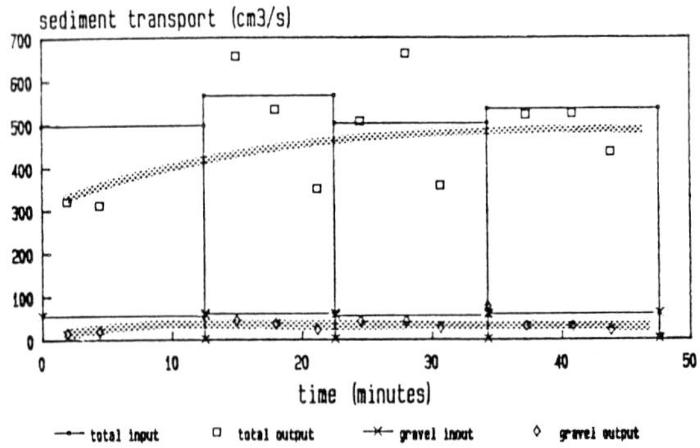

Bed and Energy Slope

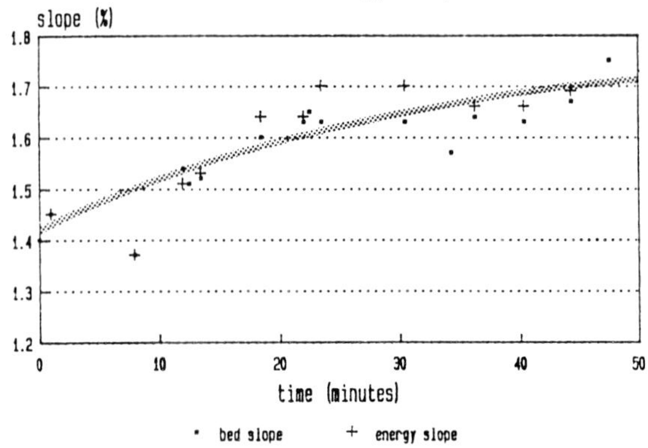

Composition of Transport and Bottom

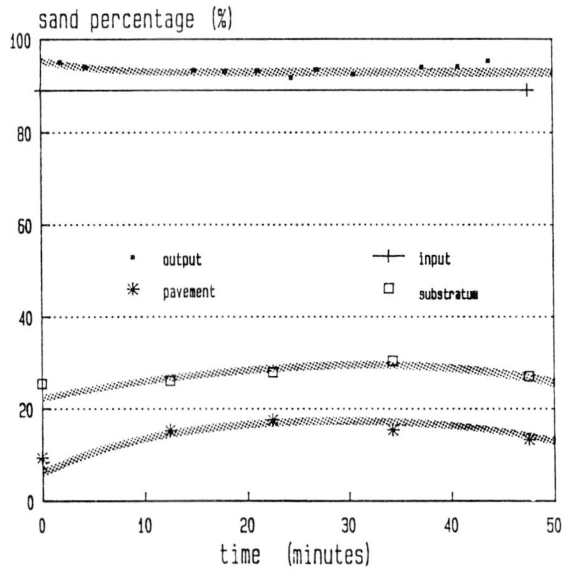

In all the experiments 1, 2, 3 and 4 (ordinary conditions) the following adaptation pattern was observed:

i) Sand is washed out from the upper layer of the bottom with the formation of a coarser surface having the thickness of the gravel diameter. Note that this coarser surface is not composed by absolutely immobile particles (the so-called "armouring" of the bed), but both the sand and the gravel are subjected to frequent, if not continuous, movement (the so-called "pavement", according to the definition of Parker [3], [4]. In any case, as the mobility of the gravel is smaller than the mobility of the sand, the pavement composition is always much coarser than the composition of the transport.

ii) Part of the material fed to the flume tends to deposit with consequent rising of the substratum and increase of the bottom slope. Note that, since gravel tends to deposit faster than sand, also the "substratum" has a composition coarser than the "transport", although finer than the "pavement".

iii) As the slope of the bottom increase, there is a progressive coarsening of the pavement and a minor adjustement of the initial substratum composition. Note that the initial composition of the flume bottom was purposely set equal to the computed equilibrium composition of the substratum.

iv) The output rate gradually increases towards the same value of the input rate. Note that the initial output composition is very rich of sand, but it becomes coarser and coarser following the evolution of the pavement.

A somewhat different adaptation pattern is observed in experiments 5 and 6, which were respectively conducted starting from the final (equilibrium) configurations of experiments 3 and 4 but with a solid discharge 10 and 7 times larger.

i) Unlike the previous experiments, the pavement is already formed, from the start and its initial composition corresponding to the ordinary conditions is too coarse with respect to the huge input of fine sediments. At the beginning of the experiments, then, the composition of the pavement tends to become finer, due to the large deposition rate of sand.

ii) While deposition goes on also the substratum tends initially to become finer, although less conspicuously and with a slight delay with respect to the pavement.

iii) As the slope of the bottom become steeper, however, there is an inversion of tendence in the bottom composition: both the pavement and the substratum start to become coarser, again with a certain delay of the latter.

iv) During the experiments, the output rates of both sand and gravel continue to increase towards the same values of the input. At the same time, the sand percentage in the output (always larger than in the input) shows very slight variations, apparently connected with the above mentioned evolution of the bottom.

7. Conclusions.

On the basis of the experiments described in the previous paragraph, some conclusion can be drawn regarding: (a) the vertical structure of the bottom in mountain rivers; (b) the relation between bottom and transport composition; and, (c), the transport formula to be used for sediment mixtures.

(a) The bottom of mountain streams conveying sand and gravel mixtures generally presents a well defined upper layer ("pavement") much coarser than the "substratum" underneath (Fig. 1); the substratum composition, in its turn, is generally coarser than the composition of the transported material.

Although, at any instant of time, the pavement and the substratum are composed by particles at rest (a moving particle, by definition, belongs to the transport), all or many of these particles are susceptible to be entrained by the waterflow; in this way both layers continuously exchange material between themselves and with the sediment transported by the stream. From this point of view, pavement and substratum are respectively similar to the "mixing layer" and the "intrusion layer" in streams with large dunes [1].

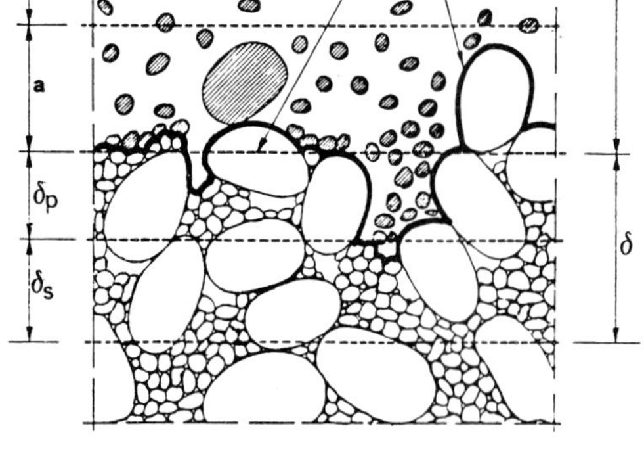

Fig. 1 - *Transport of a mixture of sand and gravel by a waterstream. In the upper layer (h-a) only suspended transport takes place, while in the layer below (a) bed-load transport is dominant. In the bottom is clearly visible a coarser pavement (δ_P) and a finer substratum (δ_S). At any given instant, the particles of both the pavement and the substratum are at rest; yet these particles are subjected to more (in the pavement) or less (in the substratum) frequent detatchments by the waterflow. The four layers can be approximately aggregated in a " transport layer" (h) and in a "storage layer" (δ), as described in [1].*

Note that, as pointed out by Parker [3], [4], bed pavement has nothing to do with bed armouring (in which all the particles are absolutely immobile under the action of flow); yet it may evolve towards this last situation by progressive removal of finer material (e.g., downstream of a dam).

(b) In equilibrium conditions, as there is no net vertical flux of sediments, the percentage of a certain fraction in the substratum should have a precise relationship with the percentage of the same fraction in both the transport and the pavement. This relationship is controlled by the exchange mechanisms among the three layers and thus will probably depend upon the specific range of granulometric and hydraulic characteristics of the stream; in the situation described here, for example, the percentage of any fraction (sand or gravel) in the substratum (β^S) is given, in equilibrium conditions, by a weighted average between the percentage in the pavement (β^P) and in the transport (β^T):

$$\beta^S = g\,\beta^P + (1 - g)\,\beta^T \tag{1}$$

where the weight is quite constant (g = 0.77) for all the experiments. It should be mentioned, in this respect, that completely different situations have been reported: the substratum composition of the Oak Creek [4] has been found to be practically coincident with the composition of the measured transport; while in some laboratory tests [3] the substratum resulted to be even coarser than the transported material. The difference is striking, but it is to be noted that in these situations bedload was the dominant mode of transport; moreover, due to the measuring technique, the finer particles were probably underrepresented in the measured transport.

In unsteady conditions, at any rate, the composition of the three layers is continuously variable as different size of sediments move from one to another. In principle, then, there is no fixed relationship among their granulometries and to describe their evolution, the sediment balance of the three layers should be made, provided that a suitable expression can be found for the net vertical fluxes [1].

As a first approximation, however, it seems possible to assume that the relationship among the composition of the three layer remains the same as in equilibrium conditions. In this way, pavement and substratum can be aggregated in one single layer (storage layer) having an appropriate thickness and an average composition. The fluxes through the upper and lower boundaries of the "storage layer" are provided by the erosion/deposition rates, while the relation between its composition and the composition of the transported material is expressed by the sediment transport formula for each fraction. This approximation has been used in the aggradation model presented in this Workshop [2].

(c) As far as the transport formula is concerned, some indications are provided by the measurements performed in various conditions during the experiments. Being the flume relatively short, the transport rate is assumed to be the aritmetic average between input and output. As expected, the transport rate of a given fraction, T_i, results to increase with the water discharge, Q, with the bottom slope, I, and with the percentage β_i of that fraction present in the bottom (the bottom may alternatively be defined as the pavement, the substratum or the storage layer). Under the same conditions, moreover, the transport rate of the finer fraction ($d_1 = 0.120$ mm) is definely larger than the transport rate of the coarser fraction ($d_2 = 7$ mm). In other words, a monomial expression as the one used in [2]:

$$T_i = \alpha\, \beta_i \frac{Q^m\, I^n}{b^p\, d_i^q} r_i \qquad (2)$$

seems to be a practical way to express the transport rate of a mixture of sand and gravel by a mountain stream having the width b. In eq. (2), the coefficient α and , to a certain extent, the exponents m, n, p, and q, should be appropriately chosen according to the specific range of granulometric and hydraulic characteristics of the stream (basically, plain or mountain river).

The hiding-and-exposure coefficient, r_i, accounts for the stronger (weaker) action exerted by the flow on the larger (smaller) diameters of a mixture, with respect to the corresponding uniform-size material. As it has been observed by Parker [4], the hiding-and-exposure effect has a different physical meaning, depending upon the definition of "bottom". If β_i is defined as the percentage of a given fraction in the pavement, the numerical effect of ri should be relatively less important as it accounts only for the "horizontal" protection exerted by larger particles over smaller ones. If bi is defined as the percentage in the substratum (or in the storage layer), the effect should be larger, as ri includes the "vertical" protection given by the coarser diameters of the pavement .

In any case, as the relationship between transport and bed composition is controlled by this coefficient, an appropriate expression for r_i under different circumstances becomes crucial. The hiding-and-exposure coefficient r_i is usually expressed as a function of the ratio between the diameter of the fraction, d_i, and a representative diameter of the mixture (e.g., the mean diameter $\bar{d} = \Sigma\, \beta_i\, d_i$).It seems reasonable, however, that the flow action on a certain grain-size fraction would be relatively stronger when these grain-sizes are relatively scarcer (for examples, few sand particles over a gravel bed). The expression of r_i, therefore, should also include, at the denominator, the very same percentage of that fraction, β_i.

8. References.

[1] Armanini, A. and Di Silvio, G.: "On the coexistence of bedload and suspended transport for a uniform grainsize material". Int. Symp. on Sediment Transport Modeling, ASCE, Hydr. Div., New Orleans, Aug. 14-18, 1989.

[2] Di Silvio, G. and Peviani, A.: "Modelling short- and long term evolution of mountain rivers: an application to the torrent Mallero (Italy), presented to the same Workshop.

[3] Parker, G.; Dhamotharan, S. and Stefan, H.: "Model Experiments on Mobile, Paved and Gravel Bed Streams". Water Resources Research, vol. 18, n. 5, October 1982.

[4] Parker, G. and Klingeman, P.C.: On Why the Gravel Bed Streams Are Paved". Water Resources Research, vol. 18, n. 5, October 1982.

VARIATION OF BED AND TRANSPORT MEAN DIAMETERS IN NON-EQUILIBRIUM CONDITIONS

Aronne Armanini

Department of Civil and Environmental Engineering
University of Trent, Trent, Italy

ABSTRACT

The paper presents a mathematical scheme able to model the evolution of deposition-erosion processes in a one-dimensional stream, also taking account of non uniform grain-size mixtures and possibly non-equilibrium of sediment transport rate.

Instead of describing the sediment transport for fractions, that is, each fraction by means of its continuity equation, the model uses mean diameter and second-order (and possibly third-) moment of grain-size distribution. As a consequence the number of equations and variables is reduced.

INTRODUCTION

The problem of river aggrading and degrading has been strongly emphasized in the recent years. The interest in solving this problem is related not only to a certain unresolved problems in the field of the research, but also to a better insight of long and short time consequences of river training engineering.

A number of numerical models has been proposed in the literature in order to describe the sediment transport of non uniform grain-size material, some of them even able to account for non-equilibrium transport rate (Ashida and Michiue,1971 ; Karim and Kennedy, 1982 ; Armanini and Di Silvio, 1982 and 1988 ; Rahuel and Holly, 1989 ; Andrews and Parker, 1987).

The general approach adopted in order to consider the different behaviour of different grain-size materials is based on the idea of dividing the grain-size distribution curve into a discrete number of classes and describe the behaviour of each class by means of separate mass- or volume- balances. Following the procedure adopted by Einstein(1950), the actual transport rate of each class is related to the percentage present in the bed sediment transport.

Aim of this paper is to approach the problem in terms of first-, second- and possibly third-order moments of the grain-size distribution.

GENERAL FORMULATION OF THE PROBLEM

The continuity equation for sediment transport in a quasi one-dimensional stream is given by :

$$\frac{\partial T}{\partial x} + \frac{\partial Ch}{\partial t} = -\frac{\partial Z}{\partial t} \qquad 1)$$

where h and Z are water depth and bottom elevation; $T = \int ucdy$ is the actual transport rate and $C = \int cdy$ is the depth-averaged concentration of sediments.

In natural rivers, but also in many laboratory settings, the bed material is far different from uniform. In this case the transport rate of each individual grain fraction depends on its size. Nevertheless the continuity within each grain-size class must be established.

Following Einstein (1950), the ratio between the actual transport rate T_j for each grain-size fraction j, and its transport capacity T_{c_j} (that is the transport rate in case of steady flow and uniform grain-size material) is equal to the percentage β_j of the same class present in the bed. That is :

$$T_j = \beta_j T_{c_j} \qquad 2)$$

This assumption, referred as *similarity hypothesis*, exhibits at least two limitations. The first one is that it refers to (local) equilibrium condition; that is, it does not account for different adaptation processes due to unsteadiness of sediment transport. The second limitation lies in the fact that it does not account for hiding effect, that is the sheltering exerted by the coarser materials over the finer fraction.

In order to account for the unsteadiness of the sediment transport, it is generally possible to describe the process by means of a linear adaptation model, in the following eq. 3) (Armanini and Di Silvio 1982):

$$T_j = \beta_j T_{c_j} - L_j^* \frac{\partial}{\partial x} T_j - \tau_j^* \frac{\partial}{\partial t} T_j \qquad 3)$$

L_j^* and τ_j^* represent, respectively, the adaptation length and the adaptation time for the same class. L_j^* accounts for the space lag between the actual transport rate T_j and the transport capacity $\beta_j T_{c_j}$; this effect is more pronounced the smaller is the grain-size of the class. For suspended load the following expression has been proposed by Armanini and Di Silvio (1988):

$$\frac{L^* w}{hU} = \frac{a}{h} + (1 - \frac{a}{h}) \exp\left[-1.5(\frac{a}{h})^{-1/6} \frac{w}{u_*}\right] \qquad 4)$$

where w is the particles falling velocity, a is the thickness of the bottom layer, h and U water depth and mean velocity.

In eq.3) τ_j^* accounts for time lag between the same quantities as above. For suspended load one can assume:

$$\tau^* \simeq \frac{L^*}{U} \qquad 5)$$

For bed load, adaptation length and adaptation time are small (e.g. of order of magnitude of bed forms), so in general it is possible to assume immediate adaptation of bed load to equilibrium condition. This assumption is generally accepted in case of gravel bed rivers where the bed load prevails over suspended load.That is:

$$L^* \sim \tau^* \sim 0 \qquad 6)$$

It should be remarked that β_j is the percentage of j class present on the bed surface. Experimental evidence in fact shows, under the layer where sediment transport occurs, the existence of a *surface layer*, where the longitudinal transport is negligible, but where the vertical exchange of sediments with the transport layer is determinant. This exchange is in fact responsible for the sorting process of the bed material.

The thickness of such a mixing layer (or surface layer) δ is order of magnitude of roughness elements. In duned beds it can be assumed as half the dune height Δ:

$$\delta = \frac{1}{2}\Delta \tag{7)}$$

In order to account for possible hiding effect, it is possible to introduce a proper hiding factor. This factor reduces the transport capacity of finer fraction of the mixture and increases the transport capacity of the coarser fraction. Different expression for the hiding factor have been proposed in the literature, suitable to be applied to different kind of sediment transport formulas (Eghiazaroff,1965 ; Day,1980 ; Ranga Raju 1985 ; Andrews and Parker, 1987). In general the hiding factor is a function of the ratio between the diameter d_j of j class and the mean diameter $\bar{D}$ of the mixture:

$$\xi_j = \xi_j(\frac{D_j}{\bar{D}})^m$$

The mixing process among the classes constituting the bed material may be described by means of the continuity equation inside the mixing bed layer δ (where just mixing but not longitudinal transport takes place), as described by the following eq.8).

$$\frac{\partial}{\partial t}\delta\beta_j = \beta_j^*\frac{\partial(\delta - Z)}{\partial t} - (\frac{\partial T_j}{\partial x} + \frac{\partial}{\partial t}C_j h) \tag{8)}$$

In the erosion processes, β_j^* represents the percentage in the subsurface layer, (the volume below the mixing layer) of the material that belongs to the j-th class ; while during the deposition phase $\beta_j^* = \beta_j$.

Finally the total transport T is the summation of the transport rates of each class j , that is :

$$T = \sum_{j=1}^{m} T_j \tag{9)}$$

If m is the number of classes, the set formed by eqs. 1), and 9), and m eqs.3) (or eq.2) and 8), in the (2 m + 2) unknowns : $Z, T, m \cdot T_j$, $m \cdot \beta_j$, (coupled with the appropriate hydrodynamics equations), are sufficient to describe the behaviour of each class of sediments.

The relationship between C_j and T_j is given by the transport formula and concentration distribution; it is generally possible to put:

$$C_j = \frac{T_j}{U_j}$$

being U_j the celerity of the $j - th$ class of the grain-size distribution.

The set of equation can be numerically solved for a discrete number of classes. Many numerical models have been recently proposed in literature (Andrews and Parker, 1987 ; Armanini and Di Silvio, 1988 ; Rahuel and Holly, 1989; Bouchard et alii., 1989). The models differ for the sediment transport formula, for the expression of adaptation quantities L_j^* and τ_j^*, for the definition of the mixing layer, as well as for numerical scheme and number of classes. Since now the number of classes adopted in literature ranges from 2 to 5.

From a numerical point of view the models are quite complex owing to the number of (non linear) equations: one possibility to simplify the problem is to write the equations in terms of mean values instead of single grain-size quantities.

EQUATIONS IN TERMS OF MEAN DIAMETERS

Adding up to m classes all the terms in eq.3), one obtains:

$$T=\sum_{j=1}^{m}\beta_j T_{c_j}-\sum_{j=1}^{m}L_j^*\frac{\partial T_j}{\partial x}-\sum_{j=1}^{m}\tau_j^*\frac{\partial T_j}{\partial t} \tag{10)}$$

If α_j is the percentage of class j present in the transport load, by definition it is :

$$T_j=\alpha_j T \tag{11)}$$

Eq. 10) can then rewritten as follows:

$$T=\sum_{j=1}^{m}\beta_j T_{c_j}-\sum_{j=1}^{m}L_j^*\frac{\partial}{\partial x}T\alpha_j-\sum_{j=1}^{m}\tau_j^*\frac{\partial}{\partial t}T\alpha_j \tag{12)}$$

Given that both α_j and β_j are distribution functions, they vanish as when the diameter tends to zero and infinity the summation in eq.10) can be calculated expanding in Taylor series T_{c_j}, L_j^* and τ_j^* .

If transport capacity for each class j is expressed by a formula linking it to the power of respective diameter D_j, one obtains:

$$T_{c_j}=AD_j^{-n} \tag{13)}$$

Expanding 13) in Taylor series, the first right hand term of eq. 12) can be written as:

$$\sum_{j=1}^{m}\beta_{c_j}T_{c_j}=T_{cb}[1+c_2\sigma_b+c_3\gamma_b+O(\kappa_b)] \tag{14)}$$

where T_{cb} is the transport capacity related of the mean diameter of the bed material :

$$D_b=\sum_{j=1}^{m}\beta_j D_j \tag{15)}$$

σ_b , γ_b and κ_b are second-, third- and fourth-order dimensionless moments of the β_j distribution (variance, skewness and kurtosis):

$$\sigma_b=\sum_{j=1}^{m}\frac{\beta_j(D_j-D_b)^2}{D_b^2}\quad;\quad\gamma_b=\sum_{j=1}^{m}\frac{\beta_j(D_j-D_b)^3}{D_b^3}\quad;\quad\kappa_b=\sum_{j=1}^{m}\frac{\beta_j(D_j-D_b)^4}{D_b^4} \tag{16)}$$

The coefficients c_2 and c_3 are, respectively:

$$c_2=\frac{n(n+1)}{2!} \tag{17)}$$

$$c_3=\frac{-n(n+1)(n+2)}{3!} \tag{18)}$$

In the same way, upon expansion of L_j^* and τ_j^* it is possible to write the second third and forth right hand terms of eq. 12) :

$$\sum_{j=1}^{m}L_j^*\frac{\partial}{\partial x}T\alpha_j=L_t^*\frac{\partial}{\partial x}T[1+d_2\sigma_t+d_3\gamma_t] \tag{19)}$$

and

$$\sum_{j=1}^{m}\tau_j^*\frac{\partial}{\partial t}T\alpha_j=\tau_t^*\frac{\partial}{\partial t}T[1+d_2\sigma_t+d_3\gamma_t] \tag{20)}$$

where: L_t^* and τ_t^* are the adaptation length and time related to the mean transport diameter, D_t , defined as follows:

$$D_t = \sum_{j=1}^{m} \alpha_j D_j \qquad 21.a)$$

and σ_t , γ_t and κ_t are second-, third- and fourth-order dimensionless moments of the α_j distribution (variance, skewness and kurtosis):

$$\sigma_t = \sum_{j=1}^{m} \frac{\alpha_j (D_j - D_t)^2}{D_t^2} \quad ; \quad \gamma_t = \sum_{j=1}^{m} \frac{\alpha_j (D_j - D_t)^3}{D_t^3} \quad ; \quad \kappa_t = \sum_{j=1}^{m} \frac{\alpha_j (D_j - D_t)^4}{D_t^4} \qquad 21.b)$$

If eq.4) can be approximated by a power low (e.g. $L^* \propto D^{-l}$), the coefficients d_2 and d_3 are, respectively:

$$d_2 = \frac{l(l+1)}{2!} \quad and \quad d_3 = \frac{-l(l+1)(l+2)}{3!} \qquad 21.c)$$

if complete eq.4) is used instead, one will obtain a more complex expression for coefficients d_2 and d_3 .

After substitution of eqs. 14), 19) and 20) in eq.12) one obtains:

$$T = T_{bc}[1 + c_2\sigma_b + c_3\gamma_b + \ldots] - L_t^* \frac{\partial}{\partial x} T[1 + d_2\sigma_t + d_3\gamma_b + \ldots] - \tau_t^* \frac{\partial}{\partial t} T[1 + d_2\sigma_t + d_3\gamma_b + \ldots] \qquad 22)$$

The terms in eq.3) are now multiplied by the respective diameter D_j, and summed up to the m classes, so that:

$$\sum_{j=1}^{m} T_j D_j = \sum_{j=1}^{m} \beta_j D_j T_j - \sum_{j=1}^{m} L_j^* D_j \frac{\partial T_j}{\partial x} - \sum_{j=1}^{m} \tau_j^* D_j \frac{\partial T_j}{\partial t} \qquad 23)$$

Introducing 21.a) into the left hand term of eq.23), one has:

$$\sum_{j=1}^{m} T_j D_j = T \sum_{j=1}^{m} \alpha_j D_j = T D_t \qquad 24)$$

By means of a procedure similar to that used in eq.12), it is possible to obtain for the left hand terms of eq. 23):

$$\sum_{j=1}^{m} \beta_j D_j T_j = D_b T_{cb}[1 + c_1\sigma_b + c_2\gamma_b + \ldots] \qquad 25)$$

$$\sum_{j=1}^{m} L_j^* D_j \frac{\partial T_j}{\partial x} = L_t^* D_t \frac{\partial}{\partial x} T[1 + d_1\sigma_t + d_2\gamma_t + \ldots] \qquad 26)$$

$$\sum_{j=1}^{m} \tau_j^* D_j \frac{\partial T_j}{\partial t} = \tau_t^* D_t \frac{\partial}{\partial t} T[1 + d_1\sigma_t + d_2\gamma_t + \ldots] \qquad 27)$$

where :

$$c_1 = \frac{n(n-1)}{2!} \qquad and \qquad d_1 = \frac{l(l-1)}{2!} \qquad 27.b)$$

After substitution of eq. 24), 25), 26) and 27) into eq.23) one has:

$$TD_t = D_b T_{cb}[1+c_1\sigma_b+c_2\gamma_b+\ldots]-L_t^*\frac{\partial}{\partial x}D_t T[1+d_1\sigma_t+d_2\gamma_t+\ldots]-\tau_t^*\frac{\partial}{\partial t}D_t T[1+d_1\sigma_t+d_2\gamma_t+\ldots] \quad 28)$$

Finally multiplying eq.8) by the respective diameter D_j and summing up the m classes, gives:

$$\frac{\partial}{\partial t}\delta D_b = D_b^*\frac{\partial}{\partial t}(\delta - Z) - \frac{\partial}{\partial x}TD_t - \frac{1}{U_t}\frac{\partial}{\partial t}TD_t \quad 29)$$

Assuming that the variance and the skewness of α_j and β_j distributions remains unchanged, the set formed by eqs. 1), 22), 28) and 29), in the unknowns T, Z, D_t,D_b is complete and can be solved numerically or, in some simplified cases, analytically. If the latter hypothesis proves unsatisfactory, an equation for the second moment can be obtained just multiplying each term of eq.3) by D_j^2 and repeating the same procedure as above.

IMMEDIATE ADAPTATION HYPOTHESIS

When the bed load prevails over suspended load, as in gravel bed rivers, it reasonable to assume immediate adaptation of the transport rate to equilibrium condition:

$$L^* \sim t^* \sim 0$$

for the same reason it is:

$$\frac{\partial T}{\partial x}D_t \gg \frac{1}{U_t}\frac{\partial TD_t}{\partial t} \quad 30)$$

The complete set of equations it this case will be the following:

$$\frac{\partial T}{\partial x} = -\frac{\partial Z}{\partial t} \quad 31.a)$$

$$T = T_{cb}[1 + c_2\sigma_b + c_3\gamma_b \ldots] \quad 31.b)$$

$$TD_t = D_b T_{cb}[1 + c_1\sigma_b + c_2\gamma_b + \ldots] \quad 31.c)$$

$$\frac{\partial}{\partial t}\delta D_b = D_b^*\frac{\partial}{\partial t}(\delta - Z) - \frac{\partial}{\partial x}TD_t \quad 31.d)$$

Combining eqs. 31.b and 31.c) one has:

$$\frac{D_t}{D_b} = \frac{1 + c_1\sigma_b + c_2\gamma_b + \ldots}{1 + c_2\sigma_b + c_3\gamma_b + \ldots} \quad 32)$$

Eq. 32) states that the ratio between the mean diameter of the transport load and mean diameter of the surface bed material , is function of the moments of the bed load distribution and of the transport formula, (trough the coefficients c_i).

If transport formula can be expressed as a power law (see eq. 13), the above ratio depends just on the exponent n . In fact substituting eqs. 17), 18) and 27.b) into eq. 32) (neglecting $O(\gamma_b)$ terms) gives:

$$\frac{D_t}{D_b} = \frac{2+\sigma_b n(1-n)}{2+\sigma_b n(1+n)} \qquad 33)$$

In fig. 1, eq. 33) is compared to some experimental data.

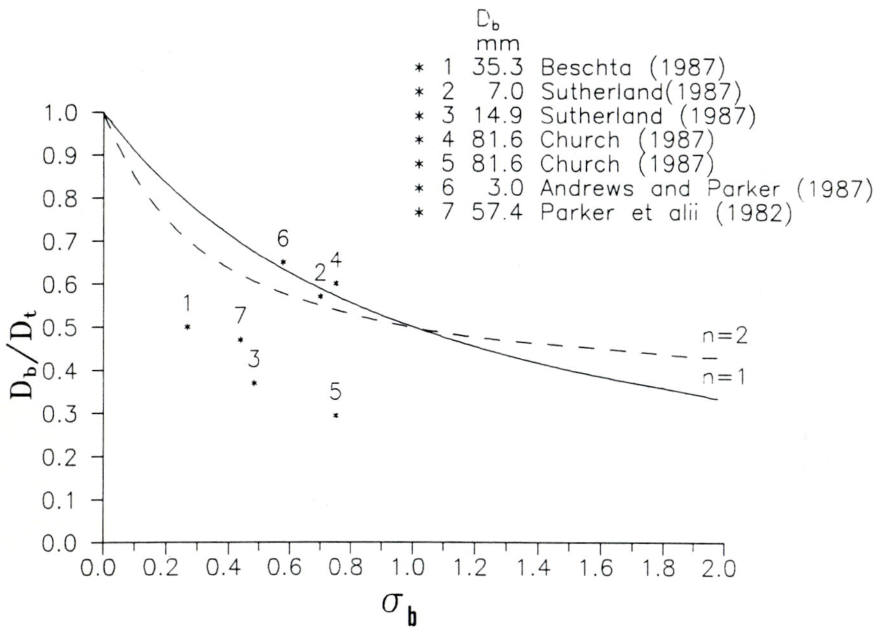

Fig.1 The ratio between mean diameters of bed-load and surface bed material are compared to eq.33)

The value $n = 0$ corresponds to equal mobility condition; in this case, in fact, $D_t/D_b \to 1$; the same result is obtained when $n \to \infty$ (absence of sediment transport).

The value $n = 1$ for the exponent in eq. 13) is a reliable value accounting for hiding effect.

The two curves in Fig.1 correspond respectively to $n = 1$ and $n = 2$: for small values of σ_b , the difference between the two curves is not striking.

In the same figure almost 3 points lie definitely above the curves. All these points correspond to a quasi-static armouring situation. From eq. 33) one has that no real values for exponent n can match these experimental points. On the other hand, it must be remarked that the power expression (eq. 13) for sediment transport formula can be used if the gradation curve is not very wide, and , above all, if sediment transport is far from incipient motion condition.

Instead of eq.13, it is possible to use a binomial expression (e.g. Mayer-Peter Müller type equation):

$$T_{c_j} = const_1\sqrt{g}D_j^{1.5}\left(\frac{u_*^2}{gD_j} - \xi_j \cdot const_2\right)^{1.5} \qquad 34)$$

Eq. 34 can be generalized in the following form.

$$T_{c_j} = (B_2 - A_2 D_j^{n_1})^s \qquad 35)$$

where B_2 and A_2 depend on hydrodynamic conditions, but not on D_j.
In this case the coefficients c_2 and c_2 will be:

$$c_1 = -\frac{n_1 s A_2(1-n_1)D_b^{n_1}}{2!(B_2 - A_2 D_b^{n_1})^2}(-B_2(1+n_1) + A_2(1+n_1 s)) \tag{36.a}$$

$$c_2 = c_1 - \frac{n_1 s A_2}{B_2 D_b^{n_1} - A_2} \tag{36.b}$$

Eqs. 36.a) and 36.b), inserted in eq. 32), allow for any solution for $D_t/D_b < 1$, provided that the correct sediment transport formula is given.

Unfortunately eqs. 36.a) and 36.b) are more difficult to be used, because they require the knowledge of the sediment transport formula (this quantity does not appear in eq. 33). Comparison between theory and literature experiments is under way.

Similar conclusions can be drawn for the coefficient c_3. More in general c_1, c_2 and c_3 are function of D_b.

From eqs. 31.a) - 31.d), algebraic manipulation (and neglecting skewness terms) yields:

$$\delta dt D_b = D_b^* \frac{\partial}{\partial x} T_{cb}[1 + c_2\sigma_b] - \frac{\partial}{\partial x} D_b T_{cb}[1 + c_1\sigma_b] + (D_b^* - D_b)\frac{\partial \delta}{\partial t} \tag{37}$$

In aggrading processes, (deposition) D_b^* is equal to D_b, and eq.37) leads to:

$$\frac{\partial}{\partial t}\delta D_b + T_{cb}[1 + c_1\sigma_b]\frac{\partial D_b}{\partial x} = D_b \frac{\partial}{\partial x} T_{cb}\sigma_b[c_2 - c_1] \tag{38}$$

and:

$$\frac{\partial}{\partial t}\delta D_b + [T_{cb}(1 + c_1\sigma_b) - nD_b\sigma_b\frac{\partial T_{cb}}{\partial D_b}]\frac{\partial D_b}{\partial x} = nD_b\sigma_b \frac{\partial U}{\partial x}\frac{\partial T_{cb}}{\partial U} \tag{39}$$

Eq.39) gives the following celerity, for the propagation for the mean diameter:

$$\tilde{c}_{D_b} = \frac{[T_{cb}(1 + c_1\sigma_b) - nD_b\sigma_b\frac{\partial T_{cb}}{\partial D_b}]}{\delta} = \frac{T_{cb}}{\delta}[1 + \sigma_b\ (c_1 + n^2)] \tag{40}$$

In nonuniform flow, the celerity is obviously non constant, and increases with increasing velocity. The thickness δ of the mixing layer reduces then the celerity.

CONCLUSIONS

This mathematical model can describe the behaviour, in one-dimensional channels, of a mixture of materials having non uniform grain-size distribution.

Compared to complete schemes, where each fraction of the gradation curve is described by two partial differential equations, the model is based on mean diameters and variance of grain-size distribution (and if necessary, of the other moments of the distribution).

From a numerical point of view the model, reducing the number of variables, represents a relevant simplification.

Some questions are still open in particular concerning:

-the role of the hiding factor (this is a open point also for the other models);

-the possibility of describing the armouring processes (static armouring);

- the legitimacy of keeping variance (and other moments) constant in space an time;

- the legitimacy of neglecting higher order moments.

Comparison between the proposed scheme and literature experiments, as well as comparison with complete models, are under way.

Finally an improvement in the model can be obtained, writing the moments in terms of logarithm of the particles diameter instead of the diameter itself.

This research was made in the framework of the Italian Groups for Disaster Prevention (CNR-GNDCI) and for Sediment Transport Research (MURST 40 %)

NOTATIONS

The following symbols have been used in this paper:.

A_2, B_2 parameters of sediment transport formula
C concentration
c_1, c_2, c_3 coefficients
$\tilde{c}_{D_b}$ celerity of transported mean diameter
D_j mean diameter of class j-th
D_b mean diameter in the mixing layer
D_t mean diameter of the transported material
h water depth
L^* adaptation length
m number of grain-size classes
n, n_1, l exponents
x longitudinal coordinate
t time
T_j transport rate of j-th grain-size class
T_c transport capacity of the j-th grain-size class
U depth averaged velocity
$U_j = \frac{T_j}{C_j}$
u_* friction velocity
w particle fall velocity
Z bed elevation
α_j percentage of the j-th class inside the transport layer
β_j percentage of the j-th class into the exchange layer
δ exchange layer thickness
Δ height of the dunes
σ_b, γ_b second and third moments of grain-size distribution inside the mixing layer
σ_t, γ_t second and third moments of grain-size distribution of transported material
τ^* adaptation time
ξ_j hiding factor for j-th grain-size class

REFERENCES

Andrews,E.D., Parker,G.(1987). Formation of Coarse Surface Layer as the Response to Gravel Mobility, *Sediment Transport in Gravel- bed Rivers*, John Wiley & S. Ltd.,p.269-300.

Armanini,A.,Di Silvio,G.(1982).Sudden Morphological Modifications along Mountain River Simulated by Mathematical Model. 3rd Congress of the APD of IAHR,Bandung,Indonesia.

Armanini, A.,Di Silvio, G.(1988). A One-dimensional Model for the Transport of Sediment Mixture in Non-equilibrium Condition, Juorn.of IAHR, vol26,n.3.

Ashida,K.,Michiue,M.(1971),An Investigation of River Bed Degradation of Dam, Proc. 14th IAHR Congr., vol.3, pp 247-255,1971.

Beschta,R.L.(1987). Conceptual Models of Sediment Transport in Streams, *Sediment Transport in Gravel-bed Rivers*, John Wiley & S. Ltd.,p.387-420.

Bouchard,J-P.,Cordelle,M.,Labadie,G.,Lorin,J.(1989) Numerical Simulation of Mud Erosion in Reservoirs by Floods Application to Reservoir of the Durance River. Proc. of XXIII IAHR Congr., Vol.B,Ottawa.pp. 575-581.

Church,M.A.(1987) Discussion,*Sediment Transport in Gravel-bed Rivers*, John Wiley & S. Ltd.,p.314-322.

Day,T.J.(1980). A Study of the Transport of Graded Sediments. HRS Wallingford Rep. No.IT 190,April.

Eghiazaroff,I.V.(1965). Calculation of Non-Uniform Sediment Concentrations. Proc.A.S.C.E. J.Hydr. Div., 91(HY4)

Einstein,H.A.(1950). The Bed-Load Function for Sediment Transport in Open Channel Flows. Tech. Bull. No.1026, U.S.Dept. of Agr., Soil Cons.Serv., Sept.

Karim,M.F.,and Kennedy,J.F.(1982).Computer Based Predictors for Sediment Discharge and Friction Factor of Alluvial Streams. IIHR report no.242,The University of Iowa.USA.

Parker,G.,Klingeman,P.C. and McLean,D.G.(1982), Bed-load and size distribution in paved gravel-bed stream. *Proc. A.S.C.E., J. Hydr. Div.,HY4*,108,544-571

Ranga Raju,K.G.(1985). Transport of Sediment Mixture, Ippen Lecture, XXI IAHR Congr., Melbourne, Australia.

Rahuel,J.,Holly,F.M.(1989), Numerical Simulation of Sediment Mixture Dynamics. Proc. XXIII-IAHR Congr., vol.B, Ottawa, pp.315-320.

Sutherland,A.J.(1987). Static Armour Layers by Selective Erosion, *Sediment Transport in Gravel-bed River*, John Wiley & S. Ltd.,p.243-267.